NONLINEAR SCIENCE THEORY AND APPLICATIONS

Numerical experiments over the last thirty years have revealed that simple nonlinear systems can have surprising and complicated behaviours. Nonlinear phenomena include waves that behave as particles, deterministic equations having irregular, unpredictable solutions, and the formation of spatial structures from an isotropic medium.

The applied mathematics of nonlinear phenomena has provided metaphors and models for a variety of physical processes: solitons have been described in biological macromolecules as well as in hydrodynamic systems; irregular activity that has been identified with chaos has been observed in continuously stirred chemical flow reactors as well as in convecting fluids; nonlinear reaction diffusion systems have been used to account for the formation of spatial patterns in homogeneous chemical systems as well as biological morphogenesis; and discrete-time and discrete-space nonlinear systems (cellular automata) provide metaphors for processes ranging from the microworld of particle physics to patterned activity in computing neural and self-replicating genetic systems.

Nonlinear Science: Theory and Applications will deal with all areas of nonlinear science – its mathematics, methods and applications in the biological, chemical, engineering and physical sciences.

Nonlinear science: theory and applications

Series editor: Arun V. Holden, *Reader in General Physiology, Centre for Nonlinear Studies, The University, Leeds LS2 9NQ, UK*

Control and optimization J. E. Rubio

Chaos A. V. Holden (*Editor*)

Automata networks in computer science F. Fogelman Soulié, Y. Robert and M. Tchuente (*Editors*)

Simulation of wave processes in excitable media V. S. Zykov

Oscillatory evolution processes I. Gumowski

Mathematical models of chemical reactions P. Erdi and J. Tóth

Introduction to the algebraic theory of invariants of differential equations K. S. Sibirsky

Almost periodic operators and related nonlinear integrable systems V. A. Chulaevsky

Soliton theory: a survey of results A. P. Fordy (Editor)

Fractals in the physical sciences H. Takayasu

Other volumes are in preparation

Soliton theory: a survey of results

Edited by Allan P. Fordy

Manchester University Press

Manchester and New York

Distributed exclusively in the US and Canada by St. Martin's Press

Published by Manchester University Press
Oxford Road, Manchester M13 9PL, UK
and Room 400, 175 Fifth Avenue,
New York, NY 10010, USA

Distributed exclusively in the USA and Canada
by St. Martin's Press, Inc.,
175 Fifth Avenue, New York, NY 10010, USA

British Library cataloguing in publication data
Soliton theory.
1. Nonlinear evolution equations. Solution. Spectral transforms & solitons
I. Fordy, Allan P. II. Series
515.3′53

Library of Congress cataloguing in publication data
Soliton theory: a survey of results/edited by Allan P. Fordy.
p. cm. — (Nonlinear science)
ISBN 0–7190–1491–3
1. Solitons. I. Fordy, Allan P. II. Series.
QC174.26.W28S62 1990
530.1′4–dc20 89–39338

ISBN 0 7190 1491 3 *hardback*

Typeset in Hong Kong
by Graphicraft Typesetters Ltd

Printed in Great Britain by
Biddles Ltd, Guildford and King's Lynn

Contents

List of contributors

M. Antonowicz Institute of Theoretical Physics, Warsaw University, Hoza 69, 00–681 Warsaw, Poland

P. J. Caudrey Mathematics Department, University of Manchester Institute of Science and Technology, PO Box 88, Sackville Street, Manchester M60 1QD, UK

R. K. Dodd School of Mathematics, Trinity College Dublin, Dublin 2, Republic of Ireland

A. P. Fordy Department of Applied Mathematical Studies and Centre for Nonlinear Studies, University of Leeds, Leeds LS2 9JT, UK

J. D. Gibbon Department of Mathematics, Imperial College, London SW7 2BZ, UK

P. G. Grinevich Landau Institute for Theoretical Physics, 117940, Kosygina 2, Moscow, USSR

I. M. Krichever Institute for Problems in Mechanics, Academy of Science USSR, pr. Vernadskogo 101, 117526 Moscow, USSR

P. S. Lomdahl Theoretical Division and Center for Nonlinear Studies, Los Alamos National Laboratory, Los Alamos, New Mexico 87545, USA

T. Miwa Research Institute for Mathematicl Sciences, Kyoto University, Kyoto 606, Japan

J. J. C. Nimmo Department of Mathematics, University Gardens, University of Glasgow, Glasgow G12 8QW, UK

A. R. Osborne Istituto di Cosmo-Geofisica del CNR, Corso Fiume 4, Turin 10133, Italy

C. Rogers Department of Applied Mathematics, University of Waterloo, Waterloo, Ontario, Canada

M. Tabor Department of Applied Mathematics and Nuclear Engineering, Columbia University, New York, New York 10027, USA

S. Rauch-Wojciechowski Department of Mathematics, LiTH, Linköping University, S–581 83 Linköping, Sweden

Part I

Introduction

1 *A. P. Fordy*

Soliton theory: a brief synopsis

1.1 Historical background

It is customary to start a book on soliton theory with John Scott Russell's description of 'the great wave of translation'. This description has such charm that I am not inclined to break with this tradition. The story of the soliton thus starts in 1834 [42] on the Union Canal near Edinburgh:

> I believe I shall best introduce this phenomenon by describing the circumstances of my own first acquaintance with it. I was observing the motion of a boat which was rapidly drawn along a narrow channel by a pair of horses, when the boat suddenly stopped – not so the mass of water in the channel which it had put in motion; it accumulated round the prow of the vessel in a state of violent agitation, then suddenly leaving it behind, rolled forward with great velocity, assuming the form of a large solitary elevation, a rounded, smooth and well defined heap of water, which continued its course along the channel apparently without change of form or diminution of speed. I followed it on horse back, and overtook it still rolling on at a rate of some eight or nine miles an hour, preserving its original figure some thirty feet long and a foot to a foot and a half in height. Its height gradually diminished, and after a chase of one or two miles I lost it in the windings of the channel. Such, in the month of August 1834, was my first chance interview with that singular and beautiful phenomenon which I have called the Wave of Translation, a name which it now very generally bears.

It was another 60 years before Korteweg and de Vries [21] were to derive their (now famous) equation describing the propagation of waves on the surface of a shallow channel. After performing a Galilean and a variety of scale transformations the Korteweg–de Vries (KdV) equation can be written in simplified form:

$$u_t = u_{xxx} + 6uu_x. \tag{1.1.1}$$

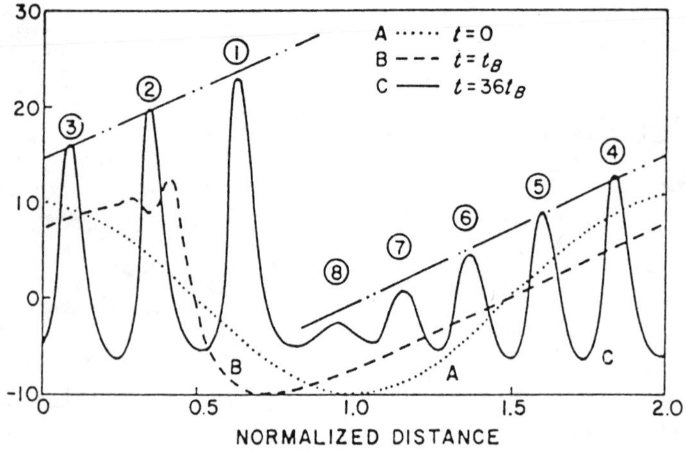

Figure 1.1 Solution of KdV equation at three different times (from [50])

It is easy to find a travelling wave solution of the form $u(x, t) = \phi(x + ct)$:

$$u(x, t) = 2a^2\text{sech}^2 a(x + 4a^2t + \varepsilon) \tag{1.1.2}$$

where ε is the phase and the amplitude $2a^2$ is half of the speed.

However, the KdV equation was not famous, or even widely studied, until the remarkable discoveries of Zabusky and Kruskal (ZK) in 1965 [50]. They were studying the Fermi–Pasta–Ulam (FPU) problem of recurrence on a nonlinear lattice. The KdV equation arises after taking the continuum limit. They found that initial data of a cosine function (periodic boundary conditions) evolved into a series of pulses, each of which resembled the solitary wave solution (1.1.2). Since the speed of the wave (1.1.2) is directly proportional to the amplitude the larger pulses travelled faster than the smaller so that they eventually lined themselves up in order of magnitude, as seen in Figure 1.1 (where they have wrapped themselves around by periodicity). The most suprising result came next. Continuing the evolution from the state depicted by Figure 1.1, the larger pulses overtake the smaller ones. Even though they undergo a fully nonlinear interaction, they reappear afterwards unscathed. These pulses retain their height, width and speed, undergoing nothing but a phase shift as a result of their interaction. Because of the particle-like nature of these interacting solitary waves, ZK coined the name 'soliton' to describe the pulses.

Stimulated by this numerical experiment Kruskal and his co-workers carried out their analytic study of the KdV equation, which culminated in a brief announcement [13] and a series of papers [12, 14, 23, 34, 36, 43].

The importance of these papers cannot be overemphasised since they comprise the origin of a whole new theory: the theory of solitons. These articles are very enjoyable to read; they convey a remarkable sense of wonder and discovery. I would thoroughly recommend them, together with the reviews by Zabusky [49], Kruskal [22] and Miura [35]. The latter includes some marvellous anecdotes regarding their calculation of conservation laws.

The remainder of this chapter gives a brief synopsis of some of the more important and elementary features of soliton theory. I also give a brief guide to the remaining chapters of the book.

1.2 Conservation laws

Let $u(x, t)$ be a solution of an evolution equation:

$$u_t = K[u] \tag{1.2.1}$$

where $K[u]$ is a locally defined function of u and its x-derivatives. The KdV equation (1.1.1) is our prime example of such an equation. A local conservation law for equation (1.2.1) is any equation of the form

$$\mathcal{T}_t + \mathcal{X}_x = 0 \tag{1.2.2}$$

where $\mathcal{T}[u]$ and $\mathcal{X}[u]$ are local expressions in u and its x-derivatives (terms involving u_t, u_{tx}, etc. being eliminated through equation (1.2.1)).

Remark

$\mathcal{T}_t$ and $\mathcal{X}_x$ denote total derivatives with respect to t and x and are given by the chain rule as:

$$\mathcal{T}_t = \frac{\partial \mathcal{T}}{\partial u} u_t + \frac{\partial \mathcal{T}}{\partial u_x} u_{xt} + \ldots \tag{1.2.3}$$

and similarly for $\mathcal{X}_x$. Upon substitution for u_t, u_{tx}, and so on from equation (1.2.1), equation (1.2.2) reduces to an identity.

The importance of local conservation laws is that with appropriate boundary conditions they lead to conserved quantities (constants of the motion). Integrating (1.2.2) with respect to x we get:

$$\frac{\mathrm{d}}{\mathrm{d}t} \int_A^B \mathcal{T} \mathrm{d}x + [\mathcal{X}]_A^B = 0 \tag{1.2.4a}$$

Under periodic boundary conditions with $(B-A)$ an integer multiple of the period or with $u(x, t)$ rapidly vanishing as $x \to \pm\infty$ and $(A, B) = (-\infty, \infty)$ the square bracket in (1.2.4a) vanishes. Thus, with these or any other appropriate boundary conditions (1.2.2) gives rise to the constant of motion $\int_A^B \mathcal{T} \mathrm{d}x$.

Remark
Such conserved quantities are only defined up to exact x-derivatives, since we can alter $\mathscr{T}$ and $\mathscr{X}$ by:

$$\mathscr{T} \to \mathscr{T} + f_x, \mathscr{X} \to \mathscr{X} - f_t \tag{1.2.4b}$$

without changing (1.2.2). This defines an equivalence relation '$\sim$' by:

$$\mathscr{T} \sim \mathscr{T}' \text{ iff } \mathscr{T} - \mathscr{T}' = f_x \tag{1.2.4c}$$

for some f.

For the KdV equation (1.1.1) the first three conservation laws are given by:

$$\mathscr{T}_0 = u, \mathscr{X}_0 = -u_{xx} - 3u^2, \tag{1.2.5a}$$

$$\mathscr{T}_1 = \tfrac{1}{2}u^2, \mathscr{X}_1 = -uu_{xx} + \tfrac{1}{2}u_x^2 - 2u^3, \tag{1.2.5b}$$

$$\mathscr{T}_2 = u^3 - \tfrac{1}{2}u_x^2, \mathscr{X}_2 = u_x u_{xxx} - \tfrac{1}{2}u_{xx}^2 - 3u^2 u_{xx} + 6uu_x^2 - \tfrac{9}{2}u^4. \tag{1.2.5c}$$

The first of these is just the equation itself, while the second results from multiplying the equation by u and integrating. Inspired by their numerical results, Zabusky and Kruskal calculated a further two conservation laws and then Miura another six. These are all presented in [36]. In fact, these authors present a proof of the existence of an infinite number of such conservation laws for the KdV equation (1.1.1). The method of proof uses a 'remarkable explicit nonlinear transformation', which is itself a very interesting mathematical object.

Following his calculations with the KdV equation. Miura considered the more general equation:

$$u_t = u_{xxx} + 6u^n u_x \tag{1.2.6a}$$

and found that for $n = 1$ and $n = 2$ (and *only* these values) there existed many conservation laws. With a slight change in notation the second of these, called the modified KdV (MKdV) equation, can be written:

$$v_t = v_{xxx} - 6v^2 v_x. \tag{1.2.6b}$$

The first four conservation laws are given by:

$$\tilde{\mathscr{T}}_{-1} = v, \tilde{\mathscr{X}}_{-1} = -v_{xx} + 2v^3, \tag{1.2.7a}$$

$$\tilde{\mathscr{T}}_0 = v^2, \tilde{\mathscr{X}}_0 = -2vv_{xx} + v_x^2 + 3v^4, \tag{1.2.7b}$$

$$\tilde{\mathscr{T}}_1 = \tfrac{1}{2}(v_x^2 + v^4), \tilde{\mathscr{X}}_1 = -v_x v_{xxx} + \tfrac{1}{2}v_{xx}^2 - 2v^3 v_{xx} + 6v^2 v_x^2 + 2v^6, \tag{1.2.7c}$$

$$\tilde{\mathcal{T}}_2 = -\left(v^6 + 5v^2v_x^2 + \tfrac{1}{2}v_{xx}^2\right),$$

$$\tilde{\mathcal{X}}_2 = -\left(v_{xx}v_{xxxx} - \tfrac{1}{2}\, v_{xxx}^2 - 10v^2v_xv_{xxx} + 8v^2v_{xx}^2 + 10vv_x^2v_{xx}\right.$$
$$\left. - 6v^5v_{xx} + \tfrac{1}{2}v_x^4 + 45v^4v_x^2 + \tfrac{9}{2}v^8\right). \tag{1.2.7d}$$

Miura then noticed a remarkable fact: by substituting

$$u = -v_x - v^2 \tag{1.2.8}$$

into the conserved densities (1.2.5) of the KdV equation they were transformed into those of the MKdV equation ($\tilde{\mathcal{T}}_{-1}$ is not included this way). For example,

$$\mathcal{T}_0 = u = -v_x - v^2 \sim -\tilde{\mathcal{T}}_0,$$

$$\mathcal{T}_1 = \frac{1}{2}u^2 = \frac{1}{2}(v_x^2 + v^4 + 2v^2v_x) \sim \tilde{\mathcal{T}}_1,$$

where '$\sim$' refers to the equivalence relation (1.2.4c). The substitution (1.2.8) is now referred to as the Miura transformation (more properly, Miura map). Such maps play a very important role in the Hamiltonian theory of soliton equations, as will be seen in Chapter 11. Furthermore, (1.2.8) gives a direct relation between equations (1.1.1) and (1.2.6b) since

$$u_t = -(\partial + 2v)v_t = -(\partial + 2v)(v_{xxx} - 6v^2v_x) \tag{1.2.9a}$$

$$= (-v_x - v^2)_{xxx} + 6(-v_x - v^2)(-v_{xx} - 2vv_x) = u_{xxx} + 6uu_x \tag{1.2.9b}$$

where $\partial \equiv \mathrm{d}/\mathrm{d}x$. Thus if v satisfies the MKdV equation (1.2.6b), then u satisfies the KdV equation (1.1.1). The remarkable occurrence in this calculation is that the mess of terms in v and its x-derivatives, resulting from the right-hand side of (1.2.9a), should collect together to give an expression entirely in terms of the combination $(-v_x - v^2)$ and thus u.

Remark

This is a one-way passage since we cannot deduce that if u satisfies the KdV equation then v satisfies the MKdV equation.

In order to prove the existence of an infinite number of conservation laws for the KdV equation we need Gardner's generalisation of the Miura map [36]. First, we exploit the Galilean symmetry of the KdV equation (and the lack of such for the MKdV equation) to introduce a parameter into (1.2.8). Specifically, we add a constant onto the function u, so that

$$u - \frac{1}{4\varepsilon^2} = -v_x - v^2. \tag{1.2.10a}$$

Now define w by

$$w = \frac{v}{\varepsilon} + \frac{1}{2\varepsilon^2} \text{ or } v = \varepsilon w - \frac{1}{2\varepsilon} \tag{1.2.10b}$$

so that

$$u = w - \varepsilon w_x - \varepsilon^2 w^2. \tag{1.2.11a}$$

Carrying out a manipulation similar to (1.2.9) we find

$$u_t = (1 - 2\varepsilon^2 w - \varepsilon\partial) w_t \tag{1.2.11b}$$

and that if w satisfies

$$\begin{aligned} w_t &= w_{xxx} + 6ww_x - 6\varepsilon^2 w^2 w_x \\ &= (w_{xx} + 3w^2 - 2\varepsilon^2 w^3)_x \end{aligned} \tag{1.2.11c}$$

then u satisfies the KdV equation (1.1.1). Gardner's equation (1.2.11c) reduces to the KdV equation and Gardner's map (1.2.11a) reduces to the identity when $\varepsilon = 0$. Maps with these characteristics are now widespread and referred to as deformations [11, 24, 25]. To prove the existence of an infinite number of conservation laws, Gardner expanded w asymptotically:

$$w = u + \sum_{k=1}^{\infty} \varepsilon^i w_i \tag{1.2.12a}$$

On substituting this into (1.2.11a) we can solve recursively for w_i, which are written in terms of u and its x-derivatives. The first few are

$$\begin{aligned} &w_1 = u_x,\ w_2 = u_{xx} + u^2,\ w_3 = (u_{xx} + 2u^2)_x, \\ &w_4 = (u_{xxx} + 6uu_x)_x + 2\left(u^3 - \tfrac{1}{2}u_x^2\right). \end{aligned} \tag{1.2.12b}$$

Since Gardner's equation is in conservation form, w, and therefore *each of* w_i, is conserved (in particular $w_2 \sim 2\mathcal{T}_1$, $w_4 \sim 2\mathcal{T}_2$). It is easy to show that the odd part of this series is an exact derivative and therefore trivial as a conserved density (but essential for constructing the series!) and that the resulting conserved quantities are independent.

Remark

If $\{\mathcal{T}_n[u]\}_{n=0}^{\infty}$ is the sequence of non-trivial conserved densities for the KdV equation, then $\{\hat{\mathcal{T}}_n[v]\}_{n=0}^{\infty}$ and $\{\hat{\mathcal{T}}_n[w]\}_{n=0}^{\infty}$, where $\hat{\mathcal{T}}_n[v] = \mathcal{T}_n[-v_x - v^2]$ and $\hat{\mathcal{T}}_n[w] = \mathcal{T}_n[w - \varepsilon w_x - \varepsilon^2 w^2]$, are corresponding infinite sequences of non-trivial conserved densities for the MKdV and Gardner's equation respectively.

1.3 Inverse spectral transform

The Miura map (1.2.10a) can be viewed as a Riccati equation for v and thus linearised by the substitution $v = \psi_x/\psi$, giving

$$\psi_{xx} + u\psi = \lambda\psi \tag{1.3.1a}$$

where we have written $\lambda = 1/4\varepsilon^2$. This is the time-independent Schrödinger equation (well known from quantum theory) with $u(x, t)$ playing the role of potential and λ the energy. It is important to realise that *t is not the time of the time-dependent Schrödinger equation*. We think of x as the spatial variable and t as a parameter. It is natural to ask how λ and ψ change with t as $u(x, t)$ evolves from some initial state according to the KdV equation. Gardner, Green, Kruskal and Miura (GGKM) discovered [13, 14] the remarkable fact that the (discrete part of the) spectrum necessarily remains constant in 'time' while the corresponding wave functions ψ evolve according to a very simple *linear differential equation*.

Today (following Lax [30]) we usually take the opposite route. We postulate that ψ evolves through a linear differential equation:

$$\psi_t = \mathbf{P}\psi \tag{1.3.1b}$$

and that the nonlinear evolution equation can be represented by the *Lax equations*:

$$\mathbf{L}_t = [\mathbf{P}, \mathbf{L}] \tag{1.3.1c}$$

where (in the present case) $\mathbf{L} = \partial^2 + u$ is the Schrödinger operator. A consequence of (1.3.1c) is that all eigenvalues corresponding to the function $u(x, t)$ remain constant. Any nonlinear evolution equation defined by such a Lax equation is thus called an *isospectral flow*. For the case of the KdV equation (1.1.1), $\mathbf{P}$ is given by

$$\mathbf{P} = 4\partial^3 + 6u\partial + 3u_x + C \tag{1.3.1d}$$

where C is a constant.

Remark
For the purpose of most algebraic treatments we may set $C = 0$ since this commutes with $\mathbf{L}$ and thus contributes nothing to the evolution $\mathbf{L}_t$. However, for the purpose of scattering theory (described below and in Chapter 2) this constant should not be taken to be zero.

The identification of the KdV equation as an isospectral flow of the Schrödinger operator enabled GGKM to devise a method of solving the KdV equation, called the inverse scattering method or inverse spectral transform (IST). This is a direct generalisation of the Fourier transform used to solve linear equations. The important inversion formulae had been derived by Gel'fand and Levitan in 1955 [16] in the context of quantum-mechanical scattering theory. These enable the potential u to be constructed out of the spectral or scattering data. Details of this can be found in Chapter 2 but a brief exposition is given below.

Consider the Schrödinger equation (1.3.1a) with potential $u(x)$ (suppressing t dependence for the moment) belonging to a class of functions which vanish *rapidly* as $x \to \pm\infty$. For instance, we can impose

$$\int_{-\infty}^{\infty} |u(x)|(1 + |x|)\mathrm{d}x < \infty \tag{1.3.2}$$

to ensure that only a finite number of bound states exist. The latter correspond to the discrete eigenvalues $\lambda_n = \kappa_n^2$, $n = 1, \ldots, N$. For each such eigenvalue the asymptotic form of the solution $\psi_n(x)$ must be a linear combination of $e^{\pm\kappa_n x}$ (since $u(x) \to 0$ as $x \to \pm\infty$). Let us impose boundary conditions:

$$\psi_n(x) \sim e^{\kappa_n x} \text{ as } x \to -\infty, \tag{1.3.3a}$$

$$\psi_n(x) \sim b_n e^{-\kappa_n x} \text{ as } x \to +\infty. \tag{1.3.3b}$$

Next we have the scattering states, corresponding to the continuous spectrum, with $\lambda = -k^2$. In this case we may impose boundary conditions:

$$\psi_k(x) \sim T(k)e^{-ikx} \text{ as } x \to -\infty, \tag{1.3.4a}$$

$$\psi_k(x) \sim e^{-ikx} + R(k)e^{ikx} \text{ as } x \to +\infty, \tag{1.3.4b}$$

where $R(k)$ and $T(k)$ are called the reflection and transmission coefficients, in accordance with their interpretation in quantum mechanics. For a given potential $u(x)$ we can (in principle) calculate κ_n, b_n, $R(k)$ and $T(k)$ although explicit expressions can only be calculated for special potentials. The scattering data is defined to be

$$S = \{(\kappa_n, b_n)_{n=1}^{N}, R(k), k \text{ real}\}, \tag{1.3.5}$$

and from this we can also calculate $T(k)$. Gel'fand and Levitan [16] proved that this scattering data is enough to characterise and reconstruct the potential function $u(x)$. This reconstruction requires the solution of a *linear integral equation*.

The inverse scattering method of solving the KdV equation requires three steps:

(1) Construct the scattering data S_0 at $t = 0$ from the initial condition $u(x, 0) = u_0(x)$.
(2) Calculate the scattering data $S(t)$ at a later time, corresponding to $u(x, t)$.

(3) Construct the potential $u(x, t)$ from the scattering data $S(t)$.

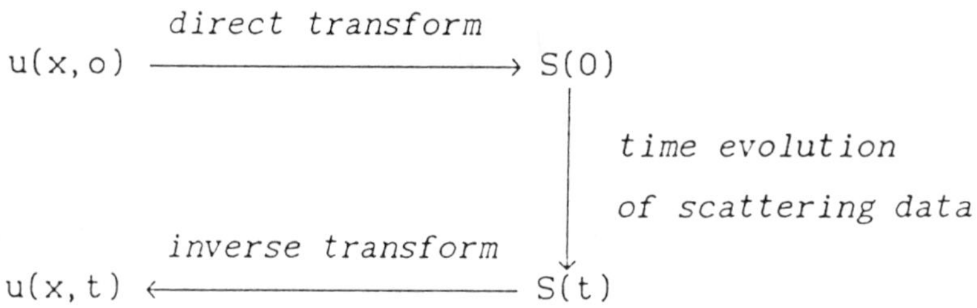

Figure 1.2 Inverse spectral transform

The first step is called the *direct spectral problem* while the third is the *inverse spectral problem*. Both of these steps can in principle be carried out. The remarkable discovery of GGKM was that the scattering data evolves in an extremely simple way while the potential function evolves according to the KdV equation. Specifically, the linear evolution (1.3.1b) for $\psi(t)$ implies a *linear constant coefficient* evolution for $S(t)$. Indeed, the asymptotic form of the linear evolution (1.3.1b) has constant coefficients:

$$\psi_t = 4\psi_{xxx} + C\psi. \tag{1.3.6a}$$

Thus:

(i) (1.3.3a) $\Rightarrow C_n = -4\kappa_n^3$,
(ii) (1.3.3b) $\Rightarrow b_{nt} = -8\kappa_n^3 b_n$,
(iii) (1.3.4b) $\Rightarrow C(k) = -ik^3,\ R_t = -2ik^3 R$,
(iv) (1.3.4a) $\Rightarrow T_t = 0$.

And, of course, $\kappa_{nt} = 0$. Thus, if

$$S_0 = \{(\kappa_n, b_n)_{n=1}^N, R_0(k)\} \tag{1.3.6b}$$

then

$$S(t) = \{(\kappa_n, b_n e^{-8\kappa_n^3 t})_{n=1}^N, R_0(k)e^{-2ik^3 t}\} \tag{1.3.6c}$$

and $T(k, t) = T_0(k)$.

This inverse spectral transform can be depicted schematically as in Figure 1.2. It will be seen in Chapter 2 that the 'hard part' of the inversion formula corresponds to the reflection coefficient. Remarkably, there exist 'reflectionless potentials' for which an *explicit* formula can be constructed for $u(x, t)$. This is the *N-soliton solution*, where N is the number of discrete eigenvalues.

Before ending this section we should consider the Burgers equation:

$$u_t = u_{xx} + 2uu_x \tag{1.3.7a}$$

which can be *linearised* by the Hopf–Cole transformation [7, 18]:

$$u = \frac{h_x}{h} \tag{1.3.7b}$$

where $h(x, t)$ satisfies the heat equation:

$$h_t = h_{xx}. \tag{1.3.7c}$$

Thus, to solve the initial-value problem for the Burgers equation we carry out the following three steps:

(1) direct transform: $h(x, 0) = \exp(\int u(x, 0)\mathrm{d}x)$,
(2) calculate $h(x, t)$ using Fourier transform,
(3) inverse transform: $u(x, t) = h_x/h$,

which can be represented by a diagram analogous to Figure 1.2.

The remarkable thing here is that (1.3.7b) transforms the Burgers equation into a *constant coefficient* linear equation, thus giving a very simple method of solution. This should be contrasted with (1.3.1a) which can be considered as a transformation between the KdV equation and the linear equation (1.3.1b,d). The latter has coefficients which *depend upon the unknown function* $u(x, t)$. It is for this reason that we need the much more complicated IST to solve the KdV equation.

1.4 Lax hierarchy

The operator (1.3.1d) is just one example of **P** given by Lax. In [30] he showed that there exists an infinite sequence of such operators, one of each odd order. The first three of these are:

$$\mathbf{P}_1 = \partial, \tag{1.4.1a}$$

$$\mathbf{P}_3 = 4\partial^3 + 6u\partial + 3u_x, \tag{1.4.1b}$$

$$\mathbf{P}_5 = 16\partial^5 + 40u\partial^3 + 60u_x\partial^2 + (50u_{xx} + 30u^2)\partial + 15u_{xxx} + 30uu_x, \tag{1.4.1c}$$

corresponding to isospectral flows:

$$u_{t_1} = u_x, \tag{1.4.2a}$$

$$u_{t_3} = u_{xxx} + 6uu_x, \tag{1.4.2b}$$

$$u_{t_5} = u_{xxxxx} + 10uu_{xxx} + 20u_xu_{xx} + 30u^2u_x, \tag{1.4.2c}$$

where I have labelled the times in an obvious manner.

Remark

Since the left-hand side of equation (1.3.1c) is a differential operator of *order zero*, so is the right-hand side. This gives precisely the number of equations required to determine the coefficients of an operator **P** of the form:

$$\mathbf{P} = \partial^n + b_{n-2}\partial^{n-2} + \ldots + b_0.$$

The zeroth-order part of the right-hand side of (1.3.1c), when equated with u_t, gives rise to the corresponding nonlinear evolution equation.

Thus, there exists an infinite hierarchy of isospectral flows of the Schrödinger operator, each of which is soluble by means of the IST of Section 1.3. Furthermore these flows can be shown to mutually commute so that u can be considered as a function of *all* t_i simultaneously: $u(t_1, t_3, t_5, \ldots)$. The conserved densities of Section 1.2 are constants of motion for *each* flow of this infinite hierarchy.

An alternative way of generating the hierarchy (1.4.2) is to use the recursion operator, presented in [14] but attributed to Lenard. This is an integro-differential operator:

$$\mathbf{R} = \partial^2 + 4u + 2u_x\partial^{-1} \tag{1.4.3a}$$

where ∂^{-1} denotes the formal inverse of ∂. The infinite hierarchy of isospectral flows satisfies the relation:

$$\mathbf{R}u_{t_{2n-1}} = u_{t_{2n+1}},\ n = 1, 2, \ldots \tag{1.4.3b}$$

The remarkable fact is that at each step $u_{t_{2n+1}}$ is an exact derivative so that the next integration can be carried out to give a locally defined expression. The recursion operator is associated with the squared eigenfunctions of **L**. Let $A = \psi^2$. Then,

$$A_{xxx} + 4uA_x + 2u_xA = 4\lambda A_x \tag{1.4.4a}$$

or

$$\mathbf{R}A_x = 4\lambda A_x \tag{1.4.4b}$$

so that A_x is an eigenfunction of **R** corresponding to eigenvalue 4λ when ψ is an eigenfunction of **L** corresponding to eigenvalue λ. Multiplying (1.4.4a) by A and integrating:

$$AA_{xx} - \tfrac{1}{2}A_x^2 + 2(u - \lambda)A^2 = C(\lambda) \tag{1.4.5a}$$

where $C(\lambda)$ is a constant. Putting $C(\lambda) = -4\lambda$ we can find an asymptotic expansion for A:

$$A = 2 + \frac{u}{\lambda} + \frac{\frac{1}{4}(u_{xx} + 3u^2)}{\lambda^2} + \ldots + \frac{A_n}{\lambda^n} + \ldots \tag{1.4.5b}$$

A_{n+1} can be found recursively in terms of local expressions in $A_1, \ldots, A_n$ since

$$8A_{n+1} = 4A_0A_{n+1} = \text{expression in only } A_1, \ldots, A_n. \tag{1.4.5c}$$

Equation (1.4.4a) is an example of a bi-Hamiltonian ladder, discussed in detail in Chapter 11. In the context of bi-Hamiltonian systems, such as the KdV equation, the integro-differential recursion operator is the formal quotient of two (generally matrix) differential operators:

$$\mathbf{R} = \mathbf{B}_1\mathbf{B}_0^{-1} \tag{1.4.6}$$

where $\mathbf{B}_0$ and $\mathbf{B}_1$ are Hamiltonian operators (see Chapter 11 for details). However, recursion operators appear in a more general context [39] than Hamiltonian systems, such as with the Burgers equation (1.3.7a). The integro-differential operator

$$\mathbf{R} = \partial + u + u_x\partial^{-1} \tag{1.4.7}$$

generates an infinite sequence of flows, starting from the x-translation flow, the first three of which are:

$$u_{t_1} = u_x, \tag{1.4.8a}$$

$$u_{t_2} = u_{xx} + 2uu_x, \tag{1.4.8b}$$

$$u_{t_3} = u_{xxx} + 3(uu_{xx} + u_x^2 + u^2u_x). \tag{1.4.8c}$$

Even though the Burgers equation possesses this infinite number of commuting flows, it is dissipative and does *not* possess a corresponding infinite sequence of conserved densities.

A useful characterisation of recursion operators of an evolution equation

$$u_t = K[u], \tag{1.4.9a}$$

where $K[u]$ is a locally defined function of u and its x-derivatives, is given by the equation [39]:

$$\mathbf{R}_t = [\mathbf{K}', \mathbf{R}] \tag{1.4.9b}$$

where the differential operator $\mathbf{K}'$ is the Fréchet derivative of K, defined by $\mathbf{K}'[u]v = \mathrm{d}/\mathrm{d}\varepsilon\, K[u + \varepsilon v]|_{\varepsilon=0}$ (equation (11.3.6) of Chapter 11 is the adjoint of this). Indeed this can be considered as a Lax equation representing the evolution equation (1.4.9a) [6].

Hierarchies of commuting flows, such as (1.4.2) and (1.4.8) can be considered as *generalised symmetries* of any one (and thus all) of the equations in the hierarchy. This is a generalisation of the more familiar geometric symmetry (see [40] for a clear introduction to this subject) and just refers to the fact that these flows commute. Let time τ parametrise the evolution

$$u_\tau = f[u]. \tag{1.4.10a}$$

Then, if

$$\mathbf{K}'f - \mathbf{f}'K = 0, \tag{1.4.10b}$$

$u(x, t, \tau)$ can be considered as *simultaneously* a function of t and τ, and $u_{\tau t} = u_{t\tau}$. The latter equality, together with the definition of $\mathbf{K}'$, implies a linear equation for f:

$$f_t = \mathbf{K}'f. \tag{1.4.10c}$$

This is the symmetry equation of (1.4.9a) satisfied by any right-hand side of the associated hierarchy. It is easy to see that if f satisfies (1.4.10c) and $\mathbf{R}$ satisfies (1.4.9b) then $\mathbf{R}f$ also satisfies (1.4.10c). Thus the recursion operator can be considered as mapping symmetries into symmetries. The hierarchies (1.4.2) and (1.4.8) are, in this way, generated by the x-translation symmetry $u_{t_1} = u_x$. Starting with the scale symmetry of either equation, one can generate hierarchies of complicated integro-differential equations, with x and t dependent coefficients, which commute with the KdV and Burgers equations.

I have taken a considerable amount of space in this introduction to describe the KdV equation and its associated spectral transform and hierarchy. This should act as a paradigm for the many spectral problems and associated (soluble) nonlinear evolution equations which appear in the remainder of this chapter and of the book.

1.5 Other spectral problems

Until 1971 the KdV equation (and associated hierarchy) was the only known example of a nonlinear evolution equation soluble by inverse scattering. However, in that year, Zakharov and Shabat (ZS) showed how to integrate the nonlinear Schrödinger (NLS) equation by IST [51]. The NLS equation:

$$\mathrm{i}q_t = q_{xx} + 2|q|^2 q \tag{1.5.1}$$

for complex potential q, is very important in nonlinear optics, but occurs in many other circumstances (as can be found in Chapter 6). In 1972 Wadati [45] gave the spectral transform for the MKdV equation (1.2.6b) and Ablowitz, Kaup, Newell and Segur (AKNS) solved the sine-Gordon equation [1]. All of these spectral problems (including (1.1.1)) were included in a general scheme by AKNS [2]. This is a first-order differential equation for a two-component vector $(\psi_1, \psi_2)^{\mathrm{T}}$:

$$\begin{bmatrix}\psi_1\\ \psi_2\end{bmatrix}_x = \begin{bmatrix}\lambda & q\\ r & -\lambda\end{bmatrix}\begin{bmatrix}\psi_1\\ \psi_2\end{bmatrix} \equiv \mathbf{F}\psi \tag{1.5.2a}$$

where $q(x, t)$, $r(x, t)$ are a pair of potential functions and λ is the spectral parameter. They considered time evolutions of this vector of the form:

$$\begin{bmatrix}\psi_1\\ \psi_2\end{bmatrix}_t = \begin{bmatrix}A & B\\ C & -A\end{bmatrix}\begin{bmatrix}\psi_1\\ \psi_2\end{bmatrix} \equiv \mathbf{G}\psi \tag{1.5.2b}$$

where A, B and C are functions of q, r and their x-derivatives and of the spectral parameter λ. Then integrability conditions $\psi_{xt} = \psi_{tx}$ imply

$$\mathbf{F}_t - \mathbf{G}_x + [\mathbf{F}, \mathbf{G}] = 0, \tag{1.5.3a}$$

which is written explicitly as

$$q_t = B_x - 2\lambda B + 2qA, \tag{1.5.3b}$$

$$r_t = C_x + 2\lambda C - 2rA, \tag{1.5.3c}$$

$$A_x = qC - rB. \tag{1.5.3d}$$

The (now-called) AKNS method starts by stipulating some λ-dependence of A, B and C. To obtain the hierarchy of polynomial flows, they substituted polynomial expressions for A, B and C:

$$\mathbf{G} = \sum_{i=0}^{N} \mathrm{G}_{N-i}\lambda^i. \tag{1.5.4a}$$

An immediate consequence is that

$$B_0 = C_0 = 0, \; A_0 = a_0 = \text{const.} \tag{1.5.4b}$$

It is then possible to recursively solve for all further components of the expansion, finally arriving at the nonlinear evolution equations:

$$q_t = B_{Nx} + 2qA_N, \tag{1.5.5a}$$

$$r_t = C_{Nx} - 2rA_N. \tag{1.5.5b}$$

Remark
At each step it is possible to introduce an arbitrary constant of integration, corresponding to adding lower-order isospectral flows to (1.5.5).

The second-order expansion is

$$A = a_0(\lambda^2 - \tfrac{1}{2}qr), \; B = a_0(\lambda q + \tfrac{1}{2}q_x),$$
$$C = a_0(\lambda r - \tfrac{1}{2}r_x); \tag{1.5.6a}$$

$$q_t = \tfrac{1}{2}a_0(q_{xx} - 2q^2r), \; r_t = \tfrac{1}{2}a_0(-r_{xx} + 2qr^2). \tag{1.5.6b}$$

With $a_0 = -2\mathrm{i}$ and $\mathrm{r} = -q^*$ we get the NLS equation (1.5.1).

The third-order expansion is

$$A = a_0(\lambda^3 - \tfrac{1}{2}\lambda qr + \tfrac{1}{4}(qr_x - rq_x)), \; B = a_0(\lambda^2 q + \tfrac{1}{2}\lambda q_x + \tfrac{1}{4}(q_{xx} - 2q^2 r)),$$
$$C = a_0(\lambda^2 r - \tfrac{1}{2}\lambda r_x + \tfrac{1}{4}(r_{xx} - 2qr^2)); \tag{1.5.7a}$$

$$q_t = \tfrac{1}{4}a_0(q_{xxx} - 6qrq_x), \; r_t = \tfrac{1}{4}a_0(r_{xxx} - 6qrr_x). \tag{1.5.7b}$$

When $r = q$ (and $a_0 = 4$) we get the MKdV equation (1.2.6b), while with $r = -1$ we get the KdV equation (1.1.1). Furthermore, we may set $r = -1 + \varepsilon^2 q$ to get Gardner's equation (1.2.11c).

Remark
It is also possible to use the AKNS spectral problem to solve the sine- and sinh-Gordon equations, whenever we have the MKdV reduction $r = \mp q$ respectively. This is discussed in Chapter 12.

We can obtain the general recursion operator for the AKNS system by eliminating A to get

$$\begin{bmatrix} q_t \\ -r_t \end{bmatrix} = \begin{bmatrix} \partial - 2q\partial^{-1}r & 2q\partial^{-1}q \\ -2r\partial^{-1}r & -\partial + 2r\partial^{-1}q \end{bmatrix}\begin{bmatrix} B \\ C \end{bmatrix} - 2\lambda\begin{bmatrix} B \\ C \end{bmatrix} \equiv (\Lambda - 2\lambda\mathbf{I})\begin{pmatrix} B \\ C \end{pmatrix}. \tag{1.5.8a}$$

The operator Λ is the recursion operator which both recursively generates the components $(B_i, C_i)^{\mathrm{T}}$ and the equations themselves.

Remark
By substituting $r = -1$, it is easy to obtain

$$q_t = -\tfrac{1}{2}(C_{xxx} + 4qC_x + 2q_xC) + 2\lambda^2 C_x, \tag{1.5.8b}$$

giving rise to the same recursion as (1.4.4a).

Equations (1.3.1a) and (1.5.2a) are not the only spectral problems which arise, but they are the simplest. We can generalise these in a number of ways in order to obtain larger systems of equations. The basic Lax approach can be extended to higher-order differential operators [15, 48, 52], while to generalise (1.5.2) we just need to take larger vectors and matrices (see Chapter 12). We can also allow Lax operators and matrices to have a more complicated dependence upon the spectral parameter λ (see Chapter 11).

The Schrödinger and AKNS spectral problems have been covered in considerable detail in this chapter so that from the beginning the importance of the linear spectral problem is understood by the reader. Many more examples occur throughout the book.

1.6 The organisation of this book

The remaining chapters of the book are written by a variety of authors and grouped under five headings: solution methods, physical applications, Hamiltonian theory, algebraic and geometric structures, and methods of testing for complete integrability.

Chapter 2 gives a detailed description of IST in (1 + 1)-dimensions, already touched upon in this introduction. This is extended in Chapter 3 to 'two-dimensional spectral transforms' which enable us to solve non-linear evolution equations in (2 + 1)-dimensions (usually interpreted as two space and one time). The first step is to represent such an equation as the integrability conditions of a pair of linear equations, generalising

(1.3.1) or (1.5.3). This is achieved by replacing the spectral parameter λ by the differential operator $\partial/\partial y$. Since the latter does not commute with functions of y, the ensuing manipulations are much more complicated. This is the starting point in Chapter 3, which is principally concerned with the actual spectral transform.

In 1972 Hirota [17] devised his 'direct method' of finding the N-soliton solution of the KdV equation. This is an extremely simple and effective method which leads directly to the determinantal form of N-soliton solution, known from IST, but does *not* itself use IST! Although originally *ad hoc*, Hirota's method has been placed in a deep algebraic setting by Sato's group in Kyoto (see [20]). Both the calculational and the vertex-operator approach are covered in the present book by Chapters 4 and 13.

Bäcklund devised his method of solving the sine-Gordon equation (and thus constructing surfaces of constant negative cuvature) in 1875. Even though Bäcklund transformations appeared a few times in the literature over the years, they did not play a significant role in mathematics and mathematical physics until the era of the soliton. In 1967 Lamb [26] used the Bäcklund transformation to construct multi-soliton solutions of the sine-Gordon equation, which was only solved by inverse scattering much later [1]. Wahlquist and Estabrook obtained the Bäcklund transformation for the KdV equation in 1973 [46] and in [27, 28] Lamb resurrected Clairin's method, constructing Bäcklund transformations for other equations. In [5] Chen showed the relationship between the Bäcklund transformation and the linear spectral problem. Weiss [47] uses the Painlevé method (see Chapter 16) to construct Bäcklund transformations and linear spectral problems. These topics and many applications and examples are discussed in Chapter 5.

Part III of this book deals with physical applications. Soliton equations, being (usually) not only Hamiltonian but also completely integrable, are very very rare in the 'space' of all possible equations but, nevertheless, do occur in an enormous number of physical theories. This is partly due to the ubiquitous nature of both the KdV and NLS equations, but there are numerous other soliton equations which occur in physics. In this book we have examples from laser physics, fluid dynamics, general relativity and protein dynamics. We regrettably do not have the Great Red Spot of Jupiter [32], Josephson junctions [41] and numerous other subjects (see [44] for a good survey of solitons in physics). However, it is hoped that the chapters included in Part III will be enough to convince the physicist that the remaining mathematical chapters are worth reading.

The two chapters of Part IV are concerned with Hamiltonian theory. The first covers finite-dimensional (or discrete) systems such as the Toda lattice, while the second is concerned with nonlinear evolution equations as infinite-dimensional Hamiltonian systems. This is very important and *very* large, and therefore much has been omitted. Soliton theory has stimulated a lot of interest in Hamiltonian dynamics and given rise to many new examples of completely integrable systems, both finite- and infinite-dimensional. The whole topic of bi-Hamiltonian systems grew directly out of soliton theory.

The appearance of the commutator in the integrability conditions (1.3.1c) and (1.5.3a) immediately places us in the realm of Lie algebras. Indeed, the generalisation of the AKNS system from 2×2 to larger matrices is best carried out in the framework of simple Lie algebras. An *ad hoc* generalisation from 2×2 to 3×3 matrices is possible, but to go further in a systematic way it is best to use the Cartan classification of simple Lie algebras. The spectral problem of the KdV, MKdV, NLS and other equations can be generalised in this way, as described in Chapter 12. During the present decade a deep connection was discovered between integrable equations of 'KdV' type and infinite-dimensional Lie algebras. Chapter 13 is a brief introduction to this topic (see also [20] and references therein). The IST described in Chapter 2 is suitable for potentials which rapidly decrease as $x \to \pm\infty$. However, for potentials which are periodic (in x) or finite gap type it is necessary to introduce algebraic-geometric methods. These are described in Chapter 14.

The final section of this book is concerned with the question of testing an equation for complete integrability. We suppose that a physicist or engineer (for instance) gives us an equation (or class of equations) which is suspected of being exactly soluble. How do we test the equation(s) and, where appropriate, construct the associated linear spectral problem. In this book we present two approaches to this problem. Chapter 15 describes the Wahlquist–Estabrook prolongation method which, from the start, attempts to construct a linear spectral problem. Chapter 16 introduces the Painlevé method, in which the main emphasis is on *testing* for integrability, rather than constructing the associated linear spectral problem. However, in some cases it is possible to continue the calculation so as to construct the linear problem. A third method is to search for generalised symmetries (commuting flows) [10, 19] or constants of the motion [33].

Inevitably, when editing this book I have had to be very selective. The topics have been chosen to be 'elementary' since they are intended as an introductory survey (but not as a textbook!) of the subject for postgraduate students and practising scientists. Thus an alarming amount of abstract algebraic and Hamiltonian theory has been omitted and numerous physical applications not touched upon. The reader can get some flavour of the range of current developments by reading a recent conference proceeding (see, for example, [8, 31]). However, it is hoped that this book will be a useful complement to the several textbooks available [3, 4, 9, 29, 37, 38], giving broader scope but less foundations.

The importance of soliton theory is without question: it has stimulated a tremendous amount of 'pure' mathematics and has an enormously wide range of applications. Furthermore, solitons will soon enter our everyday life when they begin to be used in telecommunications (nonlinear optics), the design of computer hardware (Josephson junctions), and many other places.

References

[1] Ablowitz, M. J., Kaup, D. J., Newell, A. C. and Segur, H. Method for solving the sine-Gordon equation, *Phys. Rev. Lett.* **30**, 1262–4 (1973).

[2] Ablowitz, M. J., Kaup, D. J., Newell, A. C. and Segur, H. The inverse scattering transform – Fourier analysis for nonlinear problems, *Stud. Appl. Math.* **53**, 249–315 (1974).

[3] Ablowitz, M. J. and Segur, H. *Solitons and the Inverse Scattering Transform*, SIAM Studies in Applied Mathematics 4, SIAM, Philadelphia (1981).

[4] Calogero, F. and Degasperis, A. *Spectral Transform and Solitons I*, North-Holland, Amsterdam (1982).

[5] Chen, H. H. General derivation of Bäcklund transformations from inverse scattering problems, *Phys. Rev. Lett.* **33**, 925–8 (1974).

[6] Chen, H. H., Lee, Y. C. and Lin, C. S. Integrability of nonlinear Hamiltonian systems by inverse scattering method, *Phys. Scr.* **20**, 490–2 (1979).

[7] Cole, J. D. On a quasilinear parabolic equation occurring in aerodynamics, *Quart. App. Math.* **9**, 225–36 (1951).

[8] Degasperis, A., Fordy, A. P. and Lakshmanan, M. (eds), *Nonlinear Evolution Equations: Integrability and Spectral Methods*, Manchester University Press, Manchester (1990).

[9] Dodd, R. K., Eilbeck, J. C., Gibbon, J. D. and Morris, H. C. *Solitons and Nonlinear Wave Equations*, Academic Press, London (1982).

[10] Fokas, A. S. A symmetry approach to exactly soluble evolution equations, *J. Math. Phys.* **21**, 1318–26 (1980).

[11] Fordy, A. P. Projective representations and deformations of integrable systems, *Proc. R. Ir. Acad.* **83A**, 75–93 (1983).

[12] Gardner, C. S. The Koretweg–de Vries equation and generalizations. IV. The Korteweg–de Vries equation as a Hamiltonian system, *J. Math. Phys.* **12**, 1548–51 (1971).

[13] Gardner, C. S., Greene, J. M., Kruskal, M. D. and Miura, R. M. Method for solving the Korteweg–de Vries equation, *Phys. Rev. Lett.* **19**, 1095–7 (1967).

[14] Gardner, C. S., Greene, J. M., Kruskal, M. D. and Miura, R. M. The Korteweg–de Vries equation and generalisations. VI. Methods for exact solution, *Commun. Pure Appl. Math.* **27**, 97–133 (1974).

[15] Gel'fand, I. M. and Dikii, L. Fractional powers of operators and Hamiltonian systems, *Funct. Anal. and Apps.* **10**, 259–73 (1976).

[16] Gel'fand, I. M. and Levitan, B. M. On the determination of a differential equation from its spectral function, *Am. Math. Soc. Transl. Ser. 2* **1**, 259–309 (1955).

[17] Hirota, R. Exact solution of the Korteweg–de Vries equation for mutliple collisions of solitons, *Phys. Rev. Lett.* **27**, 1192–4 (1971).

[18] Hopf, E. The partial differential equation $u_t + uu_x = \mu u_{xx}$, *Commun. Pure Appl. Math.* **3**, 201–30 (1950).

[19] Ibragimov, N. J. and Shabat, A. B. Evolutionary equations with nontrivial Lie–Bäcklund group, *Funct. Anal. Apps.* **14**, 1928 (1980).
[20] Jimbo, M. and Miwa, T. (eds), *Nonlinear Integrable Systems. Classical Theory and Quantum Theory*, World Scientific, Singapore, (1983).
[21] Korteweg, D. J. and de Vries, G. On the change of form of long waves advancing in a rectangular canal, and on a new type of long stationary waves, *Philos. Mag. Ser. 5*, **39**, 422–43 (1895).
[22] Kruskal, M. D. Nonlinear wave equations. In *Dynamical Systems, Theory and Applications*, J. Moser (ed.), Lecture Notes in Physics *38*, Springer, Berlin (1975).
[23] Kruskal, M. D., Miura, R. M., Gardner, C. S. and Zabusky, N. J. Korteweg–de Vries equation and generalizations. V. Uniqueness and non-existence of polynomial conservation laws, *J. Math. Phys.* **11**, 952–60 (1970).
[24] Kupershmidt, B. A. On the nature of the Gardner transformation, *J. Math. Phys.* **22**, 449–51 (1981).
[25] Kupershmidt, B. A. Deformations of integrable systems, *Proc. R. Ir. Acad.* **83A**, 45–74 (1983).
[26] Lamb, G. L. Propagation fo ultrashort optical pulses, *Phys. Letts. A* **25**, 181–2 (1967).
[27] Lamb, G. L. Bäcklund transformations for certain nonlinear evolution equations, *J. Math. Phys.* **15**, 2157–65 (1974).
[28] Lamb, G. L. Bäcklund transformations at the turn of the century. In *Bäcklund transformations* R. M. Miura (ed.), Lecture Notes in Mathematics 515, Springer, Berlin (1976).
[29] Lamb, G. L. *Elements of Soliton Theory*, Wiley, New York (1980).
[30] Lax, P. D. Integrals of nonlinear equations of evolution and solitary waves, *Commun. Pure Appl. math.* **21**, 467–90 (1968).
[31] Leon, J. J.-P. (ed.), *Nonlinear Evolutions*, World Scientific, Singapore (1988).
[32] Maxworthy, T. and Redekopp, L. G. Theory of the Great Red Spot and other observed features of the Jovian atmosphere, *Icarus* **29**, 261–71 (1976).
[33] Mikhailov, A. V., Shabat, A. B. and Yamilov, R. I. Extension of the module of invertible transformations, *Commun. Math. Phys.* **115**, 1–19 (1988).
[34] Miura, R. M. Korteweg–de Vries equation and generalizations I. A remarkable explicit nonlinear transformation, *J. Math. Phys.* **9**, 1202–4 (1968).
[35] Miura, R. M. The Korteweg–de Vries equation: a survey of results, *SIAM Review* **18**, 412–59 (1976).
[36] Miura, R. M., Gardner, C. S. and Kruskal, M. D. Korteweg–de Vries equation and generalizations. II. Existence of conservation laws and constants of motion, *J. Math. Phys.* **9**, 1204–9 (1968).
[37] Newell, A. C. *Solitons in Mathematics and Physics*, SIAM, Philadelphia (1985).
[38] Novikov, S. P., Manakov, S. V., Pitaevskii, L. P. and Zakharov, V. E. *Theory of Solitons*, Plenum, New York (1984).
[39] Olver, P. J. Evolution equations possessing infinitely many symmetries, *J. Math. Phys.* **18**, 1212–5 (1977).
[40] Olver, P. J. *Application of Lie Groups to Differential Equations*, Springer, Berlin (1986).
[41] Pederson, N. F. Solitons in Josephson transmission lines. In *Solitons*, S. E. Trullinger, V. E. Zakharov and V. L. Pokrovsky (eds), 469–501, North-Holland, Amsterdam (1986).
[42] Scott Russell, J. Report on waves. *Fourteenth meeting of the British Associa-*

tion for the Advancement of Science, 311–90, John Murray, London (1844).

[43] Su, C. H. and Gardner, C. S. Korteweg–de Vries equation and generalizations. III. Derivation of the Korteweg–de Vries and Burgers equation, *J. Math. Phys.* **10**, 536–9 (1969).

[44] Trullinger, S. E., Zakharov, V. E. and Pokrovsky, V. L. (eds), *Solitons*, North-Holland, Amsterdam (1986).

[45] Wadati, M. The exact solution of the modified Korteweg–de Vries equation, *J. Phys. Soc. Jap.* **32**, 1681ff (1972).

[46] Wahlquist, H. D. and Estabrook, F. B. Bäcklund transformation for solutions of the Korteweg–de Vries equation, *Phys. Rev. Lett.* **23**, 1386–9 (1973).

[47] Weiss, J. The Painlevé property for partial differential equations. II: Bäcklund transformation, Lax pairs and the Schwarzian derivative, *J. Math. Phys.* **24**, 1405–13 (1983).

[48] Wilson, G. Commuting flows and conservation laws for Lax equations, *Math. Proc. Camb. Phil. Soc.* **86**, 131–43 (1979).

[49] Zabusky, N. J. A synergetic approach to problems of nonlinear dispersive wave propogation and interaction. In *Proc. Symposium on Nonlinear Partial Differential Equations*, W. F. Ames (ed.) Academic Press, New York, 1967.

[50] Zabusky, N. J. and Kruskal, M. D. Interaction of solitons in a collisionless plasma and the recurrence of initial states, *Phys. Rev. Lett.* **15**, 240–3 (1965).

[51] Zakharov, V. E. and Shabat, A. B. Exact theory of two-dimensional self focusing and one-dimensional self modulation of waves in nonlinear media, *Zh. Eksp. Teor. Fiz.* **61**, 118 (1971). English translation *Sov. Phys. JETP* **34**, 62–9 (1972).

[52] Zakharov, V. E. and Shabat, A. B. A scheme for integrating the nonlinear equations of mathematical physics by the method of the inverse scattering problem. I, *Func. Anal. and Apps.* **8**, 226–35 (1974).

Part II

Solution methods

2 *P. J. Caudrey*

Spectral transforms

2.1 Historical introduction

The history of spectral transforms (otherwise known as inverse scattering methods) began in 1967 when Gardner, Greene, Kruskal and Miura [1] (see also [2]) used one to solve the Korteweg–de Vries equation

$$u_t - 6uu_x + u_{xxx} = 0. \qquad (2.1.1)$$

What they did was to assume that $u(x, t) \to 0$ as $x \to \pm\infty$ and make it the potential in a time-independent one-dimensional Schrödinger equation

$$-\psi_{xx} + u\psi = \zeta^2\psi. \qquad (2.1.2)$$

The solutions depend on time which enters parametrically via $u(x, t)$.

The interesting solutions are the so-called scattering and bound states. The scattering states are those for which the spectral parameter $\zeta = \xi$ (ξ real) and which have the following behaviour as $x \to \pm\infty$:

$$\psi(x, \zeta) \sim e^{i\xi x} + R(\xi)e^{-i\xi x} \quad \text{as} \quad x \to -\infty \qquad (2.1.3)$$

and

$$\psi(x, \zeta) \sim T(\xi)e^{i\xi x} \quad \text{as} \quad x \to +\infty \qquad (2.1.4)$$

If u satisfies (2.1.1) it turns out that the reflection coefficient $R(\xi)$ satisfies

$$R_t = 8i\xi^3 R \qquad (2.1.5)$$

and the transmission coefficient $T(\xi)$

$$T_t = 0. \qquad (2.1.6)$$

The bound states satisfy

$$\psi(x, \zeta) \to 0 \quad \text{as} \quad x \to \pm\infty. \qquad (2.1.7)$$

Non-trivial solution satisfying this condition exist for only a finite number of values of ζ:

$$\zeta = \mathrm{i}\kappa_j, \ j = 1, 2, \ldots, n \tag{2.1.8}$$

where (assuming u is real) κ_j is real and positive. Then

$$\psi_j(x) = \psi(x, \mathrm{i}\kappa_j) \sim C_j \mathrm{e}^{\kappa_j x} \quad \text{as} \quad x \to -\infty. \tag{2.1.9}$$

In this case

$$\kappa_{jt} = 0 \tag{2.1.10}$$

and if we adopt the slightly unusual normalisation

$$\int_{-\infty}^{\infty} \psi_j^{\,2}(x)\, \mathrm{d}x = 1 \tag{2.1.11}$$

then C_j satisfies

$$C_{jt} = 4\kappa_j^{\,3} C_j. \tag{2.1.12}$$

The quantities

$$S = \{R(\xi), -\infty < \xi < \infty; \kappa_j, C_j, j = 1, 2, \ldots, n\} \tag{2.1.13}$$

are known as the spectral or scattering data and it had been shown by Gel'fand and Levitan in 1955 [3] that this data is sufficient to reconstruct the potential u. Since all this data evolves in a particularly simple way with respect to time the initial value problem for the Korteweg–de Vries equation is found as follows:

(1) Equation (2.1.2) is solved for time $t = 0$ and the spectral data $S(0)$ is found.
(2) The evolution equations (2.1.5), (2.1.10), (2.1.12) are used to find $S(t)$.
(3) The potential $u(x, t)$ is reconstructed from $S(t)$.

The remarkable thing is that each of these three stages involves solving a linear problem and yet the end result is a solution of the nonlinear equation (2.1.1).

Until 1971 this ingenious piece of work was regarded as a freak result applicable only to the Korteweg–de Vries equation. Then Zakharov and Shabat [4] demonstrated that the nonlinear Schrödinger equation

$$q_t - \mathrm{i}q_{xx} \pm 2\mathrm{i}|q|^2 q = 0 \tag{2.1.14}$$

could be solved in a similar way by using the spectral problem

$$\left.\begin{aligned} \psi_{1x} + \mathrm{i}\zeta\psi_1 &= q\psi_2 \\ \psi_{2x} - \mathrm{i}\zeta\psi_2 &= r\psi_1 \end{aligned}\right\}. \tag{2.1.15}$$

Soon afterward the modified Korteweg–de Vries equation [5]:

$$u_t - 6u^2u_x + u_{xxx} = 0 \tag{2.1.16}$$

and the sine-Gordon equation [6]:

$$u_{xx} - u_{tt} = \sin x \tag{2.1.17}$$

fell to this technique and in 1974 Ablowitz, Kaup, Newell and Segur published their now famous paper [7] in which they effectively defined the classes of equations which can be solved by using these two spectral problems and pointed out that the derivation of the spectral data and the reconstructions of the potentials are just nonlinear versions of a Fourier transform and its inverse.

Since then many other spectral problems have been used. First by inserting matrices into (2.1.2), (2.1.15) [8] and then by introducing third- [9, 10] and higher-order [11, 12] problems.

In this chapter we shall examine a general N^{th}-order spectral problem, taking it through each of the three stages needed to solve a nonlinear differential equation. This will be followed by some examples, including those mentioned above and one in which $N \to \infty$.

2.2 The direct spectral problem

We shall write our spectral problem in the form of N first-order coupled homogeneous linear ordinary differential equations

$$\psi_x = U(x, t, \zeta)\, \psi \tag{2.2.1}$$

where ψ is an N-dimensional column vector and U is an $N \times N$ matrix depending on x and two parameters, t, the (real) 'time' and ζ, the (complex) spectral parameter. At this stage we are not interested in the 'time' so the t-dependence will not be displayed explicitly. It is convenient to split U into two parts, A, the limit as $x \to \pm\infty$ (assumed to be the same at both ends) and B, the rest. The spectral problem now is

$$\psi_x = \{A(\zeta) + B(x, \zeta)\}\psi \tag{2.2.2}$$

where $B \to 0$ as $x \to \pm\infty$. We also assume that A and B are regular functions of ζ.

If B is identically zero then a solution of (2.2.2) is

$$\psi = \exp\{\lambda_i(\zeta)x\}\mathbf{v}_i(\zeta) \tag{2.2.3}$$

where $\lambda_i(\zeta)$, $i = 1, 2, \ldots, N$ is an eigenvalue of A and $\mathbf{v}_i(\zeta)$ is the corresponding eigenvector. For simplicity we shall assume that the eigenvalues are all distinct. (Several conditions will be assumed 'for simplicity'. None of them is necessary but the complications which arise if they are not assumed make it more appropriate to consider them in particular cases rather than in the general theory.) We expect an arbitrary solution to (2.2.2) to behave like a linear combination of these as $x \to \pm\infty$.

We define a set of Jost functions $\boldsymbol{\phi}_i(x, \zeta)$, $i = 1, 2, \ldots, N$ to be the solutions of (2.2.2) which satisfy

$$\exp\{-\lambda_i(\zeta)x\}\ \boldsymbol{\phi}_i(x, \zeta) \to \mathbf{v}_i(\zeta) \quad \text{as} \quad x \to -\infty \tag{2.2.4}$$

and $\exp\{-\lambda_i(\zeta)x\}\boldsymbol{\phi}_i(x, \zeta)$ is bounded for $-\infty < x < \infty$.

By putting

$$\boldsymbol{\phi}_i(x, \zeta) = \exp\{\lambda_i(\zeta)x\}\ \boldsymbol{\Phi}_i(x, \zeta) \tag{2.2.5}$$

the problem of finding these solutions can be expressed as a matrix Fredholm equation

$$\boldsymbol{\Phi}_i(x, \zeta) = \mathbf{v}_i(\zeta) + \int_{-\infty}^{\infty} K_i(x, y, \zeta)\ \boldsymbol{\Phi}_i(y, \zeta)\ \mathrm{d}y \tag{2.2.6}$$

with kernel

$$K_i(x, y, \zeta) = \sum_{j=1}^{N} \exp\{(\lambda_j - \lambda_i)(x - y)\} \times \{\theta(x - y) - \theta(\lambda_j - \lambda_i)\}\mathbf{v}_j\tilde{\mathbf{v}}_j\ B(y, \zeta) \tag{2.2.7}$$

where $\tilde{\mathbf{v}}_j(\zeta)$ is the reciprocal vector satisfying

$$\tilde{\mathbf{v}}_j(\zeta)\ \mathbf{v}_i(\zeta) = \delta_{ji} \tag{2.2.8}$$

and $\theta(\alpha)$ is the Heaviside function defined for complex α by

$$\theta(\alpha) = \begin{cases} 1 & \text{if} \quad \mathrm{Re}\ \alpha > 0 \\ 0 & \text{if} \quad \mathrm{Re}\ \alpha \leq 0 \end{cases} \tag{2.2.9}$$

It should be noted that this kernel is a regular function of ζ except on the curves where $\mathrm{Re}\ \lambda_i(\zeta) = \mathrm{Re}\ \lambda_j(\zeta)$ for some $j \neq i$. These curves divide the complex ζ-plane into regions. The kernel $K_i(x, y, \zeta)$ is regular within each region but suffers a jump as ζ crosses a boundary from one region to another.

One way of solving (2.2.6) is to use Fredholm's method which is easily extended to cover matrix kernels. The Fredholm determinant is given in the usual way as a series

$$f_i(\zeta) = \sum_{m=0}^{\infty} f_i^{(m)}(\zeta) \tag{2.2.10}$$

where

$$f_i^{(m)} = \frac{(-1)^m}{m!} \sum_{j_1, j_2, \ldots, j_m = 1}^{N} \int_{-\infty}^{\infty}\int_{-\infty}^{\infty} \cdots \int_{-\infty}^{\infty} K_i^{(m)}\begin{pmatrix} x_1, j_1, x_2, j_2, \ldots, x_m, j_m \\ x_1, j_1, x_2, j_2, \ldots, x_m, j_m \end{pmatrix} \mathrm{d}x_1\mathrm{d}x_2 \ldots \mathrm{d}x_m \tag{2.2.11}$$

$$f_i^{(0)} = 1$$

and the integrand is a special case of the determinant

$$K_i^{(m)}\begin{pmatrix} x_1, j_1, x_2, j_2, \ldots, x_m, j_m \\ y_1, k_1, y_2, k_2, \ldots, y_m, k_m \end{pmatrix} = \begin{vmatrix} [K_i(x_1, y_1)]_{j_1k_1} & [K_i(x_1, y_2)]_{j_1k_2} & \cdots \\ [K_i(x_2, y_1)]_{j_2k_1} & [K_i(x_2, y_2)]_{j_2k_2} & \cdots \\ \vdots & \vdots & \end{vmatrix}_{m\times m} \tag{2.2.12}$$

The Fredholm minor is a matrix and is also given as a series

$$F_i(x, y, \zeta) = \sum_{m=0}^{\infty} F_i^{(m)}(x, y, \zeta) \tag{2.2.13}$$

where the jk^{th} element of $F_i^{(m)}$ (x, y, ζ) is

$$[F_i^{(m)}(x, y, \zeta)]_{jk} = \frac{(-1)^m}{m!} \sum_{j_1, j_2, \ldots, j_m=1}^{N} \int_{-\infty}^{\infty}\int_{-\infty}^{\infty} \cdots \int_{-\infty}^{\infty} K_i^{(m)}\begin{pmatrix} x, j, x_1, j_1, x_2, j_2, \ldots, x_m, j_m \\ y, k, x_1, j_1, x_2, j_2, \ldots, x_m, j_m \end{pmatrix} dx_1 dx_2 \ldots dx_m \tag{2.2.14}$$

If there exists a function $M(y, \zeta)$ such that

$$|[K_i(x, y, \zeta)]_{jk}| < M(y, \zeta) \tag{2.2.15}$$

and

$$\int_{-\infty}^{\infty} M(y, \zeta)\, dy < \infty. \tag{2.2.16}$$

then both of these series converge absolutely. In general this condition is satisfied if

$$\int_{-\infty}^{\infty} |[B(y, \zeta)]_{jk}| dy < \infty. \tag{2.2.17}$$

Under these circumstances both the Fredholm determinant and the minor are regular functioons of ζ except for jumps on the boundaries between the regions. They satisfy the equations

$$\begin{aligned} F_i(x, y, \zeta) - f_i(\zeta)K_i(x, y, \zeta) &= \int_{-\infty}^{\infty} F_i(x, z, \zeta)\, K_i(z, y, \zeta)\, dz \\ &= \int_{-\infty}^{\infty} K_i(x, z, \zeta)\, F_i(z, y, \zeta)\, dz \end{aligned} \tag{2.2.18}$$

and provided $f_i(\zeta) \neq 0$ the solution to (2.2.6) is given uniquely by

$$\boldsymbol{\Phi}_i(x, \zeta) = \mathbf{v}_i(\zeta) + \frac{1}{f_i(\zeta)} \int_{-\infty}^{\infty} F_i(x, y, \zeta)\, \mathbf{v}_i(\zeta)\, dy. \tag{2.2.19}$$

If we assume for simplicity that $f_i(\zeta)$ does not approach zero as ζ approaches a boundary between the regions then the only singularities of $\Phi_i(x, \zeta)$ and hence of $\boldsymbol{\Phi}_i(x, \zeta)$ as functions of ζ are poles where $f_i(\zeta) = 0$ and jumps on the regional boundaries. It is certain pieces of information about these singularities which constitute the spectral data.

Let us examine the the poles first. For simplicity we shall suppose that they are simple. Let the poles of $\boldsymbol{\Phi}_i(x, \zeta)$ occur where $\zeta = \zeta_i^{(k)}$, $k = 1, 2, \ldots, m_i$. Then $f_i(\zeta)$ has a simple zero at each of these points and

$$\text{Res}\, \boldsymbol{\Phi}_i(x, \zeta_i^{(k)}) = \frac{1}{f'(\zeta_i^{(k)})} \int_{-\infty}^{\infty} F_i(x, y, \zeta_i^{(k)})\, \mathbf{v}_i(\zeta_i^{(k)})\, dy \tag{2.2.20}$$

where

$$f_i'(\zeta) = \frac{d}{d\zeta} f_i(\zeta). \tag{2.2.21}$$

Thus from (2.2.18)

$$\text{Res}\, \boldsymbol{\Phi}_i(x, \zeta_i^{(k)}) = \int_{-\infty}^{\infty} K_i(x, y, \zeta_i^{(k)})\, \text{Res}\, \boldsymbol{\Phi}_i(y, \zeta_i^{(k)})\, dy. \tag{2.2.22}$$

Now from (2.2.7)

$$\begin{aligned} K_i(x, y, \zeta) = {} & \exp\{(\lambda_j(\zeta) - \lambda_i(\zeta))(x - y)\}\{K_j(x, y, \zeta) \\ & - \mathbf{v}_j(\zeta)\, \tilde{\mathbf{v}}_j(\zeta)\, B(y, \zeta)\} \end{aligned} \tag{2.2.23}$$

where j is such that $\text{Re}\{\lambda_j(\zeta) - \lambda_i(\zeta)\}$ achieves its smallest strictly positive value (if $\text{Re}\{\lambda_j(\zeta) - \lambda_i(\zeta)\} \leq 0$ for all j then (2.2.6) is a Volterra equation and $\boldsymbol{\Phi}_i(x, \zeta)$ cannot have a pole) and hence

$$\begin{aligned} & \exp\{(\lambda_i(\zeta_i^{(k)}) - \lambda;(\zeta_i^{(k)})x\}\text{Res}\, \Phi_i(x, \zeta_i^{(k)}) \\ & \quad = \int_{-\infty}^{\infty} K_j(x, y, \zeta_i^{(k)}) \exp\{(\lambda_i(\zeta_i^{(k)}) - \lambda_j(\zeta_i^{(k)}))y\}\, \text{Res}\, \boldsymbol{\Phi}_i(y, \zeta_i^{(k)})\, dy \\ & \qquad + \gamma_i^{(k)}\, \mathbf{v}_j(\zeta_i^{(k)}) \end{aligned} \tag{2.2.24}$$

where

$$\begin{aligned} \gamma_i^{(k)} &= -\int_{-\infty}^{\infty} \exp\{(\lambda_i(\zeta_i^{(k)}) - \lambda_j(\zeta_i^{(k)}))y\}\tilde{\mathbf{v}}_j(\zeta_i^{(k)}) \\ & \qquad \times B(y, \zeta_i^{(k)})\, \text{Res}\, \boldsymbol{\Phi}_i(y, \zeta_i^{(k)})\, dy \\ &= -\int_{-\infty}^{\infty} exp\{-\lambda_j(\zeta_i^{(k)})y\}\, \tilde{\mathbf{v}}_j(\zeta_i^{(k)})\, B(y, \zeta_i^{(k)}) \\ & \qquad \times \text{Res}\, \boldsymbol{\Phi}_i(y, \zeta_i^{(k)})\, dy. \end{aligned} \tag{2.2.25}$$

It follows that

$$\exp\{(\lambda_i(\zeta^{(k)}) - \lambda_j(\zeta^{(k)}))x\} \text{ Res } \mathbf{\Phi}_i(x, \zeta_i^{(k)}) = \gamma_i^{(k)} \mathbf{\Phi}_j(x, \zeta_i^{(k)}) \tag{2.2.26}$$

That is,

$$\text{Res } \phi_i(x, \zeta_i^{(k)}) = \gamma_i^{(k)} \phi_j(x, \zeta_i^{(k)}). \tag{2.2.27}$$

The constants $\zeta_i^{(k)}$, $\gamma_i^{(k)}$ for $k = 1, 2, \ldots, m_i$ and $i = 1, 2, \ldots, N$ form the discrete part of the spectral data.

Now let us look at the jump singularities on the regional boundaries. We assume first of all that the boundary is such that there is only one j such that Re λ_i = Re λ_j. We call such a boundary simple and label the two sides ±ve according to the sign of Re$(\lambda_i - \lambda_j)$. The superfix ± will be used to denote the limit as ζ approaches the boundary from the ±ve side. From (2.2.7) we have

$$K_i^+(x, y, \zeta) - K_i^-(x, y, \zeta) = \exp\{(\lambda_j - \lambda_i)(x - y)\}\mathbf{v}_j\tilde{\mathbf{v}}_j\, B(y, \zeta) \tag{2.2.28}$$

and

$$K_i^+(x, y, \zeta) = \exp\{(\lambda_j - \lambda_i)(x - y)\}K_j^-(x, y, \zeta). \tag{2.2.29}$$

Hence

$$\begin{aligned}\exp\{(\lambda_i - \lambda_j)x\}&\{\mathbf{\Phi}_i^+(x, \zeta) - \mathbf{\Phi}_i^-(x, \zeta)\} \\ &= \int_{-\infty}^{\infty} K_j^-(x, y, \zeta) \exp\{(\lambda_i - \lambda_j)y\}\{\mathbf{\Phi}_i^+(y, \zeta) - \mathbf{\Phi}_i^-(y, \zeta)\}\, dy \\ &\quad + Q_{ij}(\zeta)\, \mathbf{v}_j(\zeta)\end{aligned} \tag{2.2.30}$$

where

$$\begin{aligned}Q_{ij}(\zeta) &= \int_{-\infty}^{\infty} \exp\{(\lambda_i - \lambda_j)y\}\tilde{\mathbf{v}}_j(\zeta)\, B(y, \zeta)\, \mathbf{\Phi}_i^-(y, \zeta)\, dy \\ &= \int_{-\infty}^{\infty} \exp\{-\lambda_j(\zeta)y\}\tilde{\mathbf{v}}_j(\zeta)\, B(y, \zeta)\, \phi_i^-(y, \zeta)\, dy.\end{aligned} \tag{2.2.31}$$

It follows that

$$\exp\{(\lambda_i - \lambda_j)x\}\{\mathbf{\Phi}_i^+(x, \zeta) - \mathbf{\Phi}_i^-(x, \zeta)\} = Q_{ij}(\zeta)\, \mathbf{\Phi}_j^-(x, \zeta). \tag{2.2.32}$$

That is,

$$\phi_i^+(x, \zeta) - \phi_i^-(x, \zeta) = Q_{ij}(\zeta)\, \phi_j^-(x, \zeta). \tag{2.2.33}$$

Non-simple boundaries can be regarded as superpositions of simple boundaries. Labelling the ±ve sides arbitrarily we get

$$\phi_i^+(x, \zeta) - \phi_i^-(x, \zeta) = \sum_j Q_{ij}(\zeta)\, \phi_j^{\mp}(x, \zeta) \tag{2.2.34}$$

where for each term of the right-hand side the superfix is chosen to give the limit as ζ approaches the boundary from the side on which $\mathrm{Re}(\lambda_i - \lambda_j) < 0$. The quantities $Q_{ij}(\zeta)$ where ζ runs along all the boundaris constitute the continuum part of the spectral data.

$$S = \{\zeta_i^{(k)}, \gamma_i^{(k)}, Q_{ij}(\zeta); i, j = 1, 2, \ldots, N, i \neq j, k = 1, 2, \ldots, m_i \text{ and } \zeta \text{ runs over all the boundaries}\}. \tag{2.2.35}$$

One extra piece of information will be needed. This concerns the behaviour of $\mathbf{\Phi}_i(x, \zeta)$ as $\zeta \to \infty$. In general this depends on the particular case but usually equation (2.2.11) and (2.2.14) ensure either directly or with the help of the Riemann–Lebesgue lemma that $f_i^{(m)}(\zeta)$ $(m \neq 0)$ and $F_i^{(m)}(x, y, \zeta)$ tend to zero as $\zeta \to \infty$, so that

$$\mathbf{\Phi}_i(x, \zeta) - \mathbf{v}_i(\zeta) \to 0. \tag{2.2.36}$$

2.3 Time dependence

Since the matrix $U(x, t, \zeta)$ in the spectral problem (2.2.1) depends on 'time' t it is clear that the Jost functions and hence the spectral data must also depend on t. It is more convenient, however, to introduce 'time' rather more directly by requiring the wave function ψ to satisfy an 'auxiliary spectral equation'

$$\psi_t = V(x, t, \zeta)\,\psi. \tag{2.3.1}$$

The condition for compatibility between this and (2.2.1) is

$$U_t - V_x + [U, V] = 0 \tag{2.3.2}$$

and this is the nonlinear partial differential equation to be solved. The problem of expressing a given equation as a 'Lax pair', U and V is examined in Chapter 15.

The boundary conditions which make

$$U(x, t, \zeta) \to A(\zeta) \tag{2.3.3}$$

as $x \to \pm\infty$ will in general make

$$V(x, t, \zeta) \to L(\zeta). \tag{2.3.4}$$

Clearly

$$[A, L] = 0 \tag{2.3.5}$$

and the eigenvectors $\mathbf{v}_i(\zeta)$ of A are also eigenvectors of L.

$$A(\zeta)\,\mathbf{v}_i(\zeta) = \lambda_i(\zeta)\,\mathbf{v}_i(\zeta) \tag{2.3.6}$$

$$L(\zeta)\,\mathbf{v}_i(\zeta) = \mu_i(\zeta)\,\mathbf{v}_i(\zeta). \tag{2.3.7}$$

It is easy to check that the Jost functions $\phi_i(x, t, \zeta)$ as defined in (2.2.4) do not satisfy (2.3.1) but the

$$\exp\{\mu_i(\zeta)t\}\ \phi_i(x,\,t,\,\zeta) \tag{2.3.8}$$

do. Thus

$$\frac{\partial}{\partial t}\,\phi_i(x,\,t,\,\zeta) = \{V(x,\,t,\,\zeta) - \mu_i(\zeta)I\}\phi_i(x,\,t,\,\zeta) \tag{2.3.9}$$

and it follows quite easily that

$$\frac{\mathrm{d}}{\mathrm{d}t}\zeta_i^{(k)} = 0, \tag{2.3.10}$$

$$\frac{\mathrm{d}}{\mathrm{d}t}\gamma_i^{(k)} = (\mu_j - \mu_i)\ \gamma_i^{(k)} \tag{2.3.11}$$

and

$$\frac{\partial}{\partial t}Q_{ij}(\zeta) = (\mu_j - \mu_i)\ Q_{ij}(\zeta). \tag{2.3.12}$$

2.4 The inverse spectral problem

The reconstruction of the matrix $B(x,\,\zeta)$ from the spectral data S can easily be done by substitution in (2.2.2) if $\Phi_i(x,\,\zeta)$ and hence $\phi_i(x,\,\zeta)$, $i = 1, 2, \ldots, N$ are known. Now $\mathbf{\Phi}_i(x,\,\zeta)$ has the following properties:

(i) $\mathbf{\Phi}_i(x,\,\zeta) - \mathbf{v}_i(\zeta) \to \mathbf{0}$ as $\zeta \to \infty$.

(ii) It has simple poles at $\zeta = \zeta_i^{(k)}$, $k = 1, 2, \ldots, m_i$ with residues

$$\text{Res}\ \Phi_i(x,\,\zeta_i^{(k)}) = \gamma_i^{(k)}\exp\{(\lambda_j(\zeta_i^{(k)}) - \lambda_i(\zeta_i^{(k)}))x\}\ \Phi_j(x,\,\zeta_i^{(k)})$$

(iii) It has jump singularities on the regional boundaries

$$\mathbf{\Phi}_i^+(x,\,\zeta) - \mathbf{\Phi}_i^-(x,\,\zeta) = \sum_j Q_{ij}(\zeta)\exp\{(\lambda_j(\zeta) - \lambda_i(\zeta))x\}\ \mathbf{\Phi}_j^{\mp}(x,\,\zeta)$$

(iv) Elsewhere it is regular.

This is a Riemann–Hilbert problem. The existence and uniqueness of a solution in the general case is an open question. However, provided the boundaries are piecewise smooth and the quantities $Q_{ij}(\zeta)$ are Hölder-continuous there is no difficulty. (See [13] for a nice discussion of this.)

The actual procedure for finding $\mathbf{\Phi}_i(x,\,\zeta)$ is as follows. Properties (i)–(iv) enable us to write

$$\mathbf{\Phi}_i(x,\,\zeta) = \mathbf{v}_i(\zeta) - \sum_{k=1}^{m_i}\frac{\gamma_i^{(k)}}{\zeta_i^{(k)} - \zeta}\exp\{(\lambda_j(\zeta_i^{(k)}) - \lambda_i(\zeta_i^{(k)}))x\}\mathbf{\Phi}_j(x,\,\zeta_i^{(k)})$$
$$+ \frac{1}{2\pi\mathrm{i}}\int\sum_j\frac{Q_{ij}(\zeta')}{\zeta' - \zeta}\exp\{(\lambda_j(\zeta') - \lambda_i(\zeta'))x\}\mathbf{\Phi}_j^{\mp}(x,\,\zeta')\,\mathrm{d}\zeta' \tag{2.4.1}$$

where the integral is along all the boundaries, the direction of integration being so that the positive side is on the left. By choosing appropriate values or limits for ζ the left-hand side can be made into any of the Φ's appearing on the right-hand side. This gives a set of linear matrix/Fredholm equations which can be solved to find all the Φ's on the right-hand side of (2.4.1) and hence $\Phi_i(x, \zeta)$ for general values of ζ.

2.5 The Schrödinger spectral problem

The Schrödinger spectral equation (2.1.1) can be rewritten in the form (2.2.2) simply by putting

$$\psi = \begin{pmatrix} \psi \\ \psi_x \end{pmatrix}, \quad A = \begin{pmatrix} 0 & 1 \\ -\zeta^2 & 0 \end{pmatrix}, \quad B = \begin{pmatrix} 0 & 0 \\ u(x) & 0 \end{pmatrix} \tag{2.5.1}$$

The eigenvalues of A are

$$\lambda_1 = \mathrm{i}\,\zeta, \quad \lambda_2 = -\mathrm{i}\,\zeta \tag{2.5.2}$$

and hence the complex ζ-plane is divided into just two regions, the boundary where Re λ_1 = Re λ_2 being the real axis.

Quite a lot of labour can be saved by noticing first that we need only find the top element $\phi_1(x, \zeta)$ of $\phi_i(x, \zeta)$ since the bottom one is just the x-derivative and secondly that

$$\phi_1(x, \zeta) = \phi_2(x, -\zeta). \tag{2.5.3}$$

Hence we need only find $\phi_1(x, \zeta)$. Fortunately $K_1(x, y, \zeta)$ has the form

$$K_1(x, y, \zeta) = \begin{pmatrix} K(x, y, \zeta) & 0 \\ K_x(x, y, \zeta) & 0 \end{pmatrix} \tag{2.5.4}$$

and so $\Phi_1(x, \zeta) = \exp(-\mathrm{i}\zeta x)\phi_1(x, \zeta)$ satisfies the scalar Fredholm equation

$$\Phi_1(x, \zeta) = 1 + \int_{-\infty}^{\infty} K(x, y, \zeta)\, \Phi_1(y, \zeta)\, \mathrm{d}y \tag{2.5.5}$$

where

$$K(x, y, \zeta) = -\frac{\mathrm{i}}{2\zeta}\{\theta(x-y)(1 - \mathrm{e}^{-2\mathrm{i}\zeta(x-y)}) + \theta(-\mathrm{i}\zeta)\, \mathrm{e}^{-2\mathrm{i}\zeta(x-y)}\}\, u(y) \tag{2.5.6}$$

There is a minor difficulty as $\zeta \to 0$ because of the factor ζ^{-1}. This arises because the assumption $\lambda_1 \neq \lambda_2$ is violated here. If we restrict ourselves to the lower half-plane, $K(x, y, \zeta)$ remains finite as $\zeta \to 0$ and it is sufficient to strengthen (2.2.17) to

$$\int_{-\infty}^{\infty} (1 + |y|)|u(y)|\, \mathrm{d}y < \infty. \tag{2.5.7}$$

In the upper half-plane $K(x, y, \zeta) + (i/2\zeta)u(y)$ remains finite as $\zeta \to 0$ and so, assuming that (2.5.7) is satisfied we can find $\psi(x, \zeta)$

$$\begin{aligned}\Psi(x, \zeta) &= 1 + \int_{-\infty}^{\infty} \{K(x, y, \zeta) + \frac{i}{2\zeta}u(y)\}\, \Psi(y, \zeta)\, dy \\ &= R + \int_{-\infty}^{\infty} K(x, y, \zeta)\, \Psi(y, \zeta)\, dy \end{aligned} \tag{2.5.8}$$

where

$$R = 1 + \frac{i}{2\zeta}\int_{-\infty}^{\infty} u(y)\, \Psi(y, \zeta)\, dy. \tag{2.5.9}$$

$\psi(x, \zeta)$ is well behaved as $\zeta \to 0$ and hence so is $\Phi_1(x, \zeta)$ since

$$\Phi_1(x, \zeta) = \frac{1}{R}\, \Psi(x, \zeta) \tag{2.5.10}$$

(unless of course $R \to 0$, in which case $\Phi_1(x, \zeta)$ will have a pole which we have already assumed does not occur on a boundary).

There are no poles in the lower half-plane since (2.5.6) is a Volterra equation for $\text{Im}\ \zeta < 0$. In the upper half-plane the residues are the solutions of (2.1.2) which vanish as $x \to \pm\infty$. These are usually known as bound states and since $(-\partial^2/\partial x^2) + u(x)$ is a Hermitian operator (assuming $u(x)$ is real) its eigenvalues ζ^2 are real. Thus the poles only occur on the imaginary axis,

$$\zeta = i\kappa_j, \quad j = 1, 2, \ldots, m \tag{2.5.11}$$

where κ_j is real and positive.

In the neighbourhood of a pole put

$$\begin{aligned}\psi(x, \zeta) &= (\zeta - i\kappa_j)\, \phi_1(x, \zeta) \\ &\sim (\zeta - i\kappa_j)\, e^{i\zeta x} + P(\zeta)\, e^{-i\zeta x}\end{aligned} \tag{2.5.12}$$

as $x \to -\infty$.

Then

$$\gamma_1^{(j)} = P(i\kappa_j). \tag{2.5.13}$$

The Wronskian

$$W(\psi_\zeta, \psi) = \psi_\zeta\psi_x - \psi_{\zeta x}\psi \tag{2.5.14}$$

satisfies

$$\frac{\partial}{\partial x}\, W(\psi_\zeta, \psi) = 2\zeta\psi^2. \tag{2.5.15}$$

Integrating gives

$$\int_x^{\infty} \psi^2(y, \zeta)\, dy = -\frac{1}{2\zeta}W(\psi_\zeta, \psi) \tag{2.5.16}$$

and when $\zeta = i\kappa_j$

$$\int_{-\infty}^{\infty} \psi^2(y, i\kappa_j)\, dy = i\gamma_1^{(k)}. \tag{2.5.17}$$

The normalised bound state $\psi_j(y)$

$$\int_{-\infty}^{\infty} \psi_j^2(y)\, dy = 1 \tag{2.5.18}$$

is given by

$$\begin{aligned}\psi_j(y) &= (i\, \gamma_1^{(k)})^{-1/2}\, \psi(y, i\kappa_j)\\ &\sim C_j e^{\kappa_j x} \quad \text{as} \quad x \to -\infty.\end{aligned} \tag{2.5.19}$$

where

$$C_j^2 = -i\, \gamma_1{}^{(j)}. \tag{2.5.20}$$

On the real axis $\zeta = k$,

$$\begin{aligned}\phi_1(x, k + i0) - \phi_1(x, k - i0) &= R(k)\, \phi_2(x, k + i0)\\ &= R(k)\, \phi_1(x, -k - i0)\end{aligned} \tag{2.5.21}$$

Thus

$$Q_{12}(k) = R(k). \tag{2.5.22}$$

For the inverse transform equation (2.4.1) becomes

$$\begin{aligned}\Phi_1(x, \zeta) = 1 - i\sum_{j=1}^{m} \frac{C_j^2}{i\kappa_j - \zeta} \exp(2\kappa_j x)\Phi_1(x, -i\kappa_j)\\ + \frac{1}{2\pi i}\int_{-\infty}^{\infty} \frac{R(k)}{k - \zeta} \exp(-2ikx)\Phi_1(x, -k - i0)dk.\end{aligned} \tag{2.5.23}$$

Taking Fourier transforms we put

$$G(x, y) = \frac{1}{2\pi}\int_{-\infty}^{\infty} \{\Phi_1(x, k - i0) - 1)\}e^{ik(x-y)}\, dk. \tag{2.5.24}$$

Since this vanishes for $x < y$ (because $\Phi_1(x, \zeta)$ has no singularities in the lower half-plane) the inverse transform may be written

$$\Phi_1(x, \zeta) = 1 + \int_{-\infty}^{x} G(x, y)\, e^{-i\zeta(x-y)}\, dy \tag{2.5.25}$$

valid in the lower half-plane. Equation (2.5.23) transforms into

$$G(x, y) + F(x + y) + \int_{-\infty}^{x} G(x, z)\, F(z + y)\, dz = 0, \; x > y \tag{2.5.6}$$

where

$$F(x) = \sum_{j=1}^{m} C_j^2 e^{\kappa_j x} + \frac{1}{2\pi}\int_{-\infty}^{\infty} R(k)\, e^{-ikx} dk. \qquad (2.5.27)$$

This is called the Gel'fand–Levitan equation.

The final step to find $u(x)$ can be done by considering the asymptotic expansion of $\Phi_1(x, \zeta)$ as $\zeta \to \infty$,

$$\Phi_1(x, \zeta) = 1 + \frac{\alpha(x)}{\zeta} + \ldots \qquad (2.5.28)$$

where from (2.5.23)

$$\begin{aligned} \alpha(x) &= i\sum_{j=1}^{m} C_j^2 \exp(2\kappa_j x)\Phi_1(x, -i\kappa_j) \\ &\quad - \frac{1}{2\pi i}\int_{-\infty}^{\infty} R(k) \exp(-2ikx)\, \Phi_1(x, -k - i0)\, dk \\ &= iF(2x) + i\int_{-\infty}^{x} G(x, z)\, F(z + x)\, dz \\ &= -iG(x, x - 0). \end{aligned} \qquad (2.5.29)$$

Directly from the spectral problem we find that

$$\begin{aligned} u(x) &= 2i\alpha_x \\ &= 2\frac{d}{dx} G(x, x - 0). \end{aligned} \qquad (2.5.30)$$

It is worth noting that when $u(x)$ is small there are no bound states and

$$R(k) \simeq -\frac{i}{2k}\, \tilde{u}(-2k) \qquad (2.5.31)$$

where $\tilde{u}(k)$ is the Fourier transform of $u(x)$,

$$\tilde{u}(k) = \int_{-\infty}^{\infty} e^{-ikx}\, u(x)\, dx. \qquad (2.5.32)$$

The time dependence obviously depends on the particular nonlinear equation being solved. For the Korteweg–de Vries equation the auxiliary spectral problem is usually written in the form

$$\psi_t = \left(-4\frac{\partial^3}{\partial x^3} + 3\frac{\partial}{\partial x}u + 3u\frac{\partial}{\partial x}\right)\psi. \qquad (2.5.33)$$

Using the main spectral problem to eliminate x-derivatives of ψ higher than the first we easily find that

$$V = \begin{pmatrix} -u_x & 4\zeta^2 + 2u \\ -u_{xx} + (4\zeta^2 + 2u)(u - \zeta^2) & u_x \end{pmatrix} \qquad (2.5.34)$$

and hence

$$L(\zeta) = \begin{pmatrix} 0 & 4\zeta^2 \\ -4\zeta^4 & 0 \end{pmatrix}. \tag{2.5.35}$$

The eigenvalues are

$$\mu_1 = -4i\zeta^3, \quad \mu_2 = 4i\zeta^3, \tag{2.5.36}$$

which gives the time evolution of the spectral data already displayed in (2.1.5), (2.1.10) and (2.1.12).

Finally to round this section off we display the famous '*m*-soliton' solution. This is found simply by putting $R(k) \equiv 0$. The inverse transform now reduces to matrix algebra and the final result is

$$u(x) = -2\frac{\partial^2}{\partial x^2} \ell n\{\det M\} \tag{2.5.37}$$

where the matrix M has elements

$$M_{jk} = \delta_{jk} + \frac{C_j C_k}{\kappa_j + \kappa_k} exp\{(\kappa_j + \kappa_k)x\}. \tag{2.5.38}$$

2.6 The Zakharov–Shabat spectral problem

In this case

$$\psi = \begin{pmatrix} \psi_1 \\ \psi_2 \end{pmatrix}, \quad A = \begin{pmatrix} i\zeta & 0 \\ 0 & -i\zeta \end{pmatrix}, \quad B = \begin{pmatrix} 0 & q(x) \\ r(x) & 0 \end{pmatrix}. \tag{2.6.1}$$

The eigenvalues of A are

$$\lambda_1 = -i\zeta, \quad \lambda_2 = i\zeta \tag{2.6.2}$$

and so once more the complex ζ-plane is divided into just two regions by the real axis.

The two kernels are

$$K_1(x, y, \zeta) = \theta(x-y)\begin{pmatrix} 0 & q(y) \\ 0 & 0 \end{pmatrix} + e^{2i\zeta(x-y)}\{\theta(x-y) - \theta(i\zeta)\}\begin{pmatrix} 0 & 0 \\ r(y) & 0 \end{pmatrix} \tag{2.6.3}$$

and

$$K_2(x, y, \zeta) = e^{-2i\zeta(x-y)}\{\theta(x-y) - \theta(-i\zeta)\}\begin{pmatrix} 0 & q(y) \\ 0 & 0 \end{pmatrix} + \theta(x-y)\begin{pmatrix} 0 & 0 \\ r(y) & 0 \end{pmatrix} \tag{2.6.4}$$

where

$$\mathbf{\Phi}_1(x, \zeta) = \begin{pmatrix}1\\0\end{pmatrix} + \int_{-\infty}^{\infty} K_1(x, y, \zeta)\mathbf{\Phi}_1(y, \zeta)\,\mathrm{d}y \tag{2.6.5}$$

and

$$\mathbf{\Phi}_2(x, \zeta) = \begin{pmatrix}0\\1\end{pmatrix} + \int_{-\infty}^{\infty} K_2(x, y, \zeta)\mathbf{\Phi}_2(y, \zeta)\,\mathrm{d}y. \tag{2.6.6}$$

Although $\lambda_1 = \lambda_2$ at $\zeta = 0$ there is no problem in this case because the eigenvectors remain distinct. Condition (2.2.17) can be written

$$\int_{-\infty}^{\infty} \{|q(y)| + |r(y)|\}\,\mathrm{d}y < \infty. \tag{2.6.7}$$

For Im $\zeta > 0$ equation (2.6.5) is a Volterra equation and hence $\mathbf{\Phi}_1(x, \zeta)$ and $\phi_1(x, \zeta)$ have no poles in the upper half-plane. Similarly $\phi_2(x, \zeta)$ has no poles in the lower half-plane.

The spectral data

$$S = \{\zeta_1^{(j)}, \gamma_1^{(j)}, j = 1, 2, \ldots, m_1; \zeta_2^{(j)}, \gamma_2^{(j)}, j = 1, 2, \ldots, m_2; \\ Q_{12}(k), Q_{21}(k), -\infty < k < \infty\} \tag{2.6.8}$$

are standard (called S_- by Newell [14]).

Equation (2.4.1) for the inversion becomes

$$\mathbf{\Phi}_1(x, \zeta) = \begin{pmatrix}1\\0\end{pmatrix} - \sum_{j=1}^{m_1} \frac{\gamma_1^{(j)}}{\zeta_1^{(j)} - \zeta} \exp(2\mathrm{i}\zeta_1^{(j)}x)\mathbf{\Phi}_2(x, \zeta_1^{(j)}) \\ + \frac{1}{2\pi\mathrm{i}} \int_{-\infty}^{\infty} \frac{Q_{12}(k)}{k - \zeta} \exp(2\mathrm{i}kx)\mathbf{\Phi}_2(x, k - \mathrm{i}0)\,\mathrm{d}k \tag{2.6.9}$$

and

$$\mathbf{\Phi}_2(x, \zeta) = \begin{pmatrix}0\\1\end{pmatrix} - \sum_{j=1}^{m_2} \frac{\gamma_2^{(j)}}{\zeta_2^{(j)} - \zeta} \exp(-2\mathrm{i}\zeta_2^{(j)}x)\, \mathbf{\Phi}_1(x, \zeta_2^{(j)}) \\ + \frac{1}{2\pi\mathrm{i}} \int_{-\infty}^{\infty} \frac{Q_{21}(k)}{k - \zeta} \exp(-2\mathrm{i}kx)\, \mathbf{\Phi}_1(x, k + \mathrm{i}0)\,\mathrm{d}k. \tag{2.6.10}$$

Again we can take Fourier transforms

$$\mathbf{G}_1(x, y) = \frac{1}{2\pi} \int_{-\infty}^{\infty} \{\mathbf{\Phi}_1(x, k + \mathrm{i}0) - \begin{pmatrix}1\\0\end{pmatrix}\}\mathrm{e}^{-\mathrm{i}k(x-y)}\,\mathrm{d}k, \tag{2.6.11}$$

$$\mathbf{G}_1(x, y) = \frac{1}{2\pi} \int_{-\infty}^{\infty} \{\mathbf{\Phi}_2(x, k - \mathrm{i}0) - \begin{pmatrix}0\\1\end{pmatrix}\}\mathrm{e}^{\mathrm{i}k(x-y)}\,\mathrm{d}k. \tag{2.6.12}$$

Both of these vanish for $x < y$ and equations (2.6.9) and (2.6.10) transform into the Gel'fand–Levitan equations

$$\mathbf{G}_1(x, y) + F_1(x + y) \begin{pmatrix} 0 \\ 1 \end{pmatrix} + \int_{-\infty}^{x} \mathbf{G}_2(x, z)\, F_1(z + y)\, \mathrm{d}z = 0 \quad (2.6.13)$$

and

$$\mathbf{G}_2(x, y) + F_2(x + y) \begin{pmatrix} 1 \\ 0 \end{pmatrix} + \int_{-\infty}^{x} \mathbf{G}_1(x, z)\, F_2(z + y)\, \mathrm{d}z = 0 \quad (2.6.14)$$

where

$$F_1(x) = \mathrm{i} \sum_{j=1}^{m_1} \gamma_1^{(j)} \exp(\mathrm{i}\zeta_1^{(j)} x) - \frac{1}{2\pi} \int_{-\infty}^{\infty} Q_{12}(k)\, \mathrm{e}^{\mathrm{i}kx}\, \mathrm{d}k \quad (2.6.15)$$

and

$$F_2(x) = -\mathrm{i} \sum_{j=1}^{m_2} \gamma_2^{(j)} \exp(-\mathrm{i}\zeta_2^{(j)} x) - \frac{1}{2\pi} \int_{-\infty}^{\infty} Q_{21}(k)\, \mathrm{e}^{-\mathrm{i}kx}\, \mathrm{d}k. \quad (2.6.16)$$

The potentials are again most easily found by considering asymptotic expansions of the wave functions as $\zeta \to \infty$. These give the results

$$\begin{pmatrix} \int_{-\infty}^{x} q(y)\, r(y)\, \mathrm{d}y \\ r(x) \end{pmatrix} = 2\, \mathbf{G}_1(x, x - 0) \quad (2.6.17)$$

and

$$\begin{pmatrix} q(x) \\ \int_{-\infty}^{x} q(y)\, r(y)\, \mathrm{d}y \end{pmatrix} = 2\mathbf{G}_2(x, x - 0). \quad (2.6.18)$$

If $q(x)$ and $r(x)$ are small again there are no bound states and the spectral transform reduces to a Fourier transform

$$Q_{12}(k) \simeq \tilde{r}(2k) \quad (2.6.19)$$

and

$$Q_{21}(k) \simeq -\tilde{q}(-2k). \quad (2.6.20)$$

Three important equations which can be solved by the Zakharov–Shabat spectral problem are the nonlinear Schrödinger (NLS), modified Korteweg–de Vries (MKdV) and sine-Gordon equations.

The nonlinear Schrödinger equation

$$\mathrm{i}q_t + q_{xx} \pm 2|q|^2 q = 0 \quad (2.6.21)$$

is obtained by putting

$$r = \mp q^* \quad (2.6.22)$$

and

$$V = \begin{pmatrix} -2i\zeta^2 \pm i|q|^2 & -2\zeta q + iq_x \\ \pm 2\zeta q^* \pm iq_x^* & 2i\zeta^2 \mp i|q|^2 \end{pmatrix} \tag{2.6.23}$$

giving

$$L = \begin{pmatrix} -2i\zeta^2 & 0 \\ 0 & 2i\zeta^2 \end{pmatrix} \tag{2.6.24}$$

so that the eigenvalues are

$$\mu_1 = -\mu_2 = -2i\zeta^2. \tag{2.6.25}$$

For the modified Korteweg–de Vries equation

$$q_t \mp 6q^2 q_x + q_{xxx} = 0 \tag{2.6.26}$$

we put

$$r = \pm q \tag{2.6.27}$$

and

$$V = \begin{pmatrix} 4i\zeta^3 \pm 2i\zeta q^2 & 4\zeta^2 q - 2i\zeta q_x \pm 2q^3 - q_{xx} \\ \pm 4\zeta^2 q \pm 2i\zeta q_x + 2q^3 \mp q_{xx} & -4i\zeta^3 \mp 2i\zeta q^2 \end{pmatrix}. \tag{2.6.28}$$

Then

$$L = \begin{pmatrix} 4i\zeta^3 & 0 \\ 0 & -4i\zeta^3 \end{pmatrix} \tag{2.6.29}$$

and

$$\mu_1 = -\mu_2 = 4i\zeta^3. \tag{2.6.30}$$

The sine-Gordon equation

$$u_{xt} = \pm \sin u \tag{2.6.31}$$

requires

$$q = -r = \tfrac{1}{2}u_x \tag{2.6.32}$$

and

$$V = \pm\frac{i}{4\zeta}\begin{pmatrix} -\cos u & \sin u \\ \sin u & \cos u \end{pmatrix}. \tag{2.6.33}$$

So that

$$L = \pm\frac{i}{4\zeta}\begin{pmatrix} -1 & 0 \\ 0 & 1 \end{pmatrix} \tag{2.6.34}$$

and

$$\mu_1 = -\mu_2 = \mp\frac{\mathrm{i}}{4\zeta}. \tag{2.6.35}$$

2.7 The Boussinesq equation

One example which requires a third-order spectral problem is the Boussinesq equation. This spectral problem is usually written in the form

$$\hat{L}\psi = 4\psi_{xxx} + (1 + 6w_x)\psi_x + (3w_{xx} - \sqrt{3}\mathrm{i}w_t)\psi = \Lambda\psi \tag{2.7.1}$$

and the auxiliary spectral problem is

$$\psi_t = \hat{A}\psi = \sqrt{3}\mathrm{i}\,(\psi_{xx} + w_x\psi). \tag{2.7.2}$$

The integrability condition for these,

$$\hat{L}_t = [\hat{A}, \hat{L}], \tag{2.7.3}$$

is equivalent to

$$w_{tt} = (w_{xxx} + 3w_x{}^2 + w_x)_x \tag{2.7.4}$$

so that

$$u = w_x \tag{2.7.5}$$

must satisfy the Boussinesq equation

$$u_{tt} = (u_{xx} + 3u^2 + u)_{xx}. \tag{2.7.6}$$

The spectral problem is written in the form (2.2.2) simply by putting

$$\psi = \begin{pmatrix} \psi \\ \psi_x \\ \psi_{xx} \end{pmatrix}, \tag{2.7.7}$$

$$A = \begin{pmatrix} 0 & 1 & 0 \\ 0 & 0 & 1 \\ \frac{1}{4}\Lambda & -\frac{1}{4} & 0 \end{pmatrix} \tag{2.7.8}$$

and

$$B = \begin{pmatrix} 0 & 0 & 0 \\ 0 & 0 & 0 \\ -\frac{3}{4}w_{xx} + \frac{\sqrt{3}}{4}iw_t & -\frac{3}{2}w_x & 0 \end{pmatrix} \tag{2.7.9}$$

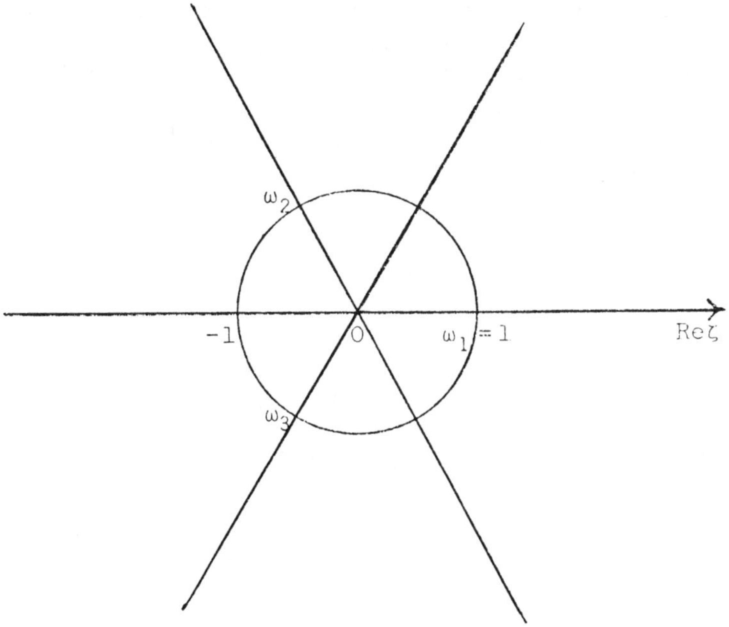

Figure 2.1

It is convenient to choose ζ as the spectral parameter where

$$\Lambda = \frac{1}{6\sqrt{3}}\left(\zeta^3 - \frac{1}{\zeta^3}\right) \tag{2.7.10}$$

then

$$\lambda_j(\zeta) = \frac{1}{2\sqrt{3}}\left(\omega_j\zeta - \frac{1}{\omega_j\zeta}\right) \tag{2.7.11}$$

where

$$\omega_1 = 1, \quad \omega_2 = -\tfrac{1}{2} + \frac{\sqrt{3}}{2}\mathrm{i}, \quad \omega_3 = -\tfrac{1}{2} - \frac{\sqrt{3}}{2}\mathrm{i} \tag{2.7.12}$$

are the cube roots of unity.

The regional boundaries are the radial lines $\arg\zeta = n\pi/3$ and the unit circle $|\zeta| = 1$ which divide the complex ζ-plane into twelve regions (see Figure 2.1)

The difficulty which occurred at the origin for the Schrödinger spectral problem occurs here at each of the six points $\zeta = \pm\mathrm{i}$, $\pm\frac{1}{2}(\sqrt{3} + \mathrm{i})$, $\pm\frac{1}{2}(\sqrt{3} - \mathrm{i})$. Fortunately, providing

$$\int_{-\infty}^{\infty} (1 + |x|)|b| \, \mathrm{d}x < \infty, \tag{2.7.13}$$

where b is either of the two non-zero elements of the matrix B, the same device can be used to overcome this difficulty.

It should also be noted that although $\lambda_i \to \infty$ as $\zeta \to 0$, $\Phi_i(x, \zeta)$ (but not $\phi_i(x, \zeta)$) remains well behaved there.

Once again labour can be saved first by considering only the top element $\phi_1(x, \zeta)$ of $\phi_i(x, \zeta)$, the other two being the first and second x-derivatives of it, and secondly by considering only $\phi_1(x, \zeta)$ since

$$\phi_1\left(x, \frac{\zeta}{\omega_1}\right) = \phi_2\left(x, \frac{\zeta}{\omega_2}\right) = \phi_3\left(x, \frac{\zeta}{\omega_3}\right). \tag{2.7.14}$$

Because of the symmetry

$$\phi_1(x, \zeta) = \phi_2\left(x, -\frac{1}{\zeta}\right) \tag{2.7.15}$$

the poles occur in pairs $\zeta_1^{(k)}$ and $\zeta_1^{(k')}$ where

$$\zeta_1^{(k)}\, \zeta_1^{(k')} = -1 \tag{2.7.16}$$

and

$$\gamma_1^{(k')} = \gamma_1^{(k)}/(\zeta_1^{(k)})^2. \tag{2.7.17}$$

Also if we take the direction of the boundaries to be away from the origin for the lines and anti-clockwise for the circle then

$$Q_{12}(\zeta) = -Q_{13}\left(-\frac{1}{\zeta}\right). \tag{2.7.18}$$

Using this symmetry, equation (2.4.1) becomes

$$\begin{aligned}\Phi_1(x, \zeta) &= 1 \\ &- \sum_{k=1}^{m} \frac{\tilde{\gamma}_1^{(k)}}{\lambda_1(\zeta_1^{(k)}) - \lambda_1(\zeta)} \exp\{(\lambda_j(\zeta_1^{(k)}) - \lambda_1(\zeta_1^{(k)}))x\}\Phi_1(x, \omega_j\zeta_1^{(k)}) \\ &+ \frac{1}{2\pi i}\int \frac{\tilde{Q}_{12}(\zeta')}{\lambda_1(\zeta') - \lambda_1(\zeta)} \exp\{(\lambda_2(\zeta') - \lambda_1(\zeta'))x\}\Phi_1(x, \omega_2\zeta')\, d\zeta'\end{aligned} \tag{2.7.19}$$

where the summation is restricted to one pole from each pair, say the one outside the unit circle, $j(= 2 \text{ or } 3)$ is chosen to give $\mathrm{Re}(\lambda_j(\zeta_1^{(k)}) - \lambda_1(\zeta_1^{(k)}))$ its smallest strictly positive value,

$$\tilde{\gamma}_1^{(k)} = \frac{1}{2\sqrt{3}}\{1 + (\zeta_1^{(k)})^{-2}\}\gamma_1^{(k)} \tag{2.7.20}$$

and

$$\tilde{Q}_{12}(\zeta) = \frac{1}{2\sqrt{3}}(1 + \zeta^{-2})Q_{12}(\zeta). \tag{2.7.21}$$

Fourier transformation is not possible in this case.

It can also be shown [12] that outside the unit circle $|\zeta| = 1$ the poles can only occur with

$$\zeta_1^{(k)} = i\omega_2\xi_k \quad or \quad -i\omega_3\xi_k \tag{2.7.22}$$

where the ξ_k's are either real or occur in conjugate pairs

$$\xi_k = \xi_{k'}^* \tag{2.7.23}$$

with $\pi/3 < \arg \xi_k < \pi/2$ or $-\pi/2 < \arg \xi_k < -\pi/3$.

For the time evolution it is easy to discover that

$$L = \sqrt{3}iA^2 \tag{2.7.24}$$

so that

$$\mu_i = \sqrt{3}i\mu_i{}^2 \tag{2.7.25}$$

and hence

$$\zeta_i^{(k)}(t) = \zeta_i^{(k)}(0) \tag{2.7.26}$$

$$\gamma_i^{(k)}(t) = \exp\{\sqrt{3}i\,[(\lambda_j(\zeta_i^{(k)}))^2 - (\lambda_i(\zeta_i^{(k)}))^2]t\}\gamma_i^{(k)}(0), \tag{2.7.27}$$

$$Q_{ij}(t, \zeta) = \exp\{\sqrt{3}i\,[(\lambda(\lambda_j(\zeta))^2 - (\lambda_i(\zeta))^2]t\}Q_{ij}(0, \zeta). \tag{2.7.28}$$

The m-soliton solution is found by putting

$$\tilde{Q}_{12}(\zeta) \equiv 0. \tag{2.7.29}$$

Then from equation (2.7.19) we get

$$\sum_{l=1}^{m} M_{kl}(x) \exp\{\lambda_{j'}(\zeta_1^{(l)})x\}\Phi_1(x, \omega_{j'}\zeta_1^{(l)}) = \exp\{\lambda_j(\zeta_1^{(k)})x\} \tag{2.7.30}$$

where

$$M_{kl}(x) = \delta_{kl} - \frac{\tilde{\gamma}_1^{(l)}}{\lambda_j(\zeta_1^{(k)}) - \lambda_1(\zeta_1^{(l)})} \exp\{(\lambda_j(\zeta_1^{(k)}) - \lambda_1(\zeta_1^{(l)}))x\}. \tag{2.7.31}$$

and the asymptotic expansion of $\Phi_1(x, \zeta)$ is

$$\begin{aligned}\Phi_1(x, \zeta) &\sim 1 + \frac{1}{\lambda_1(\zeta)} \sum_{k=1}^{m} \tilde{\gamma}_1^{(k)}\exp\{(\lambda_j(\zeta_1^{(k)}) - \lambda_1(\zeta_1^{(k)}))x\}\,\Phi_1(x, \omega_j\zeta_1{}^{(k)})\\ &= 1 - \frac{1}{\lambda_1(\zeta)} \operatorname{Tr}\{M^{-1}(x)\frac{\partial}{\partial x}M(x)\}\\ &= 1 - \frac{1}{\lambda_1(\zeta)}\frac{\partial}{\partial x}\ln\{\det M(x)\}.\end{aligned} \tag{2.7.32}$$

But from (2.7.1)

$$\Phi_1(x, \zeta) \sim 1 - \frac{w(x) - w(-\infty)}{2\lambda_1(\zeta)} \tag{2.7.33}$$

and hence the solution of the Boussinesq equation is

$$u(x, t) = w_x = 2 \frac{\partial^2}{\partial x^2} \ln\{\det M(x, t)\}. \qquad (2.7.34)$$

2.8 The Kadomtsev–Petviashvili equation

The Kadomtsev–Petviashvili equation [15]

$$(u_t - 6uu_x + u_{xxx})_x + 3\sigma^2 u_{yy} = 0 \qquad (2.8.1)$$

where $\sigma = \mathrm{i}$ (KPI) or $\sigma = -1$ (KPII) is a version of the Korteweg–de Vries equation in which an extra 'space' variable y has been introduced.

The appropriate spectral equation is

$$-\psi_{xx} + u\psi = \sigma\psi_y \qquad (2.8.2)$$

and the auxiliary spectral equation is

$$\psi_t = \left(-4\frac{\partial^3}{\partial x^3} + 3\frac{\partial}{\partial x}u + 3u\frac{\partial}{\partial x} - 3\sigma v\right)\psi. \qquad (2.8.3)$$

The integrability conditions are

$$v_x = u_y \qquad (2.8.4)$$

and

$$u_t - 6uu_x + u_{xxx} + 3\sigma^2 v_y = 0, \qquad (2.8.5)$$

which are clearly equivalent to (2.8.1).

As it stands (2.8.2) cannot be expressed in the form (2.2.2) and the use of this spectral problem requires some new techniques [16–18]. However, by 'discretising' one of the variables x or y [19] we can get an approximation to (2.8.2) which has the right form. Let us 'discretise' y. We express the y-dependence by writing ψ as an N-element column vector and replace the operator $\partial/\partial y$ by the matrix

$$D(\zeta) = \frac{1}{2h}\begin{pmatrix} 0 & 1 & 0 & \dots & 0 & -\zeta^{-N} \\ -1 & 0 & 1 & \dots & 0 & 0 \\ 0 & -1 & 0 & \dots & 0 & 0 \\ \vdots & \vdots & \vdots & & \vdots & \vdots \\ 0 & 0 & 0 & \dots & 0 & 1 \\ \zeta^N & 0 & 0 & & -1 & 0 \end{pmatrix} \qquad (2.8.6)$$

where h, the step length, is a positive real constant. The eigenvalues of D are

$$\frac{1}{2h}\left(\omega_n\zeta - \frac{1}{\omega_n\zeta}\right) \tag{2.8.7}$$

and the corresponding eigenvectors are $\mathbf{W}(\omega_n\zeta)$ where

$$\mathbf{W}(\zeta) = \begin{pmatrix} \zeta \\ \zeta^2 \\ \vdots \\ \zeta^N \end{pmatrix} \tag{2.8.8}$$

and $\omega_n = \exp(2n\pi i/N)$, $n = 0, 1, 2, \ldots, N-1$ are the Nth roots of unity. The potential $U(x, y)$ becomes the diagonal matrix

$$U = \mathrm{diag}(u_1(x), u_2(x), \ldots, u_N(x)) \tag{2.8.9}$$

which tends to zero as x tends to $\pm\infty$. The spectral problem now is

$$\psi_{xx} = (-\sigma D(\zeta) + U(x))\psi \tag{2.8.10}$$

and can be expressed in the form (2.2.2) with order $2N$ by putting

$$\boldsymbol{\phi} = \begin{pmatrix} \psi \\ \psi_x \end{pmatrix} \tag{2.8.11}$$

so that

$$A(\zeta) = \begin{pmatrix} 0 & \mathbf{I} \\ -\sigma D(\zeta) & 0 \end{pmatrix} \tag{2.8.12}$$

and

$$B(x, \zeta) = \begin{pmatrix} 0 & 0 \\ U(x) & 0 \end{pmatrix}. \tag{2.8.13}$$

Once again it is unnecessary to evaluate all the components of all the Jost functions. In fact we only need the top halves $\psi^{\pm}(x, \zeta)$ of two Jost functions. The total of $2N$ Jost functions is given by $\psi^{\pm}(x, \omega_n\zeta)$, $n = 0, 1, 2, \ldots, N-1$ and the bottom halves are just the derivatives of the tops. From now on we shall refer to these two halves as *the* Jost functions. They satisfy (2.8.10) together with the boundary conditions

$$\begin{aligned} \Psi^{\pm}(x, \zeta) &= \exp\{\pm\lambda(\zeta)x\}\psi^{\pm}(x, \zeta) \\ &\to \mathbf{W}(\zeta) \quad \text{as} \quad x \to -\infty \end{aligned} \tag{2.8.14}$$

and $\Psi^{\pm}(x, \zeta)$ remain finite as $x \to +\infty$.

$$\lambda(\zeta) = \sqrt{\left[-\frac{\sigma}{2h}(\zeta - \zeta^{-1})\right]} \tag{2.8.15}$$

is an eigenvalue of $A(\zeta)$.

The Fredholm equations are

$$\Psi^{\pm}(x, \zeta) = W(\zeta) + \int_{-\infty}^{\infty} Q^{\pm}(x, x', \zeta)\,\Psi^{\pm}(x', \zeta)\,\mathrm{d}x' \qquad (2.8.16)$$

where

$$Q^{\pm}(x, x'1, \zeta) = \sum_{\varepsilon=\pm 1} \varepsilon \sum_{n=0}^{N-1} \exp\{(\varepsilon\lambda(\omega_n\zeta) \mp \lambda(\zeta))(x - x')\}$$
$$\times \{\theta(x - x') - \theta(\varepsilon\lambda(\omega_n\zeta) \mp \lambda(\zeta))\, \frac{\mathbf{W}(\omega_n\zeta)\tilde{\mathbf{W}}(\omega_n\zeta)}{2\lambda(\omega_n\zeta)}\, U(x') \qquad (2.8.17)$$

and $\tilde{\mathbf{W}}(\omega_n\zeta)$, $n = 0, 1, 2, \ldots, N-1$ are the reciprocal vectors of $\mathbf{W}(\omega_n\zeta)$, $n = 0, 1, 2, \ldots, N-1$.

$$\mathbf{W}(\zeta) = \frac{1}{N}(\zeta^{-1}, \zeta^{-2}, \ldots, \zeta^{-N}). \qquad (2.8.18)$$

The regional boundaries where

$$\mathrm{Re}\{\varepsilon\lambda(\omega_n\zeta) \mp \lambda(\zeta)\} = 0 \qquad (2.8.19)$$

are in general simple, that is, there is only one choice of ε and ω_n for which (2.8.19) is satisfied on a given boundary. Across a boundary

$$\psi^{\pm}(x, \zeta) - \psi^{\pm}_{\mathrm{r}}(x, \zeta) = \mathrm{R}^{\pm}(\zeta)\, \psi^{\varepsilon}_{\mathrm{r}}(x, \omega_n\zeta) \qquad (2.8.20)$$

where the sides have been labelled l (left) and r (right) with $\mathrm{Re}(\varepsilon\lambda(\omega_n\zeta) \mp \lambda(\zeta)) < 0$ on the left. The coefficient $R^{\pm}(\zeta)$ is

$$R^{\pm}(\zeta) = \frac{\varepsilon}{2\lambda(\omega_n\zeta)}\, \tilde{\mathbf{W}}(\omega_n\zeta) \int_{-\infty}^{\infty} \exp(-\varepsilon\lambda(\omega_n\zeta)x)\, U(x)\, \psi^{\pm}_{\mathrm{r}}(x, \zeta)\, \mathrm{d}x. \qquad (2.8.21)$$

Now we let $N \to \infty$ and $h \to 0$ so that

$$2l = Nh \qquad (2.8.22)$$

remains finite. The continuum variable y is recovered by putting

$$y = mh - l \qquad (2.8.23)$$

where $m = 1, 2, \ldots, N$ is the element number in the vectors $\psi^{\pm}(x, \zeta)$. To prevent $\lambda(\zeta)$ becoming infinite it is necessary to choose a new spectral parameter. A convenient choice is α where

$$\zeta = \exp(-\alpha^2 h/\sigma) \qquad (2.8.24)$$

so that

$$\pm\lambda(\zeta) = \pm\sqrt{\left[\frac{\sigma}{h}\sinh(\alpha^2 h/\sigma)\right]}$$
$$\to \alpha, \qquad (2.8.25)$$

the $\pm$ being cared for in the sign of α. The matrix $D(\zeta)$ turns back into the differential operator $\partial/\partial y$ $(-l \le y \le l)$ with the pseudo-periodic boundary condition

$$f(l) = \exp(-2\alpha^2 l/\sigma)\, f(-l) \tag{2.8.26}$$

and $U(x)$ reverts to the multiplicative function $u(x, y)$. Finally it is more natural if we change the phase of the Jost function so that

$$\psi^{\pm}(x, \zeta) \to \exp(-\alpha^2 l/\sigma)\, \psi(x, y, \alpha). \tag{2.8.27}$$

The spectral problem now becomes

$$-\psi_{xx} + u\psi = \sigma\psi_y \quad (-l \le y \le l) \tag{2.8.28}$$

with the boundary condition

$$\psi(x, l) = \exp(-2\alpha^2 l/\sigma)\, \psi(x, -l). \tag{2.8.29}$$

The Jost function $\psi(x, y, \alpha)$ satisfies

$$\begin{aligned} \Psi((x, y, \alpha) &= \exp(-\alpha x + \alpha^2 y/\sigma)\, \psi(x, y, \alpha) \\ &\to 1 \quad \text{as} \quad x \to -\infty \end{aligned} \tag{2.8.30}$$

and $\Psi((x, y, \alpha)$ remains finite as $x \to +\infty$. It is given by the Fredholm equation

$$\Psi(x, y, \alpha) = 1 + \int_{-\infty}^{\infty} \int_{-l}^{l} Q(x, x', y, y', \alpha)\, \Psi(x', y', \alpha)\, dy'\, dx' \tag{2.8.31}$$

where the kernel has now become

$$\begin{aligned} &Q(x, x', y, y', \alpha) \\ &\quad = \sum_{\alpha'} \frac{1}{4l\alpha'} \exp\{(\alpha' - \alpha)(x - x') - (\alpha'^2 - \alpha^2)(y - y')/\sigma\} \\ &\quad\quad \times \{\theta(x - x') - \theta(\alpha' - \alpha)\}\, u(x, y) \end{aligned} \tag{2.8.32}$$

the sum being over all α' such that

$$\frac{l}{\pi i \sigma} (\alpha'^2 - \alpha^2) = \text{an integer}. \tag{2.8.33}$$

For $\sigma = -1$ the boundaries have become the imaginary axis together with the rectangular hyperbolae

$$\text{Re}\, \alpha\ \text{Im}\, \alpha = \frac{n\pi}{4l}, \quad n = \pm 1, \pm 2, \ldots \tag{2.8.34}$$

These are still simple and (2.8.20) and (2.8.21) have become

$$\psi(x, y, \alpha + 0) - \psi(x, y, \alpha - 0) = R(\alpha)\, \psi(x, y, \bar{\alpha} - 0) \tag{2.8.35}$$

and

$$R(\alpha) = \frac{1}{4l\bar{\alpha}} \int_{-\infty}^{\infty} \int_{-l}^{l} \exp(-\bar{\alpha}x - \bar{\alpha}^2 y)\, u(x, y)\, \psi(x, y, \alpha - 0)\, dy\, dx \tag{2.8.36}$$

where $\bar{\alpha}$ is the complex conjugate of α.

For $\sigma = \mathrm{i}$ there is again an infinite number of boundaries but they have all moved onto the imaginary axis giving a single non-simple boundary across which

$$\psi(x, y, \mathrm{i}k + 0) - \psi(x, y, \mathrm{i}k - 0) = \sum_{k'} R(k, k')\, \psi(x, y, \mathrm{i}k' - 0) \tag{2.8.37}$$

where k is real and the sum is over all real values of k' which satisfy

$$\frac{l}{\pi}(k'^2 - k^2) = \text{an integer}. \tag{2.8.38}$$

There is an expression for $R(k, k')$ similar to (2.8.36) but it involves Jost functions from regions of the α-plane which have been 'squeezed away to nothing' as the boundaries have moved onto the imaginary axis.

In both cases, $\sigma = -1$ or i, any poles suffer nothing more than a change in notation as the limit is taken. If the poles are located at $\alpha = \alpha_j$, $j = 1, 2, \ldots, m$

$$\operatorname{Res} \psi(x, y, \alpha_j) = \gamma_j \psi(x, y, \alpha_j') \tag{2.8.39}$$

where α_j' is the solution of

$$\frac{l}{\pi \mathrm{i} \sigma}(\alpha_j'^2 - \alpha_j^2) = \text{an integer} \tag{2.8.40}$$

which gives the smallest positive value to $\mathrm{Re}(\alpha_j' - \alpha_j)$.

The time dependence of the spectral data is easily found by realising that (2.8.3) is satisfied by

$$\exp(-4\alpha^2 t)\, \psi(x, y, \alpha) \tag{2.8.41}$$

so that

$$\alpha_j(t) = \alpha_j(0) \tag{2.8.42}$$

$$\gamma_j(t) = \exp\{4(\alpha_j^3 - \alpha_j'^3)t\, \gamma_j(0) \tag{2.8.43}$$

and

$$R(\alpha, t) = \exp\{4(\alpha^3 - \bar{\alpha}^3)t\}\, R(\alpha, 0) \tag{2.8.44}$$

for the case $\lambda = -1$ and

$$R(k, k', t) = \exp\{4(k'^3 - k^3)t\}\, R(k, k', 0) \tag{2.8.45}$$

for $\sigma = \mathrm{i}$.

It is interesting to see what happens as $l \to \infty$. The spectral equation (2.8.28) remains unchanged, but now $-\infty < y < \infty$ and, assuming that $u(x, y) \to 0$ as $x, y \to \pm\infty$, the boundary condition becomes that for any given x

$$\Psi(x, y, \alpha) = \exp(-\alpha x + \alpha^2 y/\sigma)\ \psi(x, y, \alpha) \tag{2.8.46}$$

is bounded as $y \to \pm\infty$. For the Jost solution we also have

$$\Psi(x, y, \alpha) \to 1 \quad \text{as} \quad x \to -\infty \tag{2.8.47}$$

and it remains bounded as $x \to +\infty$.

The only change in the Fredholm equation (2.8.31) is that the range of the integral over y is from $-\infty$ to $+\infty$ and the kernel becomes

$$\begin{aligned} Q(x, x', y, y', \alpha) &\\ = \frac{1}{2\pi i\sigma} \int_\Gamma & \exp\{(\alpha' - \alpha)(x - x') - (\alpha'^2 - \alpha^2)(y - y')/\sigma\} \\ &\times \{\theta(x - x') - \theta(\alpha' - \alpha)\}\ d\alpha'\ u(x', y'). \end{aligned} \tag{2.8.48}$$

The path of integration in the complex α'-plane is along both branches of a rectangular hyperbola which passes through α. For $\sigma = \mathrm{i}$ the asymptotes are the real and imaginary axes and the direction of integration is from $\pm\infty$ to $\pm\mathrm{i}\infty$. For $\sigma = -1$ the asymptotes are $\mathrm{Re}\,\alpha' = \pm\mathrm{Im}\alpha'$ and the path comes in in the first and third and goes out in the second and fourth quadrants. In both cases deforming the path and changing the dummy variable gives

$$\begin{aligned} Q(x, x', y, y', \alpha) = \frac{1}{2\pi\sigma} \int_{-\infty}^{\infty} & \exp\{\mathrm{i}\beta(x - x') + \beta(\beta - 2\mathrm{i}\alpha)(y - y')/\sigma\} \\ &\times \{\theta(y - y') - \theta[\beta(\beta - 2\mathrm{i}\alpha)/\sigma]\}\ d\beta\ u(x', y') \end{aligned} \tag{2.8.49}$$

(notice $\theta(y - y')$ instead of $\theta(x - x')$).

For $\sigma = \mathrm{i}$ the imaginary axis remains the only boundary. However, the behaviour of $\psi(x, y, \alpha)$ across the boundary becomes

$$\psi(x, y, \mathrm{i}k + 0) - \psi(x, y, \mathrm{i}k - 0) = \int_{-\infty}^{\infty} R(k, k')\ \psi(x, y, \mathrm{i}k' - 0)\ dk'. \tag{2.8.50}$$

This is just Manakov's non-local Riemann–Hilbert problem [16].

For $\sigma = -1$ the boundaries become dense throughout the complex α-plane as $l \to \infty$. Thus $\psi(x, y, \alpha)$ is no longer analytic anywhere in this plane. To find out what happens we rescale $\mathrm{R}(\alpha)$ before taking the limit so that (2.8.35) and (2.8.36) become

$$\psi(x, y, \alpha + 0) - \psi(x, y, \alpha - 0) = \frac{\pi\mathrm{i}}{2l\bar{\alpha}} R(\alpha)\ \psi(x, y, \bar{\alpha} - 0) \tag{2.8.51}$$

and

$$R(\alpha) = \frac{1}{2\pi i} \int_{-\infty}^{\infty} \int_{-l}^{l} \exp(-\bar{\alpha}x - \bar{\alpha}^2 y)\ u(x, y)\ \psi(x, y, \alpha - 0)\ dy\ dx. \tag{2.8.52}$$

As $l \to \infty$ the only things that alter in (2.8.52) are the limits of the inner integral. To find the limit of (2.8.51) we consider a contour C in the α-plane. Then

$$\int_C \psi(x, y, \alpha)\ d\alpha = -\frac{\pi i}{2l} \sum_n \int_{\Gamma_n} \frac{1}{\bar{\alpha}} R(\alpha)\ \psi(x, y, \bar{\alpha} + 0)\ d\alpha \tag{2.8.53}$$

where Γ_n is the section(s) of the nth boundary enclosed by C and the sum is over all such boundaries. In terms of the real parameter k, Γ_n is given by

$$\alpha = -\frac{n\pi}{2lk} - i\frac{k}{2}. \tag{2.8.54}$$

Putting

$$m = \frac{n}{l}, \quad \delta m = \frac{1}{l} \tag{2.8.55}$$

and then letting $l \to \infty$ gives

$$\begin{aligned} \int_C \psi(x, y, \alpha) d\alpha &= -\frac{\pi i}{2} \sum_m \int_{\Gamma_n} \frac{1}{k} R(\alpha)\ \psi(x, y, \bar{\alpha})\ dk\ \delta m \\ &\to -\frac{\pi i}{2} \iint \frac{1}{k} R(\alpha)\ \psi(x, y, \bar{\alpha})\ dk\ dm \end{aligned} \tag{2.8.56}$$

where the double integral is over k and m such that

$$\alpha = -\frac{m\pi}{2k} - i\frac{k}{2} \tag{2.8.57}$$

lies in the domain D enclosed by the contour C. Thus

$$\begin{aligned} \int_C \psi(x, y, \alpha)\ d\alpha &= 2i \iint_D \operatorname{sgn}(\operatorname{Im} \alpha)\ R(\alpha)\ \psi(x, y, \bar{\alpha})\ d(\operatorname{Re} \alpha)\ d(\operatorname{Im} \alpha) \\ &= -\iint_D \operatorname{sgn}(\operatorname{Im} \alpha)\ R(\alpha)\ \psi(x, y, \bar{\alpha})\ d\alpha \wedge d\bar{\alpha} \end{aligned} \tag{2.8.58}$$

and using the extension to Cauchy's theorem [20]

$$\int_C f(z)\ dz + \iint_D \frac{\partial f}{\partial \bar{z}} dz \wedge d\bar{z} = 0 \tag{2.8.59}$$

we find that

$$\frac{\partial}{\partial \bar{\alpha}} \psi(x, y, \alpha) = \operatorname{sgn}(\operatorname{Im} \alpha)\ R(\alpha)\ \psi(x, y, \bar{\alpha}). \tag{2.8.60}$$

which is the '$\bar{\partial}$' problem of Ablowitz *et al.* [18].

The behaviour of the poles is very complicated. In fact, for $\sigma = -1$ it is an unsolved problem. It seems likely, however, that there are no poles in this case. For $\sigma = \mathrm{i}$ we expand γ_j as a power series in $\delta\alpha_j = \alpha_j' - \alpha_j$,

$$\gamma_j = \gamma_j^{(0)} + \gamma_j^{(1)}\,\delta\alpha_j + \gamma_j^{(2)}\,\delta\alpha_j^{\,2} + 0(\delta\alpha_j^{\,3}). \tag{2.8.61}$$

Now

$$\begin{aligned}\psi(x, y, \alpha) &= \frac{\operatorname{Res}\psi(x, y, \alpha_j)}{\alpha - \alpha_j} + \tilde{\psi}(x, y, \alpha)\\ &= \frac{\gamma_j}{\alpha - \alpha_j}\,\psi(x, y, \alpha_j + \delta\alpha_j) + \tilde{\psi}(x, y, \alpha)\end{aligned} \tag{2.8.62}$$

where $\tilde{\psi}(x, y, \alpha)$ is regular at $\alpha = \alpha_j$. Thus

$$\begin{aligned}\tilde{\psi}(x, y, \alpha_j + \delta\alpha_j) &= \left(1 - \frac{\gamma_j}{\delta\alpha_j}\right)\psi(x, y, \alpha_j + \delta\alpha_j)\\ &= \left(-\frac{\gamma_j^{(0)}}{\delta\alpha_j} + 1 - \gamma_j^{(1)} - \gamma_j^{(2)}\,\delta\alpha_j + 0(\delta\alpha_j^{\,2})\right)\psi(x, y, \alpha_j + \delta\alpha_j)\end{aligned} \tag{2.8.63}$$

As $l \to \infty$, $\delta\alpha_j \to 0$ and so if this is to remain finite we must have

$$\gamma_j^{(0)} = 0,\ \gamma_j^{(1)} = 1 \tag{2.8.64}$$

and then

$$\tilde{\psi}(x, y, \alpha_j) = -\gamma_j^{(2)} \operatorname{Res}\psi(x, y, \alpha_j). \tag{2.8.65}$$

If we put

$$\psi(x, y, \alpha) = \frac{\operatorname{Res}\psi(x, y, \alpha_j)}{\alpha - \alpha_j} + \tilde{\psi}(x, y, \alpha) \tag{2.8.66}$$

then from (2.8.30) we find

$$\operatorname{Res}\psi(x, y, \alpha_j) = \exp(-\alpha_j z - \mathrm{i}\alpha_j^{\,2} y) \operatorname{Res}\psi(x, y, \alpha_j) \tag{2.8.67}$$

and

$$\begin{aligned}\tilde{\psi}(x, y, \alpha_j) = \exp(-\alpha_j x - \mathrm{i}\alpha_j^{\,2} y)\{&\tilde{\psi}(x, y, \alpha_j)\\ &- (x + 2\mathrm{i}\alpha_j y) \operatorname{Res}\psi(x, y, \alpha_j)\}\end{aligned} \tag{2.8.68}$$

giving

$$\tilde{\psi}(x, y, \alpha_j) = -(\gamma_j^{(2)} + x + 2\mathrm{i}\alpha_j y) \operatorname{Res}\psi(x, y, \alpha_j) \tag{2.8.69}$$

in agreement with Fokas and Ablowitz [17].

It is also possible to 'discretise' the x-variable [19]. Reversing this 'discretisation' gives the spectral problem for the Kadomtsev–Petviashvili equation with periodic boundary conditions in x.

References

[1] Gardner, C. S., Greene, J. M., Kruskal, M. D. and Miura, R. M., *Phys. Rev. Letters* **19**, 1095 (1967).

[2] Gardner, C. S., Greene, J. M., Kruskal, M. D. and Miura, R. M., *Comm. Pure Appl. Math.* **27**, 97 (1974).

[3] Gel'fand, I. M. and Levitan, B. M., *Amer. Math. Soc. Transl. Ser. 2* **1**, 253 (1955).

[4] Zakharov, V. E. and Shabat, A. B., *Zh. Eksp. Teor. Fiz.* **61**, 118 (1971). English translation, *Soviet Phys. JETP* **34**, 62 (1972).

[5] Wadati, M., *J. Phys. Soc. Japan* **32**, 1681 (1972).

[6] Ablowitz, M. J., Kaup, D. J., Newell, A. C. and Segur, H., *Phys. Rev. Letters* **30**, 1262 (1973).

[7] Ablowitz, M. J., Kaup, D. J., Newell, A. C. and Segur, H., *Studies in Appl. Math.* **53**, 249 (1974).

[8] Wadati, M. and Kamijo, T., *Prog. Theor. Phys.* **52**, 397 (1974).

[9] Kaup, D. J., *Studies in Appl. Math.* **62**, 189 (1980).

[10] Caudrey, P. J., *Phys. Lett. A* **79**, 264 (1980).

[11] Mikhailov, A. V., *Physica* **3D**, 73 (1981).

[12] Caudrey, P. J., *Physica* **6D**, 51 (1982).

[13] Muskhelishvili, N. I., *Singular Integral Equations*, Noordhoff, Groningen, Holland (1953).

[14] Newell, A. C. In *Solitons*, Topics in Current Physics 17, R. K. Bullough and P. J. Caudrey (eds), Springer, Berlin, Heidelberg, New York, ch: 6 (1980).

[15] Kadomtsev, B. B. and Petviashvili, V. I., *Soviet Phys. Dokl.* **15**, 539 (1970).

[16] Manakov, S. V., *Physica* **3D**, 470 (1981).

[17] Fokas, A. S. and Ablowitz, M. J., *Phys. Lett. A* **94**, 67 (1983).

[18] Ablowitz, M. J., Bar Yaacov, D. and Fokas, A. S., *Studies in Appl. Math.* **69**, 135 (1983).

[19] Caudrey, P. J., *Inverse Problems* **2**, 281 (1986).

[20] Hörmander, L., *An Introduction to Complex Analysis in Several Variables*, Van Nostrand, Princeton, NJ (1966).

3 *P. J. Caudrey*

Two-dimensional spectral transforms

3.1 Two-dimensional spectral transforms and Lax pairs

In this chapter we shall consider certain types of nonlinear partial differential equations or systems of equations in 2 + 1 dimensions (i.e. there are three independent variables x, y and t say, where x and y are arbitrarily regarded as spatial coordinates and t is time). If such an equation or system can be written in the form

$$U_t\left(x, y, t, \frac{\partial}{\partial y}\right) - V_x\left(x, y, t, \frac{\partial}{\partial y}\right) + \left[U\left(x, y, t, \frac{\partial}{\partial y}\right), V\left(x, y, t, \frac{\partial}{\partial y}\right)\right] = 0 \tag{3.1.1}$$

where U and V are $N \times N$ matrix functions of x, y, t and the differential operator $\partial/\partial y$ then U and V are said to form a *Lax pair*. Equation (3.1.1) is the compatibility condition for the *spectral equation*

$$\frac{\partial}{\partial x}\psi = U\left(x, y, t, \frac{\partial}{\partial y}\right)\psi \tag{3.1.2}$$

and the *auxiliary spectral equation*

$$\frac{\partial}{\partial t}\psi = V\left(x, y, t, \frac{\partial}{\partial y}\right)\psi. \tag{3.1.3}$$

The nonlinear partial differential equation (3.1.1) can be solved in three stages as follows:

(1) Certain 'Jost solution' of (3.1.2) are found for an initial time $t = t_0$ and certain pieces of information known as the *spectral data* S are extracted from these solutions.
(2) The time evolutions of the spectral data S are deduced from (3.1.3) and hence these data can be found for a later time t.
(3) The *inverse spectral problem*, that of finding $U(x, y, t, \partial/\partial y)$ from S is solved for this later time.

The amazing fact is that each of these three stages involves the solution of a *linear* problem.

The problem of finding a Lax pair, or whether one exists, for a given system of partial differential equations is not addressed here.

The rest of the chapter is arranged as follows. In Section 3.2 the spectral data S in the case when

$$U\left(x, y, t, \frac{\partial}{\partial y}\right) = A\left(\frac{\partial}{\partial y}\right) + B(x, y) \tag{3.1.4}$$

where $B(x, y) \to 0$ as $x^2 + y^2 \to \infty$ is derived. The time evolutions of S are found in Section 3.3 and the method of solution of the inverse spectral problem is described in Section 3.4. In Sections 3.5 and 3.6 two important examples, the Kadomtsev–Petviashvili equation and the Davey–Stewartson equations are considered.

3.2 The direct spectral problem

We take as our spectral equation

$$\frac{\partial}{\partial x} \psi = \left\{A\left(\frac{\partial}{\partial y}\right) + B(x, y)\right\} \psi \tag{3.2.1}$$

where ψ is an N-element column vector and $A(\partial/\partial y)$ and $B(x, y)$ are $N \times N$ matrices. The elements of A are functions of the differential operator $\partial/\partial y$ and

$$B(x, y) \to 0 \quad \text{as} \quad x^2 + y^2 \to \infty. \tag{3.2.2}$$

We introduce the eigenvalues $\lambda_i(\zeta)$ and the eigenvectors $\mathbf{v}_i(\zeta)$ of $A(\zeta)$,

$$A(\zeta)\, \mathbf{v}_i(\zeta) = \lambda_i(\zeta)\, \mathbf{v}_i(\zeta), \quad i = 1, 2, \ldots, N. \tag{3.2.3}$$

together with the reciprocal (row) vectors $\tilde{\mathbf{v}}_i(\zeta)$

$$\tilde{\mathbf{v}}_i(\zeta)\, \mathbf{v}_j(\zeta) = \delta_{ij}. \tag{3.2.4}$$

We assume that these quantities are regular throughout the cut complex ζ-plane and that as ζ crosses a cut all that happens is that the suffices $i = 1, 2, \ldots, N$ are permuted.

The Jost functions $\boldsymbol{\phi}_i(x, y, \zeta)$ are defined to be solutions of (3.2.1) such that

$$\boldsymbol{\phi}_i(x, y, \zeta) = \exp\{\lambda_i(\zeta)x + \zeta y\}\, \boldsymbol{\Phi}_i(x, y, \zeta) \tag{3.2.5}$$

where the $\boldsymbol{\Phi}_i(x, y, \zeta)$ satisfy the boundary conditions

$$\boldsymbol{\Phi}_i(x, y, \zeta) \to \mathbf{v}_i(\zeta) \quad \text{as} \quad x \to -\infty \tag{3.2.6}$$

and remain bounded for $-\infty < x < \infty$. Finding these is equivalent to solving the matrix Fredholm equations

$$\Phi_i(x, y, \zeta) = \int_{-\infty}^{\infty}\int_{-\infty}^{\infty} K_i(x, x', y, y', \zeta)\, \Phi_i(x', y', \zeta)\, dx'dy' + \mathbf{v}_i(\zeta) \tag{3.2.7}$$

where the kernels are given by

$$K_i(x, x', y, y', \zeta) = \frac{1}{2\pi}\sum_{j=1}^{N}\int_{-\infty}^{\infty} \exp\{(\lambda_j(\zeta + ia) - \lambda_i(\zeta))(x - x') + ia(y - y')\}$$
$$\times\, \phi\theta(x - x') - \theta(\mathrm{Re}\,\{\lambda_j(\zeta + ia) - \lambda_i(\zeta)\})\}$$
$$\times\, \mathbf{v}_j\{\zeta + ia)\, \tilde{\mathbf{v}}_j(\zeta + ia)\, da\, B(x', y') \tag{3.2.8}$$

with

$$\theta(\alpha) = \begin{cases} 1 & \text{if} \quad \alpha \geq 0 \\ 0 & \text{if} \quad \alpha < 0 \end{cases}$$

In general $\phi_i(x, y, \zeta)$ and $\Phi_i(x, y, \zeta)$ are not regular functions of ζ because the kernels are not regular. In fact

$$\frac{\partial}{\partial\bar{\zeta}} K_i(x, y, \zeta)$$
$$= -\frac{1}{4\pi}\sum_{k=1}^{N}\int_{-\infty}^{\infty} \exp\{(\lambda_j(\zeta + ia) - \lambda_i(\zeta))(x - x') + ia(y - y')\}$$
$$\times\, \overline{\frac{\partial}{\partial\zeta}(\lambda_j(\zeta + ia) - \lambda_i(\zeta))}\, \delta(\mathrm{Re}\{\lambda_j(\zeta + ia) - \lambda_i(\zeta)\})$$
$$\times\, \mathbf{v}_j(\zeta + ia)\, \tilde{\mathbf{v}}_j(\zeta + ia)\, da\, B(x', y') \tag{3.2.9}$$

where a bar $\overline{}$ over an expression denotes its complex conjugate.

Differentiating (3.2.7)

$$\frac{\partial}{\partial\bar{\zeta}}\Phi_i(x, y, \zeta) - \int_{-\infty}^{\infty}\int_{-\infty}^{\infty} K_i(x, x', y, y', \zeta)\frac{\partial}{\partial\bar{\zeta}}\Phi_i(x', y', \zeta)\, dx'\, dy'$$
$$= \int_{-\infty}^{\infty}\int_{-\infty}^{\infty}\left\{\frac{\partial}{\partial\bar{\zeta}} K_i(x, x', y, y', \zeta)\right\}\Phi_i(x', y', \zeta)\, dx'\, dy'$$
$$= \sum_{k=1}^{N}\int_{-\infty}^{\infty} \exp\{(\lambda_j(\zeta + ia) - \lambda_i(\zeta))x + iay\}\, \mathbf{v}_j(\zeta + ia)\, R_{ij}(a, \zeta)\, da \tag{3.2.10}$$

where

$$R_{ij}(a, \zeta) = -\frac{1}{4\pi}\int_{-\infty}^{\infty}\int_{-\infty}^{\infty} \exp\{-(\lambda_j(\zeta + ia) - \lambda_i(\zeta))x' - iay'\}$$
$$\times\, \tilde{\mathbf{v}}_j(\zeta + ia)\, B(x', y')\, \Phi_i(x', y', \zeta)\, dx'\, dy'$$
$$\times\, \overline{\frac{\partial}{\partial\zeta}(\lambda_j(\zeta + ia) - \lambda_i(\zeta))}\, \delta(\mathrm{Re}\{\lambda_j(\zeta + ia) - \lambda_i(\zeta)\}). \tag{3.2.11}$$

If we now put

$$\chi_i(x, y, \zeta) = \frac{\partial}{\partial\bar{\zeta}} \Phi_i(x, y, \zeta) - \sum_{j=1}^{N}\int_{-\infty}^{\infty} \exp\{(\lambda_j(\zeta + ia) - \lambda_i(\zeta))x + iay\} \times \Phi_j(x, y, \zeta + ia)\, R_{ij}(a, \zeta)\, da \tag{3.2.12}$$

we find that

$$\chi_i(x, y, \zeta) = \int_{-\infty}^{\infty}\int_{-\infty}^{\infty} K_i(x, x', y, y', \zeta)\, \chi_i(x', y', \zeta)\, dx'\, dy'. \tag{3.2.13}$$

Thus providing the homogeneous Fredholm equation has only the trivial solution

$$\frac{\partial}{\partial\bar{\zeta}} \Phi_i(x, y, \zeta) = \sum_{j=1}^{N} \int_{-\infty}^{\infty} \exp\{(\lambda_j(\zeta + ia) - \lambda_i(\zeta))x + iay\} \times \Phi_j(x, y, \zeta + ia)\, R_{ij}(a, \zeta)\, da. \tag{3.2.14}$$

That is

$$\frac{\partial}{\partial\bar{\zeta}} \phi_i(x, y, \zeta) = \sum_{j=1}^{N} \int_{-\infty}^{\infty} \phi_j(x, y, \zeta + ia)\, R_{ij}(a, \zeta)\, da. \tag{3.2.15}$$

The $R_{ij}(a, \zeta)$ form part of the spectral data.

A complication arises if for some value of ζ

$$\mathrm{Re}\{\lambda_j(\zeta + ia) - \lambda_i(\zeta)\} = 0 \tag{3.2.16}$$

for all (real) a. In this case $\Phi_i(x, y, \zeta)$ has a finite (jump) singularity. We label the two sides of the singularity + and − and denote the limit of any function as the singularity is approached from either side by replacing ζ by ζ^+ or ζ^- as appropriate.

It is necessary to introduce the functions $\psi_{ij}(x, y, \zeta, \zeta')$ which are defined by

$$\psi_{ij}(x, y, \zeta, \zeta') = \int_{-\infty}^{\infty}\int_{-\infty}^{\infty} K_i(x, x', y, y', \zeta)\, \Phi_{ij}(x', y', \zeta, \zeta')\, dx'\, dy' + exp\{(\lambda_j(\zeta') - \lambda_i(\zeta))x + (\zeta' - \zeta)y\}\, \mathbf{v}_j(\zeta'). \tag{3.2.17}$$

In general these functions are only well defined when both $(\lambda_j(\zeta') - \lambda_i(\zeta))$ and $(\zeta - \zeta')$ are purely imaginary. Fortunately this is sufficient. Now

$$\exp\{-(\lambda_j(\zeta + ia) - \lambda_i(\zeta))x - iay\}\, \Phi_i(x, y, \zeta^-)$$
$$- \int_{-\infty}^{\infty}\int_{-\infty}^{\infty} K_j(x, x', y, y', \zeta^+ + ia) \exp\{-(\lambda_j(\zeta + ia) - \lambda_i(\zeta))x' - iay\} \times \Phi_i(x', y', \zeta^-)\, dx'\, dy'$$

$$= \sum_{k=1}^{N} \int_{-\infty}^{\infty} P_{ijk}(\zeta^-, \zeta^+ ia, \zeta + ib) \exp\{\lambda_k(\zeta + ib) - \lambda_j(\zeta + ia))x$$
$$+ i(b - a)y\} \, \mathbf{v}_k(\zeta + ib) \, db$$
$$+ \exp\{-(\lambda_j(\zeta + ia) - \lambda_i(\zeta))x - iay\} \, \mathbf{v}_i(\zeta) \tag{3.2.18}$$

where

$$P_{ijk}(\zeta^-, \zeta^+ + ia, \zeta + ib) = \frac{1}{2\pi} \int_{-\infty}^{\infty} \int_{-\infty}^{\infty} \exp\{-(\lambda_k(\lambda + ib) - \lambda_i(\zeta))x' - iby'\}$$
$$\times \{\theta(\mathrm{Re}\{\lambda_k(\zeta^+ + ib) - \lambda_j(\zeta^+ + ia)\}) - \theta(\mathrm{Re}\{\lambda_k(\zeta^- + ib) - \lambda_i(\zeta^-)\})\}$$
$$\times \bar{\mathbf{v}}_k(\zeta + ib) \, B(x', y') \, \Phi_i(x', y', \zeta^-) \, dx' \, dy'. \tag{3.2.19}$$

Hence

$$\exp\{-(\lambda_j(\zeta + ia) - \lambda_i(\zeta))x - iay\} \, \Phi_i(x, y, \zeta)$$
$$= \sum_{k=1}^{N} \int_{-\infty}^{\infty} P_{ijk}(\zeta^-, \zeta^+ + ia), \zeta + ib) \, \psi_{jk}(x, y, \zeta^+ + ia, \zeta + ib) \, db$$
$$+ \psi_{ji}(x, y, \zeta^+ + ia, \zeta). \tag{3.2.20}$$

This matrix Fredholm equation can be solved to give

$$\psi_{ji}(x, y, \zeta^+ + ia, \zeta) = \sum_{k=1}^{N} \int_{\infty}^{\infty} Q_{ijk}(\zeta, \zeta^+ + ia, \zeta^- + ib)$$
$$\times \exp\{\lambda_k(\zeta + ib) - \lambda_j(\zeta + ia))x + i(b - a)y\} \, \Phi_k(x, y, \zeta^- + ib) \, db$$
$$+ \exp\{-(\lambda_j(\zeta + ia) - \lambda_i(\zeta))x - iay\} \, \Phi_i(x, y, \zeta^-) \tag{3.2.21}$$

where the resolvent $Q_{ijk}(\zeta, \zeta^+ + ia, \zeta^- + ib)$ satisfies

$$\sum_{k=1}^{N} \int_{-\infty}^{\infty} Q_{ijk}(\zeta, \zeta^+ + ia, \zeta^- + ib) \, P_{kjl}(\zeta^- + ib, \zeta^+ + ia, \zeta + ic) \, db$$
$$= \sum_{k=1}^{N} \int_{-\infty}^{\infty} P_{ijk}(\zeta^-, \zeta^+ + ia, \zeta + ib) \, Q_{kjl}(\zeta + ib, \zeta^+ + ia, \zeta^- + ic) \, db$$
$$= -P_{ijl}(\zeta^-, \zeta^+ + ia, \zeta + ic) - Q_{ijl}(\zeta, \zeta^+ + ia, \zeta^- + ic). \tag{3.2.22}$$

Finally from (3.2.21) by putting $j = i$, $a = 0$ and $Q_{iik}(\zeta, \zeta^+, \zeta^- + ib) = Q_{ik}(b, \zeta)$ we get

$$\Phi_i(x, y, \zeta^+) - \Phi_i(x, y, \zeta^-) = \sum_{k=1}^{N} \int_{-\infty}^{\infty} Q_{ik}(b, \zeta)$$
$$\times \exp\{(\lambda_k(\zeta + ib) - \lambda_i(\zeta))x + iby\} \, \Phi_k(x, y, \zeta^- + ib) \, db. \tag{3.2.23}$$

The quantities $Q_{ik}(b, \zeta)$ belong in the set of spectral data. This relation is neater in terms of $\phi_i(x, y, \zeta)$,

$$\phi_i(x, y, \zeta^+) - \phi_i(x, y, \zeta^-) = \sum_{k=1}^{N} \int_{-\infty}^{\infty} Q_{ik}(b, \zeta) \, \phi_k(x, y, \zeta^- + ib) \, db. \tag{3.2.24}$$

The question of what happens when the homogeneous Fredholm equation (3.2.13) has non-trivial solutions is as yet unanswered in the general case. However, in one important special case the answer is known. This is when the non-trivial solutions occur for values of ζ lying in a region of the ζ-plane for which the only integer value of j and the only real value of a such that

$$\mathrm{Re}\{\lambda_j(\zeta + ia) - \lambda_i(\zeta)\} = 0 \tag{3.2.25}$$

are $j = i$ and $a = 0$. In this case

$$\frac{\partial}{\partial\bar{\zeta}} K_i(x, x', y, y', \zeta) = 0. \tag{3.2.26}$$

That is, $K_i(x, x', y, y', \zeta)$ is a regular function of ζ throughout the region. It follows that the Fredholm determinant $f_i(\zeta)$ and Fredholm minor $F_i(x, x', y, y', \zeta)$ are also regular and that the Jost function

$$\mathbf{\Phi}_i(x, y, \zeta) = \frac{1}{f_i(\zeta)}\int_{-\infty}^{\infty}\int_{-\infty}^{\infty} F_i(x, x', y, y', \zeta)\, \mathbf{v}_i(\zeta)\, dx'\, dy' + \mathbf{v}_i(\zeta) \tag{3.2.27}$$

is analytic, with poles at the zeros of $f_i(\zeta)$. For simplicity we shall assume that the poles are simple.

If the poles occur at $\zeta = \zeta_i^{(k)}$, $k = 1, 2, \ldots, n_i$ then in a neighbourhood of $\zeta_i^{(k)}$ we can put

$$\mathbf{\Phi}_i(x, y, \zeta) = \mathbf{\Phi}_i^{(k)}(x, y)/(\zeta - \zeta_i^{(k)}) + \bar{\mathbf{\Phi}}_i^{(k)}(x, y, \zeta). \tag{3.2.28}$$

The residue

$$\mathbf{\Phi}_i^{(k)}(x, y) = \frac{1}{\dfrac{\partial}{\partial\zeta} f_i(\zeta_i^{(k)})}\int_{-\infty}^{\infty} F_i(x, x', y, y', \zeta_i^{(k)})\, \mathbf{v}_i(\zeta_i^{(k)})\, dx'\, dy' \tag{3.2.29}$$

satisfies the homogeneous Fredholm equation

$$\mathbf{\Phi}_i^{(k)}(x, y) = \int_{-\infty}^{\infty}\int_{-\infty}^{\infty} K_i(x, x', y, y', \zeta_i^{(k)})\, \mathbf{\Phi}_i^{(k)}(x', y')\, dx'\, dy' \tag{3.2.30}$$

and $\bar{\mathbf{\Phi}}_i^{(k)}(x, y, \zeta_i^{(k)})$ satisfies

$$\begin{aligned}\bar{\mathbf{\Phi}}_i^{(k)}(x, y, \zeta_i^{(k)}) = {} & \int_{-\infty}^{\infty}\int_{-\infty}^{\infty} K_i(x, x', y, y', \zeta_i^{(k)})\, \bar{\mathbf{\Phi}}_i^{(k)}(x', y', \zeta_i^{(k)})\, dx'\, dy' \\ & + \int_{-\infty}^{\infty}\int_{-\infty}^{\infty}\left\{\frac{\partial}{\partial\zeta} K_i(x, x', y, y', \zeta_i^{(k)})\right\} \mathbf{\Phi}_i^{(k)}(x', y') \\ & + \mathbf{v}(\zeta_i^{(k)}).\end{aligned} \tag{3.2.31}$$

Differentiating (3.2.8) gives

$$\frac{\partial}{\partial \zeta} K_i(x, x', y, y', \zeta) = -\left\{\frac{\partial}{\partial \zeta} \lambda_i(\zeta)(x - x') + y - y'\right\} K_i(x, x', y, y', \zeta)$$
$$+ \frac{\mathrm{i}}{2\pi} \operatorname{sgn}\left(\operatorname{Im}\left\{\frac{\partial}{\partial \zeta}\lambda_i(\zeta)\right\}\right) \mathbf{v}_i(\zeta)\, \tilde{\mathbf{v}}_i(\zeta)\, B(x', y') \quad (3.2.32)$$

and Fredholm theory gives

$$\frac{\partial}{\partial \zeta} f_i(\zeta) = -\int_{-\infty}^{\infty}\int_{-\infty}^{\infty} \operatorname{Tr}\left[f_i(\zeta) \frac{\partial}{\partial \zeta} K_i(x, x, y, y, \zeta)\right.$$
$$\left. + \int_{-\infty}^{\infty}\int_{-\infty}^{\infty} F_i(x, x', y, y', \zeta) \frac{\partial}{\partial \zeta} K_i(x', x, y', y, \zeta)\, \mathrm{d}x'\, \mathrm{d}y' \right] \mathrm{d}x\, \mathrm{d}y$$
$$= \frac{\mathrm{i}}{2\pi} \operatorname{sgn}\left(\operatorname{Im}\left\{\frac{\partial}{\partial \zeta} \lambda_i(\zeta)\right\}\right) \int_{-\infty}^{\infty}\int_{-\infty}^{\infty}\int_{-\infty}^{\infty}\int_{-\infty}^{\infty} \tilde{\mathbf{v}}_i(\zeta)\, B(x, y)$$
$$\times F_i(x, x', y, y', \zeta)\, \mathbf{v}_i(\zeta)\, \mathrm{d}x'\, \mathrm{d}y'\, dx\, \mathrm{d}y. \quad (3.2.33)$$

Combining this with (3.2.29) gives

$$\frac{\mathrm{i}}{2\pi} \operatorname{sgn}\left(\operatorname{Im}\left\{\frac{\partial}{\partial \zeta} \lambda_i(\zeta_i^{(k)})\right\}\right) \int_{-\infty}^{\infty}\int_{-\infty}^{\infty} \tilde{\mathbf{v}}_i(\zeta_i^{(k)})\, B(x, y)\, \mathbf{\Phi}_i^{(k)}(x, y)\, \mathrm{d}x = -1 \quad (3.2.34)$$

and so (3.2.31) becomes

$$\tilde{\mathbf{\Phi}}_i^{(k)}(x, y, \zeta_i^{(k)}) = \int_{-\infty}^{\infty}\int_{-\infty}^{\infty} K_i(x, x', y, y', \zeta_i^{(k)})\, \tilde{\mathbf{\Phi}}_i^{(k)}(x', y', \zeta_i^{(k)})\, \mathrm{d}x'\, \mathrm{d}y'$$
$$- \int_{-\infty}^{\infty}\int_{-\infty}^{\infty} \left\{\frac{\partial}{\partial \zeta} \lambda_i(\zeta_i^{(k)})(x - x') + y - y'\right\}$$
$$\times K_i(x, x', y, y', \zeta_i^{(k)})\, \mathbf{\Phi}_i^{(k)}(x', y')\, \mathrm{d}x'\, \mathrm{d}y' \quad (3.2.35)$$

from which it follows that

$$\tilde{\mathbf{\Phi}}_i^{(k)}(x, y, \zeta_i^{(k)}) + \left\{\frac{\partial \lambda_i}{\partial \zeta}(\zeta_i^{(k)})x + y\right\} \mathbf{\Phi}_i^{(k)}(x, y) \quad (3.2.36)$$

satisfies the homogeneous Fredholm equation (3.2.30). Since the pole is simple there is only one linearly independent solution to (3.2.30) and hence

$$\tilde{\mathbf{\Phi}}_i^{(k)}(x, y, \zeta_i^{(k)}) = \left\{\gamma_i^{(k)} - \frac{\partial \lambda_i}{\partial \zeta}(\zeta_i^{(k)})x - y\right\} \mathbf{\Phi}_i^{(k)}(x, y) \quad (3.2.37)$$

where $\gamma_i^{(k)}$ is a constant which forms another part of the spectral data.
If we put

$$\phi_i(x, y, \zeta) = \phi_i^{(k)}(x, y)/(\zeta - \zeta_i^{(k)}) + \tilde{\phi}_i^{(k)}(x, y, \zeta) \qquad (3.2.38)$$

this simplifies to

$$\tilde{\phi}_i^{(k)}(x, y, \zeta_i^{(k)}) = \gamma_i^{(k)}\phi_i^{(k)}(x, y) \qquad (3.2.39)$$

The complete set of spectral data is

$$S = \{R_{ij}(a, \zeta),\ i, j = 1, 2, \ldots, N,\ a \in \mathbb{R},\ \zeta \in \mathbb{C};$$

$$Q_{ik}(b, \zeta),\ i, k = 1, 2, \ldots, N,\ b \in \mathbb{R},$$

ζ runs along the finite singularities of $\mathbf{\Phi}_i(x, y, \zeta)$;

$$\zeta_i^{(k)},\ \gamma_i^{(k)},\ k = 1, 2, \ldots, n_i,\ i = 1, 2, \ldots, N\}. \qquad (3.2.40)$$

Finally in this section the solution of the inverse problem requires that

$$\mathbf{\Phi}_i(x, y, \zeta) - \mathbf{v}_i(\zeta) \to \mathbf{0} \quad \text{as} \quad \zeta \to \infty. \qquad (3.2.41)$$

Whether this condition is satisfied or not depends on the individual case. So far it has proved to be true in most cases of practical interest. It can usually be deduced from (3.2.7) either with or without the assistance of the Riemann–Lebesgue lemma.

3.3 The time dependence of the spectral data

The time dependence of the wave function ψ is assumed to be governed by the auxiliary spectral equation,

$$\psi_t = V\left(x, y, t, \frac{\partial}{\partial y}\right)\psi \qquad (3.3.1)$$

where

$$V\left(x, y, t, \frac{\partial}{\partial y}\right) \to L\left(\frac{\partial}{\partial y}\right). \qquad (3.3.2)$$

as $x^2 + y^2 \to \infty$ in the same way as $U(x, y, t, \partial/\partial y) \to A(\partial/\partial y)$. From (3.1.1) it follows that

$$\left[A\left(\frac{\partial}{\partial y}\right), L\left(\frac{\partial}{\partial y}\right)\right] = 0, \qquad (3.3.3)$$

that is,

$$[A(\zeta), L(\zeta)] = 0 \qquad (3.3.4)$$

so that $A(\zeta)$ and $L(\zeta)$ have common eigenvectors.

$$L(\zeta)\,\mathbf{v}_i(\zeta) = \mu_i(\zeta)\,\mathbf{v}_i(\zeta). \qquad (3.3.5)$$

By examining the behaviour as $x^2 + y^2 \to \infty$ it is easy to discover that (3.3.1) is satisfied by

$$\exp\{\mu_i(\zeta)t\}\ \phi_i(x, y, t, \zeta) = \exp\{\lambda_i(\zeta)x + \zeta y + \mu_i(\zeta)t\}\ \Phi_i(x, y, t, \zeta), \tag{3.3.6}$$

and hence the time dependence of $\phi_i(x, y, t, \zeta)$ is given by

$$\frac{\partial}{\partial t}\phi_i(x, y, t, \zeta) = \left\{V_i\left(x, y, t, \frac{\partial}{\partial y}\right) - \mu_i(\zeta)\ I\right\}\phi_i(x, y, t, \zeta). \tag{3.3.7}$$

If we assume that the $\mu_i(\zeta)$ are regular functions of ζ then (3.2.15) gives

$$\sum_{j=1}^{N}\int_{-\infty}^{\infty}\{(\mu_i(\zeta) - \mu_j(\zeta))\ R_{ij}(t, a, \zeta) + \frac{\partial}{\partial t}R_{ij}(t, a, \zeta)\}$$
$$- \phi_j\ (x, y, t, \zeta + \mathrm{i}a)\ \mathrm{d}a = 0, \tag{3.3.8}$$

from which it follows that

$$\frac{\partial}{\partial t}R_{ij}(t, a, \zeta) = \{\mu_j(\zeta + \mathrm{i}a) - \mu_i(\zeta)\}\ R_{ij}(t, a, \zeta). \tag{3.3.9}$$

A similar argument from (3.2.24) gives

$$\frac{\partial}{\partial t}Q_{ik}(t, b, \zeta) = \{\mu_k(\zeta + \mathrm{i}b) - \mu_i(\zeta)\}\ Q_{ik}(t, b, \zeta), \tag{3.3.10}$$

and from (3.2.38) and (3.2.39) we get

$$\frac{\mathrm{d}}{\mathrm{d}t}\zeta_i^{(k)} = 0 \tag{3.3.11}$$

and

$$\frac{\mathrm{d}}{\mathrm{d}t}\gamma_i^{(k)}(t) = -\frac{\partial\mu_i}{\partial\zeta}(\zeta_i^{(k)}). \tag{3.3.12}$$

3.4 The inverse spectral problem

This is the problem of finding $B(x, y)$ from the spectral data S given by (3.2.40). Besides S we also have $A(\zeta)$ and the information associated with it, the location of the cuts in the ζ-plane, the permutations of the suffices across the cuts and the locations of the curves Γ along which $\mathrm{Re}\{\lambda_j(\zeta + \mathrm{i}a) - \lambda_i(\zeta)\} = 0$ for all (real) a.

The first step in solving this problem is to find the functions $\Phi_i(x, y, \zeta)$ which have the properties given by (3.2.14), (3.2.23), (3.2.28), (3.2.37) and (3.2.41) and such that across the cuts

$$\Phi_i(x, y, \zeta^+) = \sum_{j=1}^{N}P_{ij}\ \Phi_i(x, y, \zeta^-) \tag{3.4.1}$$

where P_{ij} are the elements of the appropriate permutation matrix. Once these have been obtained the spectral equation (3.2.1) easily gives $B(x, y)$.

We use the generalisation of Cauchy's integral formula

$$w(\zeta') = \frac{1}{2\pi i}\iint_R \frac{dw/d\bar{\zeta}}{\zeta - \zeta'} d\zeta \wedge d\bar{\zeta} + \frac{1}{2\pi i}\int_C \frac{w(\zeta)}{\zeta - \zeta'} d\zeta \qquad (3.4.2)$$

where R is a region in the complex ζ-plane whose boundary is the contour C and ζ' lies in R. This gives

$$\begin{aligned}\mathbf{\Phi}_i(x, y, \zeta') = \mathbf{v}_i(\zeta') &+ \frac{1}{2\pi i}\iint_{\zeta\text{-plane}} \frac{1}{\zeta - \zeta'} \cdot \frac{\partial \mathbf{\Phi}_i(x, y, \zeta)}{\partial \bar{\zeta}} d\zeta \wedge d\bar{\zeta} \\ &+ \frac{1}{2\pi i}\int_{\Gamma+\text{cuts}} \frac{1}{\zeta - \zeta'} \{\mathbf{\Phi}_i(x, y, \zeta^+) - \mathbf{\Phi}_i(x, y, \zeta^-)\} d\zeta \\ &+ \sum_{k=1}^{n_i} \frac{1}{\zeta' - \zeta_i^{(k)}} \mathbf{\Phi}_i^{(k)}(x, y) \end{aligned} \qquad (3.4.3)$$

where the direction of integration along Γ and the cuts is such that the side labelled + is on the left. Using the properties of $\mathbf{\Phi}_i(x, y, \zeta)$ this becomes

$$\begin{aligned}\mathbf{\Phi}_i(x, y, \zeta') &= \mathbf{v}_i(\zeta') \\ &+ \frac{1}{2\pi i}\iint_{\zeta\text{-plane}} \frac{1}{\zeta - \zeta'} \sum_{j=1}^{N}\int_{-\infty}^{\infty} \exp\{(\lambda_j(\zeta + ia) - \lambda_i(\zeta))x + iay\} \\ &\times \mathbf{\Phi}_j(x, y, \zeta + ia)\, R_{ij}(a, \zeta)\, da\, d\zeta \wedge d\bar{\zeta} \\ &+ \frac{1}{2\pi i}\int_\Gamma \frac{1}{\zeta - \zeta'} \sum_{j=1}^{N}\int_{-\infty}^{\infty} Q_{ij}(a, \zeta) \exp\{(\lambda_j(\zeta + ia) - \lambda_i(\zeta))x + iay\} \\ &\times \mathbf{\Phi}_j(x, y, \zeta^- + ia)\, da\, d\zeta \\ &+ \frac{1}{2\pi i}\int_{\text{cuts}} \frac{1}{\zeta - \zeta'} \sum_{j=1}^{N} (P_{ij} - \delta_{ij})\, \mathbf{\Phi}_j(x, y, \zeta^-) d\zeta \\ &+ \sum_{k=1}^{n^i} \frac{2}{\zeta' - \zeta_i^{(k)}} \mathbf{\Phi}_i^{(k)}(x, y) \end{aligned} \qquad (3.4.4)$$

where

$$\begin{aligned}\left\{\gamma_i^{(k)} - \frac{\partial \lambda_i}{\partial \zeta}(\zeta_i^{(k)})x + y\right\} \mathbf{\Phi}_i^{(k)}(x, y) &= \mathbf{v}_i(\zeta') \\ &+ \frac{1}{2\pi i}\iint_{\zeta\text{-plane}} \frac{1}{\zeta - \zeta_i^{(k)}} \sum_{j=1}^{N}\int_{-\infty}^{\infty} \exp\{(\lambda_j(\zeta + ia) - \lambda_i(\zeta))x + iay\} \\ &\times \mathbf{\Phi}_j(x, y, \zeta + ia)\, R_{ij}(a, \zeta)\, da\, d\zeta \wedge d\bar{\zeta} \\ &+ \frac{1}{2\pi i}\int_\Gamma \frac{1}{\zeta - \zeta_i^{(k)}} \sum_{j=1}^{N}\int_{-\infty}^{\infty} Q_{ij}(a, \zeta) \exp\{(\lambda_j(\zeta + ia) - \lambda_i(\zeta))x + iay\}\end{aligned}$$

$$\times\ \Phi_j(x, y, \zeta^- + ia)\, da\, d\zeta$$

$$+ \frac{1}{2\pi i}\int_{\text{cuts}} \frac{1}{\zeta - \zeta_i^{(k)}} \sum_{j=1}^{N} (P_{ij} - \delta_{ij})\, \Phi_j(x, y, \zeta^-)\, d\zeta$$

$$+ \sum_{\substack{l=1\\ l\neq k}}^{n_i} \frac{1}{\zeta_i^{(k)} - \zeta_i^{(l)}}\, \Phi_i^{(l)}(x, y). \tag{3.4.5}$$

These equations (3.4.4), (3.4.5) form a set of matrix/Fredholm equations. Since they are singular there is an unresolved doubt as to the existence and uniqueness of their solution. However, in practical cases there is apparently no difficulty.

3.5 The Kadomtsev–Petviashvili equation

The Kadomtsev–Petviashvili equation (KP) is a two-dimensional version of the Korteweg–de Vries equation,

$$(u_t - 6uu_x + u_{xxx})_x + 3\sigma^2 u_{yy} = 0 \tag{3.5.1}$$

where $\sigma = i$ (KP–1) or $\sigma = -1$ (KP–2). In their usual forms the spectral and auxiliary spectral equations are

$$-\psi_{xx} + u\psi = \sigma\psi_y \tag{3.5.2}$$

and

$$\psi_t = \left(-4\frac{\partial^3}{\partial x^3} + 3\frac{\partial}{\partial x}u + 3u\frac{\partial}{\partial x} - 3\sigma v\right)\psi \tag{3.5.3}$$

which have the integrability conditions

$$v_x = u_y \tag{3.5.4}$$

and

$$u_t - 6uu_x + u_{xxx} + 3\sigma^2 v_y = 0. \tag{3.5.5}$$

These can easily be cast in the form of (3.2.1) and (3.3.1) with $N = 2$. However, the subsequent algebra is considerably simplified merely by interchanging the roles of x and y. Then $N = 1$ and

$$A(\zeta) = -\zeta^2/\sigma, \tag{3.5.6}$$

$$B(x, y, t) = u(x, y, t)/\sigma \tag{3.5.7}$$

and

$$V(x, y, t, \zeta) = -4\zeta^3 + 6\zeta u(x, y, t) + 3\{u_x(x, y, t) - \sigma v(x, y, t)\}. \tag{3.5.8}$$

Dropping unnecessary suffices

$$\lambda(\zeta) = -\zeta^2/\sigma \tag{3.5.9}$$

$$\mathbf{v}(\zeta) = 1, \quad \tilde{\mathbf{v}}(\zeta) = 1 \tag{3.5.10}$$

and

$$\mu(\zeta) = -4\zeta^3. \tag{3.5.11}$$

The Jost function is given by

$$\phi(x, y, \zeta) = \exp\{\zeta x - \zeta^2 y/\sigma\}\ \Phi(x, y, \zeta) \tag{3.5.12}$$

where

$$\Phi(x, y, \zeta) = \int_{-\infty}^{\infty}\int_{-\infty}^{\infty} K(x, x', y, y', \zeta)\ \Phi(x', y', \zeta)\ \mathrm{d}x'\ \mathrm{d}y' + 1 \tag{3.5.13}$$

with

$$K(x, x', y, y', \zeta) = \frac{1}{2\pi\sigma}\int_{-\infty}^{\infty} \exp\{\mathrm{i}a(x - x') + a(a - 2\mathrm{i}\zeta)(y - y')/\sigma\} \times \{\theta(y - y') - \theta(\mathrm{Re}\{a(a - 2\mathrm{i}\zeta)/\mu\})\}\ \mathrm{d}a\ u(x', y'). \tag{3.5.14}$$

Let us consider first the case $\sigma = -1$ (KP–2). This gives

$$R(a, \zeta) = R(\zeta)\ \delta(a + 2\mathrm{Im}\zeta) \tag{3.5.15}$$

where

$$R(a) = \frac{\mathrm{i}}{2\pi}\ \mathrm{sgn}(\mathrm{Im}\zeta)\int_{-\infty}^{\infty}\int_{-\infty}^{\infty} \exp\{2\mathrm{i}\mathrm{Im}\zeta x + 4\mathrm{i}\mathrm{Re}\zeta\mathrm{Im}\zeta y\} \times u(x, y)\ \Phi(x, y, \zeta)\ \mathrm{d}x\ \mathrm{d}y. \tag{3.5.16}$$

Thus

$$\frac{\partial}{\partial\bar{\zeta}}\ \Phi(x, y, \zeta) = \exp\{-2\mathrm{i}\mathrm{Im}\zeta x - 4\mathrm{i}\mathrm{Re}\zeta\mathrm{Im}\zeta y\}\ \Phi(x, y, \bar{\zeta})\ R(\zeta). \tag{3.5.17}$$

There are no values of ζ such that $\mathrm{Re}\{\lambda(\zeta + \mathrm{i}a) - \lambda(\zeta)\} = 0$ for all a. Thus $\Phi(x, y, \zeta)$ has no finite singularities.

Current belief is that the homogeneous version of the Fredholm equation (3.5.13) has only the trivial solution for all ζ. However this has yet to be proved.

If $\sigma = \mathrm{i}$ the situation is quite different. In this case $K(x, x', y, y', \zeta)$ is regular except on the imaginary axis $\mathrm{Re}\ \zeta = 0$. Thus

$$R(a, \zeta) = 0. \tag{3.5.18}$$

Across $\mathrm{Re}\ \zeta = 0$ there is a finite singularity. Putting $\zeta = \mathrm{i}\xi$ where ξ is real and labelling the sides of the imaginary axis according to the sign of Re ζ we have

$$P(\mathrm{i}\xi + \mathrm{i}a - 0, \mathrm{i}\xi + 0, \mathrm{i}\xi + \mathrm{i}b) = -\frac{\mathrm{i}}{2\pi}\int_{-\infty}^{\infty}\int_{-\infty}^{\infty} \exp\{-\mathrm{i}(a - b)(a + b + 2\xi)y' + \mathrm{i}(a - b)x'\} \times \{\theta(-b) - \theta(b - a)\}\, u(x', y')\, \Phi(x', y', \mathrm{i}\xi + \mathrm{i}a - 0)\, \mathrm{d}x'\, \mathrm{d}y' \tag{3.5.19}$$

and $Q(a, \mathrm{i}\xi)$ satisfies

$$\int_{-\infty}^{\infty} Q(a, \mathrm{i}\xi)\, P(\mathrm{i}\xi + \mathrm{i}a - 0, \mathrm{i}\xi + 0, \mathrm{i}\xi + \mathrm{i}b)\, \mathrm{d}a + P(\mathrm{i}\xi - 0, \mathrm{i}\xi + 0, \mathrm{i}\xi + \mathrm{i}b) + Q(b, \mathrm{i}\xi) = 0. \tag{3.5.20}$$

Off the imaginary axis the conditions specified in Section 3.2 are satisfied so that $\Phi(x, y, \zeta)$ is analytic and assuming that the poles are simple we can put

$$\Phi(x, y, \zeta) = \Phi^{(k)}(x, y)/(\zeta - \zeta^{(k)}) + \bar{\Phi}(x, y, \zeta) \tag{3.5.21}$$

in a neighbourhood of the pole at $\zeta = \zeta^{(k)}$, $k = 1, 2, \ldots, n$ where

$$\bar{\Phi}^{(k)}(x, y, \zeta^{(k)}) = \{\gamma^{(k)} - 2\mathrm{i}\zeta^{(k)}x + y\}\, \Phi^{(k)}(x, y). \tag{3.5.22}$$

The equatons (3.4.4), (3.4.5) for the inversion become

$$\Phi(x, y, \zeta') = 1 + \frac{1}{2\pi}\int_{-\infty}^{\infty} \frac{1}{\mathrm{i}\xi - \zeta'} \int_{-\infty}^{\infty} Q(a, \mathrm{i}\xi) \exp\{\mathrm{i}ax - \mathrm{i}a(2\xi + a)y\} \times \Phi(x, y, \mathrm{i}\xi + \mathrm{i}a + 0)\, \mathrm{d}a\, \mathrm{d}\xi + \sum_{k=1}^{n} \frac{1}{\zeta' - \zeta^{(k)}}\, \Phi^{(k)}(x, y) \tag{3.5.23}$$

and

$$\{\gamma^{(k)} - x - 2\mathrm{i}\zeta^{(k)}y\}\, \Phi^{(k)}(x, y) = 1 + \frac{1}{2\pi}\int_{-\infty}^{\infty} \frac{1}{\mathrm{i}\xi - \zeta^{(k)}} \int_{-\infty}^{\infty} Q(a, \mathrm{i}\xi) \times \exp\{\mathrm{i}ax - \mathrm{i}a(2\xi + a)y\}\, \Phi(x, y, \mathrm{i}\xi + \mathrm{i}a + 0)\, \mathrm{d}a\, \mathrm{d}\xi + \sum_{\substack{l=1 \\ l\neq k}}^{n} \frac{1}{\zeta^{(k)} - \zeta^{(l)}}\, \Phi^{(l)}(x, y). \tag{3.5.24}$$

If there are no $\zeta^{(k)}$ then we can simplify things by Fourier transforming and putting

$$K(x, y, z) = \frac{1}{2\pi}\int_{-\infty}^{\infty} \exp\{-\mathrm{i}\xi z\}\, \{\Phi(x, y, \mathrm{i}\xi + 0) - 1\}\, \mathrm{d}\xi \tag{3.5.25}$$

so that

$$\Phi(x, y, i\xi + 0) = \int_{-\infty}^{\infty} \exp(i\xi z\} \, K(x, y, z) \, dz + 1. \qquad (3.5.26)$$

Equation (3.5.23) transforms to

$$K(x, y, z) = -\theta(-z) \left\{ \int_{-\infty}^{\infty} F(x + w, y, z - w) \, K(x, y, w) \, dw \right. \\ \left. + F(x, y, z) \right\} \qquad (3.5.27)$$

where

$$F(x, y, z) = \frac{1}{2\pi} \int_{-\infty}^{\infty} \int_{-\infty}^{\infty} Q(a, i\xi) \exp\{iax - ia(2\xi + a)y - i\xi z) \, da \, d\xi. \qquad (3.5.28)$$

By expanding $\Phi(x, y, \zeta)$ as an asymptotic series

$$\Phi(x, y, \zeta) = 1 + \frac{\alpha(x, y)}{\zeta} + O(\zeta^2) \qquad (3.5.29)$$

it follows from the spectral equation (3.5.2) that

$$u(x, y) = -2 \frac{\partial}{\partial x} \alpha(x, y) \qquad (3.5.30)$$

and from (3.5.23), (3.5.25), (3.5.26) that

$$\alpha(x, y) = -K(x, y, 0 - 0) \qquad (3.5.31)$$

Thus

$$u(x, y) = 2 \frac{\partial}{\partial x} K(x, y, 0 - 0). \qquad (3.5.32)$$

Unfortunately this simplicity is lost when $\zeta^{(k)}$ do exist.

3.6 The Davey–Stewartson equations

The spectral equation

$$\frac{\partial}{\partial x} \psi = \begin{pmatrix} \alpha \frac{\partial}{\partial y} & u \\ \pm \bar{u} & -\alpha \frac{\partial}{\partial y} \end{pmatrix} \psi \qquad (3.6.1)$$

and the auxiliary spectral equation

$$\frac{\partial}{\partial t}\psi =$$

$$\begin{bmatrix} \mathrm{i}\dfrac{\partial^2}{\partial y^2} \pm \dfrac{\mathrm{i}}{2\alpha^2}|u|^2 - \dfrac{\mathrm{i}}{2}w + \dfrac{\mathrm{i}}{2}v & \dfrac{\mathrm{i}}{\alpha}u\dfrac{\partial}{\partial y} + \dfrac{\mathrm{i}}{2\alpha^2}(u_x + \alpha u_y) \\ \pm\dfrac{\mathrm{i}}{2}\bar{u}\dfrac{\partial}{\partial y} \mp \dfrac{\mathrm{i}}{2\alpha^2}(\bar{u}_x - \alpha\bar{u}_x) & -\mathrm{i}\dfrac{\partial^2}{\partial y^2} \mp \dfrac{\mathrm{i}}{2\alpha^2}|u|^2 + \dfrac{\mathrm{i}}{2}w + \dfrac{\mathrm{i}}{2}v \end{bmatrix}\psi \tag{3.6.2}$$

have as their compatibility conditions

$$\left.\begin{aligned} \mathrm{i}u_t + \frac{1}{2\alpha^2}(u_{xx} + \alpha^2 u_{yy}) &= \mp\frac{1}{\alpha^2}|u|^2 u + wu \\ w_x + \alpha v_y &= \pm\frac{2}{\alpha^2}(|u|^2)_x \\ v_x + \alpha w_y &= 0 \end{aligned}\right\} \tag{3.6.3}$$

which are the Davey–Stewartson equations when $\alpha = \mathrm{i}$ (DSI) or 1 (DSII). In this case $N = 2$,

$$A(\zeta) = \begin{pmatrix} \alpha\zeta & 0 \\ 0 & -\alpha\zeta \end{pmatrix}, \tag{3.6.4}$$

$$B(x, y, t) = \begin{pmatrix} 0 & u(x, y, t) \\ \pm\bar{u}(x, y, t) & 0 \end{pmatrix}, \tag{3.6.5}$$

and

$$L(\zeta) = \begin{pmatrix} \mathrm{i}\zeta^2 & 0 \\ 0 & -\mathrm{i}\zeta^2 \end{pmatrix}. \tag{3.6.6}$$

Thus

$$\lambda_1(\zeta) = \alpha\zeta, \quad \mathbf{v}_2(\zeta) = \begin{pmatrix} 1 \\ 0 \end{pmatrix}, \quad \tilde{\mathbf{v}}_1(\zeta) = (1 \quad 0), \quad \mu_1(\zeta) = \mathrm{i}\zeta^2 \tag{3.6.7}$$

and

$$\lambda_2(\zeta) = -\alpha\zeta, \quad \mathbf{v}_2(\zeta) = \begin{pmatrix} 0 \\ 1 \end{pmatrix}, \quad \tilde{\mathbf{v}}_2(\zeta) = (0 \quad 1), \quad \mu_2(\zeta) = -\mathrm{i}\zeta^2. \tag{3.6.8}$$

Let us consider the case $\alpha = \mathrm{i}$ first of all. The kernels have the form

$$K_i(x, x', y, y', \zeta) = \begin{pmatrix} 0 & K_{i,12} \\ K_{i,21} & 0 \end{pmatrix} \tag{3.6.9}$$

where considering first the case when $\alpha = \mathrm{i}$

$$\left.\begin{aligned} K_{1,12} &= \frac{1}{2\pi(x - x' - \mathrm{i}y + \mathrm{i}y')}\, u(x', y') \\ K_{1,21} &= \pm\, \frac{\exp\{-2\mathrm{i}\mathrm{Re}\zeta(x - x') - 2\mathrm{i}\mathrm{Im}\zeta(y - y')\}}{2\pi(x - x' + \mathrm{i}y - \mathrm{i}y')}\, \bar{u}(x', y') \\ K_{2,12} &= \frac{\exp\{2\mathrm{i}\mathrm{Re}\zeta(x - x') - 2\mathrm{i}\mathrm{Im}\zeta(y - y')\}}{2\pi(x - x' - \mathrm{i}y + \mathrm{i}y')}\, u(x', y') \\ K_{2,21} &= \pm\, \frac{1}{2\pi(x - x' + \mathrm{i}y - \mathrm{i}y')}\, \bar{u}(x', y') \end{aligned}\right\} \tag{3.6.10}$$

The spectral data are

$$R_{11}(a, \zeta) = 0 \tag{3.6.11}$$

$$\begin{aligned} R_{12}(a, \zeta) &= \mp\, \frac{\mathrm{i}}{2\pi} \int_{-\infty}^{\infty}\int_{-\infty}^{\infty} \exp\{2\mathrm{i}\mathrm{Re}\zeta x' + 2\mathrm{i}\mathrm{Im}\zeta y'\} \\ &\quad \times \bar{u}(x', y')\, \Phi_{1,1}(x', y', \zeta)\, \mathrm{d}x'\, \mathrm{d}y'\, \delta(2Im\zeta + ia) \\ &= \mp\, \frac{\mathrm{i}}{2\pi} \int_{-\infty}^{\infty}\int_{-\infty}^{\infty} \exp\{\bar{\zeta}(\mathrm{i}x' - y')\}\, \bar{u}(x', y') \\ &\quad \times \phi_{1,1}(x', y')\, \mathrm{d}x'\, \mathrm{d}y'\, \delta(2\mathrm{Im}\zeta + a) \end{aligned} \tag{3.6.12}$$

$$\begin{aligned} R_{21}(a, \zeta) &= \frac{\mathrm{i}}{2\pi} \int_{-\infty}^{\infty}\int_{-\infty}^{\infty} \exp\{-2\mathrm{i}\mathrm{Re}\zeta x' + 2\mathrm{i}\mathrm{Im}\zeta y'\} \\ &\quad \times u(x', y')\, \Phi_{2,2}(x', y', \zeta)\, \mathrm{d}x'\, \mathrm{d}y'\, \delta(2\mathrm{Im}\zeta + a) \\ &= \frac{\mathrm{i}}{2\pi} \int_{-\infty}^{\infty}\int_{-\infty}^{\infty} \exp\{-\bar{\zeta}(\mathrm{i}x' + y')\}\, u(x', y')\, \phi_{2,2}(x', y', \zeta)\, \mathrm{d}x'\, \mathrm{d}y' \\ &\quad \times \delta(2\mathrm{Im}\zeta + a) \end{aligned} \tag{3.6.13}$$

$$R_{22}(a, \zeta) = 0 \tag{3.6.14}$$

Because of the delta functions in these the integration in (3.2.15) can easily be carried out to give

$$\frac{\partial}{\partial\bar{\zeta}}\, \phi_1(x, y, \zeta) = R_1(\zeta)\, \phi_2(x, y, \bar{\zeta}) \tag{3.6.15}$$

and

$$\frac{\partial}{\partial\bar{\zeta}}\, \phi_2(x, y, \zeta) = R_2(\zeta)\, \phi_2(x, y, \bar{\zeta}) \tag{3.6.16}$$

where

$$R_1(\zeta) = \mp\, \frac{\mathrm{i}}{2\pi}\int_{-\infty}^{\infty}\int_{-\infty}^{\infty} \exp(\bar{\zeta}(\mathrm{i}x' - \mathrm{i}y')\}\, \bar{u}(x', y')\, \phi_{1,1}(x', y', \zeta)\, \mathrm{d}x'\, \mathrm{d}y' \tag{3.6.17}$$

and

$$R_2(\zeta) = \frac{\mathrm{i}}{2\pi}\int_{-\infty}^{\infty}\int_{-\infty}^{\infty} \exp(-\bar{\zeta}(\mathrm{i}x' - \mathrm{i}y')\}\ \bar{u}(x', y')\ \phi_{2,2}(x', y', \zeta)\ \mathrm{d}x'\ \mathrm{d}y' \tag{3.6.18}$$

There are no finite singularities and while it seems possible that there are values of ζ for which the homogeneous Fredholm equation has non-trivial solutions these will not satisfy the conditions in Section 3.2.

When $\alpha = 1$ the elements of the kernels are

$$\left.\begin{aligned} K_{1,12} &= \delta(x - x' + y - y')\ \{\theta(x - x') - 1\}\ u(x', y') \\ K_{1,21} &= \pm\exp\{-2\zeta(x - x')\}\ \delta(x - x' - y + y') \\ &\quad \times \{\theta(x - x') - \theta(-\mathrm{Re}\zeta)\}\ \bar{u}(x', y') \\ K_{2,12} &= \exp(2\zeta(x - x')\}\ \delta(x - x' + y - y') \\ &\quad \times \{\theta(x - x') - \theta(\mathrm{Re}\zeta)\}\ u(x', y') \\ K_{2,21} &= \pm\delta(x - x' - y + y')\ \{\theta(x - x') - 1\}\ \bar{u}(x', y') \end{aligned}\right\} \tag{3.6.19}$$

These are regular functions of ζ except on the imaginary axis. Thus the $\mathbf{\Phi}_i(x, x', y, y', \zeta)$ are analytic except on the imaginary axis where there is a finite singularity. Labelling the sides of the singularity according to the sign of Reζ we find from (3.2.19) that $P_{121}(\mathrm{i}\xi - 0, \mathrm{i}\xi + ia + 0, \mathrm{i}\xi + ib)$, $P_{122}(\mathrm{i}\xi - 0, \mathrm{i}\xi + ia + 0, \mathrm{i}\xi + ib)$ and $P_{222}(\mathrm{i}\xi - 0, \mathrm{i}\xi + ia + 0, \mathrm{i}\xi + ib)$ (and also $P_{111}(\mathrm{i}\xi - 0, \mathrm{i}\xi + ia + 0, \mathrm{i}\xi + ib)$), where ξ is real, vanish and hence from (3.2.22)

$$Q_{22}(b, \mathrm{i}\xi) = Q_{222}(\mathrm{i}\xi, \mathrm{i}\xi + 0, \mathrm{i}\xi + ib - 0) = 0 \tag{3.6.20}$$

and

$$\begin{aligned} &Q_{21}(b, \mathrm{i}\xi) \\ &= Q_{221}(\mathrm{i}\xi, \mathrm{i}\xi + 0, \mathrm{i}\xi + ib - 0) \\ &= -P_{221}(\mathrm{i}\xi - 0, \mathrm{i}\xi + 0, \mathrm{i}\xi + ib) \\ &= -\frac{1}{2\pi}\int_{-\infty}^{\infty}\int_{-\infty}^{\infty} \exp\{-\mathrm{i}(2\xi + b)x' - iby'\}\ u(x', y') \\ &\quad \Phi_{2,2}(x', y', \mathrm{i}\xi - 0)\ \mathrm{d}x'\ \mathrm{d}y'. \end{aligned} \tag{3.6.21}$$

Thus

$$\phi_2(x, y, \mathrm{i}\xi + 0) - \phi_2(x, y, \mathrm{i}\xi - 0) = \int_{-\infty}^{\infty} Q_{21}(b, \mathrm{i}\xi)\ \phi_1(x, y, \mathrm{i}\xi + ib - 0)\ \mathrm{d}. \tag{3.6.22}$$

Unfortunately $Q_{11}(b, \mathrm{i}\xi)$ and $Q_{12}(b, \mathrm{i}\xi)$ in

$$\phi_1(x, y, \mathrm{i}\xi + 0) - \phi_1(x, y, \mathrm{i}\xi - 0) = \int_{-\infty}^{\infty} \{Q_{11}(b, \mathrm{i}\xi)\, \phi_1(x, y, \mathrm{i}\xi + ib - 0) + Q_{12}(b, \mathrm{i}\xi)\, \phi_2(x, y, \mathrm{i}\xi + ib - 0)\}\, \mathrm{d}b \tag{3.6.23}$$

are not so simple. However, interchanging the roles of the two sides we find that

$$\phi_1(x, y, \mathrm{i}\xi + 0) - \phi_1(x, y, \mathrm{i}\xi - 0) = -\int_{-\infty}^{\infty} \tilde{Q}_{12}(b, \mathrm{i}\xi)\, \phi_2(x, y, \mathrm{i}\xi + ib + 0)\, \mathrm{d}b, \tag{3.6.24}$$

where

$$\tilde{Q}_{12}(b, \mathrm{i}\xi) = \mp \frac{1}{2\pi} \int_{-\infty}^{\infty} \int_{-\infty}^{\infty} \exp\{-\mathrm{i}(2\xi + b)x' - \mathrm{i}by'\}\, \bar{u}(x', y')\, \Phi_{1,1}(x', y', \mathrm{i}\xi + 0)\, \mathrm{d}x'\, \mathrm{d}y' \tag{3.6.25}$$

From (3.6.22)–(3.6.24) we obtain

$$Q_{11}(b, \mathrm{i}\xi) = -\int_{-\infty}^{\infty} \tilde{Q}_{12}(c, \mathrm{i}\xi)\, Q_{21}(b - c, \mathrm{i}\xi + ic)\, \mathrm{d}c \tag{3.6.26}$$

and

$$Q_{12}(b, \mathrm{i}\xi) = -\tilde{Q}_{12}(b, \mathrm{i}\xi). \tag{3.6.27}$$

With these data together with those associated with any non-trivial solutions to the Fredholm equations the solution to the inverse problem can be carried out as described in Section 3.4. However, in view of (3.6.26) $Q_{11}(b, \mathrm{i}\xi)$ is redundant and the task is slightly simplified by using (3.6.22) and (3.6.24), or rather the versions of them in terms of $\Phi_i(x, y, \zeta)$, namely

$$\Phi_1(x, y, \mathrm{i}\xi + 0) - \Phi_1(x, y, \mathrm{i}\xi - 0) = \int_{-\infty}^{\infty} Q_{12}(b, \mathrm{i}\xi) \times \exp(-\mathrm{i}(2\xi + b)x + \mathrm{i}by\}\, \Phi_2(x, y, \mathrm{i}\xi + ib + 0)\, \mathrm{d}b \tag{3.6.28}$$

and

$$\Phi_1(x, y, \mathrm{i}\xi + 0) - \Phi_2(x, y, \mathrm{i}\xi - 0) = \int_{-\infty}^{\infty} Q_{21}(b, \mathrm{i}\xi) \times \exp(\mathrm{i}(2\xi + b)x + \mathrm{i}by\}\, \Phi_1(x, y, \mathrm{i}\xi + ib - 0)\, \mathrm{d}b. \tag{3.6.29}$$

These lead to

$$\Phi_1(x, y, \zeta') = \begin{pmatrix}1\\0\end{pmatrix} + \frac{1}{2\pi}\int_{-\infty}^{\infty}\frac{1}{i\xi - \zeta'}\int_{-\infty}^{\infty} Q_{12}(b, i\xi)$$
$$\times \exp\{-i(2\xi + b)x + iby\}\, \Phi_2(x, y, i\xi + ib + 0)\, db\, d\xi$$
$$+ \sum_{k=1}^{n_1} \frac{1}{\zeta' - \zeta_i^{(k)}}\, \Phi_1^{(k)}(x, y) \tag{3.6.30}$$

and

$$\Phi_2(x, y, \zeta') = \begin{pmatrix}0\\1\end{pmatrix} + \frac{1}{2\pi}\int_{-\infty}^{\infty}\frac{1}{i\xi - \zeta'}\int_{-\infty}^{\infty} Q_{21}(b, i\xi)$$
$$\times \exp\{i(2\xi + b)x + iby\}\, \Phi_1(x, y, i\xi + ib - 0)\, db\, d\xi$$
$$+ \sum_{k=1}^{n_2} \frac{1}{\zeta' - \zeta_2^{(k)}}\, \Phi_2^{(k)}(x, y) \tag{3.6.31}$$

where the residues $\Phi_1^{(k)}(x, y)$ and $\Phi_2^{(k)}(x, y)$ are given by

$$\{\gamma_1^{(k)} - x + y\}\, \Phi_1^{(k)}(x, y) = \begin{pmatrix}1\\0\end{pmatrix} + \frac{1}{2\pi}\int_{-\infty}^{\infty}\frac{1}{i\xi - \zeta_1^{(k)}}\int_{-\infty}^{\infty} Q_{12}(b, i\xi)$$
$$\times \exp\{-i(2\xi + b)x + iby\}\, \Phi_2(x, y, i\xi + ib + 0)\, db\, d\xi$$
$$+ \sum_{\substack{l=1\\ l\neq k}}^{n_1} \frac{1}{\zeta_1^{(k)} - \zeta_1^{(l)}}\, \Phi_1^{(l)}(x, y) \tag{3.6.32}$$

and

$$\{\gamma_2^{(k)} + x + y\}\, \Phi_2^{(k)}(x, y) = \begin{pmatrix}0\\1\end{pmatrix} + \frac{1}{2\pi}\int_{-\infty}^{\infty}\frac{1}{i\xi - \zeta_1^{(k)}}\int_{-\infty}^{\infty} Q_{21}(b, i\xi)$$
$$\times \exp\{i(2\xi + b)x + iby\}\, \Phi_1(x, y, i\xi + ib + 0)\, db\, d\xi$$
$$+ \sum_{\substack{l=1\\ l\neq k}}^{n_2} \frac{1}{\zeta_2^{(k)} - \zeta_2^{(l)}}\, \Phi_2^{(l)}(x, y) \tag{3.6.33}$$

If there are no poles we can again Fourier transform. We put

$$\Phi_1(x, y, i\xi - 0) = \int_{-\infty}^{\infty} \exp\{-i\xi(x + z)\}\, \mathbf{K}_1(x, y, z)\, dz + \begin{pmatrix}1\\0\end{pmatrix} \tag{3.6.34}$$

and

$$\Phi_2(x, y, i\xi + 0) = \int_{-\infty}^{\infty} \exp\{i\xi(x + z)\}\, \mathbf{K}_2(x, y, z)\, dz + \begin{pmatrix}0\\1\end{pmatrix} \tag{3.6.35}$$

Then (3.6.30) and (3.6.31) become

$$\mathbf{K}_1(x, y, z) = \theta(-x - z)\{F_{12}(y - x, z - x)\begin{pmatrix}0\\1\end{pmatrix}$$
$$+ \int_{-\infty}^{\infty} F_{12}(y - w, z - w)\, \mathbf{K}_2(x, y, w)\, dw\} \tag{3.6.36}$$

and

$$\mathbf{K}_2(x, y, z) = -\theta(-x + z)\{F_{21}(y + x, z + x)\begin{pmatrix}1\\0\end{pmatrix} + \int_{-\infty}^{\infty} F_{21}(y - w, z - w)\, \mathbf{K}_1(x, y, w)\, \mathrm{d}w\} \quad (3.6.37)$$

where

$$F_{ij}(y, z) = \frac{1}{2\pi}\int_{-\infty}^{\infty}\int_{-\infty}^{\infty} Q_{ij}(b, \mathrm{i}\xi) \exp\{iby + \mathrm{i}\xi z\}\, \mathrm{d}b\, \mathrm{d}\xi. \quad (3.6.38)$$

Asymptotic expansion of $\mathbf{\Phi}_1(x, y, \zeta)$ and $\mathbf{\Phi}_2(x, y, \zeta)$ yields the results

$$u(x, y) = -2K_{2,1}(x, y, x + 0) \quad (3.6.39)$$

$$\bar{u}(x, y) = \mp 2K_{1,2}(x, y, -x - 0) \quad (3.6.40)$$

and

$$|u(x, y)|^2 = \pm 2\left(\frac{\partial}{\partial x} - \frac{\partial}{\partial y}\right)K_{1,1}(x, y, -x - 0)$$
$$= \mp 2\left(\frac{\partial}{\partial x} + \frac{\partial}{\partial y}\right)K_{2,2}(x, y, x + 0) \quad (3.6.41)$$

Reference

Leon, J. J-P. (ed.), *Nonlinear Evaluations*, World Scientific, Singapore (1988).

4 *J. J. C. Nimmo*

Hirota's method

4.1 Introduction and brief history

Since the discovery of the inverse scattering transform and its subsequent developments it has been possible to obtain N-soliton solutions to a large number of completely integrable nonlinear partial differential equations by reconstructing the potential whose spectrum consists of just N discrete eigenvalues. However, if one is only interested in obtaining these solutions, then a more straightforward and widely applicable *direct* method is available – this is Hirota's method. By direct we mean that no reference is made to an associated 'simpler' system, be that the scattering problem or a Bäcklund transformation (BT).

To introduce this approach we shall first consider what would be a classical 'linear' approach to solving a nonlinear partial differential equation by means of a perturbation expansion. Take, for example, the Korteweg–de Vries (KdV) equation

$$u_t + 6uu_x + u_{xxx} = 0. \tag{4.1.1}$$

For convenience we introduce the potential w such that $u = w_x$ and after an integration we obtain

$$w_t + 3w_x^2 + w_{xxx} = 0. \tag{4.1.2}$$

Now if we look for an expansion

$$w = \sum_{i=1}^{\infty} \varepsilon^i w_i(x, t) \tag{4.1.3}$$

in terms of a small parameter ε, we see that the w_i satisfy

$$w_{1_t} + w_{1_{xxx}} = 0, \tag{4.1.4}$$

and the recurrence relation

$$w_{n_t} + w_{n_{xxx}} + 3\left(\sum_{j=1}^{n-1} w_{j_x} w_{n-j_x}\right) = 0 \tag{4.1.5}$$

for $n \geqslant 2$. We could now attempt to obtain a general expression for the w_n, sum over all n, and hence obtain a closed form for w which will be an exact solution of (4.1.2). It is one of the remarkable properties of the KdV equation that this is in fact possible. Rosales [33] has shown that for soliton solutions

$$w = 2\frac{\partial}{\partial x} \ln \phi(x, t) \tag{4.1.6}$$

where ϕ satisfies *linear* partial differential equations in accord with what one finds in the inverse method.

Hirota's approach to this example is somewhat different [8]. Motivated by the method of Padé approximants one chooses to look for a *rational* rather than polynomial expansion of w, and to facilitate this one takes

$$w = g/f, \tag{4.1.7}$$

where we shall later expand g and f in terms of ε. When w satisfies (4.1.2), g and f are related by

$$\begin{aligned}(g_t f - g f_t + g_{xxx} f - 3g_{xx} f_x + 3g_x f_{xx} - g f_{xxx}) f^2 \\ + 3(g_x f - g f_x)(g_x f - g f_x - 2(f_{xx} f - f_x^2)) = 0,\end{aligned} \tag{4.1.8}$$

which at first sight is rather more complicated than (4.1.2). However, we may decouple (4.1.8) by means of an arbitrary function $\lambda(x, t)$, which will be determined by the boundary conditions on w, to obtain

$$g_t f - g f_t + g_{xxx} f - 3g_{xx} f_x + 3f_x g_{xx} - g f_{xxx} + 3\lambda(g_x f - g f_x) = 0 \tag{4.1.9a}$$

$$g_x f - g f_x - 2(f_{xx} f - f_x^2) = \lambda f^2. \tag{4.1.9b}$$

For soliton solutions $\lambda = 0$, whereupon (4.1.9b) gives

$$g = 2f_x \tag{4.1.10}$$

and so

$$w = 2\frac{\partial}{\partial x} \ln f, \tag{4.1.11}$$

with f satisfying

$$f_{xt} f - f_x f_t + f_{xxxx} f - 4f_{xxx} f + 3f_{xx}^2 = 0. \tag{4.1.12}$$

Equation (4.1.12) has a familiar appearance – if all the signs were positive the left-hand side would correspond to

$$\left(\frac{\partial^2}{\partial x \partial t} + \frac{\partial^4}{\partial x^4}\right) f^2$$

$$= \left[\left(\frac{\partial}{\partial x} + \frac{\partial}{\partial x'}\right)\left(\frac{\partial}{\partial t} + \frac{\partial}{\partial t'}\right) + \left(\frac{\partial}{\partial x} + \frac{\partial}{\partial x'}\right)^4\right] f(x, t) f(x', t')\Big|_{\substack{x'=x \\ t'=t}}$$

where the expression on the right is a useful way to illustrate the Leibniz rule for the derivative of a product. In fact, what we have is

$$\left[\left(\frac{\partial}{\partial x} - \frac{\partial}{\partial x'}\right)\left(\frac{\partial}{\partial t} - \frac{\partial}{\partial t'}\right) + \left(\frac{\partial}{\partial x} - \frac{\partial}{\partial x'}\right)^4\right] f(x, t) f(x', t')\Big|_{\substack{x'=x \\ t'=t}} = 0, \quad (4.1.13)$$

or, in Hirota's notation,

$$(\mathrm{D}_x \mathrm{D}_t + \mathrm{D}_x^4) f \circ f = 0 \quad (4.1.14)$$

where in general

$$\mathrm{D}_x^i \mathrm{D}_y^j \ldots D_z^k g \circ f$$

$$\equiv \left(\frac{\partial}{\partial x} - \frac{\partial}{\partial x'}\right)^i \left(\frac{\partial}{\partial y} - \frac{\partial}{\partial y'}\right)^j \ldots \left(\frac{\partial}{\partial z} - \frac{\partial}{\partial z'}\right)^k g(x, y \ldots z) f(x', y' \ldots z')\Big|_{\substack{x'=x \\ y'=y \\ z'=z}}. \quad (4.1.15)$$

Equation (4.1.14) is the *Hirota form* of the KdV equation (4.1.1) where $u = 2\partial^2/\partial x^2 \ln f$ and D_x, D_t are *Hirota derivatives* defined by (4.1.15).

Now if we express f as a power series (in ε say), and obtain the terms in the expansion recursively as we did with w, we find that rather than having an infinite series which we need to sum to obtain a solution in closed form, it is possible to *truncate* the expansion after a *finite* number of terms. The details of this will be considered in the next section.

So we see that Hirota's method is similar in spirit to the 'classical' approach of the Rosales method considered earlier but the messy business of summing an infinite series to give (4.1.6) is circumvented by taking (4.1.6) as an *Ansatz*.

In a series of papers ([8]–[12]) this method has been exploited to obtain exact N-soliton solutions to a large number of nonlinear partial differential and differential–difference equations. So successful was this approach that a jocular criterion for complete integrability was suggested: send the PDE to Hirota – it is integrable if and only if the N-soliton solution is returned within a week! Since then this bilinear approach has been adopted by many (mainly Japanese) authors and extended to obtain wider classes of solution than solitons. We cite four diverse examples; Nakamura [26] has shown that any PDE which has Hirota form

$$f(\mathrm{D}_x, \mathrm{D}_y, \ldots, \mathrm{D}_z) f \circ f = 0, \quad (4.1.16)$$

has 2-periodic wave solution given in terms of a 2-dimensional Riemann θ-function; Matsumo [25] has carried out an extensive study of solutions of the Benjamin–Ono equation; Yajima [36] has considered a perturbed KdV equation from a bilinear viewpoint and Nakamura and Hirota ([25], [16]) have obtained N–'explode–decay' solutions to the classical Boussinesq equations in terms of Hermite polynomials. More recently the discoveries of the group at Kyoto (see [19] and Chapter 13 of this volume) have brought to light the deep connections between *bilinear* equations like (4.1.14), and the algebraic properties of their related PDEs. For example the relationship between the properties of the non-linear Schrödinger equation and the Heisenberg ferromagnetic spin equations is illuminated by the fact that they are related to the same bilinear form by different changes of variable [15]. Also, the connection between the current approach and the Painlevé test is being investigated ([7], and Chapter 16).

We will see in the following sections that not only does Hirota's method provide a useful means of constructing solutions, it is also proving to be of great importance when addressing deeper questions about the algebraic properties of soliton equations and criteria for integrability.

4.2 The direct method

To illustrate Hirota's method we shall construct the N-soliton solution to the KdV equation by obtaining an expansion for f in the bilinear form (4.1.14). At the same time we shall discuss the Hirota derivatives whose properties are crucial both to Hirota's method and the more fundamental properties that we discussed at the end of Section 4.1.

Some simple properties of these bilinear operators allow us to transform from (4.1.1) to (4.1.14) very efficiently. We have

$$\begin{aligned}(\exp \varepsilon\, \mathrm{D}_x) a \circ b &= a(x+\varepsilon) b(x-\varepsilon) \\ &= \left(\exp \varepsilon \frac{\partial}{\partial x}\right) a(x) \left(\exp\left(-\varepsilon \frac{\partial}{\partial x}\right)\right) b(x), \qquad (4.2.1\text{a})\end{aligned}$$

an analogue of Taylor's theorem that, by equating powers of ε on both sides, allows one to express Hirota derivatives in terms of ordinary derivatives. Also we have

$$\mathrm{D}_x^i \mathrm{D}_y^j \ldots D_z^k a \circ b = (-1)^{i+j+\ldots+k} \mathrm{D}_x^i \mathrm{D}_y^j \ldots \mathrm{D}_z^k b \circ a, \qquad (4.2.1\text{b})$$

which implies that odd degree terms in Hirota derivatives vanish when $b = a$, and so

$$\begin{aligned}\left(2 \cosh \varepsilon \frac{\partial}{\partial x}\right) \ln a &= \ln(a+\varepsilon) + \ln(a-\varepsilon) \\ &= \ln(\cosh(\varepsilon \mathrm{D}_x) a \circ a). \qquad (4.2.1\text{c})\end{aligned}$$

From (4.2.1c) we calculate the terms in (4.1.2):

$$w_t = 2\frac{\partial^2}{\partial x \partial t}\ln f = \frac{D_x D_t f \circ f}{f^2},$$

$$w_x = \frac{D_x^2 f \circ f}{f^2},$$

$$w_{xxx} = \frac{D_x^4 f \circ f}{f^2} - 3\left(\frac{D_x^2 f \circ f}{f^2}\right)^2,$$

and we observe that the quartic term in w_{xxx} exactly balances the term $3w_x^2$ leaving only the bilinear terms appearing in (4.1.14). This balance between highest-order derivative and nonlinearity is common to all examples and reflects the balance of this kind which is said to represent the physical mechanism underlying solitonic behaviour.

To construct soliton solutions to (4.1.1) we shall proceed as proposed by expanding f as a power series in ε:

$$f = f_0 + \varepsilon f_1 + \varepsilon^2 f_2 + \ldots, \tag{4.2.2}$$

and substitute into (4.1.14). For soliton solution we expect to take linear exponential solutions for the f_k and consequently we may take $f_0 = 1$ without loss of generality since (4.1.14) is invariant under $f \to \phi f$ whenever the second logarithmic derivative of ϕ vanishes. Now, from successive powers of ε we find

$$f_{1_{xt}} + f_{1_{xxxx}} = 0, \tag{4.2.3a}$$

$$f_{2_{xt}} + f_{2_{xxxx}} = -\tfrac{1}{2}(D_x D_t + D_x^4) f_1 \circ f_1, \tag{4.2.3b}$$

$$f_{3_{xt}} + f_{3_{xxxx}} = -(D_x D_t + D_x^4) f_2 \circ f_1, \tag{4.2.3c}$$

and so on – at each stage the f_k ($k \geqslant 2$) are defined as solutions of linear PDEs with an inhomogeneity in terms of $f_1 \ldots f_{k-1}$. We shall begin by taking the simplest non-trivial solution of the required kind to (4.2.3a), namely

$$f_1 = \exp(kx - k^2 t + \phi),$$

with ϕ an arbitrary phase constant. By virtue of the identity

$$D_x^i D_t^j e^{\alpha x + \beta t} \circ e^{\gamma x + \delta t} = (\alpha - \gamma)^i (\beta - \delta)^j e^{(\alpha+\gamma)x + (\beta+\delta)t} \tag{4.2.4}$$

which comes directly from (4.2.1a), we see that with this simplest form of f_1 the right-hand side of (4.2.3b) vanishes. Consequently, we may take $f_2 = 0$ as a solution to (4.2.3b) resulting in a vanishing inhomogeneity in (4.2.3c) so that we may also take $f_3 = 0$. It is not difficult to see that for the same reasons we may also take $f_k = 0$ for $k \geqslant 4$ and hence that we have the *exact* solution

$$f = 1 + \exp(kx - k^3 t + \phi)$$

where ε has been absorbed into the phase constant. This f gives us the familiar single-soliton solution to the KdV equation

$$u = \frac{k^2}{2}\operatorname{sech}^2 \tfrac{1}{2}(kx - k^3t + \phi).$$

To obtain the 2-soliton solution we take $f_1 = \Sigma_{i=1}^2 \exp \eta_i$, where $\eta_i = k_i x - k_i^3 t + \phi_i$, $i = 1, 2$, as the solution to (4.2.3a). Here (4.2.3b) gives

$$f_{2_{xt}} + f_{2_{xxxx}} = 3k_1k_2(k_2 - k_1)^2 \exp(\eta_1 + \eta_2),$$

the particular solution to which is seen to be

$$f_2 = \exp(A_{12} + \eta_1 + \eta_2)$$

where $\exp A_{12} = [(k_2 - k_1)/(k_2 + k_1)]^2$. Again one finds that with this value for f_2 the right-hand side of (4.2.3c) vanishes, and from this that we may truncate the expansion after the third term to obtain the exact solution

$$f = 1 + \exp \eta_1 + \exp \eta_2 + \exp(A_{12} + \eta_1 + \eta_2).$$

By carrying out this procedure for f_1 with successively more terms one finds that each time the series may be truncated after a finite number of terms and one is led to conjecture the form of f corresponding to the N-soliton solution,

$$f = \sum_{\mu_1=0}^{1} \cdots \sum_{\mu_n=0}^{1} \exp\Big(\sum_{i>j} A_{ij}\mu_i\mu_j + \sum_{i=1}^{N} \mu_i\eta_i\Big). \tag{4.2.5}$$

The verification of this solution will not be given here but may be found in the original paper [8]. We see from (4.2.5) that it is possible to remove the parameter ε from the solution as we did for the single-soliton solution, by means of a phase shift, since each f_k is found to be of degree k in the exponentials. In fact, ε serves only as a device for ordering the terms in the expansion of f and should not be thought of as a 'small' parameter in the usual sense.

Next we shall briefly consider another example – the nonlinear Schrödinger (NLS) equation

$$\mathrm{i}u_t + u_{xx} + u|u|^2 = 0. \tag{4.2.6}$$

As in the case of the KdV equation we assume the form

$$u = g/f, \tag{4.2.7}$$

where here g is complex and f is real. Using the identity

$$\left(\exp \varepsilon \frac{\partial}{\partial x}\right)\frac{g}{f} = \frac{g(x + \varepsilon)f(x - \varepsilon)}{f(x + \varepsilon)f(x - \varepsilon)} = \frac{(\exp \varepsilon\, \mathrm{D}_x)g\circ f}{(\cosh \varepsilon\, \mathrm{D}_x)f\circ f} \tag{4.2.8}$$

we find that g and f must satisfy

$$\frac{(\mathrm{iD}_t + \mathrm{D}_x^2)g \circ f}{f^2} - \frac{g}{f}\left(\frac{\mathrm{D}_x^2 f \circ f - gg^*}{f^2}\right) = 0 \tag{4.2.9}$$

which we decouple (with zero decoupling function) as

$$(\mathrm{iD}_t + \mathrm{D}_x^2)g \circ f = 0, \tag{4.2.10a}$$

$$D_x^2 f \circ f = gg^*. \tag{4.2.10b}$$

Here we see that, unlike in the previous example, the decoupling of (4.2.10) does not result in a simple relation between g and f – compare (4.2.10) and (4.1.10) – and the Hirota form of (4.2.6) is a coupled system (4.2.10) in g and f. To obtain solutions to (4.2.10) we take

$$g = \varepsilon g_1 + \varepsilon^3 g_3 + \ldots \tag{4.2.11a}$$

$$f = 1 + \varepsilon^2 f_2 + \ldots \tag{4.2.11b}$$

where we find that

$$\mathrm{i}g_{1_t} + g_{1_{xx}} = 0, \tag{4.2.12a}$$

$$f_{2_{xx}} = \tfrac{1}{2} g_1 g_1^*, \tag{4.2.12b}$$

$$\mathrm{i}g_{3_t} + g_{3_{xx}} = -(\mathrm{iD}_t + \mathrm{D}_x^2) g_1 \circ f_2, \tag{4.2.12c}$$

and so on.

As in the previous example these equations may be solved recursively until an inhomogeneity is found to vanish, whereupon the term defined by that equation and all subsequent ones may be taken to be zero. By continuing in this way the N-soliton solution may be constructed [12].

These two examples serve to introduce the ideas behind the method; as we have seen both the transformation between evolution equation and bilinear equation and the construction of exact solutions rely on the use of the identities such as (4.2.1) and (4.2.3). A more extensive list of these may be found in reference [14]. Too conclude this section we offer a word of caution: the ability to reduce an evolution to Hirota form does not guarantee an N-soliton solution. Hirota [14] has shown, however, that all bilinear equations in one function (like (4.1.14)) possess at least a 2-soliton solution (cf. [26]), but very special conditions must be satisfied if such equations are to possess multi-soliton solutions. The investigation of those Hirota equations which do have N-soliton solutions is of great interest ([34], [19], [28]) and we shall consider an example of this in Section 4.5.

4.3 Bäcklund transformations

Bäcklund transformations (BT) – [32] and Chapter 5 – provide a means of constructing new solutions from known solutions to soliton equations – typically one would transform between $(N-1)$-soliton and N-soliton solutions. In the context of bilinear equations the concept of a BT was introduced by Hirota [13], and in this section we shall investigate the techniques involved in obtaining bilinear BT's by means of two examples.

First we consider the Kadomtsev–Petviashvili (KP) equation

$$(u_t + 6uu_x + u_{xxx})_x + 3u_{yy} = 0, \tag{4.3.1}$$

a weakly 2-dimensional generalisation of the KdV equation [20]. As for this latter equation, the appropriate bilinearising change of variable is

$$u = 2\,\frac{\partial^2}{\partial x^2}\ln f, \tag{4.3.2}$$

and the resulting Hirota equation is

$$(\mathrm{D}_x\mathrm{D}_t + \mathrm{D}_x^4 + 3\mathrm{D}_y^2)f\circ f = 0. \tag{4.3.3}$$

Suppose now that f and f' are two solutions of (4.3.3) which we wish to relate by means of bilinear equations. This set of equations will constitute a bilinear BT for the KP equation. Following Hirota and Satsuma [18] we consider

$$\Omega \equiv [(\mathrm{D}_x\mathrm{D}_t + \mathrm{D}_x^4 + 3\mathrm{D}_y^2)f\circ f]f'^2 - f^2[(\mathrm{D}_x\mathrm{D}_t + \mathrm{D}_x^4 + 3\mathrm{D}_y^2)f'\circ f'] = 0 \tag{4.3.4}$$

from which we see that f' is a solution of (4.3.3) if and only if f is. To transform this into a form which will give us a set of bilinear relations between f and f', we use identities which come from the expansion in terms of its parameters of what Hirota terms the exchange formula [14],

$$\begin{aligned}\exp(\mathrm{D}^{(1)})[\exp(\mathrm{D}^{(2)})a\circ b] \circ [\exp(\mathrm{D}^{(3)})c\circ d] \\ = \exp\left(\frac{\mathrm{D}^{(2)} - \mathrm{D}^{(3)}}{2}\right)\left[\exp\left(\frac{\mathrm{D}^{(2)} + \mathrm{D}^{(3)}}{2} + \mathrm{D}^{(1)}\right)a\circ c\right] \\ \circ\left[\exp\left(\frac{\mathrm{D}^{(2)} + \mathrm{D}^{(3)}}{2} - \mathrm{D}^{(1)}\right)b\circ d\right],\end{aligned} \tag{4.3.5}$$

where $\mathrm{D}^{(i)} \equiv \varepsilon_x^{(i)}\mathrm{D}_x + \varepsilon_y^{(i)}\mathrm{D}_y + \ldots + \varepsilon_z^{(i)}\mathrm{D}_z$ for $i = 1, 2, 3$. In the present example, from (4.3.4) we obtain

$$\Omega = 2\mathrm{D}_x[(\mathrm{D}_t + \mathrm{D}_x^3)f\circ f']\circ ff' + 6\mathrm{D}_y(\mathrm{D}_y f\circ f')\circ ff' - 6\mathrm{D}_x(\mathrm{D}_x^2 f\circ f')\circ(D_x f\circ f'). \tag{4.3.6}$$

From scale symmetry arguments a reasonable guess at a bilinear relation between f and f' is

$$\mathrm{D}_y f\circ f' = (\mu \mathrm{D}^2_x + \lambda) f\circ f' \tag{4.3.7}$$

which if we substitute into (4.3.6) gives

$$\begin{aligned}\Omega = {} & 2\mathrm{D}_x[(\mathrm{D}_t + 3\lambda \mathrm{D}_x + 3\mu \mathrm{D}_x \mathrm{D}_y + \mathrm{D}_x^3) f\circ f']\circ ff' \\ & + 6(\mu^2 - 1)\mathrm{D}_x(\mathrm{D}_x^2 f\circ f')\circ(\mathrm{D}_x f\circ f'),\end{aligned} \tag{4.3.8}$$

where we have used the identity

$$\mathrm{D}_y(\mathrm{D}_x^2 f\circ f')\circ ff' = \mathrm{D}_x[(\mathrm{D}_x\mathrm{D}_y f\circ f')\circ ff' + (\mathrm{D}_y f\circ f')\circ(\mathrm{D}_x f\circ f')]$$

obtained from (4.3.5).

From (4.3.8) we see that our guess (4.3.7) was correct, provided that

$$(\mathrm{D}_y \pm D_x^2) f\circ f' = \lambda ff', \tag{4.3.9a}$$

$$(\mathrm{D}_t + 3\lambda \mathrm{D}_x - 3\mathrm{D}_x\mathrm{D}_y + \mathrm{D}_x^3) f\circ f' = 0. \tag{4.3.9b}$$

The choice of signs corresponds merely to the interchangeability of f and f' and the skew-symmetric nature of Hirota derivatives, and so without loss of generality one may take just the top one. It may be shown, [6], that if f_k gives the k-soliton solution to (4.3.1) then f_N and f_{N-1} are related by (4.3.9) with $\lambda = 0$ and $f = f_{N-1}$, $f' = f_{N-1}$. It is also interesting to note that, as is always the case, not only does (4.3.9) constitute a BT for the KP equation but it must also be the Hirota form of a new equation which possesses N-soliton solutions – in this instance the modified KP equation

$$v_t - 2v_x^3 + v_{xxx} - 3v_{xy} - 6v_x v_y = 0, \tag{4.3.10}$$

where $v = \ln(f/f')$, with (4.3.9) (with top sign and $\lambda = 0$) corresponding to the lowest-order pair of equations in the 1st-modified KP hierarchy [19].

The second example we shall look at is somewhat different, making use as it does of identities not coming from the exchange formula, and illustrating the educated guesswork often required to obtain bilinear BT's. We consider the NLS equation (4.2.6) with Hirota form (4.2.10). To construct the BT we take

$$\Omega \equiv [(\mathrm{i}\mathrm{D}_t + \mathrm{D}_x^2) g\circ f] f'^2 - f^2[(\mathrm{i}\mathrm{D}_t + \mathrm{D}_x^2) g'\circ f'], \tag{4.3.11}$$

which vanishes provided (g, f) and (g', f') satisfy (4.2.10a). By means of a pair of identities [31] one may rewrite (4.3.11) as

$$\begin{aligned}\Omega = {} & \mathrm{i}\mathrm{D}_t(g\circ f' + f\circ g')ff' - (gf' + fg')\mathrm{i}\mathrm{D}_t f\circ f' \\ & + 2\mathrm{D}_x[\mathrm{D}_x(g\circ f' + f\circ g')]\circ ff' + (gf' - fg')\mathrm{D}_x^2 f\circ f' \\ & - \mathrm{D}_x^2(g\circ f' - f\circ g')\circ ff' + \tfrac{1}{2}[(gf' + fg')(gg'^* - g^*g') \\ & - (gf' - fg')(gg'^* + g^*g')],\end{aligned} \tag{4.3.12}$$

where we also assume that (g, f) and (g', f') satisfy the second half of the Hirota form (4.2.10). As in the first example we must guess one simple bilinear equation between g, f, g' and f' – in this case from consideration of relations between successive 'rungs of the soliton ladder' we take

$$\mathrm{D}_x(g \circ f' + f \circ g') = \mu(gf' - fg'), \tag{4.3.13a}$$

whereupon (4.3.12) decouples to give

$$\mathrm{iD}_t(g \circ f' + f \circ g') - (\mathrm{D}_x^2 + \lambda)(g \circ f' - f \circ g') = 0, \tag{4.3.13b}$$

$$[\mathrm{iD}_t + 2\mu \mathrm{D}_x + \mathrm{D}_x^2 + (2\mu^2 - \lambda)] f \circ f' = gg'^*. \tag{4.3.13c}$$

Equations (4.3.13) constitute a BT between (g, f) and (g', f') satisfying (4.2.10). Also one finds that if (g, f) gives the $(N - 1)$-soliton solution and (g', f') the N-soliton solution we must choose $\mu = k_N - k_N^*$, $\lambda = 2(k_N^2 + k_N^{*2}) + (k_N - k_N^*)^2$ where k_N is a constant introduced by integration of (4.2.12a); see [31], [5] for more details.

Here we end this short introduction to bilinear BT's; for more details and examples the reader should consult references [13], [17], [18] and [32].

4.4 Vertex operators and solutions to soliton equations

The connection between completely integrable nonlinear systems and infinite-dimensional Lie algebras has now been established for some time [3, 4]. It is beyond the scope of the present chapter to fully describe this relationship and the reader is referred to Chapter 13. Here we shall very briefly and very non-rigorously give the flavour of this connection and then concentrate on the properties of the objects which provide this link – the vertex operators.

Consider the operator

$$Z(a, b) = \frac{b}{a - b}[\exp(\xi(x, a) - \xi(x, b))\exp(-\xi(\hat{\partial}, a) + \xi(\hat{\partial}, b)) - 1] \tag{4.4.1}$$

with $\xi(x, a) = \sum_{i=1}^{\infty} x_i a^i$ and $\hat{\partial} = \left(\frac{\partial}{\partial x_1}, \frac{1}{2}\frac{\partial}{\partial x_2}, \frac{1}{3}\frac{\partial}{\partial x_3}, \ldots\right)$. It may be shown [3, 18] that the terms $Z_{i,j}$ such that

$$Z(a, b) = \sum_{i,j \in \mathbb{Z}} Z_{i,j}\left(x, \frac{\partial}{\partial x}\right) a^i b^{-j}, \tag{4.4.2}$$

obey the same commutation relation

$$[Z_{i,j}, Z_{k,l}] = \delta_{j,k} Z_{i,l} - \delta_{i,l} Z_{k,j} \tag{4.4.3}$$

as the infinite matrices $E_{i,j}$ with 1 in the (i, j)th place and zeros elsewhere, and consequently constitute a representation of the Lie algebra of infinite matrices $gl(\infty)$. Associated with this representation one can define functions of $x_1, x_2, \ldots$, known as τ functions, and quadratic relations between them which are infinite-dimensional analogues of the Plücher relation of Grassman manifold theory. These relations turn out to be expressible as Hirota equations of the form

$$P(\mathrm{D}_1, \mathrm{D}_2, \ldots)\tau \circ \tau' = 0, \tag{4.4.4}$$

for certain polynomial P in the Hirota derivatives $\mathrm{D}_k \equiv \mathrm{D}_{x_k}$. The simplest non-trivial example is the equation

$$(\mathrm{D}_1^4 + 3\mathrm{D}_2^2 - 4\mathrm{D}_1\mathrm{D}_3)\tau \circ \tau = 0 \tag{4.4.5}$$

which is the bilinear form of the KP equation with $x_1 = x$, $x_2 = y$ and $x_3 = -4t$. The collection of bilinear equations given by (4.4.4) with $\tau' = \tau$ is called the KP hierarchy and the others make up the modified KP hierarchies (see [18] Appendix 1 for lists of the lower-order members of these hierarchies). As mentioned earlier the above is intended merely to give a taste of a highly complex and well-developed subject and the reader is referred to the cited literature for full details.

As well as enabling one to determine the bilinear equations (4.4.4) one is also able to construct τ functions that satisfy any member of these hierarchies. This is because the vertex operator may be shown to act as a BT on the solutions of (4.4.4). To illustrate this we shall consider two examples.

First, [3], consider the operator

$$X(a, b) = \exp(\xi(x, a) - \xi(x, b))\exp(-\xi(\hat{\partial}, a^{-1}) + \xi(\hat{\partial}, b_i^{-1})) \tag{4.4.6}$$

which is closely related to $Z(a, b)$ – this is the vertex operator (when $a = -b$) that was first used, [20], to obtain the representation of the Kač–Moody algebra $A_1^{(1)}$ that inspired all subsequent work in this field. From the observation that

$$\begin{aligned}\exp\left(\xi_1 \sum_{n=1}^{\infty} \frac{1}{n}\frac{\partial}{\partial x_n} a^{-n}\right)\exp\left(\xi_2 \sum_{m=1}^{\infty} x_m b^m\right) \\ &= \exp\left(\xi_2 \sum_{m=1}^{\infty} (x_m + \xi_1 \frac{1}{m} a^{-m}) b^m\right) \exp\left(\xi_1 \sum_{n=1}^{\infty} \frac{1}{n}\frac{\partial}{\partial x_n} a^{-n}\right) \\ &= \left(1 - \frac{b}{a}\right)^{-\xi_1\xi_2} \exp(\xi(x, b))\exp(-\xi(\hat{\partial}, a^{-1}))\end{aligned}$$

where ξ_1, ξ_2 equal ± 1, we see that

$$\begin{aligned}X(a, b)X(a', b') = \frac{(a' - a)(b' - b)}{(a' - \mathrm{b})(b' - a)} \\ &\exp[\xi(x, a) - \xi(x, b) + \xi(x, a') - \xi(x, b')] \\ &\exp[-\xi(\hat{\partial}, a^{-1}) + \xi(\hat{\partial}, b^{-1}) - \xi(\hat{\partial}, a'^{-1}) + \xi(\hat{\partial}, b'^{-1})].\end{aligned}$$

Consequently, we see that $X^2(a, b) = 0$ and the 2-solition solution of the KP equation, for example, may be written as

$$\tau^{(2)} = 1 + e^{\eta_1} + e^{\eta_2} + \frac{(a_2 - a_1)(b_2 - b_1)}{(a_2 - b_1)(b_2 - a_1)} e^{\eta_1 + \eta_2}$$

$$= \exp[\alpha_1 X(a_1, b_1) + \alpha_2 X(a_2, b_2)] \cdot 1 \tag{4.4.7}$$

where $\eta_i = \xi(x, a_i) - \xi(x, b_i) + \ln \alpha_i$. Indeed one may show that

$$\exp[\alpha_N X(a_N, b_N)] \tau^{(N-1)} = \tau^{(N)} \tag{4.4.8}$$

so that the vertex operator $X(a, b)$ acts as an infinitesimal BT – generating the N-solution solution τ function from the $(N - 1)$-soliton τ function. Hence it is clear that the N-soliton solution may be written as

$$\tau^{(N)} = \exp\left[\sum_{i=1}^{N} \alpha_i X_i(a_i, b_i)\right] \cdot 1 \tag{4.4.9}$$

The second example shows how the polynomial solutions to the bilinear equations (4.4.4) and hence rational solution of the corresponding wave equations may be generated. In order to do this we shall make some definitions. We let $x_k = \frac{1}{k}(\alpha_1^k + \ldots + \alpha_r^k)$, where $\alpha_1, \ldots, \alpha_r$ are a set of dummy variables, for $k \in \mathbb{Z}^+$ and define for each partition the Schur ($S-$) functions of $\alpha_1, \ldots, \alpha_r$ as polynomials of $x_1, x_2, \ldots$

$$S_\lambda = \sum_\mu \chi_\mu^\lambda \frac{x_\mu}{\mu!} \tag{4.4.10}$$

which arises from the well-known formula, [21], due to Frobenius and Schur, relating power-sum symmetric functions to S-functions. In (4.4.10) $x_\mu = x_1^{m_1} \ldots x_k^{m_k}$ and $\mu! \equiv m_1! \ldots m_k!$ for partitions $\mu = (k^{m_k} \ldots 1^{m_1})$ with χ_μ^λ being the characters of the symmetric groups (see Section 4.5 for more on this subject). The action of the components $Z_{i,j}$ of $Z(a, b)$ on these S-functions is easily described; for $i > j$

$$Z_{i,j}(S_\lambda) = \begin{cases} (-1)^{v-1} S_{\lambda'} & \text{edge } i \text{ horizontal, edge } j \text{ vertical} \\ 0 & \text{otherwise} \end{cases}$$

where we number the edges of the Young diagram of λ as shown in Figure 4.1, and λ' is the partition whose Young diagram is obtained from that of λ by adding a strip, joining edges i and j, which occupies v rows. Jimbo and Miwa [19] indicate that for each i, $Z_{i,i-1}$ is a BT on the polynomial solutions of (4.4.4) so allowing one to construct, as indicated in Figure 4.2, all of the polynomials defined by (4.4.10) as solutions. For example, we have $S_{(1^2)} = x_2 + \frac{1}{2} x_1^2$ and

$$S_{(31)} = -x_4 - \frac{1}{2} x_2^2 + \frac{1}{2} x_2 x_1^2 + \frac{1}{8} x_1^4$$

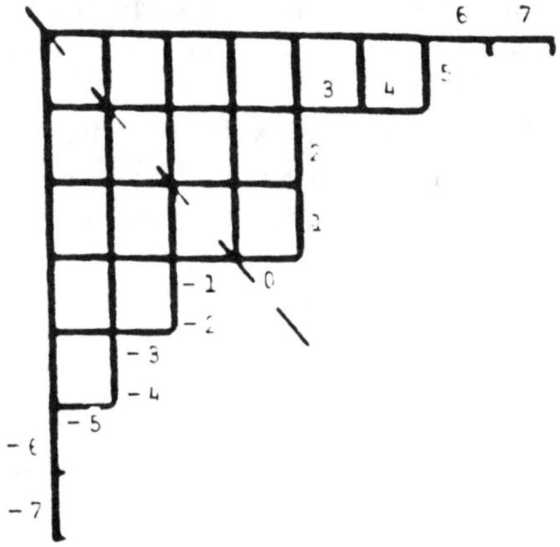

Figure 4.1 Labelling of Young diagram

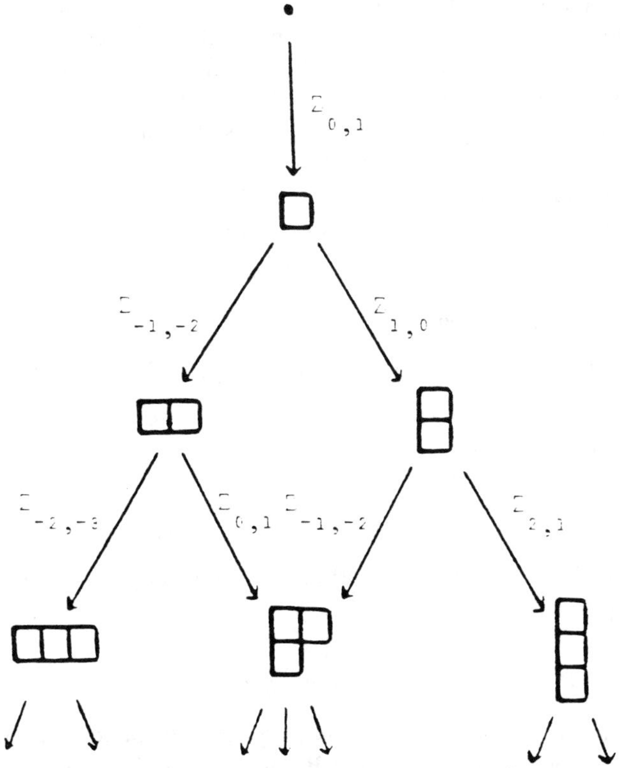

Figure 4.2 Construction of polynomial solutions

as two typical solutions. We shall use the vertex operator again in the next section when once more this description of its action on S-functions will be exploited.

We saw from the first example the very important role that is played by the 'phase-shift' term $[(a_z - a_1)(b_2 - b_1)]/[(a_2 - b_1)(b_2 - a_1)]$ in determining the underlying algebraic properties of the soliton equation. Indeed one sees from the phase-shift term for the KdV equation (see (4.2.5))

$$\left(\frac{k_2 - k_1}{k_2 + k_1}\right)^2 = \frac{(k_2 - k_1)((-k_2) - (-k_1))}{(k_2 - (-k_1))((-k_2) - k_1)}$$

that the KdV equation comes from the $a = -b$ reduction of the KP equation.

4.5 The KP hierarchy and Wronskian determinants

It was first demonstrated by Crum [2] that the potentials of the Schrödinger equation giving rise to purely discrete spectra could be written in terms of Wronskian determinants. This leads one on naturally to a formulation of the multi-soliton solutions of the KdV equation in terms of Wronskians [35] and, in a further development, these ideas were used [1] to obtain *rational* solutions of the KdV equation in this form.

An alternative way to achieve these results, and many others, which is of more interest in the present context, is the direct approach taken by Freeman and Nimmo [5, 6]. To illustrate this method we shall take the KP equation, in Hirota form as discussed in Section 4.4,

$$(\mathrm{D}_1 + 3\mathrm{D}_2^4 - 4\mathrm{D}_3^2\mathrm{D}_1)f\circ f = 0. \tag{4.5.1}$$

Writing this equation if full gives us the quadratic form

$$(f_{1^4} + 3f_{2^2} - 4f_{31})f - 4(f_{1^3} - f_3)f_1 + 3(f_{1^2}^2 - f_2^2) = 0. \tag{4.5.2}$$

We conjecture – and then prove – that the 'general' solution to (4.5.1) is given by the Wronskian

$$f = W(\phi_1, \ldots, \phi_N) = \det\left(\frac{\partial^{j-1}\phi_i}{\partial_{x_1}^{j-1}}\right) \tag{4.5.3}$$

where the ϕ_i are different but otherwise arbitrary functions of x_k, $k \in \mathbb{Z}^+$ satisfying

$$\frac{\partial\phi_i}{\partial x_j} = \frac{\partial^j\phi_i}{\partial x^j{}_1}, \tag{4.5.4}$$

the 'extra' variables x_4, x_5, ... having been introduced as they were in Section 4.4 play no role in the KP equation itself but will be of importance when we come to look at the equations in the KP hierarchy later.

The Wronskian in (4.5.3) – and its derivatives – may be completely characterised by the number of derivatives of $(\phi_1, \ldots \phi_N)^{\mathrm{T}}$ that are required to obtain each of its columns, and hence we adopt the notation

$$f = (0, 1, \ldots, N-1) \equiv (\widehat{N-1}) \qquad (4.5.5)$$

where in general k represents the column $\partial^k/\partial x_1^k\ (\phi_1, \ldots, \phi_N)^{\mathrm{T}}$ and $\widehat{N-m}$ the $N-m+1$ consecutive columns $0, \ldots, N-m$. We shall demonstrate that (4.5.5) satisfies (4.5.2) by direct substitution and so first we must calculate the various derivatives of f. This is not as difficult as might be supposed; because of the nature of a Wronskian determinant and the fact that the ϕ_i satisfy (4.5.4), each of its derivatives takes a very compact form consisting of a number of determinants – this number depending only on the derivative and not on N – each of which is obtained from f by 'shifting' one or more of its columns. For example, we may compute

$f_1 = (\widehat{N-2}), N), f_2 = (\widehat{N-2}, N+1) - (\widehat{N-3}, N-1, N)$, and so on.

When all of this calculation has been done one finds that the left-hand side of (4.5.2) reduces to just the three terms

$$2[(\widehat{N-3}, N, N+1)(\widehat{N-1}) - (\widehat{N-3}, N-1, N+1)(\widehat{N-2}, N) + (\widehat{N-3}, N-1, N)(\widehat{N-2}, N+1)] \qquad (4.5.6)$$

multiplied by a numerical factor. This collectioon of terms is the Laplace expansion (by $N \times N$ minors) of the $2N \times 2N$ determinant

$$\begin{vmatrix} \widehat{N-3} & \cdot & N-2 & N-1 & N & N+1 \\ \hline \cdot & \widehat{N-3} & N-2 & N-1 & N & N+1 \end{vmatrix}$$

$$\left(= \begin{vmatrix} \widehat{N-3} & \cdot \\ \hline \cdot & \widehat{N+1} \end{vmatrix} \right),$$

using an obvious extension of the notation introduced earlier with $\cdot$ indicating a zero matrix of the appropriate size, which, from the expression in parentheses, is equal to zero. Thus the solution is verified.

The N-soliton solution is obtained by choosing

$$\phi_i = \exp[\xi(x, p_i)] + c_i \exp[\xi(x, q_i)] \qquad (4.5.8)$$

with ξ as is Section 4.4. The resulting expression for f differs, by an exponential factor, from that obtained by Hirota's method, but since the second logarithmic derivative of this factor is zero, both give the same solution to the KP equation. Note, finally, that the N-soliton solution of the KdV equation is obtained by the reduction $q_i = -p_i$, $i = 1, \ldots, N$ as one might expect.

One deficiency of this method as it stands is that the Hirota form

(4.5.1) is not used directly – rather the expanded version (4.5.2) is used. One feels that one should be able to exploit some properties of the Hirota derivatives rather than use them simply as a convenient shorthand notation. We shall consider some new work in this area that goes some way to remedying this [30].

We begin with the observation that the derivatives of f may be computed in terms of the characters of symmetric groups – in fact

$$f_\lambda = \sum_\mu \chi_\lambda^\mu \, S_\mu. \tag{4.5.9}$$

This equation needs some explanation.

Let λ, μ be partitions $(\lambda_1, \ldots, \lambda_n)$, $(\mu_1, \ldots, \mu_m)$ respectively with parts in *descending* order; then $|\lambda| = \lambda_1 + \ldots + \lambda_m$ and

$$f_\lambda = \frac{\partial^n f}{\partial x_{\lambda_1} \ldots \partial x_{\lambda_n}}; \tag{4.5.10}$$

$$S_\mu = (\overbrace{N - 1 - m}, N - m + \mu_m, \ldots, N - 1 + \mu_1) \tag{4.5.11}$$

the determinant obtained when the final columns of the Wronskian f are shifted up by the corresponding parts of μ; χ_λ^μ are the characters of $S_{|\lambda|}$ and the summation is over all partitions of $|\lambda|$. Those familiar with representation theory or symmetric functions will recognise equation (4.5.9); it is precisely the relationship between the power-sum symmetric functions and the Schur $(S-)$ functions ([24], ch. 1) that was mentioned in Section 4.4. In what follows we shall think of f_λ as being 'power sums' and S_μ and 'S-functions' and use, with care, some of the well-known properties of these symmetric functions. Later, we will find it convenient to use Young diagram notation for the determinants S_μ. As an example of all this we have

$$f_{(3)} = S_{(3)} - S_{(21)} + S_{(1^3)} = (\overbrace{N - 2}, N + 2) - (\overbrace{N - 3}, N - 1, N + 1)$$
$$+ (\overbrace{N - 4}, N - 1, N, N + 1),$$

$$= \square\square\square - \begin{array}{c}\square\square\\\square\end{array} + \begin{array}{c}\square\\\square\\\square\end{array}$$

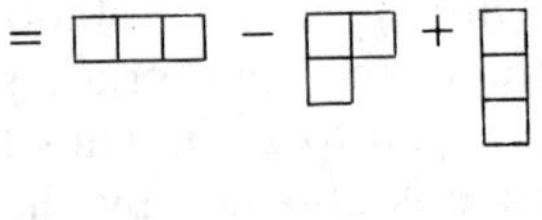

the coefficients 1, -1, 1 being read from the character table of S_3 – for example, this is the third column of Table 3 on p. 265 of Littlewood's book [23].

Also, in that book (pp. 114–18), Littlewood discusses a method of constructing 'generalised' power sums and what he calls 'generalised S-functions' $\bar{S}_\lambda$ which he shows are expressible as quadratic expressions in ordinary S-functions. It turns out that the generalised power sums correspond to Hirota derivatives in the way that ordinary derivatives correspond to power sums as described above and the formula derived by

Littlewood will enable us to express the Hirota equation *directly* as a quadratic in the S_μ. This formula is

$$\tilde{S}_\lambda = \sum_{\pi \subset \lambda} (-1)^{|\mu|} S_{\mu'} S_{\lambda\backslash\mu}, \tag{4.5.12a}$$

with

$$D_\mu f \circ f = \sum_\lambda \chi_\mu^\lambda \tilde{S}_\lambda, \tag{4.5.12b}$$

where μ' is the partition conjugate to μ (the partition with Young diagram obtained from that of μ by reflection in the diagonal) and $S_{\lambda\backslash\mu}$ is the skew S-function given by the (generalised) Jacobi–Trudi identity,

$$S_{\lambda\backslash\mu} = \det(h_{\lambda_i - \mu_j + i - j}) \tag{4.5.13}$$

where h_k are complete symmetric functions, with the sum in (4.5.12) over all partitions whose Young diagrams lie entirely within that of λ. In general the $S_{\lambda\backslash\mu}$ may be expressed as linear combinations of S-functions by means of the Littlewood–Richardson rule ([24], p. 68) but for our present purpose we shall just use the definition (4.5.13); for example

$$S_{(2^2)\backslash(1)} = \begin{vmatrix} h_{2-1} & h_3 \\ h_{1-1} & h_2 \end{vmatrix} = \begin{vmatrix} h_2 & h_3 \\ h_0 & h_1 \end{vmatrix} = S_{(21)}.$$

Finally, we use the orthogonality of group characters to invert (4.5.12b) and so express the generalised S-functions in terms of Hirota derivatives,

$$\tilde{S}_\lambda = \sum_\mu z_\mu^{-1} \chi_\mu^\lambda D_\mu f \circ f \tag{4.5.14}$$

where $z_\mu = \Pi_{i=1}^k i^{m_i} m_i!$, for partitions $\mu = (k^{m_k} \ldots 1^{m_1})$, are such that z_μ^{-1} equal $1/|\mu|!$ times the appropriate entry in the row labelled 'order' in Littlewood's tables.

Having done all this we are in a position to look at the KP equation again. We observe that

$$\tilde{S}_{(2^2)} = (D_{(1^4)} + 3D_{(2^2)} - 4D_{(31)}) f \circ f,$$

from (4.5.14) – which is the Hirota form of the KP equation – which from (4.5.12a) gives

$$\tilde{S}_{(2^2)} = 2(S_{(2^2)} S_{(\cdot)} - S_{(21)} S_{(1)} + S_{(2)} S_{(1^2)}),$$

where $(\cdot)$ is the trivial partition of zero so that $S_{(\cdot)} = f$, in agreement with (4.5.6) which vanishes as we have seen. Hirota [16] has independently made some related observations.

It is interesting to note that our equation (4.5.7) corresponds to the equation in Theorem XIX on p. 115 of Littlewood's book [23]. Littlewood also tried to determine those generalised S-functions which may be expressed as ratios of determinants (as ordinary S-functions may be) – in our terms this is equivalent to calculating those Hirota equations which have N-soliton solutions. This brings us to the main point of this section; can these ideas be used to construct the KP hierarchy?

The KP hierarchy consists of those Hirota equations of the form

$$\left(\sum_{|\lambda|=k} \alpha_\lambda D_\lambda\right) f\circ f = 0, \quad k \geqslant 4, \ \alpha_\lambda \text{ constant} \tag{4.5.15}$$

which are satisfied by the Wronskian (4.5.5). As we saw above for the KP equation itself, it is much more convenient to work in the 'basis' of generalised S-functions rather than the Hirota derivatives themselves so we shall look for those generalised S-functions satisfying

$$\sum_{|\lambda|=k} \beta_\lambda \tilde{S}_\lambda = 0 \quad k \geqslant 4. \tag{4.5.16}$$

The reason that $\tilde{S}_{(2^2)} = 0$ is that it corresponds precisely with the vanishing $2N \times 2N$ determinant (4.5.7). To begin, we shall look at higher-order analogues to this determinant.

From the basic determinant (4.5.7) we may construct a hierarchy of related determinants, all of which vanish in the same way as this determinant, and each of which give us a vanishing quadratic amongst the S_μ; we label these

$$\Delta_{\nu,\rho} \equiv \left|\begin{array}{c:c} (N-3)_\nu & \\ \hdashline & (N+1)_\rho \end{array}\right| \begin{array}{l} N \text{ rows} \\ N \text{ rows} \end{array} = 0 \tag{4.5.17}$$

where the subscripts indicate that the final columns are shifted as in the S_μ. Consequently, at each K, the number of independent quadratics in S_μ that vanish is

$$\sum_{i=0}^{k-4} p(k-i-4)p(i), \tag{4.5.18}$$

where $p(j)$ is the number of partitions of j. In turn, by inverting (4.5.9) as in (4.5.14) we obtain this number of bilinear equations in f. These are not, however, Hirota equations, nor are all of them independent of equations appearing lower down the hierarchy. The Hirota equations correspond to special linear combinations of the $\Delta_{\nu,\rho}$ that we shall obtain later; but first we shall prove by induction that the number of 'original' bilinear equations of weight k is $p(k-4)$. For $k = 4$ this is clearly true – we have just the KP equation – and for $k = m$ the number of equations derived from lower-weight ones by differentiation is $\Sigma_{i=1}^{k-4} p(k-i-4)p(i)$ – assuming that the hypothesis is true for each $4 \leqslant k < M$ – leaving $p(m-4)$ new equations. Hence the result is proved. This differs from the result of Sato and Sato [34] and Kashiwara and

Miwa [21] in which the number of equations is found to be $p(k-1)$ – this is because they count trivial (odd polynomials in D_k) as well as the non-trivial equations we have enumerated.

The number $p(k-4)$ suggests that we should be able to construct a set of operators R_m for each $m \in \mathbb{Z}^+$ that generate an Hirota equation with N-soliton solutions of weight $m+n$ from a Hirota equation with N-soliton solutions of weight n, since then each member of the hierarchy may be written as

$$\mathrm{KP}_\lambda = R_\lambda(\mathrm{KP}_{(\cdot)}) \tag{4.5.19}$$

where $R_\lambda \equiv R_{\lambda_l} \ldots R_{\lambda_m}$ for the partition $\lambda = (\lambda_1 \ldots \lambda_m)$ and $\mathrm{KP}_{(\cdot)}$ is the Hirota form of the KP equation. Such operators would also appear to correspond to those hoped for by Newell ([28], p. 126).

We have been able to construct such a set of operators which may be shown to possess a number of the appropriate necessary properties such as always producing even polynomials in which the sum of coefficients is zero, but a proof of this result requires a different approach [31]. To describe these operators we shall first look at the process of 'multiplication' of the Hirota equation by D_k.

Let $\tilde{P}_\lambda$ be the polynomial in Hirota derivatives such that $\tilde{P}_\lambda f \circ f = \tilde{S}_\lambda$; then we observe that

$$D_k\, \tilde{P}_\lambda = \sum_{i \in \mathbb{Z}} Z_{i-k,i}\left(D_j, \frac{\partial}{\partial D_j}\right) \tilde{P}_\lambda \tag{4.5.20}$$

where the $Z_{i,j}\,(x_k, \partial/\partial x_k)$ are such that $Z(p,q) = \Sigma_{i,j \in Z}\, Z_{i,j}\, p^i q^{-j}$ is the vertex operator discussed in the previous section. This is because the description of the action of the $Z_{i,j}$ on S-functions given in Section 4.4 means that (4.5.20) is the same as the formula for multiplying power-sum symmetric functions by S-functions (see, for example, [24], pp. 31–2).

We are now in a position to write down the conjectured form of the R_k:

$$R_k \tilde{P}_\lambda = \sum_{i \in \mathbb{Z}} a_i\, Z_{i-k,i}\left(D_k, \frac{\partial}{\partial D_k}\right) \tilde{P}_\lambda \tag{4.5.21}$$

where

$$a_i = \begin{cases} 1 & i > \dfrac{k-1}{2} \\ 0 & i = \dfrac{k-1}{2} \\ 1 & i < \dfrac{k-1}{2} \end{cases}, \quad k \text{ odd} \quad \text{or } a_i = \begin{cases} 1 & i \geqslant \dfrac{k}{2} \\ -1 & i < \dfrac{k}{2} \end{cases}, \quad k \text{ even,}$$

which is obtained by modifying (4.5.20) so that it satisfies the necessary conditions mentioned above. Roughly speaking we introduce an extra minus sign if we add below the diagonal of the Young diagram, for example,

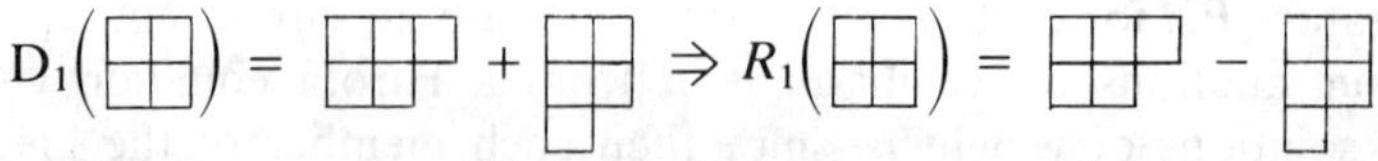

that is, $KP_{(1)} \equiv (D_{(21^3)} + 2D_{(32)} - 3D_{(41)})f \circ f$

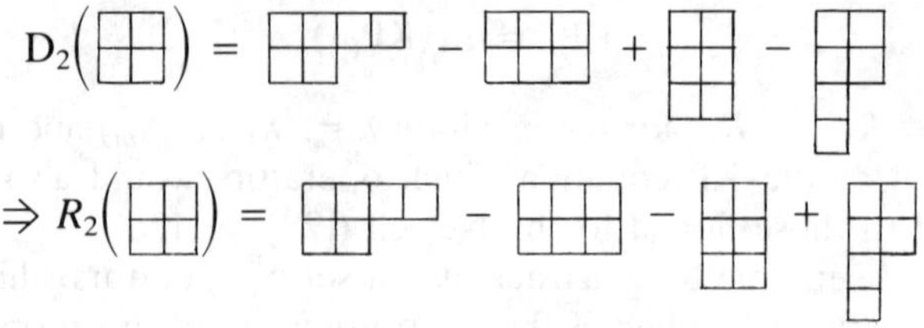

that is, $KP_{(2)} \equiv (D_{(1^6)} + 10D_{(31^3)} - 36D_{(51)} + 45D_{(42)} - 20D_{(3^2)})f \circ f = 0.$

It is possible to think of these operators entirely in terms of their actions on Young diagrams as exemplified above; however, the analytic realisation of this provided by (4.5.21) may also be utilised directly [30].

We conclude by remarking that this construction may also be performed in a similar way to obtain operators that generate the modified KP hierarchies and that it is hoped that a related approach may be taken to other hierarchies of Hirota equations.

References

[1] Adler, M. and Moser, J. On a class of polynomials connected with the Korteweg–de Vries equation, *Commun. Math.* **61**, 1–30 (1978).

[2] Crum, M. M. Associated Sturm–Liouville systems, *Quart. J. Math., ser. 2* **6**, 121–7 (1955).

[3] Date, E., Kashiwara, M. and Miwa, T. Vertex operators and τ functions. Transformation groups for soliton equations II, *Proc. Japan Acad. ser. A* **57**, 387–92 (1981).

[4] Drinfel'd, V. G. and Sokolov, V. V. Equations of Korteweg–de Vries type and simple Lie algebras, *Soviet Math. Dokl.* **23**, 457–62 (1981).

[5] Freeman, N. C. Soliton solutions of non-linear evolution equations, *I. M. A. J. App. M.* **32**, 125–45 (1984).

[6] Freeman, N. C. and Nimmo, J. J. C. Soliton solutions of the Korteweg–de Vries and Kadomtsev–Petviashvili equations: the Wronskian technique, *Proc. R. Soc. Lond. A* **389**, 319–29 (1983).

[7] Gibbon, J. D., Radmore, P., Tabor, M. and Wood, D. The Painlevé property and Hirota's method, *Stud. Appl. M.* **72**, 39–63 (1985).

[8] Hirota, R. Exact solution of the Korteweg–de Vries equation for multiple collision of solitons, *Phys. Rev. Lett.* **27**, 1192–4 (1972).

[9] Hirota, R. Exact solution of the *modified* Korteweg–de Vries equation for multiple collision of solitons, *J. Phys. Soc. Jpn.* **33**, 1456–8 (1972).

[10] Hirota, R. Exact solution of the sine-Gordon equation for multiple collision of solitons, *J. Phys. Soc. Jpn.* **33**, 1459–63 (1972).

[11] Hirota, R. Exact N-soliton solutions of a nonlinear lumped network equation, *J. Phys. Soc. Jpn.* **35**, 286–8 (1973).

[12] Hirota, R. Exact envelope-soliton solutions of a nonlinear wave equation, *J. Math. Phys.* **14**, 805–9 (1973).

[13] Hirota, R. A new form of Bäcklund transformation and its relation to the inverse scattering problem, *Prog. Theor. Phys.* **52**, 1498–1512 (1974).

[14] Hirota, R. Direct methods in solition theory, in *Solitons*, R. K. Bullough and P. J. Caudrey (eds), Springer, Berlin, Heidelberg, New York (1980).

[15] Hirota, R. Bilinearization of soliton equations, *J. Phys. Soc. Jpn.* **51**, 323–31 (1982).

[16] Hirota, R. Solutions of the classical Boussinesq equation and the spherical Boussinesq equation: the Wronskian technique, Preprint (1986).

[17] Hirota, R. and Satsuma, J. Nonlinear evolution equations generated by Bäcklund transformations for the Boussinesq equation, *Prog. Theor. Phys.* **57**, 797–807 (1977).

[18] Hirota, R. and Satsuma, J. A simple structure of superposition formula of the Bäcklund transformation, *J. Phys. Soc. Jpn.* **45**, 1741–50 (1978).

[19] Jimbo, M. and Miwa, T. Solitons and infinite dimensional Lie algebras, *Publ. RIMS, Kyoto Univ.* **19**, 943–1001 (1983).

[20] Kadomtsev, B. B. and Petviashvili, V. I. On the stability of solitary waves in weakly dispersing media, *Sov. Phys. Doklady* **15**, 539–41 (1970).

[21] Kashiwara, M. and Miwa, T. The τ function of the Kadomtsev–Petviashvili equation. Transformation groups for soliton equations I, *Proc. Japan Acad. ser. A* **57**, 342–7 (1981).

[22] Lepowsky, J. and Wilson, R. L. Construction of the affine Lie algebra $A_1^{(1)}$, *Comm. Math. Phys.* **62**, 43–63 (1978).

[23] Littlewood, D. E. *The Theory of Group Characters and Matrix Representations of Groups*, Oxford University Press, London (1950).

[24] Macdonald, I. M. *Symmetric Functions and Hall Polynomials*, Oxford University Press, New York (1979).

[25] Matsuno, Y. *Bilinear Transformation Method*, Academic Press, Orlando, Fla. (1984).

[26] Nakamura, A. A direct method of calculating periodic-wave solutions to nonlinear evolution equations. I. Exact two-periodic wave solutions, *J. Phys. Soc. Jpn.* **47**, 1701–5 (1979).

[27] Nakamura, A. and Hirota, R. A new example of explode–decay solitary waves in one dimension, *J. Phys. Soc. Jpn.* **54**, 491–9 (1985).

[28] Newell, A. C. *Solitons in Mathematics and Physics*, SIAM, Philadelphia (1985).

[29] Nimmo, J. J. C. A bilinear Bäcklund transformation for the nonlinear Schrödinger equation, *Phys. Lett.* **99a**, 279–80 (1983).

[30] Nimmo, J. J. C. Symmetric functions and the KP hierarchy, in *Nonlinear Evolutions*, J. J-P. Leon (ed.), World Scientific, Singapore (1988).

[31] Nimmo, J. J. C. Wronskian determinants, the KP hierarchy and supersymmetric polynomials, to appear in *J. Phys. A* (1989).

[32] Rogers, C. and Shadwick, W. F. *Bäcklund Transformations and their Applications*, Academic Press, New York (1982).

[33] Rosales, R. R. Exact solutions of some nonlinear evolution equations, *Stud. Appl. M.* **59**, 117–51 (1978).

[34] Sato, M. and Sato, Y. *Kokyuroku RIMS* **338**, 183 (in Japanese) (1980).

[35] Satsuma, J. A Wronskian representation of N-soliton solutions of nonlinear evolution equations, *J. Phys. Soc. Jpn.* **46**, 359–60 (1979).

[36] Yajima, N. Application of Hirota's method to a perturbed system, *J. Phys. Soc. Jpn.* **51**, 1298–1302 (1982).

5 *C. Rogers*

Bäcklund transformations in soliton theory

5.1 The importance of Bäcklund transformations in soliton theory

Bäcklund transformations originated in the study of surfaces of constant mean curvature [2–4]. The sine-Gordon equation

$$u_{xt} = \sin u \tag{5.1.1}$$

naturally arises in this connection [8]. Remarkably, it has subsequently emerged that the nonlinear hyperbolic equation (5.1.1) occurs in a wide diversity of areas of physical interest [30]. In fact, it was the relevance of (5.1.1) to the modelling of the propagation of ultra-short optical pulses in resonant laser media that, in large measure, stimulated interest in Bäcklund transformation theory among mathematical physicists. Thus, Lamb [19] reintroduced a classical Bäcklund transformation of (5.1.1) that leaves it invariant and used it to construct multi-soliton solutions.

The attractive nature of the procedure adopted by Lamb resides in the fact that this invariance allows the construction of a nonlinear superposition principle whereby multi-soliton solutions may be generated by a *purely algebraic procedure*. Since that time it has been shown that there is a wide class of important nonlinear evolution equations which admit such auto-Bäcklund transformations together with associated nonlinear superposition principles and multi-soliton solutions. It turns out, moreover, that Bäcklund transformations also have an intimate connection with the inverse scattering method whereby classes of initial value problems for privileged nonlinear evolution equations can be solved. Indeed, the linearisation of the Miura-type Bäcklund transformation linking the Korteweg–de Vries equation (KdV)

$$v_t + 6vv_x + v_{xxx} = 0 \tag{5.1.2}$$

and the modified Korteweg–de Vries (MKdV) equation

$$v_t + 6v^2v_x + v_{xxx} = 0 \tag{5.1.3}$$

leads to the linear Schrödinger equation associated with the inverse scattering method.

The derivation of the Korteweg–de Vries equation (5.1.2) as a model for the evolution of long water waves in a canal is well known. In fact, (5.1.2) serves as a canonical form for a wide class of nonlinear evolution equations. On the other hand, (5.1.3) arises in nonlinear Alfvèn wave theory and in the study of acoustic wave propagation in anharmonic lattices. Both equations reside in hierarchies which admit auto-Bäcklund transformations, nonlinear superposition principles and multi-soliton solutions.

Another important canonical equation which admits an auto-Bäcklund transformation is the cubic Schrödinger equation

$$iq_t + q_{xx} + 2|q|^2q = 0 \tag{5.1.4}$$

which describes the evolution of the wave envelope of weakly nonlinear systems with a carrier wave. Again, a nonlinear superposition theorem is available whereby solutions may, in principle, be generated.

The equations (5.1.1)–(5.1.4) all arise as particular members of the Ablowitz, Kaup, Newell and Segur (AKNS) system of nonlinear evolution equations amenable to the inverse scattering method [1]. The generation of auto-Bäcklund transformations from the AKNS formulation is straightforward. Moreover, in an interesting return to the geometric roots of the subject Chern and Tenenblat [6] have recently shown how these nonlinear evolution equations are each linked to surfaces of constant mean curvature.

In the present survey, the fundamental results for (5.1.1)–(5.1.4) as they pertain to the generation of soliton solutions via Bäcklund transformations are catalogued. The concept of reciprocal transformations is then introduced and is used to link the AKNS and WKI (Wadati–Konno–Ichikawa) scattering systems. The latter has as a member the loop soliton equation. Reciprocal transformations are then used to link the KdV and MKdV hierarchies with the Harry Dym hierarchy. The latter system raises interesting questions about the nature of integrability. In fact reciprocal transformations play an important role in the 'uniformisation' of the Harry Dym sequence as described by Weiss [38] in connection with the Painlevé test. A short account of that work and the connection with Bäcklund transformations is presented in this chapter.

5.2 The sine-Gordon equation

The classical *auto-Bäcklund transformation* for the sine-Gordon equation is embodied in the following result:

Theorem
The sine-Gordon equation

$$u_{xt} = \sin u$$

is invariant under the transformation

$$u'_x = u_x - 2\beta \sin\left\{\frac{u + u'}{2}\right\} := \mathbb{B}'_1(u, u_x; u'), \tag{5.2.1a}$$

$$u'_t = -u_t + \frac{2}{\beta} \sin\left\{\frac{u - u'}{2}\right\} := \mathbb{B}'_2(u, u_t; u'), \tag{5.2.1b}$$

$$x' = x,\ t' = t, \tag{5.2.1c}$$

where $\beta \in \mathbb{R}$ is a non-zero 'Bäcklund' parameter.

The above invariance is readily verified. Thus, application of the integrability condition

$$\frac{\partial \mathbb{B}'_1}{\partial t} - \frac{\partial \mathbb{B}'_2}{\partial x} = 0 \tag{5.2.2}$$

produces the sine-Gordon equation in u, namely (5.1.1). On the other hand, if the relations (5.2.1a) and (5.2.1b) are rewritten as

$$u_x = u'_x - 2\beta \sin\left\{\frac{u + u'}{2}\right\} := \mathbb{B}_1(u', u'_x; u'), \tag{5.2.3a}$$

$$u_t = -u'_t + \frac{2}{\beta} \sin\left\{\frac{u - u'}{2}\right\} := \mathbb{B}_2(u', u'_t; u), \tag{5.2.3b}$$

then the integrability condition

$$\frac{\partial \mathbb{B}_1}{\partial t} - \frac{\partial \mathbb{B}_2}{\partial x} = 0 \tag{5.2.4}$$

produces

$$u'_{xt} = \sin u'. \tag{5.2.5}$$

Insertion of the *trivial* 'vacuum' solution $u' = 0$ in the Bäcklund relations (5.2.1) produces the pair of equations

$$u_x = 2\beta \sin\left\{\frac{u}{2}\right\},\ u_t = \frac{2}{\beta} \sin\left\{\frac{u}{2}\right\}$$

which yield a *non-trivial* travelling wave solution of (5.1.1), namely

$$u = 4 \tan^{-1} \exp(\beta x + \beta^{-1} t + \alpha) \tag{5.2.6}$$

where α is a constant of integration.

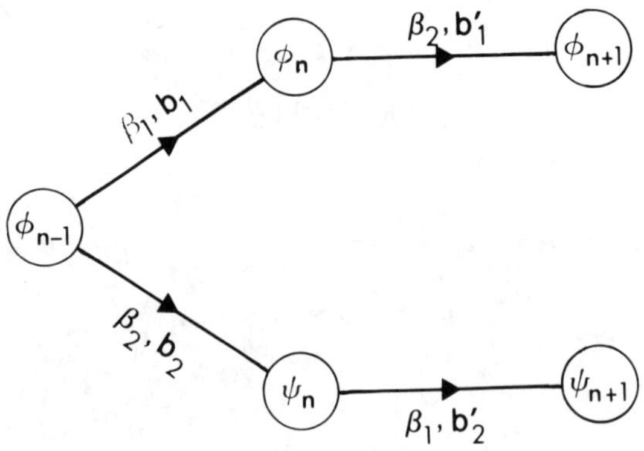

Figure 5.1 A Bianchi diagram [23]

The auto-Bäcklund transformation may now be used to generate an infinite sequence of additional multi-soliton solutions of (5.1.1) without recourse to further integration. The construction is by a purely algebraic procedure well suited to symbolic computation (MACSYMA or MAPLE).

Let ϕ_n, ψ_n be solutions of (5.1.1) produced by application of the Bäcklund transformation (5.2.1) to a known solution ϕ_{n-1} with, in turn, the Bäcklund parameters β_1 and β_2. In a similar manner, let ϕ_{n+1} and ψ_{n+1} denote solutions of (5.1.1) generated by application of the Bäcklund transformation with parameter β_2 to ϕ_n and with parameter β_1 to ψ_n, in turn. The procedure is depicted in the '*Bianchi diagram*' (Figure 5.1) wherein b_i, b_i' $i = 1, 2$ denote arbitrary constants.

The *nonlinear superposition principle* which allows the purely algebraic generation of additional multi-soliton solutions is embodied in the following 'permutability theorem'.

Theorem

There exist ϕ_{n+1} and ψ_{n+1} such that

$$\phi_{n+1} = \psi_{n+1} = \phi$$

where

$$\phi = 4\tan^{-1}\left[\left(\frac{\beta_1 + \beta_2}{\beta_1 - \beta_2}\right)\tan\left\{\frac{\phi_n - \psi_n}{4}\right\}\right] + \phi_{n-1}. \tag{5.2.7}$$

This result emerges as a consequence of the eight Bäcklund relations implied by Figure 5.1, which accordingly may be closed in the manner indicated in Figure 5.2.

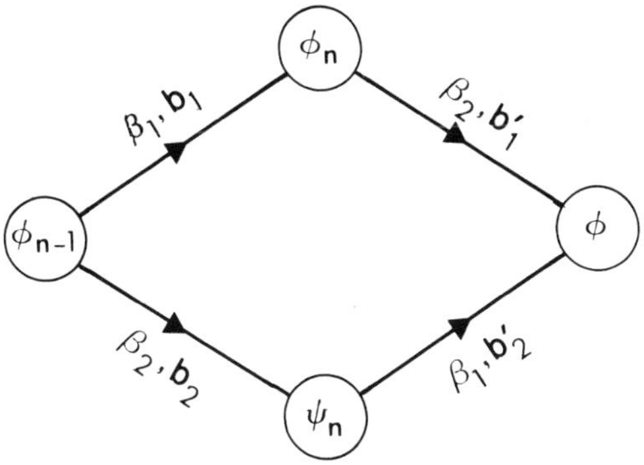

Figure 5.2 A commutative Bianchi diagram [23]

In particular, use of the permutability theorem with starting solution $\phi_0 = 0$ and first-generation solutions

$$\phi_1 = 4 \tan^{-1} e^{\nu_1}, \ \psi_1 = 4 \tan^{-1} e^{\nu_2}$$

with $\nu_i = \beta_i x + \beta_i^{-1} t$, $i = 1, 2$ produces that second-generation solution

$$\phi_2^{(1)} = 4 \tan^{-1}\left[\left(\frac{\beta_1 + \beta_2}{\beta_1 - \beta_2}\right) \tan \left\{\frac{\phi_1 - \psi_1}{4}\right\}\right]. \tag{5.2.8}$$

In the present context, ϕ_x describes a 2-soliton interaction. The procedure may be readily iterated to produce n-soliton solutions. The generation of third-generation solutions via the Bianchi lattice is indicated in Figure 5.3.

Note that a Bäcklund transformation and associated nonlinear superposition principle for an elliptic analogue of (5.1.1) have been presented by Leibbrandt [21] in connection with the propagation of magnetic flux through a long Josephson junction.

5.3 **The classical notion of Bäcklund transformation and its extension**

The invariant transformation (5.2.1) of the sine-Gordon equation (5.1.1) is of a type originally introduced by Bäcklund in connection with transformational properties of surfaces. In that context, let $u = u(x^a)$, $u' = u'(x'^a)$, $a = 1, 2$ represent two surfaces S and S' respectively in $\mathbb{R}^3$. A set of four relations

$$\mathbb{B}_i^*(x^a, u, u_a; x'^a, u', u'_a) = 0, \ i = 1, \ldots, 4; \ a = 1, 2,$$
$$\begin{pmatrix} u_a := \partial u / \partial x_a \\ u'_a := \partial u' / \partial x'_a \end{pmatrix} \tag{5.3.1}$$

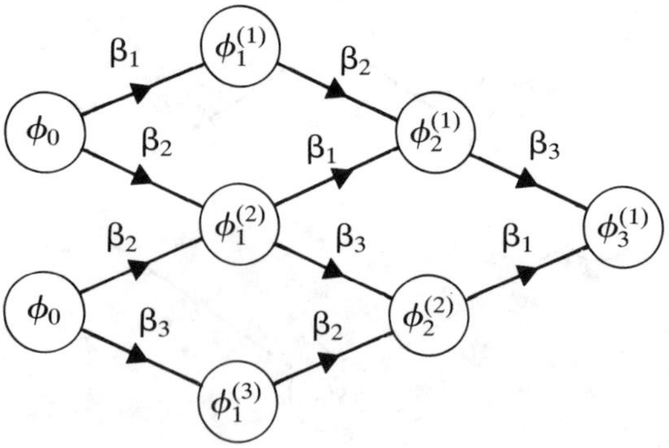

Figure 5.3 The Bianchi diagram associated with third-generation solutions

which connect the surface elements $\{x^a, u, u_a\}$ and $\{x'^a, u', u'_a\}$ of S and S' respectively is termed a Bäcklund transformation in the classical literature. If the relations (5.3.1) admit the explicit resolutions

$$u'_i = \mathrm{B}'_i(x^a, u, u_a; u') \quad i = 1, 2, \tag{5.3.2}$$

$$u_i = \mathrm{B}_i(x'^a, u', u'_a; u) \quad i = 1, 2, \tag{5.3.3}$$

$$x'^j = X(x^a, u, u_a; u') \quad j = 1, 2, \tag{5.3.4}$$

then, in order that these relations transform a surface $u = u(x^a)$ with surface element $\{x^a, u, u_a\}$to a surface $u'(x'^a)$ with surface element $\{x'^a, u', u'_a\}$ integrability requires that

$$\frac{\partial\, \mathbb{B}_1}{\partial\, x^2} - \frac{\partial\, \mathbb{B}_2}{\partial x^1} = 0, \tag{5.3.5}$$

$$\frac{\partial\, \mathrm{B}'_1}{\partial\, x'^2} - \frac{\partial\, \mathrm{B}'_2}{\partial x'^1} = 0, \tag{5.3.6}$$

It is shown in [10] that application of the integrability condition (5.3.6) to the Bäcklund relations (5.3.2)–(5.3.4) leads to a nonlinear equation of the form

$$U\{u_{11}u_{22} - u_{12}^2\} + Ru_{11} + 2Su_{12} + Tu_{22} + V = 0 \tag{5.3.7}$$

where U, R, S, T and V, in general, involve the quantities x^a, u, u_i and u' but not partial derivatives in u of second or higher order. If u' is absent in (5.3.7) then it reduces to a Monge–Ampère equation. In a similar fashion, application of the integrability condition (5.3.5) to the Bäcklund relations (5.3.2)–(5.3.4) produces a primed counterpart of (5.3.7). In particular, if the equations for u and u' derived in this way are both of the Monge–Ampère type, then the Bäcklund transformation represents a mapping between their integral surfaces. A particular case of such a transformation is that presented for the sine-Gordon equation in the preceding section.

The classical notion of Bäcklund transformation may be extended in a number of ways. Thus, Loewner [22] in an investigation of hodograph systems in gas dynamics was led to introduce a generalisation to cases involving two dependent variables $u = u(x^a)$ and $v = v(x^a)$, $a = 1, 2$. In this connection, transformations of the form

$$\mathbb{B}_i^*(x^a, u, u_a, v, v_a; x'^a, u', u'_a, v'_a) = 0, \; i = 1, \ldots, 6; \; a = 1, 2 \tag{5.3.8}$$

were presented which relate surface elements $\{x^a, u, u_a\}$, $\{x^a, v, v_a\}$ and $\{x'^a, u', u'_a\}$, $\{x'^a, v', v'_a\}$ associated with pairs of surfaces $u = u(x^a)$, $v = v(x^a)$ and $u' = u'(x'^a)$, $v' = v'(x'^a)$. Six relations were chosen in (5.3.8) so that, if the x^a are regarded as dependent variables, the number of equations is the same as the number of unknowns.

Loewner's extension of the notion of Bäcklund transformation was required for his treatment of the reduction to linear canonical form of hodograph systems in subsonic, transonic and supersonic gas dynamics. The next generalisation was motivated by consideration of the Korteweg–de Vries and modified Korteweg–de Vries equations (5.1.2) and (5.1.3). Since these are third-order equations an extension of the idea of Bäcklund transformation to incorporate higher-order derivatives was required. In this connection, Lamb [20] introduced Bäcklund transformations of the form

$$\mathbb{B}_i^*(x^a, u, u_a, u_{ab}; x'^a, u', u'_a, u'_{ab}) = 0, \; i = 1, \ldots, 4; \; a, b = 1, 2 \tag{5.3.9}$$

which, in turn, may be used to derive permutability theorems for the generation of multi-soliton solutions. These results are recorded in the next section.

Since the work of Lamb, the notion of Bäcklund transformation has been broadened considerably. In particular, a systematic extension to include higher derivatives has been developed based on a jet-bundle formalism (see [30]). Moreover, Bäcklund transformations for higher-dimensional equations such as the Kadomtsev–Petviashvili, modified Kadomtsev–Petviashvili and Davey–Stewartson equations have been developed. Bäcklund transformations have also been constructed for certain nonlinear integro-differential and differential-difference equations.

In general terms, for the purposes of the present elementary discussion, a set of relations

$$\mathbb{B}_i^*(x^a, u, u_a, u_{ab}, \ldots; x'^a, u', u'_a, u'_{ab}, \ldots) = 0 \tag{5.3.10}$$

will be called a Bäcklund transformation between two partial differential equations

$$\varepsilon(u) = 0 \tag{5.3.11}$$

and

$$\varepsilon'(u') = 0 \tag{5.3.12}$$

if the relations (5.3.10) are integrable for u' if and only if $\varepsilon(u) = 0$ and are integrable for u if and only if $\varepsilon'(u') = 0$.

5.4 The Korteweg–de Vries and modified Korteweg–de Vries equations

Wahlquist and Estabrook [35] established an *auto-Bäcklund transformation* for the Korteweg–de Vries equation (5.1.2). The result was subsequently derived by Lamb [20] who adapted a classical procedure due to Clairin for the construction of Bäcklund transformations. The result is as follows:

Theorem
The Korteweg–de Vries equation

$$v_t + 6vv_x + v_{xxx} = 0$$

is invariant under the Bäcklund transformation

$$\begin{aligned} u'_x &= \beta - u_x - \tfrac{1}{2}[u - u']^2 := \mathbb{B}'_1(u, u_x; u'), \\ u'_t &= -u_t + [u - u'][u_{xx} - u'_{xx}] - 2[u_x^2 + u_x u'_x + u'^2_x] \\ &:= \mathbb{B}'_2(u, u_x, u_{xx}, u_t; u', u'_x, u'_{xx}), \\ x' &= x,\ t' = \mathrm{t}, \end{aligned} \tag{5.4.1}$$

where

$$u := \int_{-\infty}^{x} v(\sigma, t)\mathrm{d}\sigma \tag{5.4.2}$$

is a potential and β is an arbitrary Bäcklund parameter.

The above result may be shown to generate the following *permutability theorem* whereby an infinite sequence of solutions of the Korteweg–de Vries equation may be generated from a known solution.

Theorem
If $u_{\beta_i} = \mathbb{B}_{\beta_i} u_0$, $i = 1, 2$ are solutions of the potential Korteweg–de Vries equation

$$u_t + 3u_x^2 + u_{xxx} = 0 \tag{5.4.3}$$

as generated by application of the Bäcklund transformation (5.4.1) to a known solution u_0 via, in turn, the Bäcklund parameters $\beta = \beta_i$, $i = 1, 2$ then a new solution ϕ is given by the nonlinear superposition

$$\phi = u_0 + 2\{\beta_1 - \beta_2\}/\{u_{\beta_1} - u_{\beta_2}\} \tag{5.4.4}$$

where $\phi = \mathbb{B}_{\beta_1}\mathbb{B}_{\beta_2}u_0 = \mathbb{B}_{\beta_2}\mathbb{B}_{\beta_1}u_0$.

Since, in the above, u_0 represents an arbitrary solution of (5.4.3), the general recursion relation

$$u_{(n)} = u_{(n-2)} + 2\{\beta_n - \beta_{n-1}\}/\{u_{(n-1)'} - u_{(n-1)}\},$$
$$n > 1,\ u(0) = u_0, \tag{5.4.5}$$

follows, where the subscript (n) denotes the set of n parameters $\{\beta_1, \ldots, \beta_n\}$, while the subscript $(n)'$ denotes $\{\beta_1, \ldots, \beta_{n-1}, \beta_{n+1}\}$.

Remarkably, the hierarchy of solutions of the Korteweg–de Vries equation generated by the application of the permutability theorem with starting solution $u = 0$ correspond to multi-soliton pulses. Thus, a soliton ladder is initiated by substitution of this vacuum solution into the Bäcklund relations (5.4.1) whence

$$u'_x = \beta - \tfrac{1}{2}u'^2, \tag{5.4.6}$$

$$u'_t = u'\,u'_{xx} - 2u'^2_x. \tag{5.4.7}$$

These relations yield the regular solution

$$u' = (2\beta)^{1/2}\tanh[(\beta/2)^{1/2}(x - 2\beta t)] \tag{5.4.8}$$

with associated single-soliton solution of the Korteweg–de Vries equation

$$v' = \beta\,\mathrm{sech}^2[(\beta/2)^{1/2}(x - 2\beta t)]. \tag{5.4.9}$$

Hence, it is seen that the Bäcklund parameter β gives the amplitude of the pulse while 2β gives the speed of propagation.

The pair of equations (5.4.6), (5.4.7) also admit the singular solution

$$u^* = (2\beta)^{1/2}\coth[(\beta/2)^{1/2}(x - 2\beta t)] \tag{5.4.10}$$

with associated solution of the Korteweg–de Vries equation

$$v^* = -\beta\,\mathrm{cosech}^2[(\beta/2)^{1/2}(x - 2\beta t)]. \tag{5.4.11}$$

It is interesting to note that the singular solution (5.4.10) is needed in order to construct regular second-generation solutions via the permutability theorem. Thus, the only such regular 2-soliton solution is given by

$$u_{\beta_1\beta_2} = 2\{\beta_1 - \beta_2\}/\{u_{\beta_1} - u^*_{\beta_2}\},\ \beta_1 < \beta_2, \tag{5.4.12}$$

where $u^*_{\beta_2}$ is the singular first-generation solution corresponding to the parameter β_2. The regular 3-soliton solution is constructed via the theorem of permutability with $\{u_{\beta_1}, u^*_{\beta_2}, u_{\beta_3}\}$ and $\beta_1 < \beta_2 < \beta_3$. The multi-soliton solution is readily generated.

We next turn to the modified Korteweg–de Vires equation (5.1.3) which, as will be seen, is related to the Korteweg–de Vries equation by a Bäcklund transformation. It also admits an auto-Bäcklund transformation as given below [20]:

Theorem
The modified Korteweg–de Vries equation

$$v_t + 6v^2 v_x + v_{xxx} = 0$$

is invariant under the Bäcklund transformation

$$\left.\begin{aligned} u_x &= \alpha u'_x + \beta \sin\{u + \alpha u'\}, \\ u_t &= \alpha u'_t - \beta[2\alpha u'_{xx} \cos\{u + \alpha u'\} + 2u'^2_x \sin\{u + \alpha u'\} \\ &\quad + \beta(u_x + \alpha u'_x)], \end{aligned}\right\}$$
$$x' = x,\ t' = t \quad (\alpha = \pm 1) \tag{5.4.13}$$

where

$$u = \int_{-\infty}^{x} v(\sigma, t)\mathrm{d}\sigma$$

and β is an arbitrary Bäcklund parameter.

The above Bäcklund transformation leads to the following nonlinear superposition principle:

Theorem
If u_{β_i}, $i = 1, 2$ represent solutions of the potential modified Korteweg–de Vries equation

$$u_t + 2u_x^3 + u_{xxx} = 0 \tag{5.4.14}$$

as generated by the Bäcklund transformation (5.4.13) with a starting solution u_0 and, in turn, the Bäcklund parameters $\beta = \beta_i$ $i = 1, 2$ then a new solution ϕ is given by

$$\tan\left\{\frac{\phi - u_0}{2}\right\} = \alpha\left(\frac{\beta_1 + \beta_2}{\beta_1 - \beta_2}\right) \tan\left\{\frac{u_{\beta_1} - u_{\beta_2}}{2}\right\} \tag{5.4.15}$$

where $\phi = \mathbb{B}_{\beta_1} \mathbb{B}_{\beta_2} u_0 = \mathbb{B}_{\beta_2} \mathbb{B}_{\beta_1} u_0$.

Use of the Bäcklund transformation (5.4.13) with parameters $\beta = \beta_i$, $i = 1, 2$ and initial vacuum solution $u_0 = 0$ produces the first-generation solutions

$$u_{\beta_i} = 2\tan^{-1} \mathrm{e}^{\mu_i} \tag{5.4.16}$$

where

$$\mu_i = \beta_i x - \beta_i^3 t + \gamma_i \tag{5.4.17}$$

and γ_i, $i = 1, 2$ are integration constants. Application of the permutability relation (5.4.15) now yields the second-generation solution

$$\phi = \pm 2 \tan^{-1}\left[\left(\frac{\beta_1 + \beta_2}{\beta_1 - \beta_2}\right)\frac{\sinh\{\frac{1}{2}(\mu_1 - \mu_2)\}}{\cosh\{\frac{1}{2}(\mu_1 + \mu_2)\}}\right]. \tag{5.4.18}$$

Multiple-soliton solutions of the modified Korteweg–de Vries equation are readily generated by repeated use of the permutability relation (5.4.15).

In conclusion, it is noted that the Korteweg–de Vries equation

$$v'_t + 6v'v'_x + v'_{xxx} = 0 \tag{5.4.19}$$

and the modified Korteweg–de Vries equation

$$v_t + 6v^2 v_x + v_{xxx} = 0 \tag{5.4.20}$$

are linked by the Bäcklund transformation

$$\left.\begin{aligned} v_x &= \pm\{v' + v^2\} := \mathbb{B}_1(v, v'), \\ v_t &= \mp v'_{xx} - 2\{vv'_x + v'v_x\} := \mathbb{B}_2(v, v_x; v', v'_x, v'_{xx}), \\ x' &= x,\ t' = t. \end{aligned}\right\} \tag{5.4.21}$$

The spatial part of these Bäcklund relations is commonly called a Miura transformation. It will be shown, more generally, to link Korteweg–de Vries and modified Korteweg–de Vries hierarchies in which (5.4.19) and (5.4.20) are respectively embedded.

5.5 The AKNS system and construction of Bäcklund transformations. The nonlinear Schrödinger equation

The auto-Bäcklund transformations listed for the sine-Gordon, Korteweg–de Vries and modified Korteweg–de Vries equation are readily constructed via the AKNS inverse scattering formalism [1].

Thus, consider the 2×2 eigenvalue problem

$$\left.\begin{aligned} \psi_x &= X\psi, \\ \psi_t &= T\psi, \end{aligned}\right\} \tag{5.5.1}$$

where

$$\begin{gathered} \psi = \begin{pmatrix} \psi_1 \\ \psi_2 \end{pmatrix}, \\ X = \begin{pmatrix} \lambda & q(x, t) \\ r(x, t) & -\lambda \end{pmatrix}, \quad T = \begin{pmatrix} A(x, t; \lambda) & B(x, t; \lambda) \\ C(x, t; \lambda) & -A(x, t; \lambda) \end{pmatrix}. \end{gathered} \tag{5.5.2}$$

In the above, ψ_1, ψ_2 are eigenfunctions while λ represents a time-independent eigenvalue. The compatibility condition

$$\psi_{xt} = \psi_{tx} \tag{5.5.3}$$

yields

$$X_t - T_x + [X, T] = 0 \tag{5.5.4}$$

where $[X, T] = XT - TX$ is the commutator of X and T.

In extenso, (5.5.4) produces the following set of conditions on A, B and C:

$$\left.\begin{aligned} A_x &= qC - rB, \\ B_x - 2\lambda B &= q_t - 2Aq, \\ C_x + 2\lambda C &= r_t + 2Ar. \end{aligned}\right\} \tag{5.5.5}$$

It is noted that the Lax system

$$\left.\begin{aligned} L\mathbf{v} &= \lambda\mathbf{v}, \\ M\mathbf{v} &= \mathbf{v}_t, \end{aligned}\right\} \tag{5.5.6}$$

wherein L, M are linear operators subject to the compatibility condition

$$L_t + [L, M] = 0 \tag{5.5.7}$$

produces the AKNS system if we set

$$L = \begin{pmatrix} \dfrac{\partial}{\partial x} & -q \\ r & -\dfrac{\partial}{\partial x} \end{pmatrix}, \quad M = T. \qquad (5.5.8), (5.5.9)$$

Specialisations of A, B and C in the AKNS system lead to nonlinear evolution equations amenable to the inverse scattering method for the solution of privileged initial value problems. In particular, we note the following specialisations:

The sine-Gordon equation $u_{xt} = \sin u$

$$r = -q = -u_x/2,$$

$$A = \frac{1}{4\lambda}\cos u, \; B = C = \frac{1}{4\lambda}\sin u. \tag{5.5.10}$$

The Korteweg–de Vries equation $q_t + 6qq_x + q_{xxx} = 0$

$$r = -1$$

$$\begin{gathered} A = -4\lambda^3 - 2\lambda q - q_x, \; B = -q_{xx} - 2\lambda q_x - 4\lambda^2 q - 2q^2, \\ C = 4\lambda^2 + 2q. \end{gathered} \tag{5.5.11}$$

The modified Korteweg–de Vries equation $q_t + 6q^2q_x + q_{xxx} = 0$

$$r = -q$$

$$\begin{gathered} A = -4\lambda^3 - 2\lambda q^2, \; B = -q_{xx} - 2\lambda q_x - 4\lambda^2 q - 2q^3, \\ C = q_{xx} - 2\lambda q_x + 4\lambda^2 q + 2q^3. \end{gathered} \tag{5.5.12}$$

The cubic Schrödinger equation $\mathbf{i}q_t + q_{xx} + 2|q|^2 q = 0$

$$\begin{aligned} r &= -\bar{q} \\ A = 2\mathrm{i}\lambda^2 + \mathrm{i}|q|^2,\ B &= \mathrm{i}q_x + 2\mathrm{i}\lambda q, \\ C &= \mathrm{i}\bar{q}_x - 2\mathrm{i}\lambda\bar{q}. \end{aligned} \tag{5.5.13}$$

Auto-Bäcklund transformations for the above nonlinear evolution equations are now readily constructed via the AKNS formalism. The key step in the derivation involves the introduction of $\Gamma(x, t)$ according to

$$\Gamma = \psi_1/\psi_2 \tag{5.5.14}$$

so that the AKNS system (5.5.1)–(5.5.2) produces the pair of Riccati equations

$$\Gamma_x = 2\lambda\Gamma + q - r\Gamma^2, \tag{5.5.15}$$

$$\Gamma_t = 2A\Gamma + B - C\Gamma^2. \tag{5.5.16}$$

The discussion now divides into three subclasses as follows:

Class I $\mathbf{r} = -q$

In this case, the AKNS system yields

$$\left.\begin{aligned} \psi_{xx} - \lambda^2\psi &= u\psi \\ \bar{\psi}_{xx} - \lambda^2\bar{\psi} &= \bar{u}\bar{\psi} \end{aligned}\right\} \tag{5.5.17}$$

where $\psi = \psi_1 + \mathrm{i}\psi_2$ and

$$u = -\mathrm{i}q_x - q^2. \tag{5.5.18}$$

The system (5.5.17) is invariant under the Crum-type transformation [7]:

$$\left.\begin{aligned} \psi' &= \frac{1}{\psi},\ u' = u + 2(\ln \psi')_{xx} \\ \bar{\psi}' &= \frac{1}{\bar{\psi}},\ \bar{u}' = \bar{u} + 2(\ln \bar{\psi}')_{xx} \end{aligned}\right\} \tag{5.5.19}$$

whence

$$w' = w - \mathrm{i}\ln(\psi/\bar{\psi}) = w + 2\tan^{-1}(\psi_2/\psi_1) \tag{5.5.20}$$

where w, w' are potentials given by $q = w_x$, $q' = w'_x$. Accordingly, under the transformation (5.5.19)

$$\Gamma = \cot\{\tfrac{1}{2}(w' - w)\} \tag{5.5.21}$$

and substitution in the Riccati equations (5.5.15)–(5.5.16) provides a generic Bäcklund transformation for the present subclass of the AKNS system with $r = -q$, namely

$$\left.\begin{aligned} &w_x + w'_x = 2\lambda\sin\{w - w'\}, \\ &w_t - w'_t = 2A\sin\{w' - w\} - (B + C)\cos\{w' - w\} + B - C, \\ &x' = x,\quad t' = t. \end{aligned}\right\} \tag{5.5.22}$$

If we set $\beta = \lambda/2$, $w = -u/2$ the specialisations (5.5.10) inserted in (5.5.22) produce the auto-Bäcklund transformation (5.2.1) for the sine-Gordon equation. On the other hand, if we set $\beta = 2\lambda$, $w = -u$, the specialisations (5.5.12) in (5.5.22) give the auto-Bäcklund transformation (5.4.13) with $\alpha = -1$ for the modified Korteweg–de Vries equation.

Class II $\boldsymbol{r = -1}$

In this case, the AKNS system yields the linear Schrödinger equation

$$-\psi_{2,xx} + \lambda^2\psi_2 = q\psi_2 \tag{5.5.23}$$

invariant under the Crum transformation

$$\psi_2' = \frac{1}{\psi_2}, \quad q' = q - 2(\ln \psi_2')_{xx}. \tag{5.5.24}$$

Under this transformation

$$w' - w = 2\psi_{2,x}/\psi_2 = -2(\lambda + \Gamma) \tag{5.5.25}$$

and on substitution of Γ as given by this relation into the Riccati equations (5.5.15)–(5.5.16) we obtain the generic auto-Bäcklund transformation

$$\left.\begin{aligned} w_x' + w_x &= 2\lambda^2 - \tfrac{1}{2}(w - w')^2, \\ w_t' - w_t &= 4A[\tfrac{1}{2}(w' - w) + \lambda] - 2B + 2C[\tfrac{1}{2}(w - w') - \lambda]^2, \\ x' &= x,\ t' = t, \end{aligned}\right\} \tag{5.5.26}$$

for the subclass of the AKNS system with $r = -1$. The specialisations (5.3.11) inserted in (5.5.26) give the Wahlquist–Estabrook auto-Bäcklund transformation for the Korteweg–de Vries equation.

Class III $\boldsymbol{r = -\bar{q}}$

Generic auto-Bäcklund transformations for this class were generated by Konno and Wadati [18] via Γ' and q' which leave the pair of Riccati equations (5.5.15) and (5.5.16) invariant. In particular, the specialisations (5.5.13) were shown to lead to the auto-Bäcklund transformation

$$\begin{aligned} q_x + q_x' &= \{q - q'\}\sqrt{(4\lambda^2 - |q + q'|^2)}, \\ q_t + q_t' &= \mathrm{i}\{q_x - q_x'\}\sqrt{(4\lambda^2 - |q + q'|^2)} \\ &\quad + \frac{\mathrm{i}}{2}\{q + q'\}[|q + q'|^2 + |q - q'|^2], \\ x' &= x,\ t' = t, \end{aligned} \tag{5.5.27}$$

for the nonlinear Schrödinger equation (5.1.4). This has associated nonlinear superposition principle

$$
\begin{aligned}
&(q_0 - q_1)\sqrt{(4\lambda_1^2 - |q_0 + q_1|^2)} - (q_0 - q_2)\sqrt{(4\lambda_2^2 - |q_0 + q_2|^2)} \\
&+ (q_2 - q_{12})\sqrt{(4\lambda_1^2 - |q_2 + q_{12}|^2)} - (q_1 - q_{12})\sqrt{(4\lambda_2^2 - |q_1 + q_{12}|^2)} = 0
\end{aligned}
\tag{5.5.28}
$$

where $q_i = B_{\lambda_i} q_0$, $i = 1, 2$, $q_{12} = \mathrm{B}_{\lambda_1} \mathrm{B}_{\lambda_2} q_0 = \mathrm{B}_{\lambda_2} \mathrm{B}_{\lambda_1} q_0$, and q_0 is a starting solution.

This completes the review of the elementary aspects of Bäcklund transformation theory as they relate to soliton phenomena. The interested reader is referred to the text by Rogers and Shadwick [30], for a broad account of the subject which includes the jet-bundle and Hirota bilinear operator formalisms. The remainder of the present survey will deal with selected recent developments.

5.6 **Reciprocal transformations. The link between the AKNS and WKI scattering schemes. Loop solitons. The Dym, KdV and modified KdV hierarchies**

In what follows, we make use of the following result [14].

Theorem

The conservation law

$$\frac{\partial}{\partial t}\left\{T\left(\frac{\partial}{\partial x}; \frac{\partial}{\partial t}; u\right)\right\} + \frac{\partial}{\partial x}\left\{F\left(\frac{\partial}{\partial x}; \frac{\partial}{\partial t}; u\right)\right\} = 0 \tag{5.6.1}$$

is taken to the associated conservation law

$$\frac{\partial T'}{\partial t'} + \frac{\partial F'}{\partial x'} = 0 \tag{5.6.2}$$

by the *reciprocal* transformation

$$
\left.
\begin{aligned}
&dx' = T dx - F dt, \quad t' = t && (5.6.3a)\\
&T' = \frac{1}{T(D'; \partial'; u)}, \quad F' = -\frac{F(D'; \partial'; u)}{T(D'; \partial'; u)} && (5.6.3b)\\
&0 < |T| < \infty && (5.6.3c)
\end{aligned}
\right\} R
$$

where

$$D' := \frac{\partial}{\partial x} = \frac{1}{T'}\frac{\partial}{\partial x'}, \tag{5.6.4}$$

$$\partial' := \frac{\partial}{\partial t} = \frac{F'}{T'}\frac{\partial}{\partial x'} + \frac{\partial}{\partial t'}. \tag{5.6.5}$$

The above result is readily established. Thus, the relation (5.6.3a) shows that

$$\frac{\partial x'}{\partial x} = T, \quad \frac{\partial x'}{\partial t} = -F$$

whence, compatibility produces (5.6.1). Similarly, (5.6.3a) and (5.6.3b) yield

$$\frac{\partial x}{\partial x'} = \frac{1}{T}, \quad \frac{\partial x}{\partial t'} = \frac{F}{T}$$

so that compatibility produces (5.6.2).

The reciprocal nature of the transformation resides in the involutary property $R^2 = I$. Thus,

$$\begin{aligned} \mathrm{d}x'' &= T'\mathrm{d}x' - F'\mathrm{d}t' = T'(T\mathrm{d}x - F\mathrm{d}t) - F'\mathrm{d}t = \mathrm{d}x, \\ t'' &= t' = t, \\ T'' &= \frac{1}{T'} = T, \; F'' = -\frac{F'}{T'} = \frac{F}{TT'} = F. \end{aligned}$$

The notion of reciprocal transformation may be extended to systems of conservation laws [15], and to higher dimensions [26].

The application of reciprocal-type transformations in gasdynamics and magnetogasdynamics has been surveyed in [30]. Reciprocal transformations have also recently been used to derive the exact solutions of nonlinear boundary value problems which arise in such diverse areas as two-phase flow through porous media [9], Stefan problems [24], and nonlinear viscoelasticity [32]. Application to nonlinear Cattaneo models in heat conduction has also been noted [29]. Classes of nonlinear boundary value problems solvable by reciprocal transformations have been discussed by Rogers [25]. Here, reciprocal transformations are shown to be a key component in the link between the AKNS scattering scheme and the WKI (Wadati, Konno and Ichikawa) scheme [34].

Thus, consider the inverse scattering scheme

$$\left.\begin{aligned} \psi_x &= U(x, t; \lambda)\psi \\ \psi_t &= V(x, t; \lambda)\psi \end{aligned}\right\} \tag{5.6.6}$$

with compatibility condition

$$U_t - V_x + [U, V] = 0 \tag{5.6.7}$$

where, in the above, U and V are, in general, $N \times N$ complex matrices and typically have a rational dependence on the eigenvalue λ.

It is readily verified that (5.6.6) and (5.6.7) are invariant under the gauge transformation

$$\left.\begin{aligned}\tilde{\psi} &= g^{-1}\psi,\\ \tilde{U} &= g^{-1}Ug - g^{-1}g_x,\\ \tilde{V} &= g^{-1}Vg - g^{-1}g_t,\end{aligned}\right\} G \tag{5.6.8}$$

where g is an arbitrary invertible matrix dependent on x and t.

It is noted that the above gauge transformation is invertible with

$$\left.\begin{aligned}\psi &= g\tilde{\psi}\\ U &= g\tilde{U}g^{-1} + g_x g^{-1},\\ V &= g\tilde{V}g^{-1} + g_t g^{-1}.\end{aligned}\right\} G^{-1} \tag{5.6.9}$$

The system (5.6.6) is now specialised to the 2 × 2 AKNS scheme (5.5.1)–(5.5.2) so that $U = X$ and $V = T$, where X and T are given by (5.5.2). Moreover, g is taken as the solution of (5.5.1)–(5.5.2) with $\lambda = 0$, so that

$$g_x = \begin{pmatrix} 0 & q \\ r & 0 \end{pmatrix} g, \tag{5.6.10}$$

$$g_t = \begin{pmatrix} A(r, q; 0) & B(r, q; 0) \\ C(r, q; 0) & -A(r, q; 0) \end{pmatrix}. \tag{5.6.11}$$

Thus,

$$\tilde{U} = \lambda S, \tag{5.6.12}$$

$$\tilde{V} = g^{-1}\begin{pmatrix} A - A|_{\lambda=0} & B - B|_{\lambda=0} \\ C - C|_{\lambda=0} & -A + A|_{\lambda=0} \end{pmatrix} g, \tag{5.6.13}$$

where

$$S = g^{-1}\begin{pmatrix} 1 & 0 \\ 0 & -1 \end{pmatrix} g = \begin{pmatrix} a & b \\ c & -a \end{pmatrix} \tag{5.6.14}$$

with

$$a^2 + bc = 1. \tag{5.6.15}$$

The inverse scattering scheme

$$\left.\begin{aligned}\tilde{\psi}_x &= \tilde{U}\,\tilde{\psi}\\ \tilde{\psi}_t &= \tilde{V}\,\tilde{\psi}\end{aligned}\right\} \tag{5.6.16}$$

gauge equivalent to the AKNS system has the compatibility condition

$$S_t - \lambda^{-1}\tilde{V}_x + S\tilde{V} - \tilde{V}S = 0. \tag{5.6.17}$$

If the system (5.6.16) delivers the conservation law

$$a_t + \epsilon_x = 0 \tag{5.6.18}$$

then introduction of x', t' according to the reciprocal transformation R

$$dx' = a dx - \epsilon dt, \; t' = t \tag{5.6.19}$$

leads to the associated conservation law

$$a'_{t'} + \epsilon'_{x'} = 0 \tag{5.6.20}$$

where

$$a' = \frac{1}{a}, \tag{5.6.21}$$

$$\epsilon' = -\frac{\epsilon}{a}. \tag{5.6.22}$$

Under R, the inverse scattering scheme (5.6.16) becomes

$$\left.\begin{aligned} \tilde{\psi}_{x'} &= \tilde{U}' \, \tilde{\psi} \\ \tilde{\psi}_{t'} &= \tilde{V}' \, \tilde{\psi} \end{aligned}\right\} \tag{5.6.23}$$

where

$$\tilde{U}' = \frac{\tilde{U}}{a}, \; \tilde{V}' = \tilde{V} + \frac{\epsilon \tilde{U}}{a},$$

so that

$$\tilde{U}'' = \frac{\tilde{U}'}{a'} = \frac{\tilde{U}}{aa'} = \tilde{U},$$

$$\tilde{V}'' = \tilde{V}' + \frac{\epsilon' \tilde{U}'}{a'} = \tilde{V} + \frac{\epsilon \tilde{U}}{a} - \frac{\epsilon \tilde{U}'}{aa'} = \tilde{V}.$$

Hence, R represents an involutory transformation of the inverse scattering scheme (5.6.16).

If, following Ishimori [12] and Wadati and Sogo [33] we introduce the further transformation

$$\left.\begin{aligned} \bar{q} &= \frac{-\mathrm{i}b}{\sqrt{(1 - bc)}} \\ \bar{r} &= \frac{-\mathrm{i}c}{\sqrt{(1 - bc)}} \end{aligned}\right\} T \tag{5.6.24}$$

with inverse

$$\left.\begin{aligned} b &= \frac{\mathrm{i}\bar{q}}{\sqrt{(1 - \bar{q}\,\bar{r})}} \\ c &= \frac{\mathrm{i}r}{\sqrt{(1 - \bar{q}\,\bar{r})}} \end{aligned}\right\} T^{-1} \tag{5.6.25}$$

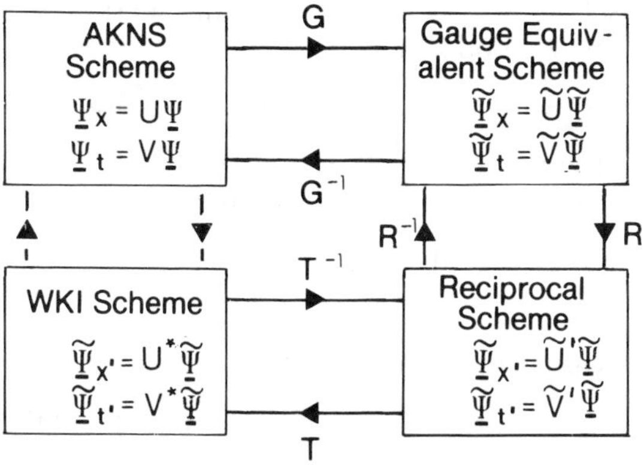

Figure 5.4 The link between the AKNS and WKI inverse scattering schemes

then the inverse scattering scheme (5.6.23) is taken to the WKI system

$$\left.\begin{aligned}\tilde{\psi}_{x'} &= U^*\tilde{\psi}\\ \tilde{\psi}_{t'} &= V^*\tilde{\psi}\end{aligned}\right\} \tag{5.6.26}$$

where

$$U^* = \begin{pmatrix}-\mathrm{i}\lambda^* & \lambda^*\bar{q}\\ \lambda^*\bar{r} & i\lambda^*\end{pmatrix}, \tag{5.6.27}$$

$$V^* = \begin{pmatrix}A^* & B^*\\ C^* & -A^*\end{pmatrix}, \tag{5.6.28}$$

with $\lambda^* = \mathrm{i}\lambda$ and A^*, B^*, C^* determined by the successive transformations G, R and T.

The link between the AKNS and WKI inverse scattering schemes may be represented as in Figure 5.4.

The above procedure may be used to link the pair of coupled nonlinear evolution equations

$$q_t + \alpha(q_{xxx} - 6qrq_x) - \beta\mathrm{i}(q_{xx} - 2q^2r) = 0, \tag{5.6.29}$$

$$r_t + \alpha(r_{xxx} - 6qrr_x) + \beta\mathrm{i}(r_{xx} - 2qr^2) = 0, \tag{5.6.30}$$

of the AKNS system with the coupled nonlinear equations

$$\bar{q}_{t'} + \alpha\left[\frac{\bar{q}_{x'}}{(1-\bar{r}\,\bar{q})^{3/2}}\right]_{x'x'} - \beta\mathrm{i}\left[\frac{\bar{q}}{(1-\bar{r}\,\bar{q})^{1/2}}\right]_{x'x'} = 0, \tag{5.6.31}$$

$$\bar{r}_{t'} + \alpha\left[\frac{\bar{r}_{x'}}{(1-\bar{r}\,\bar{q})^{3/2}}\right]_{x'x'} + \beta\mathrm{i}\left[\frac{\bar{r}}{(1-\bar{r}\,\bar{q})^{1/2}}\right]_{x'x'} = 0, \tag{5.6.32}$$

of the WKI system [12].

The specialisation $\bar{r} = -1$ together with the substitution $p = 1 + \bar{q}$ in (5.6.31) and (5.6.32) produces

$$p_{t'} - 2\alpha(p^{-1/2})_{x'x'x'} - \beta i[p^{1/2} - p^{-1/2}]_{x'x'} = 0 \qquad (5.6.33)$$

which reduces, in the case $\beta = 0$, to the Harry Dym equation.

Introduction of the reciprocal transformation

$$\left.\begin{aligned} d\bar{x} &= p dx' + [2\alpha(p^{-1/2})_{x'x'} + \beta i[p^{1/2} - p^{-1/2}]_{x'}]dt' \\ \bar{t} &= t' \end{aligned}\right\} \qquad (5.6.34)$$

produces the reciprocally associated equation

$$\bar{p}_{\bar{t}} - 2\alpha(\bar{p}^{-1/2})_{\bar{x}\bar{x}\bar{x}} - \beta i[\bar{p}^{1/2} - \bar{p}^{-1/2}]_{\bar{x}\bar{x}} = 0 \qquad (5.6.35)$$

where

$$\bar{p} = \frac{1}{p}. \qquad (5.6.36)$$

Thus, the generalised Harry Dym equation (5.6.33) is seen to be *invariant* under the reciprocal transformation given by (5.6.34). The subject of invariance under reciprocal transformations has been recently investigated by Kingston, Rogers and Woodall [16]. In particular, it will be shown that such reciprocal invariance extends to the Dym hierarchy and is linked to an auto-Bäcklund transformation for the Korteweg–de Vries hierarchy [28].

The specialisation $\bar{r} = -\bar{q}$, $\beta = 0$ in (5.6.31) and (5.6.32) produces the nonlinear evolution equation

$$\bar{q}_{t'} + \alpha\left[\frac{\bar{q}_{x'}}{(1 + \bar{q}^2)^{3/2}}\right]_{x'x'} = 0. \qquad (5.6.37)$$

If we set $\alpha = \text{sign}\,(ds/dx')$, $\bar{q} = y_{x'}$ in (5.6.37) then we obtain the equation used by Konno and Jeffrey [17] as a model for the propagation of waves along a stretched rope. In that context, y' represents the displacement normal to the undeformed axis $y' = 0$ of the rope, t' denotes time, and s measures arc length along the rope. Konno and Jeffrey obtained N-loop soliton solutions via the inverse scattering formalism. Interaction of two loop solitons is depicted in Figure 5.5. It is of interest to note the novel feature wherein the smaller loop soliton travels faster than the larger.

Introduction of the reciprocal transformation

$$\left.\begin{aligned} d\bar{x} &= \bar{q} dx' - \alpha\left[\frac{\bar{q}_{x'}}{(1 + \bar{q}^2)^{3/2}}\right]_{x'} dt' \\ \bar{t} &= t' \end{aligned}\right\} \qquad (5.6.38)$$

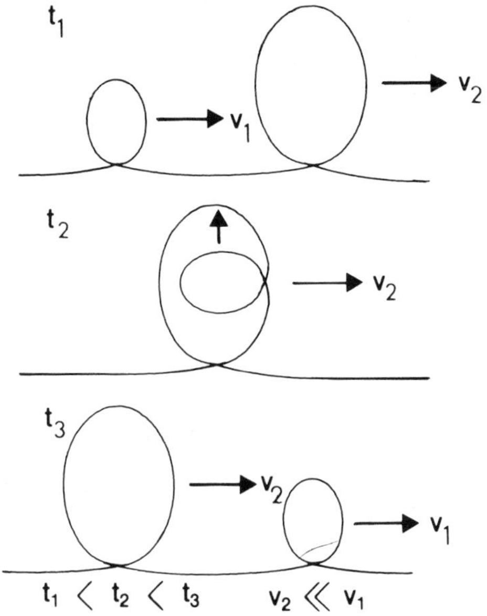

Figure 5.5 The interaction of two loop solitons [13]

leads to the reciprocally associated equation

$$\bar{q}'_{t'} + \alpha\left[\frac{\bar{q}'_{x'}}{(1 + \bar{q}^2)^{3/2}}\right]_{\bar{x}\bar{x}} = 0 \qquad (5.6.39)$$

where

$$\bar{q}' = \frac{1}{q}. \qquad (5.6.40)$$

Accordingly, it is seen that the nonlinear evolution equation (5.6.37) is *invariant* under the reciprocal transformation given by (5.6.38) and (5.6.40). In fact, such invariance may be shown to extend to a hierarchy of which (5.6.37) is the base member. This result is a consequence of an invariance property of the modified Korteweg–de Vries hierarchy [31].

To conclude this section, we show that reciprocal invariance extends to the Dym hierarchy. The spatial part of the generic auto-Bäcklund transformation for the Korteweg–de Vries hierarchy is then derived solely by consideration of reciprocal and Cole–Hopf-type transformations. The Miura transformation between the Korteweg–de Vries and modified Korteweg–de Vries hierarchies is an immediate consequence of the analysis.

The Dym hierarchy is given by [5]:

$$\rho_t = \rho^{-1}\{-\mathbf{D}^3\ \mathbf{r}\ \mathbf{I}\ \mathbf{r}\}^n \rho\rho_x, \quad n = 1, 2, \ldots \tag{5.6.41}$$

where the operators **D** and **I** are defined by

$$\mathbf{D}\phi(x, t) \equiv \phi_x, \tag{5.6.42}$$

$$\mathbf{I}\phi(x, t) \equiv \int_x^\infty \mathrm{d}y\phi(y, t), \tag{5.6.43}$$

while **r** denotes the multiplicative operator given by

$$\mathbf{r}\phi(x, t) \equiv r\phi(x, t), \tag{5.6.44}$$

with $r \equiv \rho^{-1} \neq 0$. It is assumed that $1 - \rho$ and $1 - r$ are bona fide potentials (see [5]). In particular, it is required that $\lim_{x\to\pm\infty} \rho = 1$.

It is noted that the hierarchy (5.6.41) may be rewritten as

$$\rho_t + \varepsilon_{n,x} = 0, \quad n = 1, 2, \ldots \tag{5.6.45}$$

where

$$\varepsilon_n = -\int_x^\infty \rho^{-1}[\rho^{-1}\varepsilon_{n-1}]_{xxx}\,\mathrm{d}x, \quad n = 1, 2, \ldots, \ \varepsilon_0 = 1, \tag{5.6.46}$$

so that

$$(\rho^2)_t + 2(\rho^{-1}\varepsilon_{n-1})_{xxx} = 0, \quad n = 1, 2, \ldots \tag{5.6.47}$$

It is noted that the case $n = 1$ produces the Harry Dym equation.

Under the reciprocal transformation

$$\left.\begin{aligned} \mathrm{d}x' &= \rho^2\mathrm{d}x - 2(\rho^{-1}\varepsilon_{n-1})_{xx}\mathrm{d}t, \ t' = t \\ \rho' &= \frac{1}{\rho} \end{aligned}\right\} R' \tag{5.6.48}$$

equation (5.6.47) becomes

$$(\rho'^2)_{t'} - 2[(\rho^{-1}\varepsilon_{n-1})_x]_{x'x'} = 0, \ n = 1, 2, \ldots. \tag{5.6.49}$$

In order to establish invariance of the hierarchy (5.6.47) under R', it is required to show that if we introduce ε_n', $n = 1, 2, \ldots$ according to

$$(\rho'^{-1}\varepsilon_n')_{x'} + (\rho^{-1}\varepsilon_n)_x = 0, \ n = 1, 2, \ldots \tag{5.6.50}$$

then

$$\varepsilon_n' = -\int_{x'}^\infty \rho'^{-1}[\rho'^{-1}\varepsilon_{n-1}']_{x'x'x'}\,\mathrm{d}x', \ n = 1, 2, \ldots, \ \varepsilon_0' = 1. \tag{5.6.51}$$

It is recalled that invariance of the Harry Dym equation corresponding to

$n = 1$ in (5.6.47) has already been established so that $\varepsilon_0' = 1$. Now, the relation (5.6.50) yields

$$\varepsilon_n' = \rho^{-1}\int_x^{\infty} \rho^2(\rho^{-1}\varepsilon_n)_x \, dx, \; n = 1, 2, \ldots \tag{5.6.52}$$

whence, on use of (5.2.46),

$$\varepsilon_{n,x'}' - \rho'^{-1}[\rho'^{-1}\varepsilon_{n-1}']_{x'x'x'} = \rho^{-1}(\rho^{-1})_x[\varepsilon_n' - \varepsilon_n + 2\rho^{-1}(\rho^{-1}\varepsilon_{n-1})_{xx}]. \tag{5.6.53}$$

However, (5.6.51) shows that

$$\begin{aligned} \varepsilon_n' &= -\rho^{-1}\left[\rho\varepsilon_n + 2\int_x^{\infty} \rho_x\varepsilon_n dx\right] = \varepsilon_n + 2\rho^{-1}\int_x^{\infty} \rho\varepsilon_{n,x}\, dx \\ &= \varepsilon_n + 2\rho^{-1}\int_x^{\infty} [\rho^{-1}\varepsilon_{n-1}]_{xxx}\, dx = \varepsilon_n - 2\rho^{-1}(\rho^{-1}\varepsilon_{n-1})_{xx} \end{aligned}$$

whence

$$\varepsilon_n' - \varepsilon_n + 2\rho^{-1}(\rho^{-1}\varepsilon_{n-1})_{xx} = 0. \tag{5.6.54}$$

Accordingly, equation (5.6.53) shows that

$$\varepsilon_{n,x'}' - \rho'^{-1}[\rho'^{-1}\varepsilon_{n-1}']_{x'x'x'} = 0, \; n = 1, 2, \ldots. \tag{5.6.55}$$

so that the relation (5.6.51) is established and invariance of the Dym hierarchy (5.6.41) under the reciprocal transformation (5.6.48) follows.

The above invariance of the Dym hierarchy in conjunction with another reciprocal transformation and a Cole–Hopf-type transformation may now be used to induce an auto-Bäcklund transformation for the Korteweg–de Vries hierarchy.

Thus, we now consider the action of another kind of reciprocal transformation on the Dym hierarchy based on the form (5.6.45), namely

$$\left.\begin{aligned} d\bar{x} &= \rho dx - \varepsilon_n dt, \; d\bar{t} = dt \\ \bar{\rho} &= \frac{1}{\rho} \end{aligned}\right\} \bar{R}. \tag{5.6.56}$$

Under $\bar{R}$, (5.6.45) becomes

$$\bar{\rho}_{\bar{t}} + \bar{\varepsilon}_{n,\bar{x}} = 0, \quad n = 0, 1, 2, \ldots \tag{5.6.57}$$

where

$$\bar{\varepsilon}_n = -\bar{\rho}\varepsilon_n \tag{5.6.58}$$

and equation (5.6.46) shows that the $\bar{\varepsilon}_n$ are given by the relations

$$\left.\begin{aligned} &\bar{\rho}^{-1}\frac{\partial}{\partial\bar{x}}[\bar{\rho}^{-1}\bar{\varepsilon}_n] = \frac{\partial}{\partial\bar{x}}\left[\bar{\rho}^{-1}\frac{\partial}{\partial\bar{x}}\left\{\bar{\rho}^{-1}\frac{\partial\bar{\varepsilon}_{n-1}}{\partial\bar{x}}\right\}\right] \\ &\bar{\varepsilon}_0 = -\bar{\rho}. \end{aligned}\right\} \tag{5.6.59}$$

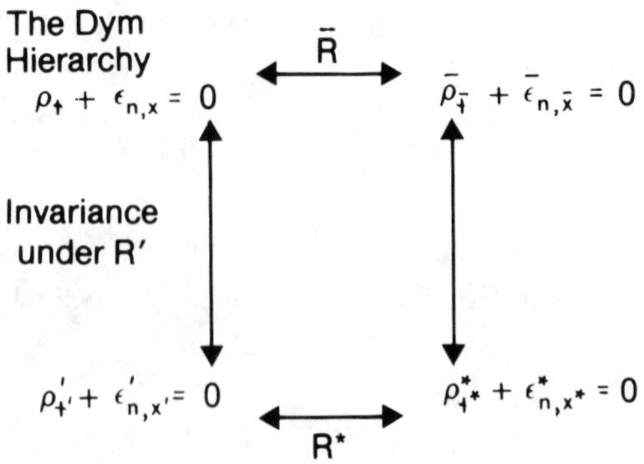

Figure 5.6 The reciprocal transformations

In a similar manner, under the reciprocal transformation

$$\left.\begin{aligned} dx^* &= \rho' dx' - \varepsilon'_n dt', \quad dt^* = dt' \\ \rho^* &= \frac{1}{\rho'} \end{aligned}\right\} \tag{5.6.60}$$

the Dym hierarchy

$$\rho'_{t'} + \varepsilon'_{n,x'} = 0, \quad n = 0, 1, 2, \ldots \tag{5.6.61}$$

reciprocally associated to the original Dym hierarchy (5.6.45) through R' is taken to

$$\rho^*_{t^*} + \varepsilon^*_{n,x^*} = 0, \quad n = 0, 1, 2, \ldots \tag{5.6.62}$$

where

$$\varepsilon^*_n = -\rho^* \varepsilon'_n \tag{5.6.63}$$

and the ε^*_n are generated by relations analogous to (5.6.59), namely

$$\left.\begin{aligned} \rho^{*-1} \frac{\partial}{\partial x^*} [\rho^{*-1} \varepsilon^*_n] &= \frac{\partial}{\partial x^*}\left[\rho^{*-1} \frac{\partial}{\partial x^*}\left\{\rho^{*-1} \frac{\partial \varepsilon^*_{n-1}}{\partial x^*}\right\}\right] \\ \varepsilon^*_0 &= -\rho^*. \end{aligned}\right\} \tag{5.6.64}$$

The situation may be represented as in Figure 5.6.

It is readily shown [28] that

$$d\bar{x} = dx^*, \quad d\bar{t} = dt^*$$

whence invariance of the Dym hierarchy under R' corresponds via $\bar{R}$ and R^* to invariance of the hierarchy (5.6.57) generated by the recurrence relations (5.6.59) under the simple inversion

$$\rho^* = \frac{1}{\rho}. \tag{5.6.65}$$

It will subsequently be seen that this invariance corresponds to an important invariance of a singularity manifold equation which arises in the Weiss version of the Painlevé test.

Introduction of the *first-order* Cole–Hopf transformation

$$u^{(1)} = \frac{(\bar{\rho}^{-1/2})_{\bar{x}}}{\bar{\rho}^{-1/2}} \tag{5.6.66}$$

into (5.6.57) produces the modified Korteweg–de Vries hierarchy

$$u^{(1)}_{\bar{t}} = \mathbf{M}^n\, u^{(1)}_{\bar{x}},\ n = 1, 2, \ldots \tag{5.6.67}$$

where **M** is the integro-differential operator defined by

$$\mathbf{M} \equiv \frac{\partial^2}{\partial \bar{x}^2} - 4u^{(1)^2} + 4u^{(1)}_{\bar{x}} \int_{\bar{x}}^{\infty} \mathrm{d}y\ u^{(1)}. \tag{5.6.68}$$

Similarly, introduction of the *first-order* Cole–Hopf transformation

$$u^{*(1)} = \frac{(\rho^{*-1/2})_{x^*}}{\rho^{*-1/2}} \tag{5.6.69}$$

into equation (5.6.62) produces another modified Korteweg–de Vries hierarchy.

The relations (5.6.66) and (5.6.69) together with (5.6.65) show that the invariance of the Dym hierarchy under the reciprocal transformation R' is associated with the invariance of the modified Korteweg–de Vries hierarchy under the simple transformation $u^{(1)} = -u^{*(1)}$.

If we next introduce the *second-order* Cole–Hopf transformation

$$u^{(2)} = \frac{(\bar{\rho}^{-1/2})_{\bar{x}\bar{x}}}{\bar{\rho}^{-1/2}} \tag{5.6.70}$$

into (5.6.57) then it may be shown that the Korteweg–de Vries hierarchy

$$u^{(2)}_{\bar{t}} = \mathbf{K}^n\, u^{(2)}_{\bar{x}}, \quad n = 1, 2, \ldots \tag{5.6.71}$$

where **K** is the integro-differential operator defined by

$$\mathbf{K} \equiv \frac{\partial^2}{\partial \bar{x}^2} - 4u^{(2)} + 2u^{(2)}_{\bar{x}} \int_{\bar{x}}^{\infty} \mathrm{d}y \tag{5.6.72}$$

results.

It is noted that relations (5.6.66) and (5.6.70) combine to produce the Miura transformation

$$u^{(2)} = u^{(1)}_{\bar{x}} + u^{(1)^2} \tag{5.6.73}$$

which links the modified Korteweg–de Vries and Korteweg–de Vries hierarchies. Thus, the Miura transformation may be regarded as a composition of *first-order* and *second-order* Cole–Hopf transformations.

An auto-Bäcklund transformation for the Korteweg–de Vries hierarchy is now readily obtained.

Thus, introduction of the *second-order* Cole–Hopf transformation

$$u^{*(2)} = \frac{(\rho^{*-1/2})_{x^*x^*}}{\rho^{*-1/2}} \tag{5.6.74}$$

into equation (5.6.62) produces another Korteweg–de Vries hierarchy

$$u^{*(2)}_{t^*} = \mathbf{K}^{*n}\, u^{*(2)}_{x^*}, \tag{5.6.75}$$

with $d\bar{x} = dx^*$, $d\bar{t} = dt^*$.

If we now set $z = \rho^{1/2}$ then (5.6.70) and (5.6.74) together with (5.6.65) show that

$$u^{(2)} = \frac{z_{\bar{x}\bar{x}}}{z}, \tag{5.6.76}$$

$$u^{*(2)} = \frac{(1/z)_{\bar{x}\bar{x}}}{1/z}. \tag{5.6.77}$$

On introduction of the potential representations

$$\phi = \int_{-\infty}^{\bar{x}} u^{(2)}(\sigma, t)d\sigma, \tag{5.6.78}$$

$$\phi^* = \int_{-\infty}^{\bar{x}} u^{*(2)}(\sigma, t)d\sigma \tag{5.6.79}$$

the relation

$$\left(z \cdot \frac{1}{z}\right)_{\bar{x}\bar{x}} = 0 \tag{5.6.80}$$

produces

$$\phi_{\bar{x}} + \phi^*_{\bar{x}} = 2\left(\frac{z_{\bar{x}}}{z}\right)^2 = 2\left[\frac{\phi - \phi^*}{2}\right]^2. \tag{5.6.81}$$

This constitutes the spatial part of an auto-Bäcklund transformation for the Korteweg–de Vries hierarchy.

5.7 The Painlevé test and Bäcklund Transformations

A partial differential equation is said to have the Painlevé property according to the work of Weiss if it admits power series solutions of the form

$$u(x, t) = \phi^{\alpha} \sum_{j=0}^{\infty} u_j \phi^j, \; u_0 \neq 0 \tag{5.7.1}$$

where

$$M = \{(x, t) : \phi(x, t) = 0\} \tag{5.7.2}$$

is the singularity manifold of the equation and $u_j(x, t)$, $\phi(x, t)$ are analytic functions in the neighbourhood of the manifold M (see Chapter 16). It is required [37] that M be non-characteristic and commonly that α be a negative integer.

In what follows, we shall indicate how the Painlevé test links with the preceding work and, in particular, how it connects with the construction of auto-Bäcklund transformations. Attention is restricted to the Korteweg–de Vries equation in the form

$$u_t + uu_x + \delta u_{xxx} = 0 \tag{5.7.3}$$

adopted by Weiss [36] in his application of the Painlevé test.

The power series expansion (5.7.1) yields

$$u_t = \sum_{j=0}^{\infty} \{(\alpha + j)u_j\phi_t\phi^{\alpha+j-1} + u_{jt}\phi^{\alpha+j}\}, \tag{5.7.4}$$

$$u_x = \sum_{j=0}^{\infty} \{(\alpha + j)u_j\phi_x\phi^{\alpha+j-1} + u_{jx}\phi^{\alpha+j}\}, \tag{5.7.5}$$

$$\begin{aligned} u_{xx} = \sum_{j=0}^{\infty} \{&(\alpha + j)(\alpha + j - 1)u_j\phi_x^2\phi^{\alpha+j-2} \\ &+ (\alpha + j)[u_j\phi_{xx} + 2u_{j,x}\phi_x]\phi^{\alpha+j-1} + u_{jxx}\phi^{\alpha+j}\}, \end{aligned} \tag{5.7.6}$$

$$\begin{aligned} u_{xxx} = \sum_{j=0}^{\infty} \{&(\alpha + j)(\alpha + j - 1)(\alpha + j - 2)u_j\phi_x^3\phi^{\alpha+j-3} \\ &+ (\alpha + j)(\alpha + j - 1)[3u_j\phi_x\phi_{xx} + 3u_{jx}\phi_x^2]\phi^{\alpha+j-2} \\ &+ (\alpha + j)[3u_{jx}\phi_{xx} + 3u_{j,xx}\phi_x + u_j\phi_{xxx}]\phi^{\alpha+j-1} \\ &+ u_{jxxx}\phi^{\alpha+j}\}. \end{aligned} \tag{5.7.7}$$

Thus, on substitution in (5.7.3) we obtain

$$\begin{aligned} \sum_{j=0}^{\infty} \{&[(\alpha + j)u_j\phi_t\phi^{\alpha+j-1} + u_{jt}\phi^{\alpha+j}] \\ &+ \left(\sum_{k=0}^{\infty} u_k\phi^{\alpha+k}\right)[(\alpha + j)u_j\phi_x\phi^{\alpha+j-1} + u_{jx}\phi^{\alpha+j}] \\ &+ \delta[(\alpha + j)(\alpha + j - 1)(\alpha + j - 2)u_j\phi_x^3\phi^{\alpha+j-3} \\ &+ (\alpha + j)(\alpha + j - 1)\{3u_j\phi_x\phi_{xx} + 3u_{jx}\phi_x^2\}\phi^{\alpha+j-2} \\ &+ (\alpha + j)\{3u_{jx}\phi_{xx} + 3u_{j,xx}\phi_x + u_j\phi_{xxx}\}\phi^{\alpha+j-1} \\ &+ u_{jxxx}\phi^{\alpha+j})\} = 0. \end{aligned} \tag{5.7.8}$$

The lowest powers of ϕ originating in the linear and nonlinear terms respectively are those in $\phi^{\alpha-3}$ and $\phi^{2\alpha-1}$. The requirement that these most singular terms balance yields $\alpha = -2$. We now set the coefficients of the powers of ϕ to be successively zero, starting with the lowest power ϕ^{-5} as follows:

$$\phi^{-5}: \quad u_0 = -12\delta\phi_x^2, \; \phi_x \neq 0$$

$$\phi^{-4}: \quad -3u_0u_1\phi_x + u_0u_{0x} - 6\delta u_1\phi_x^3 + 18\delta[u_0\phi_x\phi_{xx} + u_{0x}\phi_x^2] = 0, \tag{5.7.9}$$

so that, on insertion of the expression (5.7.9) for u_0,

$$u_1 = 12\delta\phi_{xx}. \tag{5.7.10}$$

$$\phi^{-3}: \quad \phi_x\phi_t + \phi_x^2u_2 + 4\delta\phi_x\phi_{xxx} - 3\delta\phi_{xx}^2 = 0. \tag{5.7.11}$$

$$\phi^{-2}: \quad \phi_{xt} + \phi_{xx}u_2 - \phi_x^2u_3 + \delta\phi_{xxxx} = 0. \tag{5.7.12}$$

$$\phi^{-1}: \quad u_{1t} + u_1u_{2x} + u_0u_{3x} + u_2u_{1x} + u_3u_{0x} + \delta u_{1xxx} = 0, \tag{5.7.13}$$

whence we obtain the compatibility condition

$$\frac{\partial}{\partial x}\{\phi_{xt} + \phi_{xx}u_2 - \phi_x^2u_3 + \delta\phi_{xxxx}\} = 0 \tag{5.7.14}$$

which is identically satisfied by virtue of (5.7.13).

$$\begin{aligned}\phi^0: \quad & u_{2t} + u_3\phi_t + u_0(u_4x + 3u_5\phi_x) + u_1(u_{3x} + 2u_4\phi_x)\\ & + u_2(u_{2x} + u_3\phi_x) + u_3u_{1x} + u_4(u_{0x} - u_1\phi_x) + u_5(-2u_0\phi_x)\\ & + \delta[6u_5\phi_x^3 + u_{2xxx} + u_3\phi_{xxx} + 3u_{3x}\phi_{xx} + 3u_{3xx}\phi_x +\\ & + 6u_4\phi_x\phi_{xx} + 6u_4\phi_x^2] = 0.\end{aligned} \tag{5.7.15}$$

$$\begin{aligned}\phi^1: \quad & u_{3t} + 2u_4\phi_t + u_0u_{5x} + u_1(u_{4x} + 3u_5\phi_x)\\ & + u_2(u_{3x} + 2u_4\phi_x) + u_3(u_{2x} + u_3\phi_x) + u_4u_{1x}\\ & + u_5(u_{0x} - u_1\phi_x) + \delta[u_{3xxx} + 2(u_4\phi_{xxx} + 3u_{4x}\phi_{xx} + 3u_{4xx}\phi_x)\\ & + 18(u_5\phi_x\phi_{xx} + u_{5x}\phi_x^2) = 0.\end{aligned} \tag{5.7.16}$$

It is noted that in the above (5.7.10)–(5.7.12) determine u_1, u_2, u_3 recursively in terms of u_0 as given by (5.7.9). However, at the next stage, in (5.7.13), the terms in u_4 have cancelled so that u_4 does not appear and we obtain the compatibility condition (5.7.14). A value of j such as $j = 4$ for which u_j is not given recursively in terms of its predecessors u_k, $k < j$ is called a *resonance*. At the following stage, (5.7.15) gives a recurrence relation which gives u_5 in terms of u_4 which is arbitrary. On the other hand, the terms in u_6 have cancelled in (5.7.16) so that $j = 6$ is also a resonance and (5.7.16) produces a compatibility condition which may be shown to be identically satisfied. Accordingly, u_6 is arbitrary.

The expansion (5.7.1) with $\alpha = -2$ is now specialised by setting the resonance functions

$$u_4 = u_6 = 0. \tag{5.7.17}$$

If furthermore, it is required that

$$u_3 = 0 \tag{5.7.18}$$

then (5.7.15) yields

$$u_{2t} + u_2 u_{2x} + \delta u_{2xxx} - 6\delta\phi_x^2 u_5 = 0. \tag{5.7.19}$$

Hence,

$$u_5 = 0 \tag{5.7.20}$$

if and only if

$$u_{2t} + u_2 u_{xx} + \delta u_{2xxx} = 0. \tag{5.7.21}$$

In fact, it is readily shown that $u_j = 0$, $j \geq 7$ if (5.7.17) and (5.7.18) hold along with (5.7.20).

Accordingly the Korteweg–de Vries equation (5.7.3) admits a solution of the form (5.7.1) and hence has the Painlevé property. Specifically, it has been shown to admit the solution

$$\begin{aligned} u &= \phi^{-2}[u_0 + u_1\phi + u_2\phi^2] \\ &= 12\delta \frac{\partial^2}{\partial x^2} \ln\phi + u_2 \end{aligned} \tag{5.7.22}$$

where, by virtue of (5.7.21), u_2 is likewise a solution of the Korteweg–de Vries equation (5.7.3). Moreover, elimination of u_2 between (5.7.11) and (5.7.12) produces

$$\frac{\partial}{\partial x}\left[\frac{\phi_t}{\phi_x} + \delta\left(\frac{\partial}{\partial x}\left(\frac{\phi_{xx}}{\phi_x}\right) - \frac{1}{2}\left(\frac{\phi_{xx}}{\phi_x}\right)^2\right)\right] = 0$$

whence we obtain the singularity manifold equation [36]:

$$\frac{\phi_t}{\phi_x} + \delta\{\phi; x\} = \lambda^2 \tag{5.7.23}$$

where

$$\{\phi; x\} = \frac{\partial}{\partial x}\left(\frac{\phi_{xx}}{\phi_x}\right) - \frac{1}{2}\left(\frac{\phi_{xx}}{\phi_x}\right)^2 \tag{5.7.24}$$

is the *Schwarzian derivative* of ϕ [11]. It is noted that introduction of the Galilean transformation

$$u \to u + \lambda^2,\ x \to x + \lambda^2 t,\ t \to t, \tag{5.7.25}$$

leaves the Korteweg–de Vries equation (5.7.3) invariant but reduces (5.7.23) to

$$\frac{\phi_t}{\phi_x} + \delta\{\phi; x\} = 0. \tag{5.7.26}$$

The above work is readily connected with that of the preceding section on reciprocal transformations. Thus, if we set $\bar{\rho} = \phi_{\bar{x}}$ in the notation of Figure 5.6 then, in the case $n = 1$, (5.6.70) shows that

$$u^{(2)} = \frac{(\bar{\rho}^{-1/2})_{\bar{x}\bar{x}}}{\bar{\rho}^{-1/2}} = -\frac{1}{2}\{\phi; \bar{x}\} \tag{5.7.27}$$

is a solution of the Korteweg–de Vries equation

$$u^{(2)}_{\bar{t}} + 6u^{(2)}u^{(2)}_{\bar{x}} - u^{(2)}_{\bar{x}\bar{x}\bar{x}} = 0. \tag{5.7.28}$$

If we set $u = -6\delta u^{(2)}$, $\bar{x} = x$, $\bar{t} = -\delta t$ then (5.7.27) delivers the solution

$$u = 3\delta\{\phi; x\} \tag{5.7.29}$$

of (5.7.3).

The invariance of (5.6.57) under $\bar{\rho} \to 1/\bar{\rho}$ shows that

$$u^{*(2)} = \frac{(\bar{\rho}^{1/2})_{\bar{x}\bar{x}}}{\bar{\rho}^{1/2}} = \frac{1}{2}\left[\{\phi; \bar{x}\} + \left(\frac{\phi_{\bar{x}\bar{x}}}{\phi_{\bar{x}}}\right)^2\right]$$

is also a solution of (5.7.28). Thus, if we set $u_2 = -6\delta u^{*(2)}$, $\bar{x} = x$, $\bar{t} = -\delta t$ then it is seen that

$$u_2 = -3\delta\left[\{\phi; x\} + \left(\frac{\phi_{xx}}{\phi_x}\right)^2\right] \tag{5.7.30}$$

$$= -\delta\left[-\{\phi; x\} + 4\left(\frac{\phi_{xxx}}{\phi_x} - \frac{3}{4}\left(\frac{\phi_{xx}}{\phi_x}\right)^2\right)\right] \tag{5.7.31}$$

is also a solution of (5.7.3).

Now, if we insert $\rho = \phi_{\bar{x}}$ in the reciprocal associate of the Harry Dym equation under $\bar{R}$, namely

$$\bar{\rho}_{\bar{t}} + \bar{\varepsilon}_{1,\bar{x}} = 0$$

where

$$\bar{\varepsilon}_1 = -\frac{\varepsilon_1}{\rho} = -\bar{\rho}_{\bar{x}\bar{x}} + \frac{3}{2}\frac{\bar{\rho}^2_{\bar{x}}}{\bar{\rho}},$$

then on application of a Galilean transformation of the type (5.7.25) and the change of variable $\bar{x} = x$, $\bar{t} = -\delta t$ the singularity manifold equation (5.7.26) results. Use of (5.7.26) in (5.7.31) yields

$$\phi_x\phi_t + \phi_x^2 u_2 + 4\delta\phi_x\phi_{xxx} - 3\delta\phi_{xx}^2 = 0, \tag{5.7.32}$$

which is precisely the relation (5.7.11) obtained via the Painlevé test. Moreover, this relation may be rewritten as

$$\phi_t + \phi_x u_2 + 4\delta\phi_{xxx} - 3\delta\phi_{xx}^2/\phi_x = 0,$$

whence

$$\phi_{xt} + \phi_{xx}u_2 + \delta\phi_{xxxx} + \left[\phi_x u_{2,x} + 3\delta\phi_{xxxx} - \frac{6\delta\phi_{xxx}\phi_{xx}}{\phi_x} + 3\delta\left(\frac{\phi_{xx}}{\phi_x}\right)^2\phi_{xx}\right] = 0. \tag{5.7.33}$$

But, (5.7.30) shows that

$$\phi_x u_{2,x} + 3\delta\phi_{xxxx} - \frac{6\delta\phi_{xxx}\phi_{xx}}{\phi_x} + 3\delta\left(\frac{\phi_{xx}}{\phi_x}\right)^2\phi_{xx} = 0,$$

whence (5.7.33) yields

$$\phi_{xt} + \phi_{xx}u_2 + \delta\phi_{xxxx} = 0, \tag{5.7.34}$$

which is the relation (5.7.12) with $u_3 = 0$ as derived via the Painlevé test.

The connection with the Lax equations is readily made. Thus, the Lax equations for the Korteweg–de Vries equation

$$q_t + 6qq_x + q_{xxx} = 0 \tag{5.7.35}$$

are, from (5.5.6), (5.5.8), (5.5.9) with the specialisations (5.5.11)

$$\begin{pmatrix} \dfrac{\partial}{\partial x} & -q \\ -1 & -\dfrac{\partial}{\partial x} \end{pmatrix}\begin{pmatrix}\psi_1\\ \psi_2\end{pmatrix} = \lambda\begin{pmatrix}\psi_1\\ \psi_2\end{pmatrix}, \tag{5.7.36}$$

$$\begin{pmatrix} -4\lambda^3 - 2\lambda q - q_x & -q_{xx} - 2\lambda q_x - 4\lambda^2 q - 2q^2 \\ 4\lambda^2 + 2q & -[-4\lambda^3 - 2\lambda q - q_x] \end{pmatrix}\begin{pmatrix}\psi_1\\ \psi_2\end{pmatrix} = \begin{pmatrix}\psi_1\\ \psi_2\end{pmatrix}_t. \tag{5.7.37}$$

Elimination of ψ_1 between (5.7.36) and (5.7.37) together with use of the invariance of (5.7.35) under the Galilean transformation

$$q \to q_2 = q - \lambda^2,\ x \to x - 6\lambda^2 t,\ t \to t, \tag{5.7.38}$$

yields

$$\psi_{2,xx} + q_2\psi_2 = 0, \tag{5.7.39}$$

$$\psi_{2,t} - q_{2,x}\psi_2 + 2q_2\psi_{2,x} = 0. \tag{5.7.40}$$

On application of the transformation $q_2 \to (6\delta)^{-1}u_2$, $x \to x$, $t \to \delta^{-1}t$, the pair of equations (5.7.39), (5.7.40) become

$$6\delta\psi_{2,xx} + u_2\psi_2 = 0, \tag{5.7.41}$$

$$6\psi_{2,t} - u_{2,x}\psi_2 + 2u_2\psi_{2,x} = 0, \tag{5.7.42}$$

where u_2 satisfies the Korteweg–de Vries equation (5.7.3). These relations are retrieved from the Painlevé test approach by setting

$$\bar{\rho} = \phi_x = \psi_2^2 \tag{5.7.43}$$

successively in the relation

$$u_2 = -6\delta \frac{(\bar{\rho}^{1/2})_{xx}}{\bar{\rho}^{1/2}}$$

and (5.7.34).

It remains for us to derive the relation (5.7.22) from the reciprocal transformation approach.

Invariance of (5.7.26) under the simple Möbius transformation $\phi \rightarrow 1/\phi$ shows that

$$u = -3\delta \left[\{\phi; x\} + \left(\frac{(1/\phi)_{xx}}{(1/\phi)_x} \right)^2 \right] \tag{5.7.44}$$

is also a solution of (5.7.3). Combination of (5.7.30) and (5.7.44) produces the relation (5.7.22).

Thus, the relations derived by the Painlevé test by Weiss [36] for the Korteweg–de Vries equation may be alternatively derived via the reciprocal transformation approach. Reciprocal transformations have recently been extended to 2+1-dimensions [26]. Connection with the Painlevé test for the Kadomtsev–Petviashvili equation and an associated 2+1-dimensional Harry Dym equation has been established [27].

Acknowledgement

Thanks are due to Dr Sandra Carillo of the University of Rome for background material on the Painlevé method.

References

[1] Ablowitz, M. J., Kaup, D. J., Newell, A. C. and Segur, H. Nonlinear evolution equations of physical significance, *Physical Review Letters* **31**, 125–7 (1973).

[2] Bäcklund, A. V. Ueber Flächentransformationen, *Mathematische Annalen* **9**, 297–320 (1876).

[3] Bäcklund, A. V. Zür theorie der partiellen differentialgleichungen erster ordnung, *Mathematische Annalen* **17**, 285–328 (1880).

[4] Bäcklund, A. V. Zür theorie der Flächentransformationen, *Mathematische Annalen* **19**, 387–422 (1882).

[5] Calogero, F. and Degasperis, A. *Spectral Transform and Solitons I*, North-Holland, Amsterdam (1982).

[6] Chern, S. S. and Tenenblat, K. Pseudospherical surfaces and evolution equations, *Studies in Applied Mathematics* **74**, 55–83 (1986).

[7] Crum, M. M. Associated Sturm–Liouville systems, *Oxford Journal of Mathematics* **6**, 121–7 (1955).

[8] Eisenhart, L. P. *A Treatise on the Differential Geometry of Curves and Surfaces*, Dover, New York (1960).

[9] Fokas, A. S. and Yortsos, Y. C. On the exactly solvable equation $s_t = [(\beta s + \gamma)^{-2} s_x]_x + \alpha[\beta s + \gamma]^{-2} s_x$ occurring in two-phase flow in porous media, *Society for Industrial and Applied Mathematics Journal on Applied Mathematics* **42**, 318–42 (1982).

[10] Forsyth, A. R. *Theory of Differential Equations*, Vol. 6, Dover, New York (1959).

[11] Hille, E. *Analytic Function Theory*, Vol. 2, Dover, New York (1962).

[12] Ishimori, Y. A relationship between the Ablowitz–Kaup–Newell–Segur and Wadati–Konno–Ichikawa schemes of the inverse scattering method, *Journal of the Physical Society of Japan* **51**, 3036–41 (1982).

[13] Jeffrey, A. Equations of evolution and waves, *Wave Phenomena: Modern Theory and Applications*, C. Rogers and T. B. Moodie (eds), North-Holland, Amsterdam (1984).

[14] Kingston, J. G. and Rogers, C. Reciprocal Bäcklund transformations of conservation laws, *Physics Letters A* **92**, 261–4 (1982).

[15] Kingston, J. G. and Rogers, C. Bäcklund transformations for systems of conservation laws, *Quarterly of Applied Mathematics* **51**, 423–32 (1984).

[16] Kingston, J. G., Rogers, C. and Woodall, D. Reciprocal auto-Bäcklund transformations, *Journal of Physics A: Mathematical and General* **17**, L35–8 (1984).

[17] Konno, K. and Jeffrey, A. Some remarkable properties of two loop soliton solutions, *Journal of the Physical Society of Japan* **52**, 1–3 (1983).

[18] Konno, K. and Wadati, M. Simple derivation of Bäcklund transformations from Riccati form of inverse method, *Progress in Theoretical Physics* **53**, 1652–6 (1975).

[19] Lamb, G. L. Jr. Propagation of ultrashort optical pulses, *Physics Letters* **25**, 181–2 (1967).

[20] Lamb, G. L. Jr. Bäcklund transformations for certain nonlinear evolution equations, *Journal of Mathematical Physics* **15**, 2157–65 (1974).

[21] Leibbrandt, G. Soliton-like solutions of the elliptic sine–cosine equation by means of harmonic functions, *Journal of Mathematical Physics* **19**, 960–6 (1978).

[22] Loewner, C. A transformational theory of partial differential equations of gasdynamics, National Advisory council for Aeronautics, Technical Note 2065 (1950).

[23] McLaughlin, D. W. and Scott, A. C. A restricted Bäcklund transformation, *Journal of Mathematical Physics* **14**, 1817–28 (1973).

[24] Rogers, C. Application of a reciprocal transformation to a two phase Stefan problem, *Journal of Physics A: Mathematical and General* **18**, L105–9 (1985).

[25] Rogers, C. Linked Bäcklund transformation and application to nonlinear boundary value problems, *Journal of Mathematical Physics* **26**, 393–5 (1985).

[26] Rogers, C. Reciprocal Bäcklund transformations in 2+1 dimensions, *Journal of Physics A: Mathematical and General* **19**, L491–6 (1986).

[27] Rogers, C. The Harry Dym equation in 2+1 dimensions: a reciprocal link with the Kadomtsev–Petviashvili equation, *Physics Letters A* **120**, 15–18 (1987).

[28] Rogers, C. and Nucci, M. C. On reciprocal Bäcklund tranformations and the Korteweg–de Vries hierarchy, *Physica Scripta* **33**, 289–92 (1986).

[29] Rogers, C. and Ruggeri, T. A reciprocal Bäcklund transformation: application to a nonlinear hyperbolic model in heat conduction, *Lett. Il. Nuovo Cimento* **44**, 289–96 (1985).

[30] Rogers, C. and Shadwick, W. F. *Bäcklund Transformations and their Applications*, Academic Press, New York (1982).

[31] Rogers, C., Nucci, M. C. and Kingston, J. G. On reciprocal auto-Bäcklund transformations: application to a new nonlinear hierarchy, *Il Nuovo Cimento* **96**, 55–63 (1986).

[32] Seymour, B. R. and Varley, E. Exact solutions for large amplitude waves in dispersive and dissipative systems, *Studies in Applied Mathematics* **72**, 241–62 (1985).

[33] Wadati, M. and Sogo, K. Gauge tranformations in soliton theory, *Journal of the Physical Society of Japan* **52**, 394–8 (1983).

[34] Wadati, M., Konno, K. and Ichikawa, Y. H. A generalization of the inverse scattering method, *Journal of the Physical Society of Japan* **46**, 1965–6 (1979).

[35] Wahlquist, H. D. and Estabrook, F. B. Bäcklund transformations for solutions of the Korteweg–de Vries equation, *Physical Review Letters* **31**, 1386–90 (1973).

[36] Weiss, J. The Painlevé property for partial differential equations. II: Bäcklund transformation, Lax pairs, and the Schwarzian derivative, *Journal of Mathematical Physics* **24**, 1405–13 (1983).

[37] Weiss, J. Modified equations, rational solutions, and the Painlevé property for the Kadomtsev–Petviashvili and Hirota–Satsuma equations, *Journal of Mathematical Physics* **26**, 2174–80 (1985).

[38] Weiss, J. Bäcklund transformation and the Painlevé property, *Journal of Mathematical Physics 27*, 1293–1305 (1986).

Part III

Physical applications

6 *J. D. Gibbon*

Why the NLS equation is simultaneously a success, a mediocrity and a failure in the theory of nonlinear waves

The NLS (nonlinear Schrödinger) equation has long been hailed as the canonical integrable amplitude equation governing weak oscillations in weakly nonlinear systems. In some one-dimensional examples, it is an idealisation and not truly physically applicable. However, the soliton laser provides an excellent case study for cyclical soliton propagation in a truly one-dimensional world. Transverse and longitudinal wave coupling in a spring or string can also provide a pedagogical model for understanding the behaviour of many systems which have more than one active dispersion branch. These cases are reviewed along with the circumstances when the NLS equation fails (blows up in finite time) when either higher-order nonlinearities or more spatial dimensions are introduced.

6.1 **Introduction**

Although the subject of integrable systems can trace its roots back into the last century, it only began to have any significance in 1967 when Gardner *et al.* [1] showed that the fascinating soliton behaviour demonstrated by Zabusky and Kruskal [2] was directly related to the method known as inverse scattering. John Scott Russell's solitary wave [3] was referenced in only a handful of places between 1844 and 1965 [4, 5, 6], mainly because the solitary wave plays only a minor role in fluid mechanics and hence to fluid dynamicists the soliton phenomenon has remained little more than a curiosity. Its significance outside this subject was unknown until Zabusky and Kruskal began their work on weakly nonlinear effects in nonlinear chains and plasmas in the 1960s. In theoretical physics, the soliton idea has highlighted the importance of a particular type of energy propagation: that is, localised, finite energy states or 'lumps' which are fundamentally nonlinear objects and so cannot be reached by perturbation theory from any linear state. Excluding the work on the self-dual Yang–Mills equations and its variants which stands as a separate entity on its own (see references in [7]) it would appear that for classical problems, the soliton idea is only useful if the system in question is truly one-dimensional in space. This appears to be a definite drawback as far as applications are concerned. There are one or two equations solvable by the inverse scattering transform (IST), such as the Kadomsev–Petviashvili [8] and Davey–Stewartson equations [9, 10] which involve two spatial dimensions, but even in these cases the IST only yields plane wave solutions which are one-dimensional in character. In fluid dynamics, this is an obvious drawback. Not only this, the IST is also unable to solve dissipative equations (the Burgers equation can be transformed to the heat equation). Integrable systems, as they have become known, are confined in physical importance mainly to areas outside of fluid mechanics, although peripherally, there are one or two examples which appear in weakly nonlinear limits such as surface and internal waves [6, 11], rotating baroclinic and barotropic flows [12–14] and magnetohydrodynamics [15]. All of these examples are idealised to one dimension, a situation which is not experimentally realistic in fluid dynamics except in some specifically selected cases. Nevertheless, the literature over the last 20 years has shown that certain common elements occur in general weakly nonlinear dispersive systems (see references in [16–19]). In this limit, pulse-like waves evolve according to either the Korteweg–de Vries (KdV) equation, if the leading-order nonlinearity is quadratic,

$$u_\tau + \alpha u u_\xi + u_{\xi\xi\xi} = 0. \tag{6.1.1}$$

or the MKdV equation, if the leading-order nonlinearity is cubic,

$$u_\tau + \beta v^2 v_\xi + v_{\xi\xi\xi} = 0. \tag{6.1.2}$$

In each case the slow length and time scales $\xi = \varepsilon^p(x - ct)$; $\tau = \varepsilon^{3p}t$ have values for p respectively as $p = \frac{1}{2}$ for the KdV equation and $p = 1$ for the MKdV equation. The parameter ε is 'small' in comparison to a typical length scale in the problem where the wave amplitude is $0(\varepsilon)$. The two KdV equations govern the evolution of lumps or pulses; for oscillatory wave packets, the result is different [16–19]. The typical amplitude equation to occur in this case is the nonlinear Schrödinger equation (NLS)

$$2i\frac{\partial A}{\partial T} + \left(\frac{d^2\omega}{dk^2}\right)\frac{\partial^2 A}{\partial X^2} + \gamma A|A|^2 = 0 \tag{6.1.3a}$$

where $A(X, T)$ is now the envelope of a carrier wave of wave number k and frequency ω and

$$X = \varepsilon(x - c_g t);\ T = \varepsilon^2 t. \tag{6.1.3b}$$

Equations (6.1.1)–(6.1.3), along with the sine-Gordon equation, are the most physically ubiquitous of the soliton equations. Most textbooks or review articles in applied soliton theory will provide a compendious list of references.

The restriction to one dimension (or to plane waves) means that areas other than fluid dynamics are more likely to produce realistic physical occurrences of integrable systems. Classical mechanics has only a few integrable examples such as special cases in the rigid body problem [20, 21], the 3-vortex problem [22], the Manakov rotator [23], the variants of the Calogero–Moser problem on the line [24, 25], and the Toda lattice [26–29]. Most of these examples stand out as exceptional integrable cases among more general non-integrable problems, so we must turn our thoughts to other areas of physics to find other possible physical examples where the soliton idea is fundamental. Solid state physics is one such area where crystalline or metallic structure provides a natural one-dimensional world. It is not surprising therefore that the sine-Gordon equation is useful as a solvable continuous model for pinning potentials and ferromagnetics (see references in [30]).

The newest and probably the best example of a physically important integrable system which has been built in the laboratory is the so-called 'soliton laser' designed and built by Mollenauer and Stolen [31, 32]. This device is based around several important but simple properties of the monomode optical fibre. First, it is an almost perfect one-dimensional world and, secondly it has the NLS equation as the governing amplitude equation. Solitons are produced and shaped down an optical fibre from a laser amplifier. By various means, the output from the fibre is fed back into the amplifier which reshapes and amplifies the pulses. The NLS is scale-invariant and so the equation remains the same for shorter and shorter pulses as this continuous cyclical process continues. Various other processes, which we will explain in Section 6.2, are needed to keep the process stable. Only when the fibre length is equal to the periodicity length of the $N = 2$ soliton pulse does the narrowing process stop. Pulses as narrow as 50 fm have been reported [31, 32] and the final result is a stream of remarkably narrow clean soliton pulses which have obvious applications in information technology and elsewhere. The mathematical process is interesting because it requires the mapping onto each other of the end conditions of the NLS equation over a finite domain through a nonlinear cyclical process. This process is only possible first because the optical fibre is perfectly one-dimensional and secondly because frequency space is the natural space in which an optical experimentalist works and so it is relatively easy to chop or shape a spectrum in an optical laboratory and do things in frequency space which a fluid dynamicist cannot do.

A second area of physical interest lies in the possibility of coupling systems of waves together in physical problems in which more than one process is occurring, each on different length scales. Systems which have both optical and acoustic branches are examples where coupling can occur between the two branches. The NLS equation and its variants is fundamental to this scheme. This type of wave–wave coupling on different scales, but forming a resonant coupling, has many possible applications. In Section 6.3 we will present a pedagogical model which contains most of the relevant features of these types of couplings. This model is used as an illustration of how physical systems can operate simultaneously in various modes with coupling between them. In some cases, the result is integrable. Davydov's [33] soliton model for helical proteins is an obvious example where classical acoustic oscillations in the helix are coupled to the quantum-mechanical energy states in the $C = \mathrm{O}$ bond in the amino acid groups.

Finally in Section 6.4 we will show why the NLS equation blows up in finite time once the nonlinearity is stronger than cubic or when the number of spatial dimensions is increased. In this sense, the equation fails to be of use as an amplitude equation except in those circumstances when the 'self-focusing' or blow-up can be put to physical use.

6.2 The NLS equation and the soliton laser

For a lossless medium, the dispersion relation

$$D(k, \omega) = 0 \tag{6.2.1}$$

may have several branches. The most familiar in physics are what are known as the optical and acoustic branches. Each branch represents a different dispersive process in the medium with a correspondingly different group velocity. Formally, (6.2.1) represents the linear part of the PDEs modelling the problem: any nonlinear part creates harmonics which react back on the fundamental. The competition between these two processes produces the solitary wave phenomenon. A rough explanation of why the NLS equation is generic for nonlinear media in equilibrium can be given by thinking of an optical analogy. Let us take the optical branch of (6.2.1), which reacts more strongly to high frequencies. For the moment we will assume that no coupling between this branch and any other branch can occur, although in Section 6.3 we will deal with this question. To take our optical analogy further, it is well known that when light travels through a medium, if the intensity is strong enough, the refractive index n becomes field-amplitude-dependent: $n = n_0 + n_2|A|^2$. Because of this, we will postulate that (6.2.1) is field-dependent:

$$D(k, \omega; |A|^2) = 0. \tag{6.2.2}$$

Turning (6.2.2) into operator language, it becomes the Fourier space representation of the operator

$$L\left(\frac{\partial}{\partial x}, \frac{\partial}{\partial t}; |A|^2\right)\varphi = 0 \qquad (6.2.3)$$

where φ is the dependent variable. A is related to φ by

$$\varphi = \varepsilon A(\xi, \tau) \exp[\mathrm{i}(kx - \omega t)] + \text{c.c.} \qquad (6.2.4)$$

such that it is the complex amplitude of a carrier wave (c.c. is complex conjugate). A standard multiple-scales procedure (see references [16–19]) with slow scales $\xi = \varepsilon(x - c_g t)$; $\tau = \varepsilon^2 t$ produces the NLS equation from (6.2.3)

$$2\mathrm{i}\,\frac{\partial A}{\partial \tau} + \left(\frac{\mathrm{d}^2\omega}{\mathrm{d}k^2}\right)\frac{\partial^2 A}{\partial \xi^2} + \gamma A|A|^2 = 0 \qquad (6.2.5)$$

which is the 'canonical' form of the NLS equation. It is this procedure which, when applied to individual cases, gives the NLS equation. An alternative way of writing (6.2.5) is to exchange the slow space and time variables such that

$$T = \varepsilon(t - x/c_g) = -\xi/c_g, \qquad (6.2.6a)$$

$$X = \varepsilon^2 c_g t = c_g \tau. \qquad (6.2.6b)$$

(6.2.5) now becomes ($\beta = \gamma/c_g$):

$$2\mathrm{i}\,\frac{\partial A}{\partial X} - \left(\frac{\mathrm{d}^2 k}{\mathrm{d}\omega^2}\right)\frac{\partial^2 A}{\partial T^2} + \beta A|A|^2 = 0. \qquad (6.2.7)$$

It is this latter form of the NLS equation which is relevant to the work on the soliton laser which has been built at Bell Laboratories. The sign of β is dependent on the sign of n_2 in the expansion of the field-dependent refractive index $n = n_0 + n_2|A|^2$. Generally this is positive and since we need $\beta\ \mathrm{d}^2k/\mathrm{d}\omega^2 < 0$ for soliton production, obviously we need $\mathrm{d}^2k/\mathrm{d}\omega^2 < 0$. This condition is what optical physicists call negative group velocity dispersion. The idea of the soliton laser exploits the fact that there exists a frequency range for silica fibres for which $\mathrm{d}^2k/\mathrm{d}\omega^2 < 0$. Mollenauer, Stolen and Gordon [34], in an initial experiment, passed a 7 ps soliton pulse down an optical fibre of length 700 m and confirmed the prediction of NLS behaviour. They were able to do this because of the development of mode-locked colour centre lasers (see references in [32]) which are capable of producing streams of very clean picosecond pulses of many watts peak power. The soliton laser exploits the behaviour of *oscillatory* soliton solutions of the NLS equation. To explain this, we shall take equation (6.2.7), with $\mathrm{d}^2k/\mathrm{d}\omega^2 < 0$, in a normalised form:

$$\mathrm{i}\,\frac{\partial A}{\partial X} + \frac{1}{2}\frac{\partial^2 A}{\partial T^2} + A|A|^2 = 0 \qquad (6.2.8)$$

where the space, time and electric field amplitude variables X, T and A have been rescaled to absorb various coefficients. Satsuma and Yajima [35] have used the IST on equation (6.2.8) with an initial pulse profile

$$A(0,\ T) = \frac{N}{T_0}\,\mathrm{sech}(T/T_0) \tag{6.2.9}$$

where T_0 denotes pulse width and N is an integer. They showed that this initial data produces N zeros of the transmission coefficient in the complex plane each of which has zero real part. This is caused by the fact that the initial profile has zero phase. We shall call this the 'N-soliton' case even though the solitons are not completely independent. For integer N, analytic solutions of the NLS equation exist. For instance, for $N = 2$, the solution is

$$A(x,\ t) = \frac{4\,\exp(-\tfrac{1}{2}\mathrm{i}x)\{\cosh(3t) + 3\mathrm{e}^{-4\mathrm{i}x}\cosh t\}}{\{\cosh(4t) + 4\cosh(2t) + 3\cos(4x)\}}. \tag{6.2.10}$$

The important point to note is that all the N-solitons ($N \geqslant 2$) are pulses which are periodic in X and therefore return to the same initial profile for distances

$$X_0 = nT_0^{\,2}\pi/2, \quad n \text{ integer.} \tag{6.2.11}$$

We note that X_0 scales with $T_0^{\,2}$ so a broader pulse requires a long distance to execute one period. Mollenauer and Stolen [36] have used a full-period-length fibre to demonstrate experimentally the periodicity of the higher-order N-solitons. Figure 6.1 shows an $N = 2$ pulse executing 2 cycles with $T_0 = 1$ so $X_0 = \pi/2$. It is important to note that if the periodicity length is shortened, so is the pulse width T_0. Hence every fibre of a given length has a particular pulse which 'resonates' exactly with it. Mollenauer and Stolen [31] have exploited this fact in building a soliton laser. The idea is simple in principle although in practice (as always), it is much harder to achieve experimentally or numerically. In references [31, 32], a much shorter fibre (8 cm) shapes the output from a laser oscillator. A small percentage of the pulse in the laser is fed into the fibre and temporally narrowed there. Part of this is then superimposed back on the pulse remaining in the laser cavity and, because of the temporal narrowing and hence spectral broadening induced by the fibre, extends the mode-locking region. Initially, the pulse in the fibre is far too long to have the correct periodicity length for the very short fibre. Several thousand cycles of this process continuously narrows the pulse (in time and space) and amplifies it until finally it reaches a state (for suitable initial conditions) where it is an $N = 2$ soliton whose periodicity length matches the fibre length. From equation (6.2.11) we can see that the final pulse width is proportional to the square root of the fibre length. The shorter the fibre, the shorter the final pulse. Pulse widths as narrow as

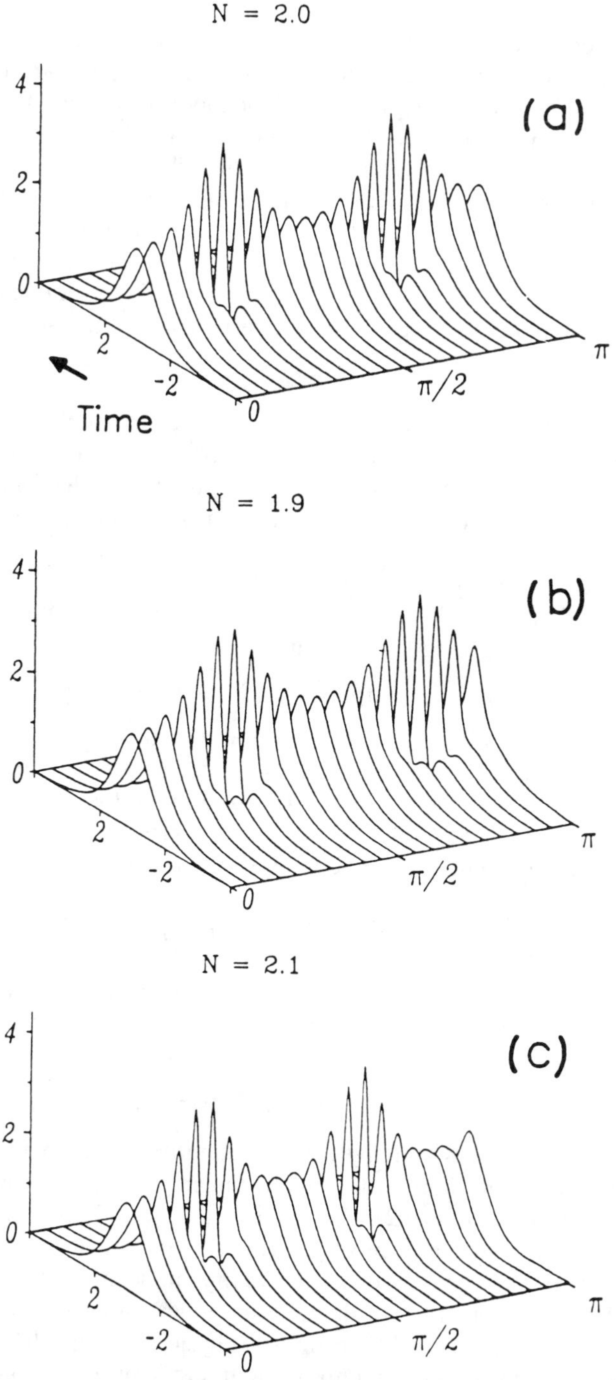

Figure 6.1 Three plots of electric field $E(x, t)$ versus t for discrete x

50 fs were reported [31, 32] with the possibility of even shorter pulses. Clearly, the possible uses for such amazingly short pulses are endless.

One can think of the system as a feedback loop which has an attracting state, which in this case is the $N = 2$ soliton, whose width is determined by the chosen length L of the fibre. In practice, the experimental set-up is more complicated: the pulse passes along the fibre, is reflected from a mirror and sent back along it. We therefore need the periodicity length X_0 for an $N = 2$ soliton to satisfy $X_0 = 2L$. The fact that this process works is dependent upon the scaling properties of the NLS equation. It is well known that the NLS equation is invariant under the scaling

$$t' = \mu t, \quad x' = \mu^2 x, \quad A = \mu A'. \tag{6.2.12}$$

The laser amplifier just 'renormalises' the pulse into the fibre onto different space and time scales on every round trip; within the fibre it is still the NLS equation with the same coefficients on *every* scale. One can look at it two ways. Either the pulse gets narrower or the pulse 'sees' a longer fibre.

Without the scaling property given in (6.2.12) the whole system would fail to work. Schematically we can think of it in the following way. Since the effective working length of the fibre is $2L$, we have

$$A(0, t) = \text{laser output}; \quad A(2L, t) = \text{laser input}. \tag{6.2.13}$$

Hence the end conditions of the fibre are mapped onto each other through the process in the laser cavity. This process cannot be represented by a discrete mapping. However, for a mode-locked laser the field inside the cavity can be broken down into discrete Fourier cavity modes $E_n(t)$. The simplest equation modelling this process is

$$\frac{\partial E_n}{\partial_t} = (\alpha - \varkappa)E_n + \beta(E_{n+1} + E_{n-1} - 2E_n) \tag{6.2.14}$$

where α is the gain (growth rate) and $\varkappa$ the cavity loss. Putting the Fourier modes back together to form the full cavity field $E(x, t)$, we can write

$$E(0, t) = A(2L, t); \quad E(l, t) = A(0, t), \tag{6.2.15}$$

where l is the laser cavity length. Consequently the 'mapping' between end conditions of the fibre is through a linear equation in cavity modes (6.2.14). Now, although (6.2.14) looks like a simple discrete form of the diffusion equation with a positive growth rate, in fact, it turns out that α and β need to be functions of n, the cavity mode number, in order to realistically model the effect of the laser amplification. As we explained in the introduction, it is possible in optical physics to shape light pulses in spectral space, a process often impossible in other systems. Since the type of laser used has gain over a large bandwidth of frequencies, birefringence plates are used in the experiment to chop off energy in the outer wings of the pulse. Also, only a small proportion of the pulse is fed back into the amplifier on every round trip. By this process, Mollenauer and Stolen [31] were able to consistently narrow the soliton pulse, without destabilising it, until its periodicity length equalled $2L$. Pulses as narrow as 50 fs were observed, which is an order of magnitude smaller than previously obtained.

To simulate this behaviour numerically is not easy. If *et al.* [37] have performed a numerical experiment using the model given here and achieved soliton pulse narrowing over 400 round trips through the fibre and amplifier. It is necessary to simulate the birefringence plates and introduce saturation into the laser model (6.2.14) in which $\varkappa$ is the cavity loss constant and α is the intensity-dependent gain, which is simplified by taking $\alpha = \alpha_0 S$ where α_0 is the linear cavity gain constant and $S = \exp(-W/W_0)$. W is the pulse energy and W_0 is an appropriate saturation energy. In this manner, we are able to include saturation effects phenomenologically, thus introducing nonlinearity into the laser model. The effect of the birefringence plates can be written into the model at this point. This is achieved by making β a function of the mode-number n. It could either be a tophat function or a Gaussian. In the latter case $\beta = \exp(-n^2/n_0^2)$ was used. The range n_0 defines the width of the coupled modes in frequency space, and hence is responsible for determining the temporal width of the (Gaussian) stationary pulses within the laser cavity. The scaling given in equation (6.2.12), which keeps the NLS equation invariant, also keeps the pulse area $\int_{-\infty}^{\infty} E(t)\mathrm{d}t$ invariant. However, the energy $W = \int_{-\infty}^{\infty} |E|^2\mathrm{d}t$ scales with μ: $W' = \mu W$. The different scaling between area and energy requires a careful choice of initial parameters, that is, of the choice of initial values of area and energy.

6.3 Coupling the NLS equation to other modes: a pedagogical model

In Section 6.2 we considered the type of system which was an intrinsically nonlinear medium, thereby enabling us to think idealistically of the dispersion relation $D(k, \omega)$ as being amplitude-dependent. In this sense, any pulse moving in a medium of this type experiences or 'sees' the nonlinearity of the medium. It is possible instead to think of nonlinearity being introduced into an intrinsically linear medium through coupling between different branches of the dispersion relation $D(k, \omega) = 0$. As in Section 6.2 it is possible to consider those systems, which are quite common, which have two or more branches to their dispersion relations; for example, an optical mode and an acoustic mode. For instance, in a plasma, the optical branch looks like $\omega^2 = 1 + k^2$ whereas the acoustic branch looks like $\omega^2 = k^2/(1 + k^2)$. To model in an idealised way how such coupling can introduce nonlinearity, we will consider an idealised model consisting

Figure 6.2 String or spring oscillating transversely (ϕ_T) and longitudinally (ϕ_L)

of a string (or a spring) as in Figure 6.2. These arguments follow those given in reference [38]. The string (or spring) is allowed to oscillate in two modes: transversely (like a guitar) with amplitude $\varphi_T(x, t)$ and longitudinally (like a spring stretching and compressing) with amplitude $\varphi_L(x, t)$.

Without transverse oscillations, the equation of motion for φ_L, which will be linear in φ_L, can be written as

$$D\left(\frac{\partial}{\partial t}, \frac{\partial}{\partial x}\right)\varphi_L = 0, \tag{6.3.1}$$

giving a dispersion relation $D(-i\omega; ik) = 0$. However, these longitudinal compressive and expansive waves have to travel around the curves in the string, once transverse oscillations have been set up. This will make the medium appear nonlinear with a dispersion relation $D(-i\omega; ik; \varphi_T) = 0$ where the amplitude dependence on the dispersion relation is in the φ_T variable. We can write the equation of motion as

$$D\left(\frac{\partial}{\partial t}; \frac{\partial}{\partial x}; \varphi_T\right)\varphi_L = 0. \tag{6.3.2}$$

For transverse oscillations we could write a dispersion relation in the reverse fashion, $M(-i\omega; ik; \varphi_L) = 0$ but in reality transverse waves do not see a medium made nonlinear by longitudinal oscillations. Rather, we can think of the transverse motion being *driven* by longitudinal oscillations which displace any given point in the string. Consequently we shall write

$$M\left(\frac{\partial}{\partial t}; \frac{\partial}{\partial x}\right)\varphi_T = \text{nonlinear driving terms in } \varphi_L \tag{6.3.3a}$$

which couples with (6.3.2). The most likely form of M is the wave operator

$$M = \frac{\partial^2}{\partial x^2} - \frac{1}{c_p^{\,2}}\frac{\partial^2}{\partial t^2} \tag{6.3.3b}$$

although it could be more complicated than this. At certain points in the following calculations it will be easiest to use (6.3.3b). The argument for the form of the driving terms is based largely on how the wave equation for a string is derived from continuity and momentum equations. When the wave equation is obtained by roughly assuming constant phase velocity, any forcing term appears as the second space derivative of an intensity of energy density. For fast oscillations in $\varphi_L \sim \varepsilon A \exp(\mathrm{i}\vartheta) + \text{c.c.}$, the main contribution to this energy density would be a term of the form $|A|^2$, in analogy with energy density in electromagnetic theory. This would only appear from a ${\varphi_L}^2$ term in which the second harmonics had been neglected, and would also be the obvious simplest nonlinear term to choose. Therefore we write the simplest forcing term on the right-hand side of (6.3.3) as

$$M\left(\frac{\partial}{\partial t};\frac{\partial}{\partial x}\right)\varphi_T = -\alpha\,\frac{\partial^2}{\partial x^2}\,({\varphi_L}^2) \tag{6.3.4a}$$

coupling with

$$D\left(\frac{\partial}{\partial t};\frac{\partial}{\partial x};\,\varphi_T\right)\varphi_L = 0 \tag{6.3.4b}$$

Along with a choice of M, there are two types of motion on different scales to consider. Furthermore the size of the coupling constant α has a considerable effect on the result. The type of argument used in constructing the r.h.s. of equation (6.3.4a) is that used by Zakharov [39] when discussing the coupling of Langmuir oscillations to ion acoustic waves in a plasma. We can now work out two limiting cases with respect to equations (6.3.4) and give some physical examples.

Long transverse waves (φ_T) and short longitudinal oscillations (φ_L)

In the usual derivation of the NLS equation, which was discussed in Section 6.2, it is normal to take 'slow' space and time scales

$$\xi = \varepsilon(x - c_g t), \quad \tau = \varepsilon^2 t, \tag{6.3.5}$$

where ε is a measure of the amplitude. We will allow these scales for φ_L in (6.3.4b)

$$\varphi = \varepsilon A(\xi,\tau)\exp(\mathrm{i}\vartheta) + \text{c.c.} \tag{6.3.6}$$

but it is then necessary to consider how φ_T depends on ε. In analogy with a purely nonlinear medium with the position of φ_T replaced by $|A|^2$ as in (6.2.3), this requires $\psi_T \sim 0(\varepsilon^2)$. Hence we take

$$\varphi_T = \varepsilon^2 n(\xi,\tau). \tag{6.3.7}$$

By this choice we have made φ_T dependent only on the slow scales (ξ, τ) with no carrier wave oscillatory part. The usual multiple-scales calculation shows that A satisfies the equation:

$$2\mathrm{i}\,\frac{\partial A}{\partial \tau} + \left(\frac{\mathrm{d}^2\omega}{\mathrm{d}k^2}\right)\frac{\partial^2 A}{\partial \xi^2} = \beta n A \tag{6.3.8}$$

where we have assumed that the contributory term to the $\beta n A$ term in (6.3.8) is the dominant linear term in n in any expansion of D.

With respect to the operator M, we use the wave operator, so we need to use the slow scales (6.3.5) on

$$\left(\frac{\partial^2}{\partial x^2} - \frac{1}{c_p^2}\frac{\partial^2}{\partial t^2}\right)\varphi_T = -\alpha \frac{\partial^2}{\partial x^2}(\varphi_L^2) \tag{6.3.9}$$

which gives

$$\left\{\varepsilon^2\left(1 - \frac{c_g^2}{c_p^2}\right)\frac{\partial^2}{\partial \xi^2} + 2\varepsilon^3 \frac{c_g}{c_p^2}\frac{\partial^2}{\partial \xi \partial \tau} - \frac{\varepsilon^4}{c_p^4}\frac{\partial^2}{\partial \tau^2}\right\}(\varepsilon^2 n)$$
$$= -\alpha\left[\frac{\partial^2}{\partial x^2} + 2\varepsilon\frac{\partial^2}{\partial \xi \partial \tau} + \varepsilon^2\frac{\partial^2}{\partial \xi^2}\right](\varepsilon^2|A|^2 + \text{second harmonics}). \tag{6.3.10}$$

To obtain a sensible matching of the scales we can identify two relevant magnitudes of the coupling constant α.

Case (i): $\alpha \sim 0(1)$

This is the strong coupling case. Ignoring the effect of the second harmonics which are too fast for the slow-scale motion and will therefore average out, we can immediately see that a balance is achieved if

$$n = -\alpha\, c_p^2(c_p^2 - c_g^2)^{-1}|A|^2. \tag{6.3.11}$$

For an optical-branch-type dispersion relation, $c_p^2 > c_g^2$, so we have exactly the NLS equation

$$2i\frac{\partial A}{\partial \tau} + \left(\frac{d^2\omega}{dk^2}\right)\frac{\partial^2 A}{\partial \xi^2} + \bar{\beta}A|A|^2 = 0,\ \bar{\beta} > 0. \tag{6.3.12}$$

In this case the coupling, via α, is strong enough to slave the transverse to the driving longitudinal modes and the NLS equation results.

Case (ii): $\alpha \sim 0(\varepsilon)$

In the weak coupling case, we can only obtain a sensible matching at a special wavelength; that is, when $c_p = c_g$. Hence, when $c_p = c_g$, we have ($\alpha = \varepsilon\bar{\alpha}$)

$$\frac{\partial n}{\partial \tau} = -\tfrac{1}{2}\bar{\alpha}\left(\frac{c_p^2}{c_g}\right)\frac{\partial}{\partial \xi}|A|^2. \tag{6.3.13}$$

Now, as expected, n is not slaved to $|A|^2$ but acts as a variable coupled to A via (6.3.13) and (6.3.8). Indeed these are the longwave–shortwave resonance (LW-SW) equations [11, 40] which have arisen in capillary-gravity waves. It has the added advantage of being integrable [41].

The condition $c_p = c_g$ has a simple interpretation. First, we note that

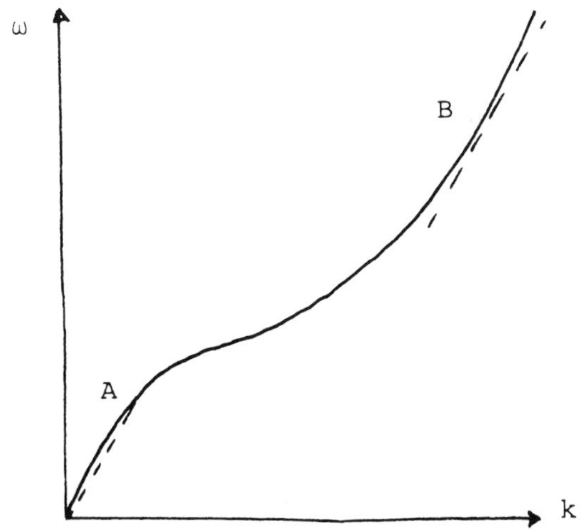

Figure 6.3 The longwave–shortwave resonance depicted on a double-branched dispersion curve. The secant at A is parallel to the tangent at B to make $c_p = c_g$.

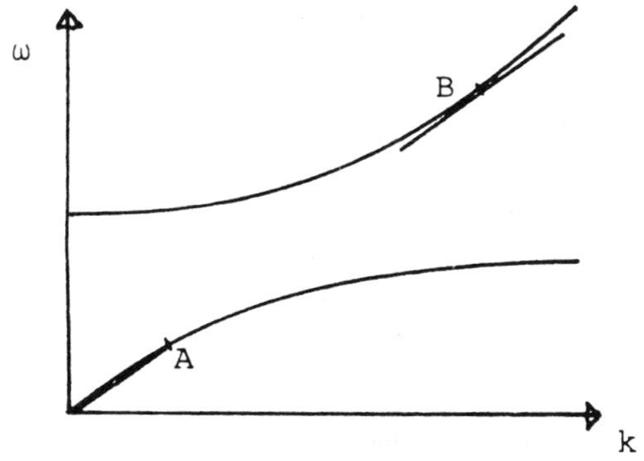

Figure 6.4 The longwave–shortwave resonance for a single-branched dispersion curve

the c_g is the group velocity from the *upper* branch, the fast or short waves, whereas the c_p is the phase velocity on the *lower* branch. In fact this condition can be shown to be a special case of the triad resonance. Take a main oscillatory wave of wave number k on the upper branch. Let it have a left sideband $k_1 = k - \varepsilon$ and a right side-band $k_2 = k + \varepsilon$ and let $k_3 = 2\varepsilon$ which is a low wave number on the lower branch. The triad resonance condition is met: $k_2 - k_1 = k_3$ but only fully provided that $\omega_2 - \omega_1 = \omega_3$. Since

$$\omega_1 = \omega(k - \varepsilon) = \omega(k) - \varepsilon c_g + \ldots, \qquad (6.3.14a)$$

$$\omega_2 = \omega(k + \varepsilon) = \omega(k) + \varepsilon c_g + \ldots, \tag{6.3.14b}$$

$$\omega_3 = (2\varepsilon)c_p, \tag{6.3.14c}$$

we can achieve the resonance condition provided that $c_p = c_g$ to $0(\epsilon^2)$. Physically we can think of this resonance condition as the two wave numbers k_1 and k_2 beating to excite a wave number 2ε on the lower branch. In the capillary-gravity wave case, there is only one branch (see Figure 6.3b) but the form of the dispersion relation still enables one to achieve the $c_g = c_p$ condition. However, it would seem to be a more natural condition to occur for the double-branched type systems (Figure 6.3a).

Case (iii): Zakharov's equations

If $c_g = 0$ and $c_p = \varepsilon\bar{c}_p$, then equation (6.3.10) allows the second T-derivative to appear and we now find

$$\left(\frac{\partial^2}{\partial\xi^2} - \frac{1}{\bar{c}_p{}^2}\frac{\partial^2}{\partial\tau^2}\right)n = -\alpha\frac{\partial^2}{\partial\xi^2}(|A|^2). \tag{6.3.15}$$

This equation and (6.3.8) are together called Zakharov's equations (1-dimensional version) and were first derived in the context of plasma physics in an attempt to couple the electron oscillations (upper dispersion branch) to the ion sound waves (lower dispersion branch) [39].

It is notable that in all three sets of equations; NLS, LW-SW and Zakharov's equations, we have coupled short modulated oscillations to a long wave $n(\xi, \tau)$. The latter is like a pulse without a carrier wave, very much like a KdV soliton, whereas the short wave oscillatory amplitude $A(\xi, \tau)$ behaves like an NLS envelope soliton. Consequently the transverse and longitudinal motions feed off one another.

Waves in φ_T and φ_L of matching wavelength

Considering our model string system

$$D\left(\frac{\partial}{\partial t}, \frac{\partial}{\partial x}; \varphi_T\right)\varphi_L = 0, \tag{6.3.16a}$$

$$M\left(\frac{\partial}{\partial t}, \frac{\partial}{\partial x}\right)\varphi_T = -\alpha\frac{\partial^2}{\partial x^2}(\varphi_L{}^2), \tag{6.3.16b}$$

let us assume now that both φ_t and φ_L are oscillatory and have amplitudes $0(\varepsilon)$:

$$\varphi_T = \varepsilon A(\xi, \tau)\exp(i\vartheta_T) + \text{c.c.} \tag{6.3.17a}$$

$$\varphi_L = \varepsilon A_L(\xi, \tau)\exp(i\vartheta_L) + \text{c.c.} \tag{6.3.17b}$$

Again, assuming that L is at least linear in φ_T and that M is at least a second-order operator, we can obtain the second harmonic resonance equations

$$\left(\frac{\partial}{\partial T} + c_L \frac{\partial}{\partial x}\right)A_L = \alpha_1 A_T A_L{}^*, \qquad (6.3.18a)$$

$$\left(\frac{\partial}{\partial T} + c_T \frac{\partial}{\partial x}\right)A_T = \alpha_2 A_L{}^2, \qquad (6.3.18b)$$

where $T = \varepsilon t$; $X = \varepsilon x$, provided the second harmonic resonance conditions are met; $2k_L = k_T$; $2\omega_L(k_L) = \omega_T(k_T)$. This means that

$$c_T(k_T) = c_L(k_L), \qquad (6.3.19)$$

where the c's are the phase velocities. Equations (6.3.18) can be fitted into the first 2×2 system [42] but the solution of the spectral problem is not yet fully understood.

Physical occurrences

Until now we have used the string idea as a convenient model for explaining how it is possible to get a coupling between the optical (upper branch) and acoustic (lower branch) of a system. While motion on each individual branch may or may not be linear, the coupling introduces nonlinearity into the problem. We will briefly list circumstances where this type of coupling can occur. In the examples referenced, by no means all the above types of wave behaviour occur in practice since real systems operate on vastly differing scales.

(i) In plasma physics, the Langmuir (electron) oscillations correspond to the longitudinal motion and the ion sound to the transverse motion. Zakharov's equations occur, of which (6.3.15) are the one-dimensional version. Zakharov [39] was the first to introduce the idea of coupling between different modes and his model has given rise to considerable work in plasma physics with respect to their relevance to plasma turbulence (see references in [43]).

(ii) Davydov's model in biophysics [33] (see also Chapter 9) couples the quantum-mechanical far-infrared spectral transitions in the $C = O$ bonds of the amino acids (longitudinal) with the acoustic oscillations in the molecular helix (transverse mode). In this case the starting equation is

$$L \equiv \mathrm{i}\frac{\partial}{\partial t} - \frac{\hbar^2}{2m}\frac{\partial^2}{\partial x^2} - n(x, t), \qquad (6.3.20a)$$

$$M \equiv \frac{\partial^2}{\partial x^2} - \frac{1}{c_p{}^2}\frac{\partial^2}{\partial t^2}, \qquad (6.3.20b)$$

thereby giving Zakharov's equations automatically before doing the multiple scaling procedure.
(iii) In water wave theory, in the consideration of capillary-gravity waves, Benney [40] and Djordjevic and Redekopp [11] have shown that the LW–SW resonance equations appear. There is no double branch to the dispersion relation but (see [11]) the dispersion curve is such that the condition $c_p = c_g$ is still achievable. Furthermore, McGoldrick [44] has shown how the second harmonic resonance equations appear also for water waves.

6.4 Circumstances when the NLS equation fails

In Section 6.2 we talked about how the NLS equation arises from a dispersive system in a nonlinear medium where the dispersion relation is field-dependent:

$$D(k, \omega; |A|^2) = 0. \tag{6.4.1}$$

We showed there that the canonical form of the NLS equation occurs:

$$2\mathrm{i}\,\frac{\partial A}{\partial \tau} + \left(\frac{\mathrm{d}^2\omega}{\mathrm{d}k^2}\right)\frac{\partial^2 A}{\partial \xi^2} + \beta A|A|^2 = 0 \tag{6.4.2}$$

when a multiple scaling procedure is used. There are two further generalisations of (6.4.2) we shall consider. The first is when the nonlinearity is stronger than cubic: say $A|A|^{2q}$, and secondly when further spatial dimensions are necessary. Putting the two together, we can write a more general form of the NLS equation as

$$2\mathrm{i}\,\frac{\partial A}{\partial \tau} + \sum_{j=1}^{n} \Delta_j \frac{\partial^2 A}{\delta \xi_j{}^2} + \beta A|A|^{2q} = 0 \tag{6.4.3}$$

where $\Delta_j = \pm 1$ (we have absorbed the $\mathrm{d}^2\omega/\mathrm{d}k_j{}^2$ into the spatial variables ξ_j for convenience). Equation (6.4.3) has arisen in a variety of forms. Its most important physical application is in self-focusing in nonlinear optics where a pencil of light can focus or defocus in a nonlinear medium [45]. In this case the τ-variable is a spatial variable (z) down the pencil or cylinder of light; $\Delta_1 = \Delta_2 = 1$ and ξ_1 and ξ_2 are the x–y radial variables. It is well known in this case that when $\beta > 0$, the NLS equation blows up in finite τ (actually z), a behaviour we will summarise later. The $\beta > 0$ case coincides with the coefficient $n_2 > 0$ in the refractive index: $n = n_0 + n_2|A|^2$. Blow-up in finite time (or z) means that the NLS fails as an amplitude equation since it breaks the scales on which it was derived in the first place. This behaviour would appear to indicate, at first sight, that the NLS is a failure as an amplitude equation and that the model is no good. If an equation blows up for much typical initial data, it would not seem to have much use. In fact, however, there are two particular physical situations where the blow-up phenomenon is either useful or tells us something about the problem.

A brief summary of the blow-up phenomenon is not easy to give in detail. It has been known for some time that initial data can collapse in finite time, in a similar way to the N-body problem in celestial mechanics. Let us, for the moment, take $\Delta_j = 1$ for all j. Equation (6.4.3) has constants of the motion which can be thought of as mass and energy:

$$M = \int |A|^2 \, d\xi \quad E = \int [\tfrac{1}{2}|\nabla A|^2 - \tfrac{1}{4}\beta|A|^{2q+2}] \, d\xi. \tag{6.4.4}$$

Defining the moment of inertia I as

$$I = \int \xi^2 \, |A|^2 \, d\xi \tag{6.4.5}$$

we find that a form of virial theorem results:

$$\ddot{I} = 2qnE + (4 - 2qn)T \tag{6.4.6}$$

where T is the positive definite kinetic energy integral which is the first term in E. Since $T > 0$ and $E =$ constant, we have

$$\ddot{I} \leqslant 0, \quad qn \geqslant 2, \tag{6.4.7}$$

provided $E < 0$, which requires $\beta > 0$. Now n is the number of dimensions in the problem. For instance, in the case $qn = 2$ we have $\ddot{I} = 4E$ and hence

$$I = 2Et^2 + \dot{I}(0) + I(0) \tag{6.4.8}$$

when initial data is chosen such that $E < 0$, then I becomes negative in finite time which indicates a singularity appearing in A. This type of behaviour is analagous to collapse of heavenly bodies into their centre of mass in astronomy. It was Zakharov [39] who first used the virial theorem idea to look at the collapse, and the idea has been extended by several others [46, 47]. While the analogy with collapse in the N-body problem has a certain elegance, it turns out that the actual collapse mechanism is different when numerical evidence is taken into account. The singularity in A appears much earlier than that predicted by the virial method [48]. Indeed the singularity is essential and has not properly been understood although the conditions on q and n seem to remain the same.

What the fact of collapse tells us in physical problems, such as self-focusing in optics, is that the beam focuses to a singularity in finite z provided $E < 0$ and $qn \geqslant 2$. This means that if $n = 2$, then the 'normal' nonlinear term $A|A|^2$ is sufficient to induce collapse. If $n = 1$, then we need a higher nonlinearity ($A|A|^4$ or higher) to induce collapse. This failure of the equation can be used in plasma physics and optics to concentrate laser beam energy at the singularity [49].

A second 'useful' example of collapse in the NLS equation is in the two-layer model for the inviscid Kelvin–Helmholtz instability. The neutral curve predicts no instability until about 11 knots for wind over water [50]. Since the problem is inviscid, small-amplitude dispersive waves can nevertheless exist below the critical point and these are governed by the NLS equation (6.4.3). In this case $q = 1$ but Δ_1, Δ_2 and β are dependent both on the two carrier wave numbers k and l and on the shear. As these vary, all three parameters can change sign. Hence blow-up can occur for some wave numbers and shears but not for others [51]. Nevertheless the fact that blow-up happens at all shows that self-focusing of waves can occur well below the critical point of 11 knots predicted by linear theory.

Finally, therefore, we can conclude that the NLS equation has a multitude of uses even though its occurrence must be treated with care. Being integrable in one dimension only means that, except in the soliton laser case or any other genuine one-dimensional situations, the soliton solutions themselves have strictly limited application. However, as we have seen above, even the occurrence of a singularity in finite time has its uses.

References

[1] Gardner, C. S., Greene, J., Kruskal, M. D. and Miura, R., *Phys. Rev. Letts.* **19**, 1095 (1967).
[2] Zabusky, N. J. and Kruskal, M. D., *Phys. Rev. Letts.* **15**, 240 (1965).
[3] Scott Russell, J. Report on Waves, *14th Meeting of British Association for the Advancement of Science, London* (1844).
[4] Stokes, G., *Camb. Trans.* **8**, 441 (1847).
[5] Boussinesq, J., *J. Math. Pures. Appl., Ser. 2* **17**, 55 (1872).
[6] Korteweg, D. J. and de Vries, B., *Phil. Mag.* **39**, 422 (1895).
[7] Atiyah, M. F. and Hitchin, N. J., *Phil. Trans. Royal Soc. Lond.* **A315**, 459–69 (1985).
[8] Kadomsev, B. B. and Petviashvili, V. I., *Kokl. Akad. Nauk SSR* **192**, 753 (1970).
[9] Davey, A. and Stewartson, K., *Proc. Royal Soc. Lond.* **A338**, 101 (1974).
[10] Anker, D. and Freeman, N. C., *Proc. Royal Soc. Lond.* **A360**, 529 (1978).
[11] Djordjevic, V. D. and Redekopp, L. G., *J. Fluid Mech.* **79**, 703 (1977).
[12] Pedlosky, J., *J. Atmos. Sci.* **29**, 680 (1972).
[13] Gibbon, J. D., James, I. N. and Moroz, I., *Proc. Royal Soc. Lond.* **A367**, 219 (1979).
[14] Redekopp, L. G., *J. Fluid Mech.* **82**, 725 (1977).
[15] Hill, N. Ph.D. thesis, Imperial College (1985).
[16] Scott, A. C., Chu, F. Y. C. and McLaughlin, D. W., *Proc. IEEE* **61**, 1443 (1973).
[17] Dodd, R. K., Eilbeck, J. C., Gibbon, J. D. and Morris, H. C. *Solitons and Nonlinear Waves Equations*, Academic Press, London (1982).
[18] Newell, A. C., *Solitons in Mathematics and Physics* SIAM, Philadelphia (1985).
[19] Gibbon, J. D., *Phil. Trans. Royal Soc. Lond.* **A315**, 335 (1985).
[20] Kovalevskaya, S., *Acta Mathematics, Stockholm* **14**, 81 (1890).

[21] Tabor, M., *Nature* **310**, 277–82 (1984).
[22] Novikov, E. A., *Sov. Phys. JETP*, **41**, 937–43 (1976).
[23] Manakov, S., *Funct. Anal. Appl.* **10**, 93 (1976).
[24] Moser, J., *Adv. Math.* **16**, 197 (1975).
[25] Gibbons, J. and Hermsen, T., *Physica* **D11**, 337 (1984).
[26] Toda, M., *Prog. Theor. Phys. Suppl.* **59**, 1 (1976).
[27] Flaschka, H., *Phys. Rev.* **B9**, 1924 (1974).
[28] Manakov, S., *Sov. Phys. JETP* **40**, 269 (1975).
[29] Henon, M., *Phys. Rev.* **B9**, 1921 (1974).
[30] Bishop, A., Krumhansl, J., and Trullinger, S., *Physica* **1D**, 1 (1980).
[31] Mollenauer, L., and Stolen, R., *Optics Letts.* **9**, 13 (1984).
[32] Mollenauer, L., *Phil. Trans. Royal Soc. Lond.* **A315**, 333 (1985).
[33] Davydov, A. S. and Kislukha, N. I., *Sov. Phys. JETP* **44**, 571 (1976).
[34] Mollenauer, L., Stolen, R., and Gordon, J., *Phys. Rev. Letts.* **45**, 1095 (1980).
[35] Satsuma, J. and Yajima, N., *Prog. Theor. Phys. Suppl.* **55**, 284 (1974).
[36] Stolen, R., Mollenauer, L. and Tomlinson, W., *Optics Lett.* **8**, 186 (1983).
[37] If, F., Christiansen, P. L., Elgin, J. N., Gibbon, J. D. and Skovgaard, O., *Optics Comm.* **57**, 350 (1986).
[38] Gibbon, J. D., Coupled wave resonance systems (in preparation).
[39] Zakharov, V. E., *Sov. Phys. JETP* **72** 908 (1972).
[40] Benney, D. J., *St. Appl. Math* **55**, 93 (1976).
[41] Yajima, N. and Oikawa, M., *Prog. Theor. Phys. Suppl.* **56**, 1719 (1976).
[42] Kaup, D. J., *St. Appl. Math.* **59**, 25 (1978).
[43] Nicholson, D. R. and Goldman, M. V., *Phys. Rev. Letts.* **41**, 406 (1978).
[44] McGoldrick, L. F., *J. Fluid Mech.* **40**, 251 (1980).
[45] Kelley, P. L. *Phys. Rev. Letts.* **15**, 1005 (1965).
Chiao, R. Y., Garmire, E. and Townes, C. H., *Phys. Rev. Letts.* **13**, 479 (1964).
[46] Zakharov, V. E. and Synakh, V. S., *Sov. Phys. JETP* **41**, 465 (1976).
[47] Gibbon, J. D. and Berkshire, F. H., *Studies in Appl. Maths.* **69**, 229 (1983).
[48] Wood, D., *Studies in Appl. Maths.* **71**, 103 (1984).
[49] Nicholson, D. and Goldman, M. *Phys. Rev. Letts.* **41**, 406 (1978).
[50] Weissman, M. A., *Phil. Trans. Royal Soc. Lond.* **290**, 639 (1979).
[51] Gibbon, J. D. and McGuinness, M. J., *Phys. Letts.* **77A**, 118 (1980).

7 *A. R. Osborne*

The generation and propagation of internal solitons in the Andaman Sea

Internal solitons with amplitudes up to 105 m have been observed in the southern Andaman Sea near the western coast of Thailand. These waves propagate beneath the free surface in density-stratified waters ranging from a few hundred metres to three kilometres in depth. The waves are evidently generated by tidal interactions with bathymetric features near island chains to the west and south-west of the measurement site. The solitons occur in packets of about six to eight waves and propagate several hundred kilometres before encountering the coastline of Thailand. The waves are normally rank-ordered by amplitude, the largest leading the rest; the leading wave in a packet has a typical crest length of 150 km. Bands of surface waves called 'rips' accompany the internal waves. The bands are 600–1200 m wide, stretch tens to hundreds of kilometres across the sea surface and are easily seen in both Apollo–Soyuz and Landsat satellite photographs. *In situ* time series measurements of local water temperature and particle velocities are analysed and evidence is presented which motivates the use of the Korteweg–de Vries equation as an appropriate model equation for these waves. We apply infinite-line scattering theory (a nonlinear generalisation of Fourier analysis for the Korteweg–de Vries equation) to gain insight into the nonlinear behaviour of long internal waves in this region.

7.1 **Preliminary discussion**

Interest in the generation and propagation of long nonlinear internal waves has been stimulated in recent years by the discovery of large *internal solitons* in the Andaman Sea (see [24] and [25]). The solitons are referred to as *internal waves* because they propagate beneath the sea surface in a medium whose density varies as a function of depth; the density stratification is determined by solar heating from above and by the local salinity content of the water. The internal solitons are long in the sense that their wavelengths (1–3 km) are much greater than the average thermocline depth (240 m). The solitons occur on rather short time scales of five to ten minutes. Thus an important experimental consideration is the rapid data recording interval of 75 seconds used for both temperature and velocity measurements. The largest of the internal solitons observed had measured amplitudes (isotherm excursions) of over 100 m. Packets of solitons occurred every 12.4 hours during the measurement programme. Each packet consisted on the average of about six to eight waves, generally rank-ordered by amplitude, with the largest leading the rest. The internal solitons are evidently tidally generated near shallow-water topographic features which are several hundred kilometres to the south and west of the measurement site (labelled E in Figure 7.1). Satellite photographs provide graphic evidence of their crest lengths which range up to 150 km. Offshore drilling operations demonstrate that the internal soliton is an ubiquitous feature of the marine environment in the southern Andaman Sea and that currents due to these waves can generate substantial hydrodynamic forces on drilling equipment [24].

Work on long internal wave motions has intensified in recent years on both theoretical and experimental fronts. Theoretical progress has been made in the discovery of inverse scattering transform solutions to several nonlinear wave equations which describe various types of long internal wave motions. Experimental progress has been made both in the laboratory and in the field. We begin a short review of recent work with a discussion of relevant theoretical aspects. For a more complete bibliography before about 1980, see [25] and cited references.

Several theories exist for long internal wave motions; all of them consist of nonlinear partial differential equations which describe unidirectional wave propagation in each of three water-depth regimes. The ratio of depth h to wavelength L separates the physics into three separate cases: (i) $h/L \gg 1$ defines deep water, (ii) $h/L \ll 1$ defines shallow water and (iii) $h/L \sim 1$ corresponds to intermediate depths. In shallow water the Korteweg–de Vries [19] (KdV) equation, accurate to second order, describes long internal motions [5, 13]; in deep water the appropriate model equation is that of Benjamin [4] and Ono [23] (BO). For intermediate water depths the second-order wave equation is given by Kubota, Ko and Dobbs [20] (KKD), an equation which reduces to the KdV equation in the shallow-water limit and the BO equation in the deep-water limit. The KKD equation is thus a theoretically reasonable generalisation of the theory for long internal wave motions to all water depths. An important consideration is that all of the above equations have rather general exact solutions which have been found using the modern mathematical method known as the scattering transform [11, 16, 17, 32] (see [1, 6, 8, 22] for reviews on scattering theory as a tool for solving integrable nonlinear wave equations). These theories solve the

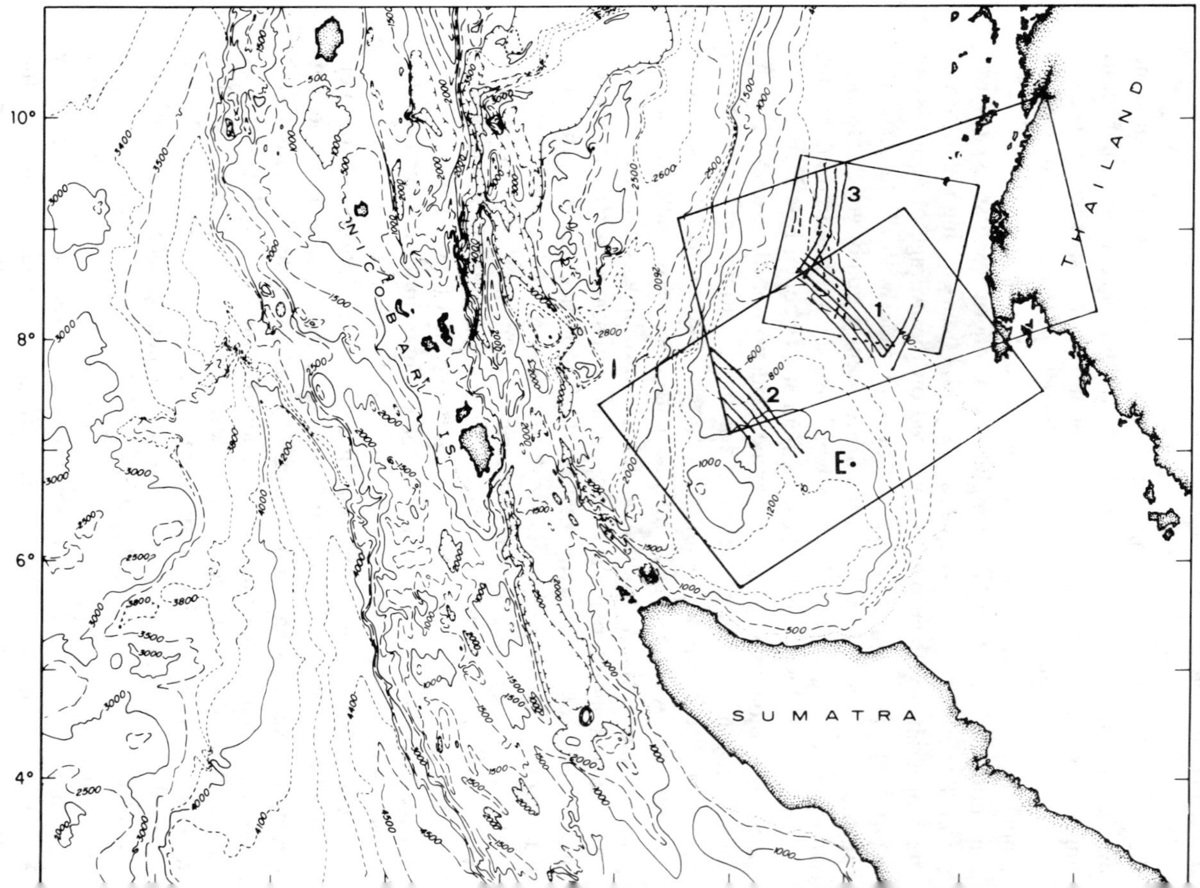

Figure 7.1 Map of the southern Andaman Sea showing the measurement site, E, and the rectangular boundaries of the Apollo–Soyuz photographs shown in Figures 7.2 and 7.3. The square boundary corresponds to a Landsat photograph [31]. Also shown are the locations of several internal wave packets observable on these photographs.

Cauchy problem: Given a sufficiently localised wave-form at some initial time $t = 0$, determine the wave motion for all future times thereafter. Since the soliton is a natural spectral component in all of these theories and because solitons have been ubiquitous in several experimental data sets (see discussion below), it therefore seems natural to consider interpretation of the data in terms of nonlinear long wave motions describable by one of KdV, KKD and BO.

We now discuss experimental results on long internal wave motions. Hodgins and Westergard [14] observed large internal solitons (amplitudes up to 45 m) in Arctic waters in the Davis Strait north of Labrador. They measured currents up to 2.3 knots; a two-layer KdV model was used to compare to the data. While their results are consistent with wave motion predicted by the KdV equation, on occasion they observe that the solitons appearing in the far field are not rank-ordered, a result which is contrary to the time-asymptotic behaviour of the KdV equation on the infinite line.

Internal solitons have been investigated experimentally in the laboratory by Koop and Butler [18] and a theoretical and experimental study has been conducted by Segur and Hammack [33]. These investigators found that in shallow water the KdV equation predicts the shapes of the measured solitons much better than the KKD equation; they found that none of the theories is applicable in deep water. Seeking better agreement Segur carried the finite-depth KKD theory to the next order of approximation and found that it becomes about as accurate as the lower-order KdV theory. The range of validity of the higher-order intermediate-depth theory is rather small, however, including only about half of the experimental data of Koop and Butler. Based upon the above studies the KdV equation apparently is applicable over a rather large depth range (extending from shallow to deeper water) while the KKD and BO equations were never found to be applicable. Segur and Hammack point out: (i) Both KdV and KKD are formally valid asymptotic equations that govern the slow evolution of long internal waves of small amplitude. However, the meaning of 'long' is different in the two theories; they describe perturbations to two different wave equations with two different linear phase speeds. (ii) The KdV equation seems to have a relatively large range of validity, and has practical predictive value even when its assumptions are satisfied only marginally. (3) The KKD equation by itself has such a small range of validity that it has been difficult even to find it experimentally.

Farmer and Smith [9] provide a thorough and detailed analysis of tidal interactions of stratified flow with a sill in Knight Inlet. Their paper opens several new avenues for future work with regard to generation mechanisms for long internal waves, particularily for situations dominated by complex topography. Chereskin [7] discusses measurements of the generation of internal waves in Massachusetts Bay; in particular tidal flow over a submarine ridge with large amplitude and asymmetry determines the type of flow response. The observations support the evolution of a solitary wave train whose properties are consistent with KdV evolution.

Fu and Holt [10] conducted an intensive study of SEASAT synthetic-aperture radar images of the Gulf of California, and in the absence of *in situ* measurements, inferred several properties of long internal waves on the basis of the KdV equation as a model.

More recently Kao, Pan and Renouard [15] investigated laboratory internal waves and they too found the KdV equation to describe long wave motions to a high degree of accuracy. These authors point out that the finite-depth KKD model is essentially a small perturbation of the BO model and is singular in the KdV limit. Thus their conclusions are consistent with those of Segur and Hammack [33]. They further point out that the scaling laws from the various models which relate the soliton amplitude to its wavelength (and to the predicted wave-form shape) have never been confirmed experimentally, either in the laboratory or the field.

The discovery of internal solitons also in the Sulu Sea (Apel, Holbrook, Liu and Tsai [2]) [2] has considerably extended the available data base for *in situ* measurements of long internal wave motions in the ocean. In particular measurements were obtained at three separate locations relative to the source region and it was observed how the waves evolved in time and space. Liu, Holbrook and Apel [21] used the KKD equation to model the internal solitons; they obtained good agreement with the data. They too found that the solitons in the far field were occasionally not rank-ordered.

Helfrich and Melville [13] present a thorough laboratory study of the propagation and stability of long nonlinear internal waves over slope–shelf topography. They use a form of the KdV equation modified to include dissipation and varying bottom topography. This model compares favourably with their data.

Several open questions face researchers in the field of long nonlinear internal wave motions. Among these are: (i) While many observers favour the KdV equation, definite determination as to which of the several possible wave equations is preferable (KdV, BO, or KKD) is still in doubt. In particular the scaling law for soliton amplitude to wavelength is still uncertain and the exact shape of the solitons has never been adequately established; in any event comparison of the KdV and BO solitons shows that the differences cannot exceed about 30%. Differences between KdV and KKD in the case of the Andaman Sea data are only of the order of 12–15 %. (ii) Occasionally measurements reveal solitons in the far field which are not rank-ordered. This violates known theoretical results about the time-asymptotic behaviour of the three model wave equations (with infinite-line boundary conditions). A major question is: Does this anomalous behaviour imply higher-order nonlinear effects, or is there a simpler explanation? (iii) Experimental verification of the presence of the radiation tail has been difficult to obtain, primarily because energetically the tail contains only a small fraction of the energy residing in the solitons. Do the radiation solutions play any major role in the dynamics of long internal waves? (iv) Are there experimental tests which can be made in order to verify whether one nonlinear wave equation is more applicable than the others? (v) What role does measurement error play in the interpretation of nonlinear experimental data? What effect can errors have on the assessment as to which wave equation may be more applicable than the others? Complete details about the resolution of these issues are more fully addressed elsewhere [31]; they are briefly discussed below.

The contents of this paper are summarised as follows. We first consider satellite photographs of internal· soliton propagation in the southern Andaman Sea (Section 7.2). Then in Section 7.3 we consider the high resolution *in situ* measurement programme conducted there in 1976 [25]. We also consider the observation that the measured particle velocities do not decay with depth and how this implies that the motion is compatible with that predicted by the KdV equation. This result, combined with the facts that (i) measurement noise obscures differences between KdV and KKD (see discussion below and [31]) and (ii) theoretically KKD is somewhat singular in the KdV limit [15, 33], leads us to select the simpler of the two models, namely KdV. This result motivates further discussion of the KdV equation both as a physical model (Section 7.4) and as a spectral description (Section 7.5) of internal soliton propagation. In Section 7.5 we also consider how the spectral theory (inverse scattering transform) for KdV may be viewed as a kind of nonlinear Fourier analysis. In Section 7.6 we use infinite-line scattering theory for the KdV equation to nonlinearly Fourier-analyse time series of measured temperature signals. The results support the soliton interpretation of the data. Remaining open questions and future work are discussed in Section 7.7.

7.2 **Satellite photographs**

Shown in Figure 7.1 is a bathymetry map of the southern Andaman Sea. The location labelled E is the position of the *in situ* measurement programme discussed herein (see Section 7.3). Also shown in Figure 7.1 are the rectangular boundaries of two photographs of the Andaman Sea surface taken during the Apollo–Soyuz programme [25] (see Figures 7.2 and 7.3) and the square boundary of a Landsat photograph [31]. The photograph of Figure 7.2 shows long striations (labelled '1' in Figure 7.1) which we intrepret as surface wave signatures ('rips') associated with

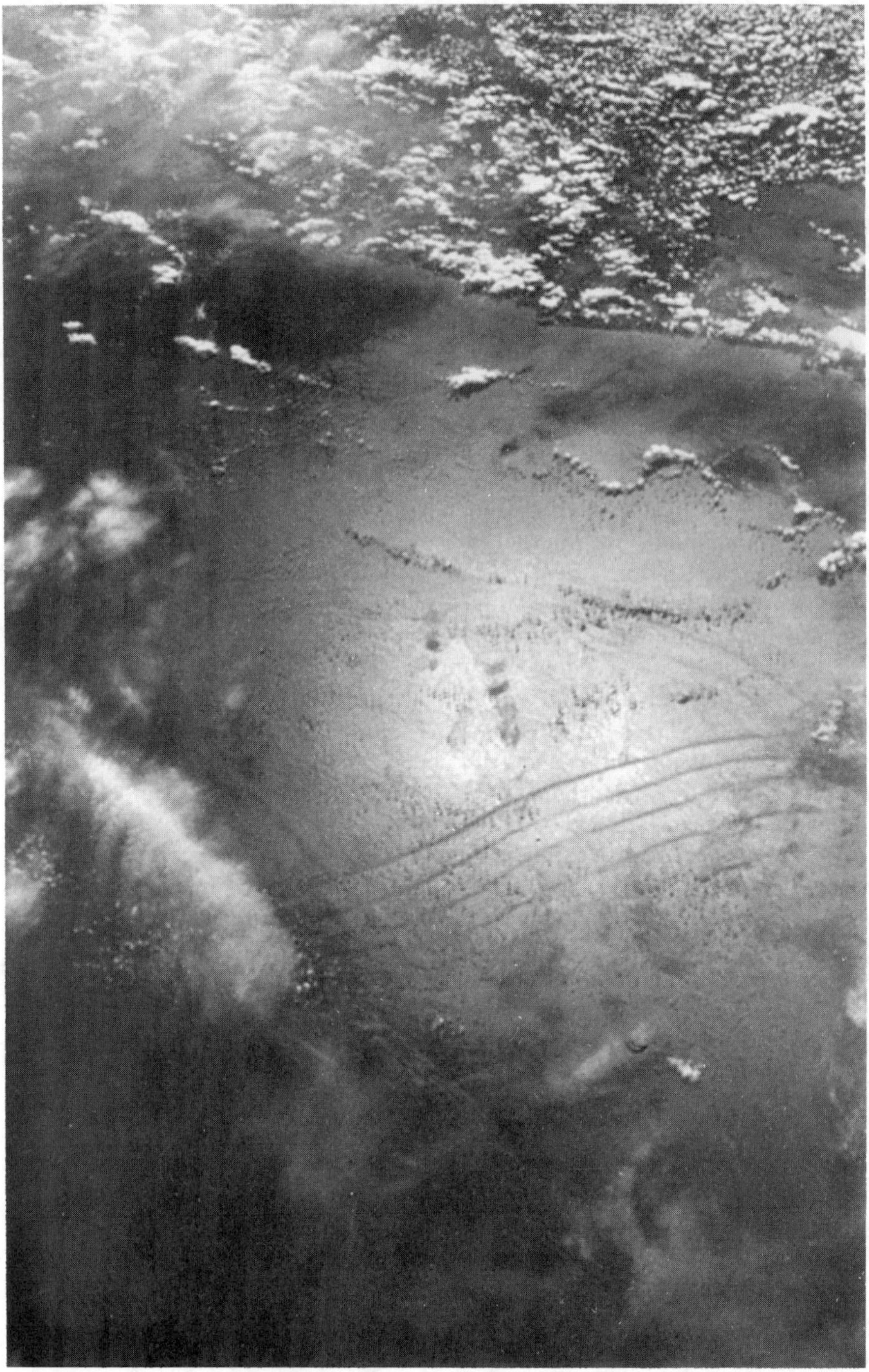

Figure 7.2 Apollo–Soyuz photograph of the southern Andaman Sea (north is to the left). Visible in the upper part of the photograph is the west coast of Thailand. The white puffy regions are clouds; the long striations in the centre of the photograph are due to the presence of internal waves propagating beneath the sea surface towards the coast of Thailand (here upwards, see packet labelled 'l' in Figure 7.1). The waves are visible from orbit because of the presence of surface wave bands ('rips') which accompany the internal waves (see Figure 7.4).

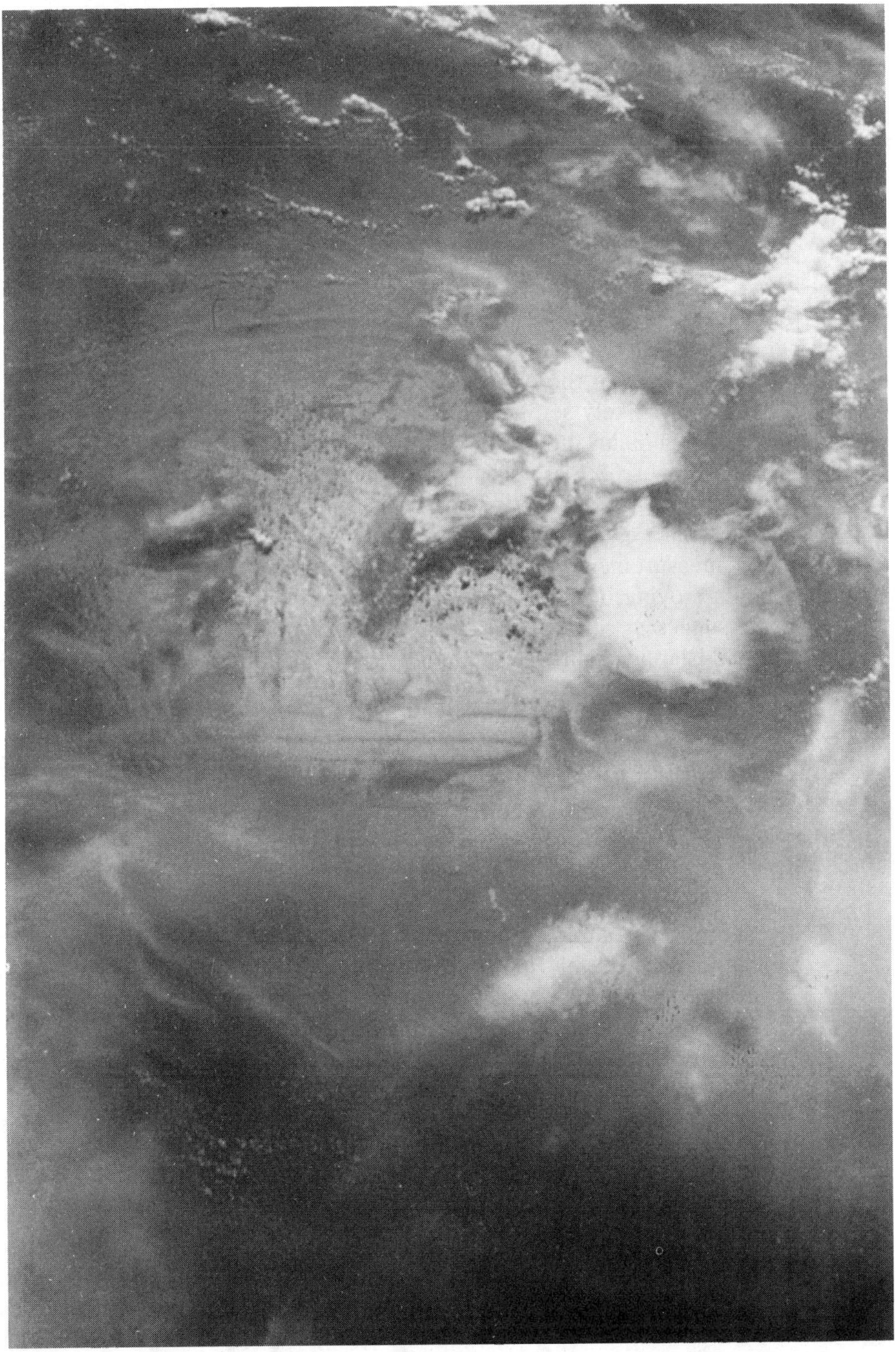

Figure 7.3 Apollo–Soyuz photograph of the southern Andaman Sea (north is to the left). This photograph was taken a few minutes before that of Figure 7.2. Two internal wave packets are visible through sparse clouds. The packet in the upper portion of the photograph is the same as that of Figure 7.2. The lower packet follows the first packet an estimated 12 hours later (see packet labelled '2' in Figure 7.1).

internal wave activity (discussed in Section 7.3 below). From orbital altitude the striations appear as packets of waves which evidently are propagating toward the coast of Thailand; to the extreme right of Figure 7.2 is a smaller packet of two waves which intersects the main packet. Another small packet may be seen at the extreme left of the main packet. Figure 7.3 is a satellite photograph taken a few minutes earlier than that shown in Figure 7.2; one also sees the packet of Figure 7.2 in the upper half of Figure 7.3; in the centre of this latter photograph is another packet (labelled '2' in Figure 7.1) which evidently trails the first by the local tidal period of about 12.4 hours.

The locations of the two larger packets of Figures 7.2 and 7.3 suggest that the most likely source region for the these waves is in the shallow-water region near the north-west coastline of Sumatra [25, 31]. Also shown in Figure 7.1 are striations obtained from a Landsat photograph (square boundary, labelled '3' in Figure 7.1) [31]. The packet in this case likely propagates from a source lying to the west of the photograph. The *in situ* measurements discussed below (Section 7.3) were dominated by currents moving in an east–west direction; we infer that these measurements suggest a source from the west, presumably somewhere near the Nicobar Islands chain, rather than a source region near the northern tip of Sumatra.

7.3 ***In situ*** **measurements of water temperature and particle velocity**

An internal wave measurement programme was conducted in the southern Andaman Sea (see location E of Figure 7.1) from 24 to 28 October 1976 [24, 25, 31]. Shipboard observations of local weather and sea conditions were made during the simultaneous deployment of subsurface instrumentation in a total depth of 1093 m. Measurements of water temperature and particle velocity were obtained with instruments placed on a taut cable (mooring) stretched vertically between an anchor deployed on the bottom and buoys suspended near, but below, the sea surface. Temperature sensors (thermistors) were placed on the mooring at approximate depths below the free surface of 53, 87, 116, 164 and 254 m. A total of ten current meters were also attached to the mooring at depths of 53, 87, 116, 121, 164, 254, 437, 635, 895 and 1001 m. The thermistors measured time series of water temperature $T(t)$ at their respective depths during the 4-day deployment period. Time series of about 3800 points each of both temperature and velocity were recorded at each of the upper five depths; time series of velocity of comparable duration were also recorded at the lower depths. Both temperature and velocity time series were each recorded at a rapid recording interval of 75 seconds.

Photographs of the sea surface were taken from the deck of the survey vessel during internal wave activity; a 35 mm camera provided the sequ-

Figure 7.4 Photographs of the surface of the Andaman Sea taken from the deck of the survey vessel during the passage of an internal soliton. The surface waves visible here are due to a long internal wave/short surface wave resonance interaction which inhances the surface wave amplitude above the convergence zones of the internal solitons (see also Figure 7.7). The bands of surface waves are referred to as 'rips' and are typically several hundred metres wide and stretch from one horizon to the other. In (a) an approaching rip band can be seen in the distance. About one minute later the rip is seen to be much closer to the survey vessel (b). The waves reach nearly 2 m maximum height about three minutes later (c). Fifteen minutes later the rip has passed and calm seas are left behind (d). The great difference in reflectivity properties of the surface waves for the rip bands and for the smooth sea surface between rips provides the contrast which allows the wave motion to be visible from satellite orbit.

ence of photographs shown in Figure 7.4. The event corresponds to the passage of an internal soliton of about 80 m amplitude recorded on 27 October, 1976; in Figure 7.4a one observes in the foreground nearby surface waves of about 0.5 m in height, while on the horizon there appears a long row of breaking waves which are approaching the observer. In panel (b) of Figure 7.4 the band of breaking waves has moved closer to the observer. In panel (c) the breaking waves of the rip band have arrived at the location of the survey vessel; the wave heights are about 2 metres. Shortly thereafter the band of breaking waves passes by the survey vessel and the sea surface is swept clean of nearly all surface wave activity (Figure 7.4d). This sequence of photographs occurred during a period of about 17 minutes. Bands of breaking waves of this type have been referred to as 'rips' in the oceanographic literature (see [25] and cited references). Sea-level observations indicate that the rips stretch from horizon to horizon and are normally 600–1200 m in width; it is thus natural to conjecture that these features are responsible for the striations on the satellite photographs shown in Figures 7.2 and 7.3. They are the result of a long internal wave/short surface wave resonance interaction [12].

Time series of temperature for the first soliton packet observed during the measurement programme are shown in Figure 7.5. Note the positive definite pulses or spikes which occur about every 40 minutes. The pulses are rank-ordered by amplitude, the largest leading the rest; the largest amplitudes occur at the station 164 m below the surface. The pulse shape and rank-ordering suggest a soliton interpretation of the data [25]. The largest wave recorded during the 4-day measurement programme was obtained with measurements made by expendable bathythermographs (XBTs, devices which measure water temperature as a function of depth). Figure 7.6 shows the isotherms constructed from these measurements. The maximum wave amplitude of 105 m occurs for the 17.5° isotherm.

The interested reader may find time series of measured particle velocities elsewhere [25, 31]. One of the more important observations about this data is that the particle velocities do not decay with increasing depth; they are relatively constant for depths below about 600 m from the free surface. This is consistent with wave motion predicted by the Korteweg–de Vries equation for a linearly varying vertical eigenfunction (see equation (7.4.4) below). KdV assumes the waves are long compared with the depth; this results in horizontal particle velocities that do not decay. The deep water (BO) theory does instead predict that the particle velocities decay with depth. Strictly speaking the soliton wave length-to-depth ratio in the Andaman Sea lies in the range 1–3; on this basis alone one might expect the KKD equation to be more applicable. However, experimental verification that the particle velocities do not decay with depth plus the fact that KKD is somewhat singular in the KdV limit, has motivated us to consider KdV as the simplest available wave equation for analysing the Andaman Sea data. KdV is simpler not only from the point of view of the partial differential equation itself but also from the point of view of its spectral theory. We are thus motivated to take a closer look at KdV and its mathematical and physical properties.

7.4 The Korteweg–de Vries equation as a model for long internal wave motions

We consider the model equation for long wave propagation given by Korteweg and de Vries [19]):

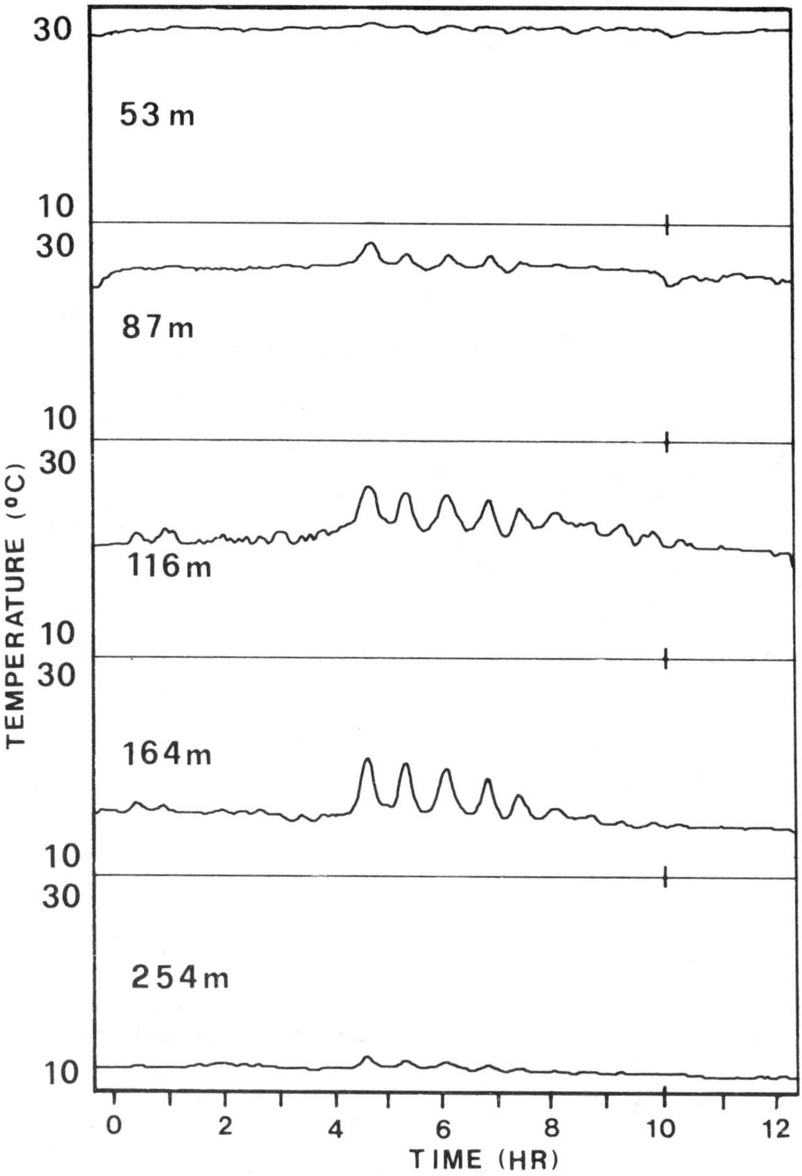

Figure 7.5 Time series of temperatures recorded at several depths below the surface at the measurement site E during the passage of a train of internal solitons on 24 October 1976. The wave amplitudes are largest for the instrument at 164 m. The waves may be viewed as moving to the left in the figure; note that they are rank-ordered by amplitude with the largest leading the rest.

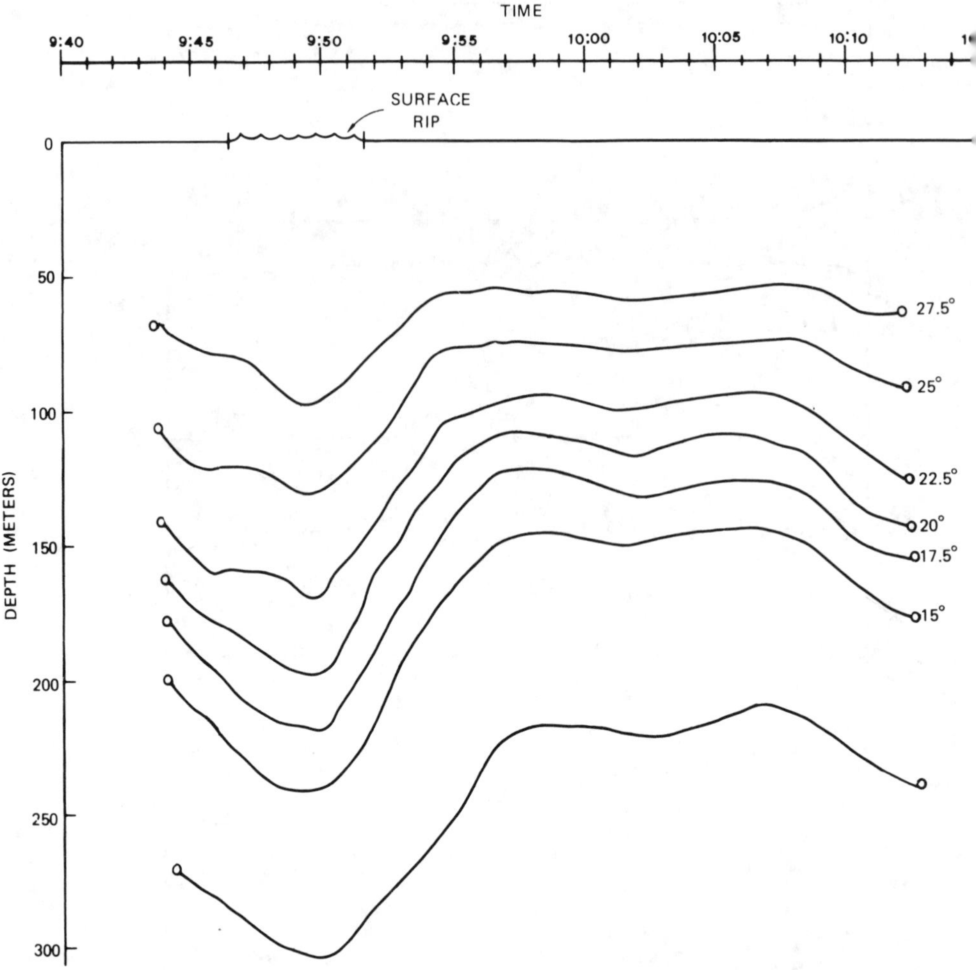

Figure 7.6 Isotherms constructed from XBT samples on 28 October 1986. This is the largest internal soliton observed in the measurement programme; the maximum excursion for the 17.5° isotherm is 105 metres.

$$\eta_t + c_0\eta_x + \alpha\eta\eta_x + \beta\eta_{xxx} = 0. \tag{7.4.1}$$

The coefficients c_0, α and β are constants ($c_0 = (gh)^{1/2}$, $\alpha = 3c_0/2h$ and $\beta = c_0h^2/6$ for surface-water waves) and the subscripts refer to partial derivatives. For long internal waves the KdV equation also describes the vertical displacement $\eta(x, t)$ of an isopycnal [5]. In this case the coefficients of the nonlinear and dispersive terms of the KdV equation are given by

$$\alpha = \frac{3c_0}{2} \frac{\int_0^h \rho_0(z)\, \phi_z^{\,3}(z)\, dz}{\int_0^h \rho_0(z)\, \phi_z^{\,2}(z)\, dz}, \tag{7.4.2}$$

$$\beta = \frac{c_0}{2} \frac{\int_0^h \rho_0(z)\, \phi^2(z)\, dz}{\int_0^h \rho_0(z)\, \phi_z^{\,2}(z)\, dz}. \tag{7.4.3}$$

The undisturbed density stratification $\rho_0(z)$ is a function of the vertical coordinate z (measured upwards from the bottom) and h is the total water depth. The constant c_0 is the linear phase speed of the wave motion; c_0 and $\phi(z)$ are the eigenvalues and eigenfunctions of the eigenvalue problem:

$$(\rho_0\phi_z)_z - (g/c_0^2)\, \rho_{0,z}\phi = 0, \tag{7.4.4}$$

where the surface and bottom boundary conditions are

$$\phi(0) = \phi(h) = 0. \tag{7.4.5}$$

The stream function associated with the flow is given by

$$\psi(x, z, t) = c_0\phi(z)\eta(x, t). \tag{7.4.6}$$

Associated horizontal and vertical particle velocities are computed from

$$u = \psi_z, \; w = -\psi_x. \tag{7.4.7}$$

When the density stratification can be approximated by a two-layer fluid, the expressions for c_0, α and β simplify; if the upper layer has density ρ_1 and thickness h_1 and the lower layer has density ρ_2 and thickness h_2 for $h_1 < h_2$ and $\rho_1 \sim \rho_2 \sim \rho$, then the coefficients of the KdV equation are given approximately by

$$c_0 = [g(\Delta\rho/\rho)h_1/(1 + r)]^{1/2}, \tag{7.4.8}$$

$$\alpha = -3c_0(1 - r)/2h_1, \tag{7.4.9}$$

$$\beta = c_0 h_1 h_2/6, \tag{7.4.10}$$

where $r = h_1/h_2$ and $\Delta\rho = \rho_2 - \rho_1 \ll 1$.

The general single-soliton solution to the KdV equation is

$$\eta(x, t) = -\eta_1 \operatorname{sech}^2[(x - c_1 t)/L_1] = -\eta_1 \operatorname{sech}^2(K_1 x - \Omega_1 t). \tag{7.4.11}$$

The depth variation in this solution is given by

$$\eta(x, z, t) = \phi(z)\, \eta(x, t). \tag{7.4.12}$$

The wavelength, phase speed and frequency of the soliton are:

$$L_1 = (12\beta/\eta_1|\alpha|)^{1/2} = K_1^{-1}, \tag{7.4.13a}$$

$$c_1 = c_0(1 + \eta_1|\alpha|/3c_0), \tag{7.4.13b}$$

$$\Omega_1 = c_1/L_1. \tag{7.4.13c}$$

In (7.4.11)–(7.4.13) the eigenfunctions, eigenvalues and coefficients of the KdV equation may be determined in general (by equations (7.4.2)–(7.4.5)) or for a two-layer fluid (by equations (7.4.8)–(7.4.10)). Note that the negative sign before the amplitude η_1 in (7.4.11) means that the soliton is a negatively (downwardly) displaced wave, a result of the assumption $h_1 < h_2$; the sign is reversed when the upper layer is thickest. The parameter α is negative for internal waves when the upper layer is thinner than the lower. In Figure 7.7 we show a typical internal soliton for a two-layer fluid (see the soliton equation (7.4.11)). The associated velocity field (isovels) are computed from (7.4.6) and (7.4.7)).

7.5 Nonlinear Fourier analysis for the KdV equation

Here we consider the space–time evolution of waves governed by the KdV equation (7.4.1). The wave motion is described by the scattering transform theory discovered by Kruskal and co-workers [11]. We consider solutions on the infinite line; here $(-\infty < x < \infty)$ in the context of a Cauchy problem: an 'initial wave' $\eta(x, 0)$ is assumed to be known at time $t = 0$ and to be localised in space (i.e. $\eta(x, 0)$ vanishes rapidly as $|x| \to \infty$). One seeks the evolution of the wave motion $\eta(x, t)$ in time thereafter. The spectral problem (which we refer to as the 'direct scattering transform', DST) is given by the Schrödinger eigenvalue problem

$$\Psi_{xx} + [\lambda\eta(x, 0) + \kappa^2]\Psi = 0, \tag{7.5.1}$$

which satisfies certain well-known asymptotic boundary conditions (see (7.5.2)–(7.5.7) below); $\lambda = \alpha/6\beta$. Note that $\eta(x, 0)$ is a 'potential' in the sense of quantum mechanics and κ is the spectral wavenumber. The problem may be formulated in terms of the 'transfer' or 'spectral matrix' $M(\kappa)$. Generally speaking $M(\kappa)$ is two-dimensional and complex. The eigenfunction solution to (7.5.1) has the asymptotic solutions

$$\left.\begin{aligned} \lim_{x\to-\infty} \Psi(x) &= A \exp(\mathrm{i}\kappa x) + B \exp(-\mathrm{i}\kappa x), \\ \lim_{x\to\infty} \Psi(x) &= E \exp(\mathrm{i}\kappa x) + F \exp(-\mathrm{i}\kappa x), \end{aligned}\right\} \tag{7.5.2}$$

where A, B, E and F are complex constants. The transfer matrix is then defined by

$$\begin{pmatrix} A \\ B \end{pmatrix} = M(\kappa)\begin{pmatrix} E \\ F \end{pmatrix}. \tag{7.5.3}$$

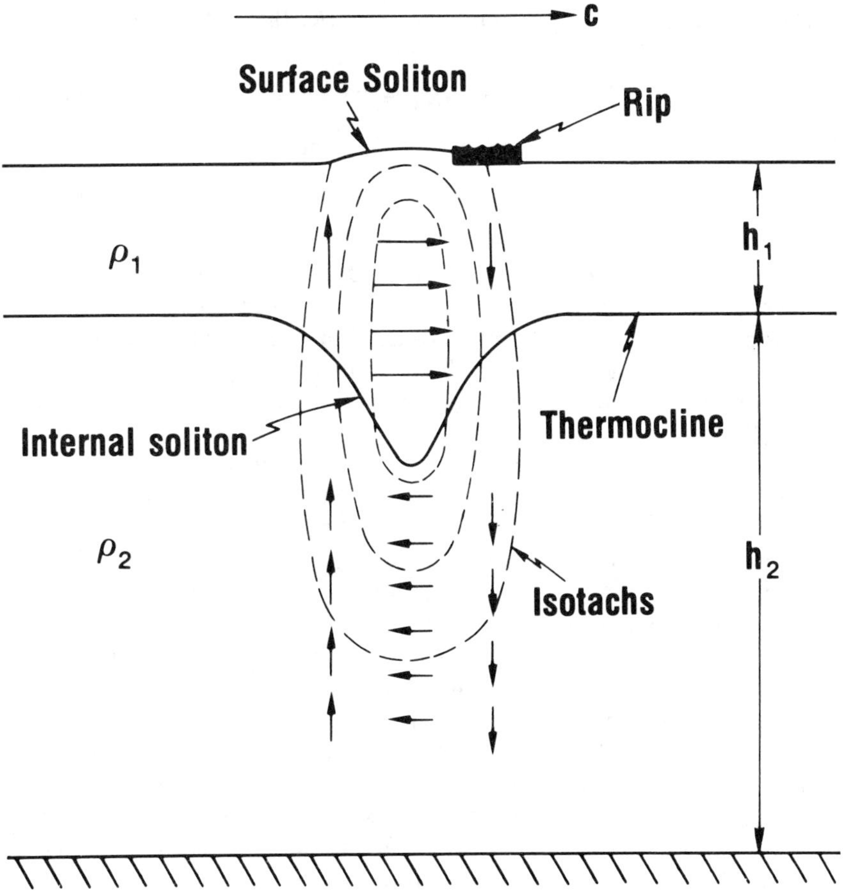

Figure 7.7 Theoretical description of an internal soliton for a two-layer fluid. The upper layer of density ρ_1 and thickness h_1 overlies a lower layer of density ρ_2 and thickness h_2 for the case $h_2 > h_1$. The soliton is a wave of depression. Note that a surface wave soliton accompanies the internal soliton; the surface soliton has an amplitude so small that it is very difficult to observe. The location of a rip band is shown to be on the leading edge (convergence zone) of the internal soliton. The isovels or isotacs (constant-velocity contours) associated with the internal soliton are also shown.

For real wavenumbers, $\kappa = k/2$, the transmission coefficient $a(k)$ and reflection coefficient $b(k)$ for (7.5.1) may then be written in terms of the components of $M(\kappa)$:

$$a(k) = 1/M_{11}(k/2), \tag{7.5.4}$$

$$b(k) = -M_{12}(k/2)/M_{11}(k/2). \tag{7.5.5}$$

The coefficient $b(k)$ is the 'radiation spectrum' of the wave train $\eta(x, 0)$.

For purely imaginary wavenumber, $\kappa = \mathrm{i}K$, the 'discrete specrum' is found from

$$M_{11}(\mathrm{i}K) = 0. \tag{7.5.6}$$

This gives the N-soliton wavenumbers, K_n, while the

$$C_n = -M_{12}(\mathrm{i}K_n) \tag{7.5.7}$$

are the soliton phase coefficients (when normalised by the usual eigenfunction normalisation condition). The DST spectrum consists of the following information: $\{K_n, C_n, N; b(k)\}$. One can reconstruct $\eta(x, 0)$ from the spectrum with the Gel'fand–Levitan–Marchenko integral equation (see [11] for a theoretical discussion, and [27] for the numerical implementation), a process we refer to as the 'inverse scattering transform' (IST).

The importance of the above formulation of the scattering transform for the KdV equation in terms of the spectral matrix M is that it is easily implemented numerically [30, 31]. This is because the spectral matrix $M(\kappa)$ may be computed by simple two-point recursion relations. Extension of the method to the KdV equation with periodic boundary conditions is also straightforward [28, 29].

In order to aid the reader in interpreting the nonlinear spectral analysis of the Andaman Sea data which follows, we show in Figure 7.8 a typical spectrum and its associated (temporally asymptotic) wave train. We graph the spectral amplitudes on the vertical axis (Figure 7.8a) and the wavenumber on the horizontal axis. The soliton amplitudes are represented as vertical arrows; the radiation spectrum ($|kb(k)|$) is shown as a continuous curve. In the present example there are three solitons in the discrete spectrum and two lobes in the continuous spectrum. In the 'far field' (asymptotically large times) the wave train tends to separate into its individual spectral components. This is seen in Figure 7.8b. The solitons become rank-ordered as they propagate to the right; the radiation trails behind and separates into packets associated with each of the lobes in the continuous spectrum.

7.6 **Nonlinear Fourier analysis of the data**

In order to compute the *nonlinear Fourier transform* of the data using the spectral theory for the KdV equation one first needs to compute the constant coefficients c_0, α and β. This requires knowledge of the undisturbed density stratification $\rho_0(z)$. Estimates of this function were computed from XBT measurements taken during 'quiet times' when large internal wave activity was not present; these temperature-versus-depth curves, when combined with supplemental data about the temperature and salinity structure, allow determination fo $\rho_0(z)$ [26, 31]. Given $\rho_0(z)$

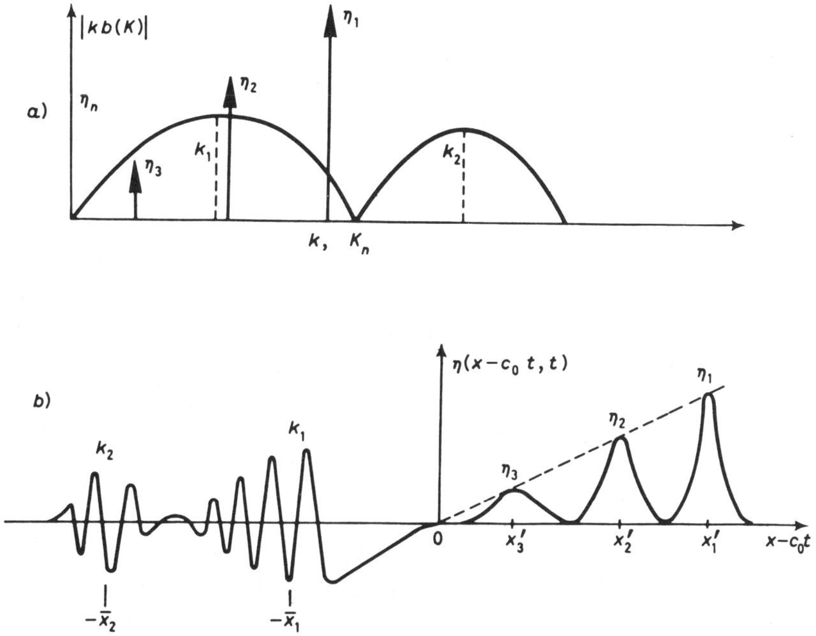

Figure 7.8 Graph of a typical soliton spectrum (a) and the associated long-time behaviour of the wave train (b). In (a) the three vertical arrows correspond to the solitons in the spectrum while the two lobes near wave numbers k_1 and k_2 correspond to the radiation spectrum. In (b) the solitons become rank-ordered after an asymptotically long time period and the radiation, trailing behind the solitons, separates into individual packets associated with each lobe of the continuous spectrum.

one numerically solves (7.4.4) for the eigenvalues c_0 and eigenfunctions $\phi(z)$. We assume only the first mode eigenfunction contributes to the motion; this is because a separate analysis reveals that about 90% of the total energy in the time series is attributable to the first mode [31]. The eigenfunction analysis for the example presented herein gives a linear phase speed of $c_0 = 2.17\,\mathrm{ms}^{-1}$ and a first mode eigenfunction with one broad maximum which occurs at a depth of $h_1 = 235\,\mathrm{m}$ in water of a total depth $H = 1093\,\mathrm{m}$. One computes the constants α and β by numerically integrating the integrals in (7.4.2) and (7.4.3); this allows computation of the ratio of nonlinearity to dispersion $\lambda = \alpha/6\beta$ (used in the eigenvalue problem (7.5.1) for the KdV equation.

The procedure for computing the spectrum is as follows. Largely because of economic considerations it is usually more convenient to measure time series in the ocean rather than space series. Thus the spectral analysis procedures for space series given in the last section must be modified for the analysis of time series [26, 30, 31]. Here one computes the spectral matrix $M(\Omega)$ (where Ω is a complex frequency) from the rescaled time series $c_0{}^2\lambda\eta(0, t)$; the soliton spectrum occurs for purely imaginary frequency, $\Omega = i\Omega_n$, and the radiation spectrum occurs for purely real frequency, $\Omega = \omega/2$. One graphs the continuous spectrum $|\omega b(\omega)|$ as a function of ω; the discrete spectrum (amplitudes of the solitons) are computed from the formula

$$\eta_n = 12\beta\Omega_n{}^2/c_n{}^2|\alpha|. \qquad (7.6.1)$$

The time series of the Andaman Sea data which we analyse herein is shown in the upper right-hand corner of Figure 7.9. The signal was recorded at a depth below the surface of 164 m on 24 October 1976. The time series of 600 points is discretised at 75 s intervals, is 12.5 hours long and begins at 10.20 of that day. While the original signal had units of temperature, the units shown in Figure 7.8 are in metres; knowledge of the density stratification was used to rescale the original temperatures to give the wave amplitude in units of length. The soliton amplitude to period relationship (7.6.1) is compared to the data in the upper left of Figure 7.9. While the agreement is rather good one notes the considerable scatter in the data points. How the inclusion of measurement error improves this comparison is discussed in [31]. The DST spectrum of the time series is shown in the main body of the figure. The vertical arrows correspond to the solitons in the spectrum. Each arrow is associated with the frequency Ω_n on the frequency axis; the length of the arrows is the soliton amplitude η_n. The continuous spectrum is the solid curve in the figure. We found a total of 12 solitons from the spectral analysis and have graphed the largest seven which have amplitudes of 57, 49, 46, 30, 14, 3 and 1.5 m. These amplitudes agree well with the amplitudes estimated directly from the time series.

The continuous DST spectrum is dominated by a large lobe extending out to near 0.0045 rad/s. The value of the frequency is close to the 'KdV cutoff frequency' of 0.0041 rad/s [26, 27]; for higher frequencies KdV is not an appropriate model equation. Thus the spectrum may be interpreted as giving rise to a single relatively large radiation tail which trails behind the solitons (the tail is not easily visible on the scale of time series in Figure 7.8). For more details of the spectral analysis see [26], [31].

7.7 Conclusions and future work

Large internal solitons have been measured in the Andaman Sea off the coast of Thailand near the northern tip of Sumatra. These waves are described to within measurement accuracy by the Korteweg–de Vries equation. Analysis of the data using KdV in the context of nonlinear Fourier analysis reveals the importance not only of the solitons but also of the radiation tail. The energy content in the radiation spectrum is surpri-

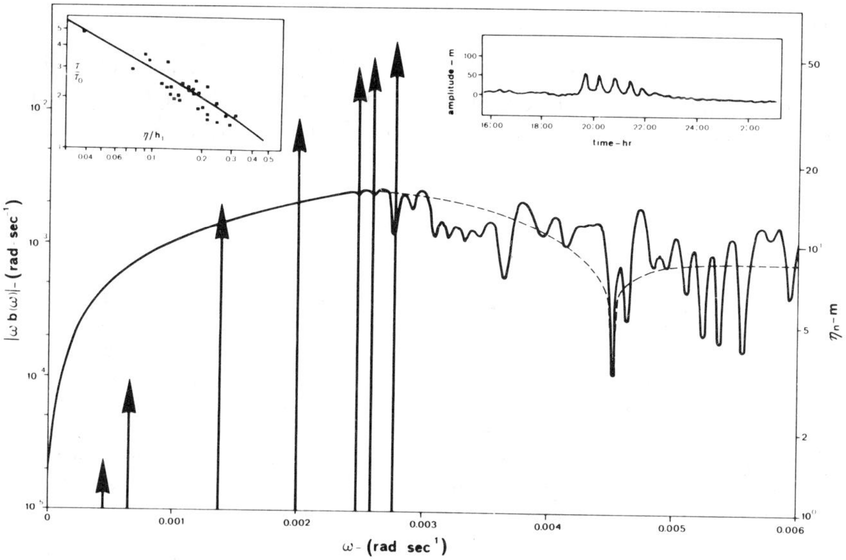

Figure 7.9 The results of a nonlinear Fourier analysis of a time series of temperature taken at a depth of 164 m below the surface during the passage of an internal soliton train. The time series is shown in the upper right-hand corner of the figure. The period of the individual solitons is compared with the soliton amplitudes (upper left); the theoretical curve is for the KdV equation. The vertical arrows in the main body of the figure correspond to the discrete eigenvalues for the solitons; the continuous curve is the radiation spectrum.

singly large; this suggests that radiative effects could play a major role in the generation and propagation of internal waves in this region.

Future work consists of using the *periodic* spectral/scattering transform for the Korteweg–de Vries equation to analyse the Andaman Sea data. This is because the physics of long internal wave motions in the Andaman Sea is basically tidally generated, approximately repeating every 12.4 hours, and therefore periodic boundary conditions are more appropriate than infinite-line boundary conditions. Analysing the data using the periodic transform should more explicitly bring out the role of the dynamics of the radiation components in the spectrum, particularly with regard to the 'asymptotic' state of the system, which technically does not exist for the periodic problem. Physically the periodic nature of nonlinear wave motion means that the radiation tail from one tidally generated initial wave could well inundate the solitons evolving from a trailing initial wave. It is conceivable that the open question regarding the fact that some of the solitons are not rank-ordered could be solved using periodic spectral analysis (see [31] for details).

Acknowledgements

Prof. C. Castagnoli provided valuable support and encouragement. Much of this chapter was adapted from earlier work and from Osborne and Segre [31]. Partial support from grants issued by the Ministero della Pubblica Istruzione is acknowledged.

References

[1] Ablowitz, M. J. and Segur, H. *Solitons and the Inverse Scattering Transform*, SIAM, Philadelphia (1981).
[2] Apel, J. R., Holbrook, J. R., Liu, A. K. and Tsai, J. J. The Sulu Sea internal soliton experiment, *J. Phys. Oceanogr.* **15**, 1625–51 (1985).
[3] Benjamin, T. B. Internal waves of finite amplitude and permanent form, *J. Fluid Mech.* **25**, 241–70 (1966).
[4] Benjamin, T. B. Internal waves of permanent form in fluids of great depth, *J. Fluid Mech.* **29**, 559–92 (1967).
[5] Benney, D. J. Long non-linear waves in fluid flows, *J. Math. Phys.* **45**, 52–63 (1966).
[6] Calogero, F. and Degasperis, A. *Spectral Transform and Solitons*, North-Holland, Amsterdam (1982).
[7] Chereskin, T. K. Generation of internal waves in Massachusetts Bay, *J. Geophys. Res.* **88**, 2649–61 (1983).
[8] Dodd, R. K., Eilbeck, J. C., Gibbon, J. D. and Morris, H. C. *Solitons and Nonlinear Wave Equations*, Academic Press, London (1982).
[9] Farmer, D. M. and Smith, J. D. Tidal interaction of stratified flow with a sill in Knight Inlet, *Deep-Sea Res.* **27A**, 239–54 (1980).
[10] Fu, L.-L., and Holt, B. Internal waves in the Gulf of California: observation from a spaceborne radar, *J. Geophys. Res.* **89**, 2053–60 (1984).
[11] Gardner, C. S., Greene, J. M., Kruskal, M. D. and Miura, R. M. Method for solving the Korteweg–de Vries equation, *Phys. Rev. Lett.* **19**, 1095–7 (1967).
[12] Gargett, A. E. and Hughes, B. A. On the interaction of surface and internal waves, *J. Fluid Mech.* **52**, 179–91 (1972).
[13] Helfrich, K. R. and Melville, W. K. On long nonlinear internal waves over slope-shelf topography, *J. Fluid Mech.* **167**, 285–308 (1986).
[14] Hodgins, D. O. and Westergard, H. H. Internal waves in Davis Strait and their measurment with a real-time system, *The Sixth International Conference on Port and Ocean Engineering under Arctic Conditions, July 27–31. Vol. 1*, 423–32 (1981).
[15] Kao, T. W., Pan F.-S. and Renouard, D. Internal solitons on the pycnocline: generation, propagation, and shoaling and breaking over a slope, *J. Fluid Mech.* **159**, 19–53 (1985).
[16] Kodama, Y., Satsuma, J. and Ablowitz, M. J. Nonlinear intermediate long-wave equation: analysis and method of solution, *Phys. Rev. Lett.* **46**, 687–90 (1981).
[17] Kodama, Y., Ablowitz, M. J. and Satsuma, J. Direct and inverse scattering problems of the nonlinear intermediate long wave equation, *J. Math. Phys.* **23**, 564–76 (1982).
[18] Koop, C. G. and Butler, G. An investigation of internal solitary waves in a two-fluid system, *J. Fluid Mech.* **112**, 225–51 (1981).

[19] Korteweg, D. J. and de Vries, G. On the change of form of long waves advancing in a rectangular canal, and on a new type of long stationary waves, *Philos. Mag. Ser.* **5**, 39, 422–43 (1895).
[20] Kubota, T., Ko, D. R. S. and Dobbs, L. D. Weakly-nonlinear, long internal gravity waves in stratified fluids of finite depth, *J. Hydronautics* **12**, 157–65 (1978).
[21] Liu, A. K., Holbrook, J. R. and Apel, J. R. Nonlinear internal wave evolution in the Sulu Sea, *J. Phys. Oceanogr.* **15**, 1613–24 (1985).
[22] Newell, A. C. *Solitons in Mathematics and Physics*, SIAM, Philadelphia (1985).
[23] Ono, H. Algebraic solitary waves in stratified fluids, *J. Phys. Soc. Japan* **39**, 1082–91 (1975).
[24] Osborne, A. R., Burch, T. L. and Scarlet, R. I. The influence of internal waves on deep-water drilling, *J. Petrol. Tech.* **30**, 1497–1504 (1978).
[25] Osborne, A. R. and Burch, T. L. Internal solitons in the Andaman Sea, *Science* **208**, 451–60 (1980).
[26] Osborne, A. R., Provenzale, A. and Bergamasco, L. The nonlinear Fourier analysis of internal solitons in the Andaman Sea, *Lett. Nuovo Cimento* **36**, 593–9 (1983).
[27] Osborne, A. R. The spectral transform: methods for the Fourier analysis of nonlinear wave data, in *Proceedings of a Workshop at the Ettore Majorana Centre, Erice, Italy, Statics and Dynamics of Nonlinear Systems*, G. Benedek, H. Bilz and R. Zeyher (eds), Springer, Berlin (1983).
[28] Osborne, A. R. and Bergamasco, L. The small-amplitude limit of the spectral transform for the periodic Korteweg–de Vries equation, *Il Nuovo Cimento* **85B**, 229–43 (1985).
[29] Osborne, A. R. and Bergamasco, L. The solitons of Zabusky and Kruskal revisited: perspective in terms of the periodic spectral transform, *Physica* **18D**, 26–46 (1986).
[30] Osborne, A. R. Nonlinear Fourier analysis for the infinite-interval Korteweg–de Vries equation I: An algorithm for the direct scattering transform, *J. Comput. Phys.* (in press).
[31] Osborne, A. R. and Segre, E. The nonlinear Fourier analysis of long internal wave motions in the Andaman Sea (in preparation).
[32] Satsuma, J., Ablowitz, M. J. and Kodama, Y. On an internal wave equation describing a stratified fluid with finite depth, *Phys. Lett.* **73A**, 283–6 (1979).
[33] Segur, H. and Hammack, J. L. Soliton models of long internal waves, *J. Fluid Mech.* **118**, 285–304 (1982).

8 *R. K. Dodd*

General relativity

8.1 The Geroch and Cosgrove groups

Prior to the middle of the 1970s the principal method for obtaining exact solutions to Einstein's field equations was to impose sufficient symmetries to reduce the number of equations and render them more amenable to analysis. The original Schwarzschild solution was obtained in this way. During the 1960s the number of exact solutions to the empty field equations proliferated due to the development of techniques which restricted attention to spacetimes with specific algebraic properties (see, for example, the excellent compendium [1]).

At the beginning of the 1970s in two remarkable papers Geroch [2, 3] showed that for empty spacetimes admitting a two-parameter commutative group of motions it appeared possible to generate all such spacetimes from flat space by the action of a hidden symmetry group associated with the field equations. In order to understand Geroch's argument and for the subsequent developments due principally to Kinnersley and Chitre we shall introduce the notation which will be used throughout the chapter.

Let M be a four-dimensional manifold (local coordinates $\{x^\alpha\}$, $\alpha = 1, \ldots, 4$) equipped with a pseudo-Riemannian structure (metric) $g_{\alpha\beta}$ which satisfies Einstein's empty field equations $R^{\alpha\beta} = 0$. If the spacetimes admit a commutative two-parameter group of motions then locally the generators of the group, the Killing vectors, form an Abelian Lie algebra. In this case the metric can be put into block diagonal form

$$\mathrm{d}s^2 = f_{AB}\, \mathrm{d}x^A\, \mathrm{d}x^B - h_{MN}\, \mathrm{d}x^M\, \mathrm{d}x^N, \tag{8.1.1}$$

where $A, B = 1, 2,\ M, N = 3, 4$ and the metric functions only depend upon x^3 and x^4. In this representation the Killing vectors are adapted to the coordinates

$$X^1 = \delta^{1\alpha} \frac{\partial}{\partial x^\alpha},\ X_1 = g_{1\alpha}\, \mathrm{d}x^\alpha,$$

$$X^2 = \delta^{2\alpha} \frac{\partial}{\partial x^\alpha}, \; X_2 = g_{2\alpha} \, dx^\alpha.$$

Then we find that the vector-valued one-form $X = (X_1, X_2)^{\mathrm{T}}$ is given by

$$X = F \begin{bmatrix} dx^1 \\ dx^2 \end{bmatrix}, \quad F = (f_{AB}) \tag{8.1.3}$$

The spacetimes (8.1.1) can be classified by the conditions det $F < 0$ or det $F > 0$. In the first case one of the Killing vectors is time-like and the spacetimes are called stationary axially symmetric empty spacetimes. If d(det F) $\neq 0$ then the metric for this case can be written in a form due to Lewis [4]:

$$ds^2 = f\,(dt - \omega\, d\phi)^2 - f^{-1}(e^{2\Gamma}\,(d\rho^2 + dz^2) + \rho^2\, d\varphi^2), \tag{8.1.4}$$

where $(t, \varphi, \rho, z) \equiv (x^1, x^2, x^3, x^4)$ by imposing the condition det $F = -\rho^2$ (signature $g_{\alpha\beta} = (+, -, -, -)$). The second case, det $F > 0$, corresponds to gravitational waves. These can be obtained from the metric (8.1.4) by imposing either of the transformations:

(i) $\quad t = i\tilde{z}, \quad z = i\tilde{t}, \quad \omega = i\tilde{\omega};$

(ii) $\quad t = i\tilde{\rho}, \quad \rho = i\tilde{t}, \quad \phi = i\tilde{\phi}. \qquad (8.1.5)$

We shall mainly consider the stationary axially symmetric empty spacetimes.

In order to explore the symmetry groups associated with the field equations we shall follow the ideas of Kinnersley [5] who was the first to systematically work out the consequences of Geroch's insights. For the Lewis metric (8.1.4) the field equations are block diagonal:

$$\begin{bmatrix} R^{AB} & 0 \\ 0 & R^{MN} \end{bmatrix} = 0. \tag{8.1.6}$$

Then since

$$R^A{}_C = -\tfrac{1}{2}\, \rho^2\, \nabla.(\rho^{-2} f^{AB} f_{BC}), \tag{8.1.7}$$

where $\nabla.V = \rho^{-1}(\rho V_1)_{,\rho} + V_{2,z}$ and $V = (V_1, V_2)$ it follows that $R^{AB} = 0$ can be written as

$$\nabla.(\rho^{-2} f^{AB}\, \nabla f_{BC}) = 0,$$

where
$$F = (f_{AB}) = \begin{bmatrix} f & -f\omega \\ -f\omega & f\omega^2 - \rho^2 f^{-1} \end{bmatrix}. \tag{8.1.8}$$

The indices on f_{AB} are raised by

$$\varepsilon = (\varepsilon^{AB}) = \begin{bmatrix} 0 & 1 \\ -1 & 0 \end{bmatrix} \tag{8.1.9}$$

The remaining field equations $R^{MN} = 0$ constitute a first-order completely integrable system for Γ:

$$\Gamma_{,\rho} = \tfrac{1}{4}\,\rho\, f^{-2}(f_{,\rho}^2 - f_{,z}^2) - \tfrac{1}{4}\,\rho^{-1}\, f^2(\omega_{,\rho}^2 - \omega_{,z}^2),$$
$$\Gamma_{,z} = \tfrac{1}{2}\,\rho f^{-2}\, f_{,\rho}\, f_{,z} - \tfrac{1}{2}\,\rho^{-1}\, f^2\, \omega_{,\rho}\, \omega_{,z}. \qquad (8.1.10)$$

The field equations (8.1.8) and (8.1.10) can be written succintly in terms of the matrices $V_M = \rho^{-1}F_{,x^M}\, F^{-1}$ (or $V_M^T = \rho^{-1}F^{-1}F_{,x^M}$), $M = 3, 4$ and the function $l = f^{-1}e^{2\Gamma}$:

$$(V_3)_{,\rho} + (V_4)_{,z} = 0 \qquad (8.1.11)$$

$$(\ln l)_{,\rho} = -\rho^{-1} + \tfrac{1}{4}\rho\, \mathrm{Tr}(V_3^2 - V_4^2) \qquad (8.1.12a)$$

$$(\ln l)_{,z} = \frac{1}{2\rho}\, Tr(V_3\, V_4) \qquad (8.1.12b)$$

The Geroch (Kinnersley–Chitre) group is associated with the equations (8.1.11) whereas the Cosgrove group which we discuss later in this section is connected with equations (8.1.12).

Two symmetry groups can be associated with the field equations (8.1.11). The first group is the Matzner–Misner symmetry group G of ignorable coordinate transformations which preserves the vector space spanned by the Killing vectors. The group is isomorphic to SL(2, $\mathbb{R}$) and is defined by

$$x'^A = g_B^A x^B, \quad \det(g_B^A) = 1. \qquad (8.1.13)$$

Under the action of G, f_{AB} transforms as an SL(2, $\mathbb{R}$) tensor

$$f'_{AB} = g_A^C\, g_B^D\, f_{CD} \qquad (8.1.14)$$

It is easy to investigate the effect of various subgroups of G. In particular in the Bruhat decomposition of G, $\varepsilon \in G$, (8.1.9), belongs to the Weyl group and its action is to interchange the Killing vectors.

The second group of symmetries is non-local and is derived by prolonging the system (8.1.11) to include additional variables, the 'twist potentials' ψ_{AB}, and searching for symmetries of the prolonged system. Their existence follows from the field equation (8.1.11) ($\equiv$ (8.1.8)), $\nabla.(\rho^{-2} f^{AB}\, \nabla f_{BC}) = 0$, since for any scalar $U = U(\rho, z)$,

$$\nabla.(\rho^{-1}\, \tilde{\nabla} U) = \rho^{-1}U_{,\rho z} - \rho^{-1}U_{,z\rho} = 0, \quad \tilde{\nabla} := (\partial_z, -\partial_\rho), \quad (8.1.15)$$

and consequently there exist ψ_{AB} such that

$$\tilde{\nabla}\, \psi^A{}_B = \rho^{-1}\, f^{AC}\, \nabla f_{CB}. \qquad (8.1.16)$$

If we drop G-covariance for the moment then the field equations (8.1.11) can be written in terms of the functions f and $\psi := \psi_{11}$ where from (8.1.16)

$$\nabla\psi = -\rho^{-1} f^2 \tilde{\nabla} \omega \tag{8.1.17}$$

The remaining linearly independent field equations in (8.1.11) can be chosen to be

$$\begin{aligned} &\nabla.(f^{-1} \nabla f + \rho^{-2} f^2\omega \nabla\omega) = 0, \\ &\nabla.(\rho^{-2} f^2 \nabla\omega) = 0. \end{aligned} \tag{8.1.18}$$

A straightforward calculation using (8.1.17) shows that these are equivalent to

$$\begin{aligned} \nabla.(f^{-2}(f \nabla f + \psi \nabla \psi)) &= 0, \\ \nabla.(f^{-2} \nabla\psi) &= 0. \end{aligned} \tag{8.1.19}$$

The equations (8.1.19) can be expressed in terms of a single complex potential function $E = f + i\psi$ called the Ernst potential [6]:

$$(\text{Re } E) \nabla^2 E = \nabla E.\nabla E \tag{8.1.20}$$

Neugebauer and Kramer [7] showed that the mapping I,

$$I\colon f \to \rho f^{-1}, \ \omega \to i\psi \tag{8.1.21}$$

transforms equations (8.1.18) directly into equations (8.1.19). Since the original field equations admitted the symmetry group G this means that the equations (8.1.19) admit an internal symmetry group H which is isomorphic to SL(2, $\mathbb{R}$). Under the action of I the tensor f_{AB} is equivalent to the SL(2, $\mathbb{R}$) tensor

$$m = (m_{AB}) = \begin{bmatrix} f^{-1} & f^{-1}\psi \\ f^{-1}\psi & f^{-1}(\psi^2 + f^2) \end{bmatrix}. \tag{8.1.22}$$

In particular, equations (8.1.19) are just the (12) and (22) components of the H convariant equations

$$\nabla.(m^{AB} \nabla m_{BC}) = 0. \tag{8.1.23}$$

The remaining equation arising from the (11) component is an algebraic consequence of the other two equations. Since H is isomorphic to SL(2, $\mathbb{R}$) it admits a Bruhat decomposition which is analogous to that for G. In particular, the subgroup of unipotent upper triangular matrices (1's on the diagonal) generate the Ehlers transformation,

$$\begin{aligned} f &\to f[1 + 2a\psi + a^2(\psi^2 + f^2)]^{-1}, \\ \psi &\to (\psi + a(\psi^2 + f^2))[1 + 2a\psi + a^2(\psi^2 + f^2)]^{-1}. \end{aligned} \tag{8.1.24}$$

The group H is called the Ehlers group [8].

Geroch showed that it was possible to generate from a given solution, solutions containing an arbitrary number of parameters by the successive application of the symmetry groups G and H. Starting from a time-like

Killing vector the group H generates a one-parameter family of metrics. A new Killing vector is then obtained by the action of G on this metric. A further application of H then results in a metric with an additional arbitrary parameter. The process can be continued to obtain solutions with an arbitrary number of parameters. Geroch [2, 3] suggested that it would be possible in this fashion to generate any exact solution of the stationary axially symmetric empty spacetimes by acting with a suitable element of the group so formed upon flat space. Hauser and Ernst [9] have proved a restricted version of this Geroch conjecture. We shall denote the symmetry group introduced above by K. Formally the Geroch group K has the direct product decomposition

$$K = \ldots\ldots H \times G \times H \times G \times H. \tag{8.1.25}$$

The group K has a nonlinear action on an infinite hierarchy of potentials which can be associated with the field equations. This hierarchy was introduced by Kinnersley and Chitre [5, 10]. Introduce the complex Ernst tensor potential

$$H_{AB} = f_{AB} + \mathrm{i}\,\psi_{AB}, \tag{8.1.26}$$

then using the definition of ψ_{AB}, (8.1.16), we get

$$\nabla H_{AB} = -\mathrm{i}\,\rho^{-1} f_A{}^C\, \tilde{\nabla}\, H_{CB}. \tag{8.1.27}$$

Equations (8.1.27) are the self-dual form of the field equations investigated by Hauser and Ernst [11]. Kinnersley and Chitre [5, 10] showed that the Geroch group acts on a hierarchy of potentials $\overset{n}{H}_{AB}$ which are defined by

$$\overset{n+1}{H}_{AB} = \mathrm{i}(\overset{1,n}{N}_{AB} + H_{AC}\,\overset{n}{H}{}^C_B) \tag{8.1.28}$$

where the potentials $\overset{m,n}{N}_{AB}$ are given by

$$\nabla\,\overset{m,n}{N}_{AB} = \overset{\overline{m}}{H}_{CA}\,\nabla\,\overset{n}{H}{}^C{}_B,\ m,\, n = 0,\, 1,\, 2,\, \ldots$$

with $\overset{0}{H}_{AB} = \mathrm{i}\,\varepsilon_{AB}$ and $\overset{0,n}{N}_{AB} = -\mathrm{i}\,\overset{n}{H}_{AB}$. (8.1.29)

The potentials $\overset{n}{H}_{AB}$ satisfy the equation (8.1.27):

$$\nabla\,\overset{n}{H}_{\mathrm{AB}} = -\mathrm{i}\,\rho^{-1} f_A{}^C\,\tilde{\nabla}\,\overset{n}{H}_{CB}. \tag{8.1.30}$$

The easiest way of investigating this hierarchy of potentials is to introduce a generating function for them,

$$R_{AB}(k) = \sum_{n=0}^{\infty} k^n\,\overset{n}{H}_{AB}. \tag{8.1.31}$$

The generating function satisfies the linear equation

$$\nabla R_{AB}(k) = \mathrm{i}k\, S(k)^{-2}\{(1 - 2kz)\, \nabla H_{AX} - 2k\rho\, \tilde{\nabla} H_{AX}\}\, R^{X}{}_{B},$$

where
$$S(k)^2 = (1 - 2kz)^2 + 4k^2\rho^2. \tag{8.1.32}$$

This is easy to verify if we use the fact that

$$H_{AB} + \overline{H}_{BA} = 2f_{AB} + 2\mathrm{i}z\, \varepsilon_{AB}. \tag{8.1.33}$$

The matrix version of (8.1.32) is

$$\mathrm{d}R = \mathrm{k}\, \mathrm{S}(k)^{-2}[1 - 2k(z - \rho^*)]\, \mathrm{d}H\, \Omega\, R,$$

where
$$\Omega := \mathrm{i}\varepsilon, \tag{8.1.34}$$

and * is the Hodge star operator

$$^*\mathrm{d}\rho = \mathrm{d}z,\ ^*\mathrm{d}z = -\mathrm{d}\rho. \tag{8.1.35}$$

The linear equation (8.1.34) is of fundamental importance in obtaining solutions to the field equations. We observe that if $\mathrm{d}k = 0$ the system is completely integrable provided H_{AB} satisfies (8.1.27). It is an example of a linear deformation problem for the field equations (8.1.27). The derivation of this and other deformation problems for the field equation (8.1.11) is presented in Section 8.2.

If the hierarchy of potentials $\overset{m,n}{N}_{AB}$ are considered, Kinnersley and Chiltre showed that the generators $\{\overset{l}{\gamma}_{AB}\}$ of the Lie algebra of K have the following action:

$$\overset{l}{\gamma}_{XY}\overset{m,n}{N}_{AB} = \varepsilon_{A(X}\overset{m+l,n}{N}_{Y)B} - \overset{m,n+l}{N}_{A(X}\varepsilon_{Y)B} + \sum_{s}^{l} \overset{m,s}{N}_{A(X}\overset{l-s,n}{N}_{Y)B}. \tag{8.1.36}$$

The indices in (8.1.36) are positive and the summation involves only positive values. Negative indices can be admitted by putting $\overset{p,-p}{N}_{AB} = -\overset{-p,p}{N}_{AB} = \varepsilon_{AB}$ where $p \geq 1$ and all other quantities with $m,\, n \leq 0$ equal to zero. If the new basis $\{\alpha^l = 1/\sqrt{2}\,(\gamma^l_{11} - \gamma^l_{22}),\ \beta^l = 1/\sqrt{2}(\gamma^l_{11} + \gamma^l_{22}),\ \gamma^l = \gamma^l_{12}\colon l \in \mathrm{Z}\}$ is introduced the commutation relations for the Lie algebra of K are given by

$$[\alpha^j, \beta^l] = \gamma^{j+l},\ [\gamma^j, \alpha^l] = 2\alpha^{j+l},$$
$$[\gamma^j, \beta^l] = -2\beta^{j+l}, \tag{8.1.37}$$

with all other commutators zero. The algebra is therefore isomorphic to the loop algebra $\hat{a}_1^{(1)} := sl(2, \mathbb{R}) \otimes \mathbb{C}[k^{-1}, k]$ with the correspondence given by

$$\alpha^j \approx e \otimes k^j,\quad \beta^j \approx f \otimes k^j,\quad \gamma^j \approx h \otimes k^j,\quad j \in \mathbb{Z}. \tag{8.1.38}$$

The Chevalley generators of $sl(2, \mathbb{R})$ are

$$e = \begin{bmatrix} 0 & 1 \\ 0 & 0 \end{bmatrix}, \quad f = \begin{bmatrix} 0 & 0 \\ 1 & 0 \end{bmatrix}, \tag{8.1.39}$$

$h = [e, f]$ and the bracket in $\hat{a}_1{}^{(1)}$ is given by

$$[a_1 \otimes k^j, a_2 \otimes k^l] = [a_1, a_2] \otimes k^{j+l}, \; a_i \in sl(2, \mathbb{R}). \tag{8.1.40}$$

The loop algebra $\hat{a}_1^{(1)}$ is a graded Lie algebra with the gradation defined on homogeneous elements by deg $(a_1 \otimes k^j) = j, j \in \mathrm{Z}$. It is clear that the whole algebra is generated by the elements of degree 1, 0 and -1. The action of the corresponding elements $\overset{1}{\gamma}_{AB}, \overset{0}{\gamma}_{AB}, \overset{-1}{\gamma}_{AB}$ on the field equation is to respectively generate infinitesimal Ehlers, ignorable coordinates and gravitational gauge transformations.

For empty space the Ernst tensor is easily seen to be

$$H = \begin{bmatrix} 1 & 2iz \\ 0 & -\rho^2 \end{bmatrix}. \tag{8.1.41}$$

The infinitesimal action of K on empty space can be shown to be given by [10, 12]:

$$\begin{aligned} \overset{k}{\gamma}_{11} &: H_{11} \to 1 + i\gamma\,(2r)^{k+1}\,P_{k+1}\,(\cos\theta), \\ \overset{k}{\gamma}_{22} &: H_{11} \to 1 - 2\gamma\,(2r)^k\,P_k(\cos\theta), \\ \overset{k}{\gamma}_{22} &: H_{11} \to 1 - i\gamma\,(2r)^{k-1}\,P_{k-1}(\cos\theta),\; k \geq 1, \end{aligned} \tag{8.1.42}$$

where $P_k(\cos\theta)$ is the kth Legendre polynomial and $r = (\rho^2 + z^2)^{1/2}$, $\cos\theta = z/r$. The fields created are therefore multipole gravitational fields of the linearised field equations, two of which are multipoles of angular momentum and the other is a mass mutlipole. As Kinnersley and Chitre [12] have pointed out, although these solutions are not asymptotically flat the corresponding exponentiated solutions of the nonlinear field equations may have this property. Notice in particular though that the infinitesimal transformation corresponding to $\overset{l}{\gamma}_{11} + \overset{l+2}{\gamma}_{22}$ is asymptotically flat. Thus the Abelian subgroup $B \subset K$ corresponding to the commutative Lie algebra generated by $\{\beta^k = \overset{k}{\gamma}_{11} + \overset{k+2}{\gamma}_{22} : k \in Z\}$ preserves asymptotic flatness. Kinnersley and Chitre in [13] show how asymptotically flat spacetimes can be obtained by exponentiation of the action of this Lie algebra. Another type of infinitesimal transformation which can be successfully exponentiated is the HKX (Hoenselaers–Kinnersley–Xanthopoulos) transformation [14–16], which arises from exponentiation of the action of a linear combination of the generators $\overset{k}{\gamma}_{22}$. As mentioned earlier, there is an alternative and much more powerful method for obtaining these solutions which employs the linear problem (8.1.34) and which provides a faithful representation of the group K. For this reason we will defer a detailed discussion of the solutions until later in the chapter.

So far we have not investigated symmetries which are associated with the field equations (8.1.12).

In fact there exist two further symmetry groups not contained in K. They were denoted Q and $\tilde{Q} = IQI$, where I is the Neugebauer–Kramer map (8.1.21), by Cosgrove [17] who discovered them. These groups are essentially equivalent to the Neugebauer Bäcklund transformations I_1 and I_2 [18] which are considered in Section 8.3. The group Q is locally isomorphic to SL(2, $\mathbb{R}$) and the action of $\delta \in Q$ is given by

$$\rho' = 4\rho(\delta_3^2\,\rho^2 + (\delta_3 z - 2\delta_4)^2)^{-1},$$

$$z' = \frac{\delta_1\delta_3\,\rho^2 + (\delta_1 z - 2\delta_2)(\delta_3 z - 2\delta_4)}{(\delta_3^2\,\rho^2 + (\delta_3 z - 2\delta_4)^2},$$

$$\Gamma'(\rho', z') = \Gamma(\rho, z),$$

where
$$\delta = \begin{bmatrix} \delta_1 & \delta_2 \\ \delta_3 & \delta_4 \end{bmatrix}. \qquad (8.1.43)$$

The functions Γ, Γ' are the coefficients appearing in the Lewis metric which satisfy equations (8.1.10). The action of the subgroup Q^0 generated by elements in which $\delta_1 = 1 = \delta_4$ and $\delta_2 = 0$ is non-trivial. The infinitesimal action of the group $\tilde{Q}^0$ on the hierarchy of potentials $\overset{m,n}{N}_{AB}$ is given by [19]:

$$\overset{m,n}{N}_{AB} \to \overset{m,n}{N}_{AB} + \varepsilon\,\{(m+1)\overset{m+1,n}{N}_{AB} + n\overset{m,n+1}{N}_{AB}\} \qquad (8.1.44)$$

In fact, Cosgrove demonstrates that I_1 and I_2 are the groups $H \times Q$ and $G \times Q$. The group Q has the property that it preserves asymptotic flatness, whereas $\tilde{Q}$ does not. These symmetry groups were omitted from the Kinnersley–Chitre analysis because that analysis was restricted to the field equations (8.1.11). The Cosgrove and Geroch groups together appear to constitute the full symmetry group for the hierarchy of fields associated with the spacetimes. Thus any symmetry ϕ of the hierarchy $\overset{m,n}{N}_{AB}$ (which includes the fields H^n_{AB}, equation (8.1.29)) can be decomposed formally into a product $\phi = ck$ where c is an element of the Cosgrove groups and $k \in K$. The question arises whether it is possible to find elements of the full symmetry group which will transform between the hierarchy of potentials for any two given spacetimes. This question is another version of the Geroch conjecture mentioned earlier. A restricted form of this conjecture proved by Hauser and Ernst will be presented in Section 8.3.

8.2 Linear deformation problems

Equation (8.1.34) is an example of a linear deformation problem for the field equations (8.1.11). Linear deformation problems are a particular type of Bäcklund tranformation. These were introduced as a generalisation of Lie's theory of contact transformations [20]. Contact transformations provided a theory of transformations for first-order partial differential equations but their extension to second-order equations resulted in only a trivial theory (the associated equations were differential prolongations of first-order systems).

The case of interest here is transformations between second-order equations in two independent variables ($(x^3, x^4) \equiv (\rho, z)$). Suppose that the two equations which are to be transformed into one another involve the variables x^M, $V = (v_1, \ldots, v_n)$, $V_M := V,_{x^M}$, $V_{MN} := V,_{x^M x^N}$ and $\hat{x}^M$, $Y = (y_1, \ldots, y_n)$, $Y_M := Y,_{\hat{x}^M}$, $Y_{MN} := Y,_{\hat{x}^M \hat{x}^N}$, $M, N = 3, 4$, respectively. We are interested in the existence of a Bäcklund transformation (type III in Clairin's classification [21]) which has the form

$$\begin{aligned} \hat{x}^M &= \mathrm{X}^{\mathrm{M}}(x^N, V, \mathrm{V}_{\mathrm{N}}, Y). \\ Y_M &= \Omega_M(x^N, V, V_N, Y),\ M, N = 3, 4. \end{aligned} \tag{8.2.1}$$

The equations are defined on the jet-bundle $J^2(\mathbb{R}^2, \mathbb{R}^n)$ which for the first equation can be taken to have the local coordinates (x^N, V, V_N, V_{NM}). Associated with each equation is the contact module I_2 on $J^2(\mathbb{R}^2, \mathbb{R}^n)$ which is generated locally by either of the sets of one-forms,

$$\mathrm{d}V = V_M\, \mathrm{d}x^M, \quad \mathrm{d}V_M = V_{MN}\, \mathrm{d}x^N, \tag{8.2.2}$$

$$\mathrm{d}Y = Y_M\, \mathrm{d}x^M, \quad \mathrm{d}Y_M = Y_{MN}\, \mathrm{d}x^N. \tag{8.2.3}$$

Provided $\mathrm{d}^2 Y = 0$, equations (8.2.1) are integrable and can be solved to define transformations (maps) on J^2 $(\mathbb{R}^2, \mathbb{R}^n)$. Because obtaining Y involves solving a differential equation the maps obtained in this way are not unique. The integrability condition can be written as

$$\mathrm{d}\Omega_M \wedge dX^M = 0$$

or
$$(\mathrm{D}_{x^N}\, \Omega_M + \Omega_{M,Y}\, \Omega_P\, \mathrm{D}_x N\, X^P)\, \mathrm{d}x^N \wedge (\mathrm{D}_x Q\, X^M + {} + X^M{}_Y\, \Omega_P\, \mathrm{D}_x Q\, X^P)\, \mathrm{d}x^Q = 0,$$

where
$$\mathrm{D}_{x^N} = \partial_{x^N} + V_N\, \partial_V + V_{MN}\, \partial_{V_M}. \tag{8.2.4}$$

Equation (8.2.4) is a second-order differential equation and provided it is equivalent to the second-order equation satisfied by the V variables equation (8.2.1) defines a Bäcklund transformation for it. The inverse transformations are obtained by inverting the relations (8.2.1), whereupon transformation of the same functional form are obtained but with the roles of x^M, V, V_M and $\hat{x}^M$, Y, Y_M interchanged. Analogous considerations to those above lead to a second-order equation for Y which is required to be equivalent to the given Y equation. Simplifications occur if we have the point transformation $\hat{x}^M = X^M \equiv x^M$ when equations (8.2.4) become

$$\mathrm{D}_{x^3}\,\Omega_4 - \mathrm{D}_{x^4}\,\Omega_3 + [\Omega_3, \Omega_4] = 0,$$

where
$$[\Omega_3, \Omega_4] := \Omega_{3,Y}\,\Omega_4 - \Omega_{4,Y}\,\Omega_3. \tag{8.2.5}$$

If in addition Ω_M is linear in Y, $\Omega_M = P_M\,Y$ then (8.2.5) becomes

$$\mathrm{D}_{x^3}\,P_4 - \mathrm{D}_{x^4}\,P_3 + [P_3, P_4] = 0, \tag{8.2.6}$$

where the bracket in (8.2.6) is the matrix commutator.

The particular cases of (8.2.1) which have been especially useful in other areas of soliton theory are when (i) V and Y satisfy the same differential equation – equation (8.2.1) is then termed an auto-Bäcklund transformation; (ii) the situation corresponding to (8.2.6) occurs – the equations (8.2.1),

$$Y_{,x^M} = P_M Y \quad M = 3, 4 \tag{8.2.7}$$

are then called a linear deformation problem for the equation in the V variables. In this section we shall consider the various deformation problems which have been derived in the literature and leave for Section 8.3 the detailed consideration of auto-Bäcklund transformations. The main reason for this is that the techniques for obtaining solutions to the field equations differ in the two cases although clearly the existence of auto-Bäcklund transformations and linear deformation problems are closely related. In fact given either transformation the other corresponding transformation can be easily derived.

The Belinsky–Zakharov deformation problem [22, 23]
In [22, 23] Belinsky and Zakharov derived a linear deformation problem for the field equations (8.1.11) in the case of a pair of space-like Killing vectors and when one of the Killing vectors was space-like. Since the Wick transformation (8.1.5) maps between the two problems we shall only consider the stationary axially symmetric vacuum case. Belinsky and Zakharov showed that the field equations (8.1.11) and their integrability conditions (obtained by re-expressing them in terms of the matrix F)

$$\begin{aligned} V_{3,\rho} + V_{4,z} &= 0 \\ V_{3,z} - V_{4,\rho} + \rho[V_3, V_4] - \rho^{-1}\,V_4 &= 0 \end{aligned} \tag{8.2.8}$$

were the conditions for the complete integrability of the linear system

$$\mathrm{D}_1 Y = \frac{1}{(\rho^2 + \lambda^2)}\,(\rho V_4 - \lambda V_3)Y,$$

$$\mathrm{D}_2 Y = \frac{1}{(\rho^2 + \lambda^2)} (\rho V_3 + \lambda V_4) Y,$$

$$D_1 := \partial_z - \frac{2\lambda^2\, \partial_\lambda}{(\rho^2 + \lambda^2)}, \quad D_2 := \partial_\rho + \frac{2\lambda\rho\, \partial_\lambda}{(\rho^2 + \lambda^2)}, \tag{8.2.9}$$

where $Y(\lambda, \rho, z)$ is a 2×2 matrix function which satisfies $Y(0, \rho, z) = F(\rho, z)$.

Harrison's deformation problem [24]

The deformation problem due to Harrison was discovered by applying the Wahlquist–Estabrook technique (see Chapter 15 and later in this section) to the Ernst equation (8.1.20).

Put $x = \frac{1}{2}(\rho + \mathrm{i}z)$, then the Ernst equation can be written as

$$E_{,x\bar{x}} + \rho^{-1}(E_{,x} + E_{,\bar{x}}) = f^{-1}\, E_{,x}\, E_{,\bar{x}} \tag{8.2.10}$$

Harrison found the one-form

$$\sigma = -\,\mathrm{d}q + \tfrac{1}{2}[s(q + q^2\gamma^{1/2}) - v(q + \gamma^{1/2})]\, \mathrm{d}x + \tfrac{1}{2}[w(q + q^2\gamma^{-1/2}) - u(q + \gamma^{-1/2})]\, \mathrm{d}\bar{x},$$

where $s = f^{-1}\, E_{,x} - \rho^{-1},\ u = f^{-1}\, E_{,\bar{x}} - \rho^{-1},$

$v = f^{-1}\, \bar{E}_{,x} - \rho^{-1},\ w = f^{-1}\, \bar{E}_{,x} - \rho^{-1},$

and
$$\gamma = \frac{1 - 4\mathrm{i}k\, \bar{x}}{1 + 4\mathrm{i}k\, x}, \quad k \in \mathrm{C}. \tag{8.2.11}$$

The one-form σ is annulled when restricted to a solution manifold $q = q(x, \bar{x})$, $w = w(x, \bar{x})$, $u = u(x, \bar{x})$, $v = v(x, \bar{x})$ and $s = s(x, \bar{x})$. It then defines a nonlinear deformation problem which can be linearised by introducing the variables y_1, y_2 such that $q = y_1/y_2$. The resulting linear deformation problem can be written in the form

$$Y_{,x} = \begin{bmatrix} \frac{1}{2}\,\mathrm{i}f^{-1}\, \psi_{,x} & -\frac{1}{2}(f^{-1}\, \bar{E}_{,x} - \rho^{-1})\gamma^{1/2} \\ -\frac{1}{2}(f^{-1}\, E_{,x} - \rho^{-1})\gamma^{1/2} & -\frac{1}{2}\mathrm{i}\, f^{-1}\, \psi_{,x} \end{bmatrix} Y$$

$$Y_{,x} = \begin{bmatrix} -\frac{1}{2}\,\mathrm{i}f^{-1}\, \psi_{,\bar{x}} & -\frac{1}{2}(f^{-1}\, E_{,\bar{x}} - \rho^{-1})\gamma^{-1/2} \\ -\frac{1}{2}(f^{-1}\, \bar{E}_{,x} - \rho^{-1})\gamma^{-1/2} & \frac{1}{2}\mathrm{i}f^{-1}\, \psi_{,\bar{x}} \end{bmatrix} Y \tag{8.2.12}$$

Hauser–Ernst deformation problem [10, 11]

In Section 8.1 we saw how this linear deformation problem was constructed from the conservation laws associated with the field equations (8.1.11) by defining a generating function (8.1.31). The linear deformation problem is given by

$$\mathrm{d}R = k\, S(k)^{-2}[1 - 2k(z - \rho^*)]\, \mathrm{d}H\, \Omega R,$$

where
$$\Omega := \mathrm{i}\varepsilon. \tag{8.2.13}$$

The operator * is the duality operator

$$* \, d\rho = dz, \; * \, dz = -d\rho. \qquad (8.2.14)$$

Instead of using the generating function the linear problem (8.2.13) can be obtained directly from the self-dual form of the field equations (8.2.11) [11].

Maison deformation problem [25]
This is an alternative formulation of the Kinnersley–Chitre/Ernst–Hauser problem. The recursion relations (8.1.30) are written as

$$\begin{aligned} \nabla\mu_{n+1} &= -\nabla(\mu\, \omega_n) - \nabla\omega\, \mu_n + \mu\nabla\, \omega_n + 2z\, \nabla\mu_n, \\ \nabla\omega_{n+1} &= \nabla(\mu\mu_n) - \nabla\omega\omega_n + \mu\, \nabla\mu_n + 2z\, \nabla\omega_n, \\ n &= 0,\, 1,\, 2,\, \ldots \end{aligned}$$

with $\mu_0 = 0,\ \mu_1 = \mu,\ \omega_1 = \omega,$ (8.2.15)

where

$$\mu^A{}_B = \varepsilon^{AC} f_{CB} \text{ and } \omega^A{}_B = \varepsilon^{AC}\, \psi_{CB}. \qquad (8.2.16)$$

The generating functions

$$V(k) = \sum_{n=0}^{\infty} k^n\, \mu_n,\; U(k) = \sum_{n=0}^{\infty} k^n\, \omega_n \qquad (8.2.17)$$

enable the equations (8.2.15) to be written concisely as

$$\begin{aligned} k^{-1}\, \nabla V(k) &= -\nabla(\mu U) - \nabla\omega V - \mu\, \nabla U + 2z\, \nabla V, \\ k^{-1}\, \nabla U(k) &= \nabla(\mu V) - \nabla\omega U + \mu\nabla V + 2z\, \nabla U. \end{aligned} \qquad (8.2.18)$$

If we impose the relations $V = \beta\mu(1 + \rho^2\, \beta^2)^{-1}\, Y$ and $U = (1 + \rho^2\, \beta^2)^{-1} Y$, then with the new independent variables $x = \rho + iz,\ \bar{x}$, equations (8.2.18) yield the system

$$\begin{aligned} Y_{,x} &= -\tfrac{1}{2}(1 - \gamma^{-1})\, \rho^{-2}\, \mu\, \mu_{,x}\, Y, \\ Y_{,\bar{x}} &= -\tfrac{1}{2}(1 - \gamma)\, \rho^{-2}\, \mu\, \mu_{,\bar{x}}\, Y, \end{aligned}$$

where $\gamma^2 = \dfrac{1 - 2ikx}{1 + 2ik\bar{x}},$

and $\beta = \tfrac{1}{2}\, k^{-1}\, \rho^{-2}\{(1 - 2kz) - [(1 - 2kz)^2 + 4k^2\rho^2]^{1/2}\}$

(8.2.19)

Neugebauer deformation problem [18]
The Neugebauer deformation problem was obtained from symmetry transformations of the equations (8.1.11) expressed in terms of the Ernst potential. Neugebauer used the non-canonical form of the metric for stationary axially symmetric empty spacetimes:

$$ds^2 = f(dt - \omega\, d\phi)^2 - f^{-1}(e^{2\gamma}\, dx\, d\bar{x} + V^2\, d\phi^2), \qquad (8.2.20)$$

where $x = \rho + iz$ and V is an arbitrary harmonic function $V_{,x\bar{x}} = 0$. The role of canonical coordinates is taken over by the pair (V, Z) where

$$\nabla V = \tilde{\nabla} Z. \tag{8.2.21}$$

Define

$$\begin{aligned} M_1 &= (2f)^{-1}\, E_{,x} \quad M_2 = (2f)^{-1}\, \bar{E}_{,x} \quad M_3 = V^{-1}\, V_{,x} \\ N_1 &= (2f)^{-1}\, \bar{E}_{,\bar{x}} \quad N_2 = (2f)^{-1}\, \bar{E}_{,x} \quad N_3 = V^{-1}\, V_{,\bar{x}} \end{aligned} \tag{8.2.22}$$

then the field equations can be written as the first-order system:

$$\begin{aligned} M_{1,\bar{x}} &= -M_1N_1 + M_1N_2 - \tfrac{1}{2}\, M_3N_2 - \tfrac{1}{2}\, M_1N_3, \\ M_{2,\bar{x}} &= -M_2N_2 + M_2N_1 - \tfrac{1}{2}\, M_3N_1 - \tfrac{1}{2}\, M_2N_3, \\ N_{1,x} &= -N_1M_1 + N_1M_2 - \tfrac{1}{2}\, M_2N_3 - \tfrac{1}{2}\, N_1M_3, \\ N_{2,x} &= -N_2M_2 + N_2M_1 - \tfrac{1}{2}\, N_3M_1 - \tfrac{1}{2}\, N_2M_3, \\ M_{3,\bar{x}} &= -M_3N_3 \quad N_{3,x} = -N_3M_3. \end{aligned} \tag{8.2.23}$$

New solutions are generated from known solutions by applying the symmetry transformation I_1,

$$I_1\colon M_i' = (\alpha_i^{\ k})M_k,\ N_i' = (\alpha_i^k)^{-1}N_k$$

where

$$\alpha_i^{\ k} = \begin{bmatrix} \alpha & 0 & 0 \\ 0 & \gamma\alpha^{-1} & 0 \\ 0 & 0 & \gamma \end{bmatrix}, \tag{8.2.24}$$

provided α and γ satisfy the Riccati equations:

$$\begin{aligned} d\gamma &= \gamma(\gamma - 1)\, M_3\, dx + (\gamma - 1)\, N_3\, d\bar{x}, \\ d\alpha &= [\alpha(\alpha - 1)\, M_1 + (\alpha - \gamma)M_2 + \tfrac{1}{2}\alpha(\gamma - 1)M_3]dx \\ &\quad + [(\alpha - 1)\, N_1 + \alpha\gamma^{-1}(\alpha - \gamma)N_2 + \tfrac{1}{2}\alpha\gamma^{-1}(\gamma - 1)N_3]d\bar{x}. \end{aligned} \tag{8.2.25}$$

The solution of the first equation in (8.2.23) is

$$\gamma = \frac{1 - 2ik(V - iZ)}{1 + 2ik(V + iZ)}, \tag{8.2.26}$$

where k is a constant of integration. The linear deformation problem is obtained from the nonlinear representation in (8.2.25) by the substitution $\alpha = -\gamma^{1/2}\, y_1/y_2$.

$$\begin{aligned} Y_{,x} &= \tfrac{1}{2}f^{-1} \begin{bmatrix} \bar{E}_{,x} & \gamma^{1/2}\, \bar{E}_{,x} \\ E_{,x} & \gamma^{1/2}\, E_{,x} \end{bmatrix} Y \\ Y_{,\bar{x}} &= \tfrac{1}{2}f^{-1} \begin{bmatrix} \bar{E}_{,\bar{x}} & \gamma^{-1/2}\, \bar{E}_{,\bar{x}} \\ E_{,\bar{x}} & \gamma^{-1/2}\, E_{,\bar{x}} \end{bmatrix} Y \end{aligned} \tag{8.2.27}$$

There is a simple relationship between the Harrison and Neugebauer deformation problems. In canonical coordinates ($V = \rho$, $Z = z$) it is given by

$$s = 4M_1 - 2M_3, \quad u = 4N_2 - 2N_3,$$
$$v = 4M_2 - 2M_3, \quad w = 4N_1 - 2N_3. \tag{8.2.28}$$

The prolongation variables q and α are related by

$$\alpha = \frac{\gamma^{1/2} + \gamma q}{\gamma^{1/2} + q}. \tag{8.2.29}$$

This somewhat bewildering proliferation of deformation problems for the same set of field equations can be connected by a general approach which is more in the spirit of soliton theory. This approach also reveals the role played by the parameter k (or the variable λ in the case of the Belinsky–Zakharov problem) which appears in the above presentation as the parameter in a generating function or as an integration constant. It is essentially the indeterminate in the loop group/algebra realisation of the Geroch group/algebra which was briefly mentioned in Section 8.1.

The Wahlquist–Estabrook method [26] provides a route for obtaining linear deformation problems of solvable equations. The basic idea has already been outlined at the beginning of this section. This is the determination of the one-form

$$\sigma = \mathrm{d}Y + FY\,\mathrm{d}\rho + GY\,\mathrm{d}z \tag{8.2.30}$$

for the field equations (8.1.11). It is convenient to use the Ernst equation (8.1.20). This can be written as the first-order system:

$$\rho^{-1} P_1 + P_{1,\rho} + P_{2,z} = - Q_1^2 - Q_2^2,$$
$$\rho^{-1} Q_1 + Q_{1,\rho} + Q_{2,z} = Q_1P_1 + Q_2P_2,$$
where $$P_1 = f^{-1} f_{,\rho},\ P_2 = f^{-1} f_{,z},\ Q_1 = f^{-1} \psi_{,\rho},$$
$$Q_2 = f^{-1} \psi_{,z}. \tag{8.2.31}$$

The integrability conditions for the P_i, Q_i, $i = 1, 2$, are

$$P_{1,z} = P_{2,\rho} \quad , Q_{1,z} - Q_{2,\rho} = P_1Q_2 - P_2Q_1. \tag{8.2.32}$$

The equations (8.2.31) and (8.2.32) can be used to generate an exterior system $E(M)$ on the manifold $M \subset \mathbb{R}^8$ with local coordinates (ρ, z, f, ψ, P_1, P_2, Q_1, Q_2) which is equivalent to the Ernst equation. This method is discussed in detail in Chapter 15. It is equivalent to solving for F and G the equation

$$\mathrm{D}_zF - \mathrm{D}_\rho G + [F, G] = 0, \tag{8.2.33}$$

where D_z, D_ρ are the total differentiation operators on M suitably restricted by the relations (8.2.31) and (8.2.32) [27]. As shown in [28] a Wahlquist–Estabrook prolongation exists provided

$$F = X_0 + P_1X_1 + P_2X_2 + Q_1X_3 + Q_2X_4,$$
$$G = X_5 - P_1X_2 + P_2X_1 - Q_1X + Q_2X_3, \qquad (8.2.34)$$

where the X_i's, $i = 0, \ldots, 5$, are matrix-valued functions of ρ and z. It turns out that X_1 and X_2 are linearly dependent. The X_i's satisfy the Lie-algebraic structure:

$$X_{5,\rho} - X_{0,z} + [X_0, X_5] = 0,$$
$$X_{1,z} + X_{2,\rho} - \rho^{-1}X_2 + [X_0, X_2] - [X_1, X_5] = 0,$$
$$X_{3,z} + X_{4,\rho} - \rho^{-1}X_4 + [X_0, X_2] - [X_3, X_5] = 0,$$
$$X_{2,z} - X_{1,\rho} - [X_0, X_1] - [X_2, X_5] = 0,$$
$$X_{4,z} - X_{3,\rho} - [X_0, X_3] - [X_4, X_5] = 0, \qquad (8.2.35)$$
$$[X_1, X_2] = 0, \; X_4 + [X_1, X_4] + [X_3, X_2] = 0,$$
$$X_3 - [X_2, X_4] - [X_1, X_3] = 0, \; X_2 - [X_3, X_4] = 0, \qquad (8.2.36)$$

where [,] is the matrix commutator. There is a 2×2 matrix representation of the algebra (8.2.36) in terms of a basis for $su(2)$,

$$Z_0 = -\tfrac{1}{2}i \begin{bmatrix} 1 & 0 \\ 0 & -1 \end{bmatrix}, \quad Z_1 = -\tfrac{1}{2} \begin{bmatrix} 0 & -1 \\ 1 & 0 \end{bmatrix}, \quad Z_2 = -\tfrac{1}{2} \begin{bmatrix} 0 & 1 \\ 1 & 0 \end{bmatrix}.$$
$$(8.2.37)$$

It is given by

$$X_1 = i \frac{(1 - v^2)}{(1 + v^2)} Z_0, \quad X_2 = \frac{-2iv}{(1 + v^2)} Z_0,$$

$$X_3 = Z_1 + i \frac{(1 - v^2)}{(1 + v^2)} Z_4,$$

$$X_4 = \frac{-2iv}{(1 + v^2)} Z_2. \qquad (8.2.38)$$

The parameter v in (8.2.38) is a function of ρ and z and its specific form is determined from (8.2.35). In fact there are two possible solutions. If $X_0 \neq 0 \neq X_5$ then we have a new type of realisation of the prolongation structure since it only has an infinite-dimensional realisation. It is necessary to assume that Y is a function of an additional variable λ. Then in this case the realisation is given by

$$X_0 = \frac{2v}{(1 + v^2)} \partial_\lambda, \quad X_5 = \frac{-2}{(1 + v^2)} \partial_\lambda \quad \text{and}$$

$$v = \rho/\lambda. \qquad (8.2.39)$$

The deformation problem defined by (8.2.38) and (8.2.39) is given by [29]:

$$\left(\partial_\rho + \frac{2\nu}{(1+\nu^2)}\,\partial_\lambda\right)Y = -\frac{f^{-1}}{(1+\nu^2)}\,\{i[(1-\nu^2)f_{,\rho} - 2\nu\, f_{,z}]Z_0$$
$$+ (1+\nu^2)\psi_{,\rho}\, Z_1 + i[(1-\nu^2)\psi_{,\rho} - 2\nu\,\psi_{,z}]Z_2\}\,Y$$
$$\left(\partial_z - \frac{2}{(1+\nu^2)}\,\partial_\lambda\right)Y = -\frac{f^{-1}}{(1+\nu^2)}\,\{i[2\nu f_{,\rho} + (1-\nu^2)f_{,z}]Z_0$$
$$+ (1+\nu^2)\psi_{,z}\, Z_1 + i[2\nu\psi_{,\rho} + (1-\nu^2)\psi_{,z}]\, Z_2\}\,Y \qquad (8.2.40)$$

The operators on the left-hand sides of (8.2.40) are the same as the differential operators in the deformation problem of Belinsky–Zakharov. Alternatively we can take $X_0 = 0 = X_5$ and in this case we find with $\tau = \pm 1$ that,

$$\tilde{\nu} = n + \tau(1+n^2)^{1/2} \quad \text{where} \quad n(\rho, z) = \frac{1}{2k\rho}\,(2kz - 1). \qquad (8.2.41)$$

We have used a tilde '˜' to distinguish this solution from that of the previous case. The parameter k is a constant of integration. In this case if we introduce the complex variables $x = \rho + iz$, $\bar{x}$ then the linear deformation problem can be written as

$$Y_{,x} = -\,\tau f^{-1}\,\{i\gamma^{1/2}(f_{,x}\, Z_0 + \psi_{,x}\, Z_2) + \tau\psi_{,x}\, Z_1\}Y$$
$$Y_{,\bar{x}} = -\,\tau f^{-1}\,\{i\gamma^{1/2}(f_{,x}\, Z_0 + \psi_{,\bar{x}}\, Z_2) + \tau\psi_{,x}\, Z_1\}Y$$

where $$\gamma = \frac{1 - 2ik\,\bar{z}}{1 + 2ik\,z} \qquad (8.2.42)$$

If Y is rescaled, $Y \to f^{1/2}Y$, and the $su(2)$ basis is changed,

$$Z_0 \to \tau\, Z_2, \quad Z_1 \to -Z_0, \quad Z_2 \to \tau\, Z_1, \qquad (8.2.43)$$

then the problem (8.2.42) becomes that introduced by Neugebauer. Alternatively the gauge transformation $Y \to \exp(-hZ_0)Y$ where $h = i\ln[(1+\tau\gamma^{1/2})/(1-\tau\gamma^{1/2})]$ and the change of basis,

$$Z_0 \to \tau\, Z_2, \quad Z_1 \to Z_0, \quad Z_2 \to \tau\, Z_1, \qquad (8.2.44)$$

applied to (8.2.42) gives the Harrison deformation problem. The relationship between the matrix representation and the infinite-dimensional realisation of the prolongation structure can be achieved by writing (8.2.40) in the form:

$$L_1 Y = (H\rho + \lambda J)Y, \; L_2 Y = (M^{-1}\rho + \lambda^{-1}N)Y,$$

where $$L_1 := \rho\,\partial_\rho + 2\lambda\partial_\lambda - \lambda\partial_z \;,\; L_2 := \rho^{-1}\,\partial_\rho + \lambda^{-1}\,\partial_z$$

and
$$H = \begin{bmatrix} \frac{1}{2}f^{-1} f_{,\rho} & 0 \\ f^{-1} \psi_{,\rho} & -\frac{1}{2}f^{-1} f_{,\rho} \end{bmatrix}$$
$$N = \begin{bmatrix} \frac{1}{2}f^{-1} f_{,z} & 0 \\ f^{-1} \psi_{,z} & -\frac{1}{2}f^{-1} f_{,z} \end{bmatrix}$$
$$M = -H^T, \quad J = N^T. \tag{8.2.45}$$

The solution of the homogeneous problem

$$L_1 u = 0, \quad L_2 u = 0, \tag{8.2.46}$$

obtained by the method of characteristics is

$$u = u(k) \quad \text{where} \quad k = \lambda(\lambda^2 + 2z\lambda - \rho^2)^{-1/2}. \tag{8.2.47}$$

Solving for λ we obtain the relationship between the functions v and $\tilde{v}$,

$$\tilde{v}(\rho, z, k) = -v(\rho, \lambda(\rho, z, k))$$
$$\lambda(\rho, z, k) = \tfrac{1}{2}k^{-1}((1 - 2kz) + \tau[(1 - 2kz)^2 + 4k^2\rho^2]^{1/2}) \tag{8.2.48}$$

If we introduce the vector-valued function

$$W(x, \bar{x}, k) = Y(\rho, z, \lambda(\rho, z, k)), \tag{8.2.49}$$

then equation (8.2.45) is transformed into (8.2.42). The Maison and Belinsky–Zakharov deformation problems are linked by a similar change of variables:

$$Y(\rho, z, k) = -F^{-1}\, Y\, (\rho, z, \lambda(\rho, z, k)). \tag{8.2.50}$$

Maison's problem can be obtained directly from (8.2.42) with the representation (8.2.37) by the following series of gauge transformations, $Y \to \left(\frac{1 - \gamma^{1/2}}{1 + \gamma^{1/2}}\right)^{-1/2} Y$, $Y \to G_\tau Y$, where $\tau = \pm 1$ and

$$G_\tau^{-1} = -\tfrac{1}{2}f^{1/2}\,\rho^{-1} \begin{bmatrix} f\omega(\tau - 1) + i\rho(\tau + 1) & -f(\tau - 1) \\ f(\tau + 1) - i\rho(\tau - 1) & -f(\tau + 1) \end{bmatrix} \tag{8.2.51}$$

Finally $Y \to HY$,

$$H = -2\rho^{-1} \frac{(1 - \gamma^{1/2})}{(1 + \gamma^{1/2})} \begin{pmatrix} 0 & 1 \\ 1 & 0 \end{pmatrix}, \tag{8.2.52}$$

and then introduce $\mu = \varepsilon F$ into the equation. The Hauser–Ernst problem can be converted into the Maison problem by observing that $U + iV = \varepsilon(F + i\Psi)$. The generating functions are easily seen to be related by $\varepsilon R = (1 + \rho^2\beta^2)^{-1}(\beta\mu + iI)\, Y$. The Belinsky–Zakharov eigenfunction is related to the Hauser–Ernst function by $Y_{BZ}(\lambda) = k^{-1}S(k)P(k)$, $R(k) = P(k) + Q(k)$.

8.3 Solution techniques

In the previous section we discussed the linear deformation problems which can be associated with the field equations and commented on the fact that they were examples of Bäcklund transformations. Linear deformation problems provide one route for obtaining solutions to the field equations, but as we indicated then, the auto-Bäcklund transformations provide another. They are particularly useful for obtaining the soliton solutions associated with the field equations. These are solutions which have an explicit analytic form. For both the linear deformation and the auto-Bäcklund transformation it is the presence of an additional parameter (k or λ in Section 8.2), which is absent from the field equations, that provides the method of solution. In the case of the auto-Bäcklund transformations it enables a nonlinear superposition principle to be formulated whereby new solutions can be obtained algebraically from a given solution and its Bäcklund transforms.

Because of the difference in the techniques used to obtain solutions from the Bäcklund transformations and the linear deformation problems we shall treat them separately. The last two examples in the subsection on Bäcklund transformations, the Belinsky–Zakharov and Dodd–Morris Bäcklund transformations are derived by a method which provides a bridge between the two approaches.

8.3.1 *Bäcklund transformations*

The Wahlquist–Estabrook method briefly outlined in the previous section is equivalent to a method due to Clairin [21] for constructing auto-Bäcklund transformations. This approach has been implemented by Omote and Wadati [30] for the Ernst equation

$$E_{,x\bar{x}} + \tfrac{1}{4}\rho^{-1}(E_{,x} + E_{,\bar{x}}) - f^{-1}\, E_{,x}E_{\bar{x}} = 0, \qquad (8.3.1)$$

where $x = \rho + iz$. Then from (8.2.1) if E' is a new Ernst potential which satisfies (8.3.1) the associated Bäcklund transformation is given by

$$\begin{aligned} E'_{,x} &= P(x, \bar{x}, E, \bar{E}, E_{,x}, E_{,\bar{x}}, \bar{E}_{,x}, \bar{E}_{,\bar{x}}, E'), \\ E'_{,x} &= Q(x, \bar{x}, E, \bar{E}, E_{,x}, E_{,\bar{x}}, \bar{E}_{,x}, \bar{E}_{,\bar{x}}, E'). \end{aligned} \qquad (8.3.2)$$

The functions P and Q are determined from the complete integrability condition on (8.3.2) and the requirements that E and E' satisfy (8.3.1). The authors restricted themselves to the cases when P and Q are linear in $E_{,x}$ and $E_{,\bar{x}}$ respectively. They obtained three types of transformation which are as follows:

(i) Neugebauer I_1 transformation with $\gamma = 1$

$$P = f^{-1} f'\, E_{,x} y \quad Q = f^{-1} f'\, E_{,x} y$$

and y is a solution of the Riccati system

$$\left.\begin{aligned} y_{,x} &= \frac{(y-1)}{2f}(E_{,x}+\bar{E}_{,x}) \\ y_{,\bar{x}} &= \frac{(y-1)}{2f}(E_{,\bar{x}}+\bar{E}_{,\bar{x}}) \end{aligned}\right. \quad \text{and} \quad y\,\bar{y} = 1$$

(ii) Neugebauer I_2 transformation, with $\gamma = 1$

$$P = f^{-1} f' E_{,x}y + \tfrac{1}{2}\rho f'(1-y)$$
$$Q = f^{-1} f' y^{-1} + \tfrac{1}{2}\rho f'(y-1)y^{-1}$$

and y is real and satisfies

$$y_{,x} = -\tfrac{1}{4} f^{-1}(y-1)[(y+1)(E_{,x}+\bar{E}_{,x}) - (y-1)(E_{,x}-\bar{E}_{,x})] + \tfrac{1}{4}\rho^{-1}\tfrac{3}{4}y - 1)(y+1)$$
$$y_{,\bar{x}} = -\tfrac{1}{4} f^{-1}(y-1)[(y+1)(E_{,\bar{x}}+\bar{E}_{,\bar{x}}) + (y-1)(E_{,\bar{x}}-\bar{E}_{,\bar{x}})] + \tfrac{1}{4}\rho^{-1}(y-1)(y+1)$$

(iii) Harrison tranformation

$$P = -\frac{f'y(\gamma^{1/2}y+1)}{f(y+\gamma^{1/2})}E_{,x} + \tfrac{1}{2}\rho^{-1} f'(\gamma^{1/2}y+1)$$

$$Q = -\frac{f'y^{-1}(\gamma^{1/2}y+1)}{f(y+\gamma^{1/2})}E_{,\bar{x}} + \tfrac{1}{2}\rho^{-1}\gamma^{-1/2}y^{-1}(\gamma^{1/2}y+1)$$

and y satisfies

$$y_{,x} = \tfrac{1}{2} f^{-1}[y(\gamma^{1/2}y+1)E_{,x} - (y+\gamma^{1/2})\bar{E}_{,x}] - \tfrac{1}{4}\rho^{-1}\gamma^{1/2}(y-1)(y+1)$$
$$y_{,\bar{x}} = \tfrac{1}{2} f^{-1}[-(\gamma^{1/2}y+1)E_{,\bar{x}} + y(y+\gamma^{1/2})\bar{E}_{,\bar{x}}] - \tfrac{1}{4}\rho\gamma^{1/2}(y-1)(y+1)$$

where $\gamma = \dfrac{1-2i\bar{x}}{1+2ikx}$ and k is constant.

We shall discuss these particular Bäcklund transformations in detail under their respective headings.

Harrison Bäcklund transformation [24]

We have already discussed the Harrison deformation problem in the previous section. The auto-Bäcklund transformation is obtained by requiring the new solution of the Ernst equation defined by the variables s', u', v', w', (cf. equation (8.2.11)) to belong to the exterior system generated by σ and the one-forms corresponding to the Ernst equation (8.2.10) written in terms of the variables s, u, v, w. Harrison assumed that the functional form of s' was linear in s and depended only upon x, $\bar{x}$ and q.

The other primed variables were defined similarly. Harrison found that

$$\begin{aligned}
u' &= \frac{-1}{(q + \gamma^{1/2})} [q^{-1}(q\gamma^{1/2} + 1)u + \rho^{-1}\gamma^{-1/2}(\gamma - 1)], \\
v' &= -\frac{1}{(q\gamma^{1/2} + 1)} [q^{-1}(q + \gamma^{1/2})v + \rho^{-1}(1 - \gamma)], \\
s' &= \frac{-q}{(q + \gamma^{1/2})} [(q\gamma^{1/2} + 1)s + \rho^{-1}(1 - \gamma)], \\
w' &= \frac{-q}{(q\gamma^{1/2} + 1)} [(q + \gamma^{1/2})v + \rho^{-1}\gamma^{-1/2}(\gamma - 1)], \qquad (8.3.4)
\end{aligned}$$

where q is a solution of the Riccati equations obtained by restricting the one-form σ (8.2.11) to a solution manifold. Many of the properties of the Harrison Bäcklund transformation are more easily obtained from the Neugebauer Bäcklund transformation.

Neugebauer Bäcklund transformation

We shall use the notation introduced in the discussion of the Neugebauer deformation problem equations (8.2.20)–(8.2.27). The Bäcklund transformations I_1 and I_2 are defined by

$$\begin{aligned}
&I_1\colon M_i' = \alpha_1^k M_k \quad, N_i' = (\alpha_i^k)^{-1} N_k, \\
&I_2\colon I_2 = SI_1S,
\end{aligned}$$

where $S\colon M_i' =$ $\quad S\colon M_i' = \mu_i^k M_k \quad, N_i' = \nu_i^k N_k,$

$$(\mu_i^k) = \begin{bmatrix} 0 & -1 & \frac{1}{2} \\ -1 & 0 & \frac{1}{2} \\ 0 & 0 & 1 \end{bmatrix}, \quad (\nu_i^k) = \begin{bmatrix} -1 & 0 & \frac{1}{2} \\ 0 & -1 & \frac{1}{2} \\ 0 & 0 & 1 \end{bmatrix}. \qquad (8.3.5)$$

The primed variables define new solutions to the field equations and the matrices (α_i^k), M and N are given in equations (8.2.24). Each transformation is characterised by the solutions α, γ to the Riccati equations (8.2.25) and are therefore denoted by $I_i(\alpha, \gamma)$, $i = 1, 2$. The group properties of the $I_i(\alpha, \gamma)$ are $I_i(\alpha_1, \gamma_1)\, I_i(\alpha_2, \gamma_2) = I_i(\alpha_1\alpha_2, \gamma_1\gamma_2)$ and $I_i^{-1}(\alpha_1, \gamma_1) = I_i(\alpha^{-1}, \gamma^{-1})$ where α_i, γ_i, $i = 1, 2$ are solutions of the Riccati equations with different integration constants. Neugebauer found that the sequence of Bäcklund transformations shown in Figure 8.1 constituted the identity transformation.

The Harrison transformation can be expressed as a product of Neugebauer transformations [19],

$$\begin{aligned}
H(\alpha, \gamma) &= I_2\left(\frac{\alpha - \gamma}{\gamma(\alpha - 1)}, \gamma^{-1}\right) I_1(\alpha, \gamma) \\
&= I_1(\alpha\,\gamma^{-1}, \gamma^{-1})\, I_2\left(\frac{\alpha - \gamma}{(\alpha - 1)}, \gamma\right). \qquad (8.3.6)
\end{aligned}$$

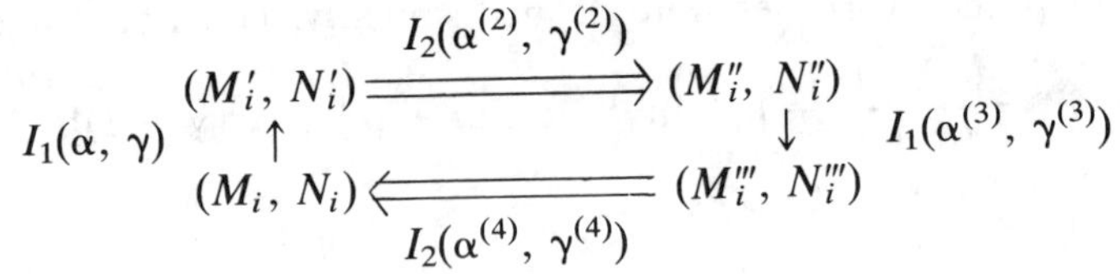

$$\alpha^{(2)} = \frac{\alpha - \gamma}{\gamma(\alpha - 1)} \qquad \gamma^{(2)} = \gamma^{-1}$$

$$\alpha^{(3)} = \gamma\alpha^{-1} \qquad \gamma^{(3)} = \gamma$$

$$\alpha^{(4)} = \frac{\alpha - 1}{\alpha - \gamma} \qquad \gamma^{(4)} = \gamma^{-1}$$

Figure 8.1 Neugebauer's identity transformation

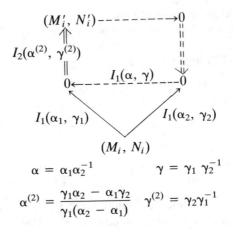

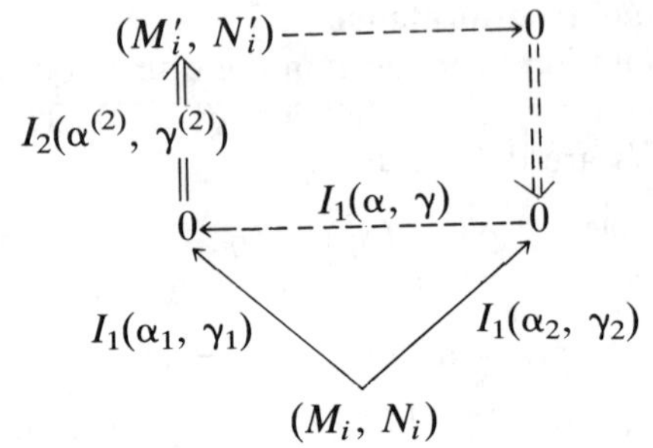

$$\alpha = \alpha_1\alpha_2^{-1} \qquad \gamma = \gamma_1\,\gamma_2^{-1}$$

$$\alpha^{(2)} = \frac{\gamma_1\alpha_2 - \alpha_1\gamma_2}{\gamma_1(\alpha_2 - \alpha_1)} \qquad \gamma^{(2)} = \gamma_2\gamma_1^{-1}$$

Figure 8.2 Neugebauer's commutation formula

The relation between the variables in the two problems was given in (8.2.28) and between q and α in (8.2.29). A nonlinear superposition principle for a product of Harrison transformations can easily be obtained from Neugebauer's commutation formula, see Figure 8.2. Thus for any two solutions obtained by applying the transforms $I_1(\alpha_1, \gamma_1)$, $I_1(\alpha_2, \gamma_2)$ to known solution (M_i, N_i), a new solution can be found by applying the transform $I_2(\alpha^{(2)}, \gamma^{(2)})\ I_1(\alpha_1, \gamma_1)$ to the known solution where $I_2(\alpha^{(2)}, \gamma^{(2)})$ is determined algebraically by the transforms $I_1(\alpha_1, \gamma_1)$ and $I_1(\alpha_2, \gamma_2)$. The Neugebauer commutation formula gives the following superposition principle for Harrison tranformations. If $H(\alpha_1, \gamma_1)$, $H(\alpha_2, \gamma_2)$ are two Harrison transformations of a known solution then a new solution is obtained by applying the Harrison transformation $H(\alpha_{12}, \gamma_{12})$, where $\gamma_{12} = \gamma_2$

and $$\alpha_{12} = \frac{\gamma_2(\gamma_1 - 1)\alpha_1 + \gamma_1(1 - \gamma_2)\alpha_2 + (\gamma_2 - \gamma_1)\alpha_1\alpha_2}{[\alpha\,(\gamma_1 - \gamma_2) + (\gamma_2 - 1)\alpha_1 + (1 - \gamma_1)\alpha_2]} \qquad (8.3.7)$$

to the known solution. The remarkable feature of this superposition principle is that the equations for the new fields obtained by applying (8.3.7) to a known solution can be integrated and the result expressed as an algebraic relation between the new Ernst potential E, the old Ernst potential E_0 and the variables γ_i, α_i, $i = 1, 2$:

$$E = E_0 + \frac{(\gamma_2 - \gamma_1)(E_0 + E_1)}{(\alpha_1 - \gamma_2) + (\gamma_2 - 1)\alpha_1 + \alpha_2(1 - \gamma_2)}. \qquad (8.3.8)$$

Neugebauer [18] showed that this could be written concisely as a determinantal expression which was easily generalisable to the case of n such superpositions involving $2n$ functions γ_j, α_j, $j = 1, \ldots, 2n$:

$$E = E_0\,\frac{D\begin{bmatrix}\alpha_0 & \alpha_1 & \alpha_2 & \ldots\alpha_{2n}\\ 1 & \gamma_1 & \gamma_2 & \ldots\ \gamma_{2n}\end{bmatrix}}{D\begin{bmatrix}1 & \alpha_1 & \alpha_2 & \ldots\ \alpha_{2n}\\ 1 & \gamma_1 & \gamma_2 & \ldots\ \gamma_{2n}\end{bmatrix}}, \quad \alpha_0 := -\bar{E}_0/E_0,$$

where

$$D\begin{bmatrix}x_1 & x_2 & \ldots & x_m\\ y_1 & y_2 & \ldots & y_m\end{bmatrix} := \det\left[\exp\left\{\frac{1 + (-1)^r}{2}\ln x_s + (r - 1)\ln y_s\right\}\right]$$
$$r, s = 1, \ldots, m. \qquad (8.3.9)$$

Belinsky–Zakharov Bäcklund transformations

In the papers [22, 23] Belinsky and Zakharov first introduced the techniques which are used in the analysis of all the linear problems which arise in general relativity. The method for generating soliton solutions is analogous to the methods used for other field equations [31, 32] and involves using λ dependent gauge transformations, or Darboux transformations as they are known. In general such transformations will not leave the field equations invariant so that additional symmetries have to be imposed on the transformations to preserve the equations. This symmetry group is called a reduction group for the linear problem [33]. For the Belinsky–Zakharov problem given in Section 8.2 suppose $(V_3^{(1)}, V_4^{(1)}, Y^{(1)})$, $(V_3^{(0)}, V_4^{(0)}, Y^{(0)})$ satisfy the linear problem (8.2.9) and are Bäcklund-related by the Darboux transformation $Y^{(1)}(\lambda) = \chi(\lambda)\, Y^{(0)}(\lambda)$. Consistency requires that

$$\frac{\rho\, V_4^{(1)} - \lambda\, V_3^{(1)}}{\lambda^2 + \rho^2} = (D_1\chi)\chi^{-1} + \frac{\chi(\rho V_4^{(0)} - \lambda\, V_3^{(0)})\chi^{-1}}{\lambda^2 + \rho^2},$$
$$\frac{\rho\, V_3^{(1)} + \lambda\, V_4^{(1)}}{\lambda^2 + \rho^2} = (D_2\chi)\chi^{-1} + \frac{\chi(\rho V_3^{(0)} - \lambda\, V_4^{(0)})\chi^{-1}}{\lambda^2 + \rho^2}. \qquad (8.3.10)$$

The reduction group requires the imposition of the following constraints on χ:

$$\overline{\chi(\bar{\lambda})} = \chi(\lambda), \quad F^{(1)} = \chi(-\rho^2/\lambda)\, F^{(0)}\, \chi^{\mathrm{T}}(\lambda). \tag{8.3.11}$$

Then since $F = Y^{\mathrm{T}}(0) = \mathrm{Y}(0)$ it follows that

$$F^{(1)} = \chi(0)\, F^{(0)}, \tag{8.3.12}$$

and then from (8.3.11) we get $\chi(\infty) = I$. The soliton solutions correspond to requiring χ to be meromorphic in the λ-plane. In particular, define

$$\chi^{(j)} = I + \frac{\mu_j^2 + \rho^2}{\mu_j(\lambda - \mu_j)}\, P^{(j)},$$

where $$(P^{(j)})_{AB} = \frac{m^{(j)C}\,(F^{(j-1)})_{CA}\, m_B^{(j)}}{m^{(j)D}\,(F^{(j-1)}))_{DF}\, m^{(j)F}}$$

and $$m_A^{(j)} = d^{(j)C}\,([Y^{(j-1)}\,(\mu_j, \rho, z)]^{-1})_{CA}, \tag{8.3.13}$$

where the $d^{(j)C}$ are arbitrary and the μ_j satisfy, with $\tau = \pm 1$,

$$D_1\, \mu_j = 0 = D_2\, \mu_j,$$
$$\mu_j = \frac{1}{2k_j}\,((1 - 2k_j z) + \tau[(1 - 2k_j z)^2 + 4k_j^2\, \rho^2]^{1/2}). \tag{8.3.14}$$

The matrix $P^{(j)}$ is a projection operator with unit trace and zero determinant. The transformation $\chi^{(1)}$ defines a Bäcklund transformation of the type (8.3.12). An alternative form of the Bäcklund transformation can be obtained from (8.3.10). The application of n such Bäcklund transformtions results in the expression

$$F^{(n)} = \prod_{j=1}^{n} \chi^{(j)}(0)\, F^{(0)}. \tag{8.3.15}$$

Notice that from (8.3.12) each succeeding $\chi^{(j)}$ is determined in terms of the preceding solution $Y^{(j-1)}$. The determinant of $F^{(n)}$ is no longer $-\rho^2$ but this can be corrected by multiplication by the factor $(-\rho^{-n}\, \Pi_{j=1}^{n}\, \mu_j)$. This works provided n is even. However, if n is odd, in particular $n = 1$, we have to start from a non-physical metric, det $F^{(0)} = \rho^2$. The function $\chi^{(j)}$ in (8.3.13) is derived from the analytic properties of equations (8.3.10) upon assuming that χ has a pole at μ_j. The n-Bäcklund transformations (8.3.15) can be applied as one transformation by deriving from (8.3.10) the form of a χ which has n poles μ_j, $j = 1, \ldots, n$ and which satisfies the boundary conditions. Alekseev [34] has obtained a determinantal expression for $F^{(n)}$, equation (8.3.15). A one-soliton solution is equivalent to a Harrison transformation followed by I. In fact it is not hard to show the equivalence of the $2n$-transformation (8.3.15) with the Neugebauer formula (cf. Cosgrove [19]).

Dodd–Morris transformations

We include these transformations [35, 36] as examples of singular solutions which can be generated by Bäcklund transformations. Unfortunately they are not asymptotically flat so that they would only serve as intermediate solutions. The Bäcklund transformation is derived from the equations (8.2.46). In order to get real solutions the poles in the Darboux transformation must occur as complex conjugate pairs. The simplest transformation with the pole at the origin of the λ-plane can be shown to give the following result:

$$f^{(1)} = \frac{(1-\Delta)f^{(0)}}{n^2(1+q^2)}, \quad \text{where} \quad \Delta = n^2\,\rho^2(1+q^2)^2$$
$$\omega^{(1)} = \frac{-n}{(1-\Delta)f^{(0)}} \tag{8.3.16}$$

where $q = f^{(0)-1}\,(C_0 - \psi^{(0)}) \quad C_0 - \text{const.}$

and
$$n_{,z} = f^{(0)-1}f^{(0)}_{,z}q - \rho n^2[2f^{(0)-1}f^{(0)}_{,\rho}q + (1-q^2)f^{(0)-1}\psi^{(0)}_{,\rho}]$$
$$n_{,\rho} = f^{(0)-1}f^{(0)}_{,\rho}q + \rho n^2[2f^{(0)-1}f^{(0)}_{,z}q + (1-q^2)f^{(0)-1}\psi^{(0)}_{,z}] \tag{8.3.17}$$

The indices (i), $i = 0, 1$ refer to the untransformed and transformed metric functions f and ω (8.1.4) and twist potential ψ (8.1.17).

8.3.2 *Linear deformation problems*

In the above discussion we omitted the Hauser–Ernst Bäcklund transformations. In this subsection we shall principally consider their deformation problem and refer only briefly to the Belinsky–Zakharov problem. The soliton solutions will be considered as part of this analysis. The reason for this approach is that the Hauser–Ernst problem provides a faithful representation of a large subgroup of K which is independent of the field equations whereas the Belinsky–Zakharov problem does not.

The construction of new solutions to the Hauser–Ernst problem from a given seed solution is achieved by solving a homogeneous Riemann–Hilbert problem. This technique was first applied to integrable systems by Zakharov and Shabat [31]. First, though, we observe from (8.2.13) that a solution of the linear deformation problem is related to a given H by ($R^\dagger(k) = \overline{R^{\mathrm{T}}(\bar{k})}$):

$$R(0) = \Omega, \quad \dot{R}(0) = H \quad \text{where} \quad \dot{R}(\mathrm{O}) = \partial_k R(k)|_{k=0}$$
$$S(k)\,\det R(k) = -1$$
$$R^\dagger(k)\{\Omega - k\,\Omega(H + H^\dagger)\Omega\}\,R(k) = \Omega$$

or
$$S(k)\bar{R}(k) = 2ikF\varepsilon R(k) - (1-2kz)R(k). \tag{8.3.18}$$

The first two relations in (8.3.18) are obtained from the definition of $R(k)$, (8.1.31) whereas the next two conditions are first integrals of (8.2.13). The last first integral is written in two alternative but equivalent forms ($\bar{R}(k) = R(\bar{k})$) corresponding to the formulation (8.1.32) or (8.2.13). It is clear that (8.3.18) only defines $R(k)$ up to a gauge function $G(k)$ multiplying $R(k)$ on the right. This function, which depends only upon k, must satisfy certain conditions which are easily deduced from (8.3.18). The gauge freedom can be used to minimise the singularities of $R(k)$ in the k-plane. In particular, Hauser–Ernst prove that their exists a unique gauge, the HE special gauge, such that provided $E = H_{11}$ is an analytic function of (ρ, z) in a neighbourhood of $(0, 0)$ for which $f = \text{Re}\, E \neq 0$ then $R(k)$ is analytic in the whole k-plane and

$$R(k) \begin{bmatrix} 1 & 0 \\ 0 & k \end{bmatrix} \tag{8.3.19}$$

is analytic at $k = \infty$ apart from branch points with index $-\frac{1}{2}$ given by the zeros of

$$S(k) = \{(1 - 2kz)^2 + 4k^2\rho^2\}^{1/2} \tag{8.3.20}$$

and the cut joining them which does not pass through $k = 0$.

The transformation group $K_L \subset K$ is the Lie group of complex 2×2 matrices $u(k)$ which satisfy the conditions

$$\det u(k) = 1, \quad u^\dagger(k)\, \varepsilon\, u(k) = \varepsilon,$$

$$u(k) \text{ is analytic in an annulus containing } L$$

$$\begin{bmatrix} 1 & 0 \\ 0 & k^{-1} \end{bmatrix} u(k) \begin{bmatrix} 1 & 0 \\ 0 & k \end{bmatrix} \text{ is analytic at } k = \infty. \tag{8.3.21}$$

The label L refers to a simple contour in the complex k-plane which contains the singularities of $u(k)$ and is symmetric about the real axis. The regions L_+ and L_- are respectively the interior and exterior of the contour L. The relation between K_L and K is established by solving a Riemann–Hilbert problem.

A 2×2 matrix-valued function $\chi(k)$ defined in the k-plane is sectionally holomorphic if (i) it is analytic in each of the regions L_+, L_- except perhaps at ∞, (ii) $\lim_{t \to k \varepsilon L} \chi(t)$ exists and is denoted by $\chi_+(k)$, $\chi_-(k)$ respectively, depending upon whether L is approached along a curve in L_+ or L_-. Here $\chi(k)$ has finite degree if in a neighbourhood of ∞ there exist $-m$, $-n \;\varepsilon\; \mathbb{Z}$ such that

$$M = \lim_{k \to \infty} \begin{bmatrix} k^m & 0 \\ 0 & k^n \end{bmatrix} \chi(k) < \infty. \tag{8.3.22}$$

The homogeneous Riemann–Hilbert problem is to construct the sectionally holomorphic matrix-valued function $\chi(k)$ which has finite degree and satisfies the boundary condition

$$\chi_-(k) = \chi_+(k)\,\Gamma(k) \tag{8.3.23}$$

for $k \in L$ where $\Gamma(k)$ is a given matrix function non-singular on L. In general a solution to this problem exists if Γ satisfies a Hölder condition [43]. In the special HE gauge for a given $H^{(0)}$, the function $R^{(0)}$ is unique. It then follows from (8.3.21) that the function $\Gamma(k) := R^{(0)}(k)\ u(k)\ R^{(0)}(k)^{-1}$ is analytic on L. If in addition $\chi(k)$ satisfies the boundary conditions

$$\chi_+(0) = I, \quad \chi_-(k) \text{ is analytic at } k = \infty, \tag{8.3.24}$$

then the solution (the function χ) is unique. We shall require $R^{(0)}(k)$ to be analytic in $L_+ \cup L$ so that the zeros of $S(k)$ and the cut joining them must be in L_-. For a given L this can be done by suitably restricting (ρ, z) sufficiently close to the origin. The function $R^{(1)}(k) := \chi_+(k)\ R^{(0)}(k)$ then satisfies (8.3.18) and defines a solution of (8.2.13) which is in special HE gauge provided $u(k)$ is analytic in L_-. Let $\mathcal{K}_L \subset K_L$ be the subgroup of matrices which are analytic in L_-.

The analyticity at ∞ (component indices $m = 0$, $n = 0$) for u and χ_- is too strong to encompass all possible solutions of the Riemann–Hilbert problem. However, K_L does define a representation of a large subgroup of K. The representation is faithful and it acts as a multitransitive transformation group, a fact which is clearly seen from the formula [9]:

$$[kE^{(1)},\ \mathrm{i}]\ u(k) \begin{bmatrix} -\mathrm{i}k^{-1} \\ E^{(0)} \end{bmatrix} = 0. \tag{8.3.25}$$

This defines a set of $u(k)$ matrices which relate two given K_L-transformable spacetimes. The $E^{(i)}$ in (8.3.25) are evaluated at $\rho = 0$, $z = (2k)^{-1}$. If $E^{(0)}$ is taken as flat space then (8.3.25) is a version of the Geroch conjecture stated in Section 8.1. Hauser and Ernst prove this conjecture for stationary axially symmetric spacetimes under the assumptions that the metric of the transformed spacetime along with the Killing vectors are C^4 in a domain which includes at least one point of the axis defined by the time-like Killing vector.

The solution of the Riemann–Hilbert problem given above is equivalent to that of the Fredholm equation

$$R^{(1)}(k) - \frac{1}{2\pi \mathrm{i}} \int_L \mathrm{d}s\ \mathrm{R}^{(1)}(s) K(s, k) = R^{(0)}(k),$$

where $$K(s, k) := \frac{k}{s(s-k)}\,[M(s, k) - u(k) M(s, k) u(k)^{-1}]$$

and $$M(s, k) := R^{(0)}(s)^{-1} R^{(0)}(k). \tag{8.3.26}$$

The group K_L defined above is a loop group and by posing a Riemann–Hilbert problem for arbitrary $u(k)$ we obtain their Birkhoff decomposition. Thus, put $u(k) = u_+(k)u_-(k)$ where $u(k)$ satisfies the conditions (8.3.21) but is not necessarily analytic in L_-. Require u_+ to be analytic in L_+, u_- to be analytic in L_- and $u_+(0) = I$, then a solution to this Riemann–Hilbert problem exists. Similarly we have the decomposition $u(k)G(k) = u_+(k)u_-(k)$ for a given $u(k)$ and arbitrary element $G(k)$ of the gauge group of (8.3.18). This shows that the condition of analyticity on $u(k)$ required for $u(k) \in \mathscr{K}_L$ is not restrictive. Thus for $u(k) = u_+(k)u_-(k)$ it follows that $R^{(1)}(k)u_+(k)$ is in special HE gauge and $u_-(k)$ satisfies (8.3.21). The required $R^{(1)}(k)$ is recovered by applying a gauge transformation $u_+(k)^{-1}$. Similarly if $R^{(0)}(\mathrm{k})$ is not in special HE gauge but $R^{(0)}(k)G(k)$ is, then we can repeat this argument for $u(k)G(k) = u_+(k)u_-(k)$. However, the condition satisfied by $u(k)$ at ∞ is a restriction on the group. To obtain all the interesting $R^{(1)}(k)$ it is necessary to go outside the Hauser–Ernst formalism by applying gauge transformations which mean that $R^{(1)}(k)$ is no longer analytic in L_+.

It is interesting to consider a few $u(k)$ matrices which define transformations which have been obtained by other means.

Harrison transformation

$$u(k) = (1 - sk^{-1})^{-1/2}\begin{bmatrix} 1 & -csk^{-1} \\ -c^{-1} & 1 \end{bmatrix} \quad s,\ c \text{ real constants}$$

HKX transformations

$$u(k) = I + \sum_{i=1}^{N} \frac{\alpha_i k}{k - s_i}\begin{bmatrix} 0 & 1 \\ 0 & 0 \end{bmatrix} \quad s_i,\ \alpha_i \text{ real constants}$$

B transformations

$$u(k) = \begin{bmatrix} \cos\alpha(k) & k^{-1}\sin\alpha(k) \\ -k\sin\alpha(k) & \cos\alpha(k) \end{bmatrix} \alpha(k) \text{ analytic in } L_+ \cup L \text{ and at } \infty.$$

The $B \subset K$ subgroup was defined in Section 8.1 as the subgroup of K with infinitesimal generators $\gamma_{11}^k + \gamma_{22}^{k+2}$. It has the property that it leaves flat space invariant. The HKX transformations [15] are generated by exponentiating linear combinations of the infinitesimal generators γ_{22}^k. We discuss these transformations in the next section.

The relationship between the Belinsky–Zakharov and Hauser–Ernst deformation problems has been investigated in [37]. The connection between the matrix functions representing elements of K which appear in the homogeneous Riemann–Hilbert problem for the two methods ($u(k)$ in HE, $G_0(\lambda)$ in Belinsky–Zakharov) is explicitly established.

8.4 Solutions

In this review space only permits a brief discussion of the solutions obtained by using the techniques discussed in the previous sections. Which technique is used depends to some extent upon the type of solution being generated. For example, the Neugebauer determinantal expression given in Section 8.3.1 is extremely efficient for obtaining solutions containing an additional $2n$-solitons, but it is not available for obtaining $2n+1$-soliton augmented solutions. In this section we restrict ourselves to the main results and include important additional references.

8.4.1 *Soliton solutions*

We took as a working definition for a soliton that it could be obtained in explicit analytic form. This definition is too weak and we modify it by requiring that the solution should also be asymptotically flat. In the literature soliton solutions have been defined as the solutions generated from flat space by the one-pole Zakharov–Belinsky transformation which is essentially equivalent to a Harrison transformation. This is not an invariant definition, however, since the Harrison transformation in the Hauser–Ernst formalism is of a different type.

The Belinsky–Zakharov one-pole transformation is equivalent to a Harrison transformation followed by the discrete Neugebauer–Kramer map I which is apparently outside K. The double Harrison and Belinsky–Zakharov transformations are in K but do not form a subgroup. They have the property that they preserve asymptotic flatness and consequently add in soliton solutions [19]. These double transformations involve four parameters and a similar transformation can be constructed from the eight-parameter Neugebauer transformation $I_2 I_1 I_2 I_1$.

All the known solutions can be generated by applying soliton transformations to the Weyl [38] or Cosgrove–Tomimatsu–Sato solutions [39, 40, 43] as seed solutions.

Weyl solutions (static axially symmetric solutions)

$$E = \bar{E} = e^{2V}, \quad \nabla^2 V = 0$$

contains as a subclass the Zipoy–Voorhees spacetimes [41, 42], $\rho = \sigma(x^2 - 1)^2(1 - y^2)^{1/2}$, $z = \sigma xy$, σ constant,

$$E = \left[\frac{x - 1}{x + 1}\right]^{\delta} \quad \delta \text{ a positive constant.}$$

$\delta = 1$ is the Schwarzschild solution.

Tomimatsu–Sato solutions

Put $\xi = \dfrac{1 - E}{1 + E}$ and $\rho = (mp/\delta)(x^2 - 1)^{1/2}$, $z = (mp/\delta)\, xy$, then the solutions have the form of a ratio of polynomials in x and y, $\xi = \beta/\alpha$:

$$\delta = 1 \quad \alpha = px - iqy, \; \beta = 1 \quad \text{Kerr solution}$$

$$\delta = 2 \quad \alpha = p^2(x^4 - 1) - 2ipqxy(x^2 - y^2) - q^2(1 - y^4),$$
$$\beta = 2px(x^2 - 1) - 2iqy(1 - y^2),$$

where $p^2 + q^2 = 1$. A solution exists for each integer δ. When $q = 0$ Tomimatsu–Sato solutions reduce to the Zipoy–Voorhees solutions.

The Cosgrove–Tomimatsu–Sato solutions involve three parameters (δ is a continuous parameter) [43, 44].

One-pole Zakharov–Belinsky and the onefold Harrison transformations were investigated in [45, 46]. The $(2n + 1)$-fold transformations can be generated from these by applying the $2n$-fold Harrison, Neugebauer, or Belinsky–Zakharov transformation. Belinsky–Zakharov generated the Kerr–NUT solution from flat space by a two-pole transformation. In fact, the $2n$-fold Harrison, $2n$-pole Zakharov–Belinsky transformations add in n Kerr particles to any seed solution. Belinsky and Fargion [47] have investigated a two-pole transformation of a Kasner solution and obtained a cosmological solution (two space-like Killing vectors) which can be interpreted as a pair of colliding universes. A $2n$-parameter static solution was derived in [22, 23]. It was shown to reduce to a subset of the Zipoy–Voorhees metrics if the poles in the transformation, labelled odd (even), were allowed to coalesce to a single distinct pair. Since these were identical with the static limits of the Tomimatsu–Sato (TS) metrics it is not suprising that a similar result holds for the n-Kerr particle solution. In [61, 62] Neugebauer and Kramer showed that if the odd(even) parameters in their n-Kerr solution were allowed to coalesce they obtained the TS $\delta = n$ solution. It turns out that some of the HKX transformations we treat next can also be obtained from multiple-pole Belinsky–Zakharov transformations.

8.4.2 *Solutions obtained from HKX transformations*

In [13–15] it was shown that it was possible to exponentiate sums of the form

$$\sum_{k=-\infty}^{\infty} s^k \, q^{XY} \gamma^k_{XY}. \tag{8.4.1}$$

The original (rank zero) HKX transformation was defined for only positive values of k with $q^{XY} = h^X h^Y$ and $h^X = (1, 0)$. The $u(k)$ matrix for N such transformations was listed in the previous section. The generalised null HKX transformation results from exponentiating (8.4.1) with h^X an arbitrary null vector and restricting the sum to positive values of k. The extended transformations are defined in the same way but with the full range of k values. It has not proved possible to exponentiate the non-null case when q^{XY} is an arbitrary non-null symmetric tensor. By allowing the poles in the $u(k)$ matrix representation of the N-HKX transformation to coalesce we obtain a transformation which involves transformations of rank $0, \ldots N - 1$. Under the generalised null HKX transformation the Hauser–Ernst Riemann–Hilbert problem gives a new $R(k)$ potential which is a gauge-transformed potential for an extended HKX transformation [52]. The extended HKX transformation can be obtained by allowing the poles to coalesce in a two-pole Zakharov–Belinsky transformation [19].

The HKX transformation applied to flat space creates an extreme Kerr-NUT solution. The twofold HKX transformation of any Weyl or Cosgrove–Tomimatsu–Sato (CTS) solution can be obtained from the twofold Harrison transformation of some other Weyl or CTS solution [19] and since the extended generalised HKX transformation is a repeated-pole Zakharov–Belinsky transformation this too can be obtained from the HKX transformations alone. The non-null HKX transformations have still not been thoroughly investigated.

8.4.3 *Solutions generated by the* B-*group*

The subgroup $B \subset K$ was introduced in Section 8.1 and its $u(k)$ matrix was given in the previous section. It maps flat space into itself. A class of solutions which are generalisations of the TS metrics (with δ an integer) were obtained from the action of the B group on the Zipoy–Voorhees metrics and are now known as the KC (Kinnersley–Chitre) solutions [13].

Further relationships between the different methods of generating solutions and the solutions themselves abound in the literature. In particular it appears that all the asymptotically flat spacetimes can be generated by the HKX transformations alone acting on the Weyl–CTS metrics as seed solutions. It follows that if any new solutions are to be discovered or a transformation between the Weyl and CTS solutions then a detailed study of the integral equation of the last section is required.

So far we have not mentioned the Dodd–Morris solutions which are not asymptotically flat and admit singularities. The solutions could play the role of intermediate solutions; however, the Ernst equation also arises as defining solutions of the Bogomolny equations which define magnetic monopoles, and it is precisely solutions of this type which are of interest [35].

8.5 **Other solvable gravitational field equations**

In this review we have concentrated on vacuum spacetimes. In fact, in their original work Kinnersley and Chitre developed the theory for the stationary axially symmetric spacetimes as a special case of the axially symmetric Einstein–Maxwell equations. Apparently the simplicity of the transformation laws associated with the Maxwell equations greatly assisted the discovery of the corresponding laws for the gravitational equations. We have concentrated on the vacuum case (i) because the results are more complete, (ii) because of limitations of space. The Kinnersley–Chitre group K' which contains K is investigated in the papers [5, 10, 12, 13, 15]. A homogeneous Riemann–Hilbert problem for this case was found by Hauser and Ernst [49]. A Bäcklund transformation was derived using Clairin's method by Omote, Michihiro and Wadati [50], and another linear deformation problem was determined by Kramer and Neugebauer [51] by using the Wahlquist–Estabrook method. Cosgrove [52] obtained an electrovac generalisation of the double Harrison transformation. This Bäcklund transformation has been used to obtain a family of new solutions [53–56].

Belinsky [59] has considered the case when space is filled with a perfect fluid with a super-rigid equation of state. The paper [60] obtains a linear deformation problem for the Jordan model and two soliton solutions which are asymptotically flat are examined for the stationary axially symmetric case. A comprehensive review of the latest results on solvable gravitational field equations is contained in [58].

References

[1] Kramer, D., Stephani, M., Herlt, E., MacCallum, M. and Schmutzer, E. *Exact Solutions of Einstein's Field Equations*, Cambridge University Press Cambridge (1980).

[2] Geroch, R. A method for generating solutions of Einstein's equations, *J. Math. Phys.* **12**, 918 (1971).

[3] Geroch, R. A method for generating solutions of Einstein's field equations II, *J. Math. Phys.* **13**, 394 (1972).

[4] Lewis, T. Some special solutions of the equations of axially symmetric gravitational fields, *Proc. Roy. Soc. Lond.* **A136**, 176 (1932).

[5] Kinnersley, W. Symmetries of the stationary Einstein–Maxwell field equations I, *J. Math. Phys.* **18**, 1529 (1977).

[6] Ernst, F. J. A new formulation of the axially symmetric gravitational field problem, *Phys. Rev.* **167**, 1175 (1968).

[7] Neugebauer, G. and Kramer, D. Eine exacte stationare Losung der Einstein–Maxwell Gleichungen, *Ann. Physik (Leipzig)* **24**, 62 (1969).

[8] Ehlers, J. In *Les Théories Rélativistes de la Gravitation*, CNRS, Paris (1959).

[9] Hauser, I. and Ernst, F. J. Proof of a Geroch conjecture, *J. Math. Phys.* **22**, 1051 (1981).

[10] Kinnersley, W. and Chitre, D. M. Symmetries of the stationary Einstein–Maxwell field equations II, *J. Math. Phys.* **19**, 1926 (1978).

[11] Hauser, I. and Ernst, F. J. A homogeneous Hilbert problem for the Kinnersley–Chitre transformations, *J. Math. Phys.* **21**, 1126 (1980).

[12] Kinnersley, W. and Chitre, D. M. Symmetries of the stationary Einstein–Maxwell field equations III, *J. Math. Phys.* **19**, 1926–31 (1978).

[13] Kinnersley, W. and Chitre, D. M. Symmetries of the stationary Einstein–Maxwell equations IV. Transformations which preserve asymptotic flatness, *J. Math. Phys.* **19**, 2037 (1978).
[14] Hoenselaers, C., Kinnersley, W. and Xanthopoulos, B. C. Generation of asymptotically flat, stationary spacetimes with any number of parameters, *Phys. Rev. Lett.* **42**, 481 (1979).
[15] Hoenselaers, C., Kinnersley, W. and Xanthopoulos, B. C. Symmetries of the stationary Einstein–Maxwell equations VI. Transformations which generate asymptotically flat spacetimes with arbitrary multiple moments, *J. Math. Phys.* **20**, 2530 (1979).
[16] Xanthopoulos, B. C. An axisymmetric stationary solution of Einstein's equations calculated by computer, *J. Phys. A.* **14**, L427 (1981).
[17] Cosgrove, C. M. Stationary axially symmetric gravitational fields. An asymptotic flatness preserving transformation. *Lecture Notes in Physics* **124**, Springer, New York (1979).
[18] Neugebauer, G. Bäcklund transformations of axially symmetric stationary gravitational fields, *J. Phys. A.* **12**, L67 (1979).
[19] Cosgrove, C. M. Relationships between the group theoretic and soliton theoretic techniques for generating stationary axisymmetric gravitational solutions, *J. Math. Phys.* **21**, 2417 (1980).
[20] Forsyth, A. R. *Theory of Differential Equations*, Vols 5 and 6. Dover, New York, (1959).
[21] Clairin, J. Sur quelques équations aux derivées partielles du second ordre, *Ann. Fac. Sci. Univ. Toulouse*, 2nd Ser., **5**, 437 (1903).
[22] Belinsky, V. A. and Zakharov, V. E. Integration of the Einstein equations by means of the inverse scattering technique and construction of exact soliton solutions, *Sov. Phys. JETP* **48**, 985 (1978).
[23] Belinsky, V. A. and Zakharov, V. E. Stationary gravitational solitons with axial symmetry, *Sov. Phys. JETP* **50**, 1 (1979).
[24] Harrison, B. K. Bäcklund transformations for the Ernst equation of general relativity, *Phys. Rev. Lett.* **41**, 1197 (1978).
[25] Maison, D. On the complete integrability of the stationary axially symmetric Einstein equations, *J. Math. Phys.* **20**, 871 (1979).
[26] Wahlquist, M. D. and Estabrook, F. B. Prolongation structure of nonlinear evolution equations, *J. Math. Phys.* **16**, 1 (1975).
[27] Dodd, R. K. and Fordy, A. P. The prolongation structures of quasi-polynomial flows, *Proc. R. Soc. Lond.* **A385**, 389 (1983).
[28] Dodd, R. K. and Morris, H. C. Deformation problems for the Ernst equations, *J. Math. Phys.* **23**, 1131 (1982).
[29] Morris, H. C. and Dodd, R. K. A two connection and operator bundles for the Ernst equation for axially symmetric gravitational fields, *Phys. Lett.* **75A**, 20 (1979).
[30] Omote, M. and Wadati, M. Bäcklund transformations for the Ernst equation, *J. Math. Phys.* **22**, 961 (1981).
[31] Zakharov, V. A. and Shabat, A. Integration of the nonlinear equations of mathematical physics by the method of the inverse scattering problem, *Funct. Anal. Appl.* **13**, 160 (1979).
[32] Zakharov, V. A. and Mikhailov, A. V. Relativistically invariant two dimensional models of field theory which are integrable by means of the inverse scattering problem, *Sov. Phys. JETP* **47**, 1017 (1978).
[33] Mikhailov, A. V. The reduction problem and the inverse scattering method, In *Proc. Joint US–USSR Symposium on Soliton Theory Kiev 1979*, V. E. Zakharov and S. V. Manakov (eds), North-Holland, Amsterdam (1981).

[34] Alekseev, G. A. On soliton solutions of the Einstein equations in vacuum, *Sov. Phys. Doklady*. **26**, 158 (1981).

[35] Morris, H. C. and Dodd, R. K. Singular solutions of the axially symmetric Bogolmolny equations, *J. Math. Phys.* **25**, 240 (1984).

[36] Dodd, R. K. Kinoulty, J. and Morris, H. C. Bäcklund transformations for the Ernst equation. In *Nonlinear Waves*, Lokenath Debnath (ed.), Cambridge University Press, Cambirdge (1983).

[37] Cosgrove, C. Relationship between the inverse scattering techniques of Belinsky–Zakharov and Hauser–Ernst in general relativity, *J. Math. Phys.* **23**, 615 (1982).

[38] Weyl, H. Zur Gravitations Theorie, *Annalen Physik* **54**, 117 (1917).

[39] Tomimatsu, A. and Sato, H. New exact solution for the gravitational field of a spinning mass, *Phys. Rev. Lett.* **29**, 1344 (1972).

[40] Tomimatsu, A. and Sato, H. New series of exact solutions for gravitational fields of spinning masses, *Prog. Theor. Phys.* **50**, 95 (1973).

[41] Voorhees, B. H. Static axially symmetric gravitational fields, *Phys. Rev.* **D2**, 2119 (1970).

[42] Zipoy, D. M. Topology of some spheroidal metrics, *J. Math. Phys.* **7**, 1137 (1966).

[43] Cosgrove, C. M. New family of exact stationary axisymmetric gravitational fields generalising the Tomimatsu–Sato solutions, *J. Phys. A.* **10**, 141 (1977).

[44] Cosgrove, C. M. In *Proceedings of the second Marcel Grossmann Meeting on recent developments of general relativity* and references therein, Trieste, Italy 5–11 July 1979, R. Ruffini (ed.), North Holland, Amsterdam (1979).

[45] Harrison, B. K. New large family of vacuum solutions of the equations of general relativity, *Phys. Rev. D.* **21**, 1695 (1980).

[46] Verdaguer, E. Stationary axisymmetric one-soliton solutions of the Einstein equations, *J. Phys. A.* **15**, 1261 (1982).

[47] Belinsky, V. and Fargion, D. Two soliton waves in anisotropic cosmology, *Il Nuovo Cim.* **59B**, 143 (1980).

[48] Alekseev, C. A. and Belinsky, V. A. Static gravitational solitons, *Sov. Phys. JETP* **51**, 655 (1980).

[49] Hauser, I. and Ernst, F. J. A homogeneous Hilbert problem for the Kinnersley–Chitre transformations of electrovac spacetimes, *J. Math. Phys.* **21**, 1418 (1980).

[50] Omote, M., Michihiro, Y. and Wadati, M. Bäcklund transformation of the axially symmetric stationary Einstein–Maxwell equations, *Phys. Lett.* **79A**, 141 (1980).

[51] Kramer, D. and Neugebauer, G. Prolongation structure and linear eigenvalue equations for the Einstein–Maxwell fields, *J. Phys. A.* **14**, L333 (1981).

[52] Cosgrove, C. Bäcklund transformations in the Hauser–Ernst formalism for stationary axisymmetric spacetimes, *J. Math. Phys.* **22**, 2624 (1981).

[53] Guo, D. S. and Ernst, F. J. Electrovac generalisation of Neugebauer's $N = 2$ solution of the Einstein vacuum field equations, *J. Math. Phys.* **23**, 1359 (1982).

[54] Wang, S. K., Guo, H. Y. and Wu, K. The N-fold Kerr family and charged Kerr family solutions, *Commun. Theor. Phys. (China)* **2**, 921 (1983).

[55] Chen, Y., Guo, D. S. and Ernst, F. J. Charged spinning mass field involving rational functions, *J. Math. Phys.* **24**, 1564 (1983).

[56] Guo, D. S. Noniterative method for constructing many-parameter solutions of the Einstein–Maxwell field equations, *J. Math. Phys.* **25**, 2284 (1984).

[57] Muskevishli, N. I. *Singular Integral Equations*, Noordhoff, Groningen, Holland (1946).

[58] Morris, H. C. and Dodd, R. K. (eds) *Solitons in General Relativity*, Plenum Press, New York, (1989).
[59] Belinsky, V. A. One-soliton waves, *Sov. Phys. JETP* **50**, 623 (1980).
[60] Belinsky, V. A. and Ruffini, R. On axially symmetric solutions of the coupled scalar–vector–tensor equations in general relativity, *Phys. Lett.* **89B**, 195 (1980).
[61] Neugebauer, G. and Kramer, D. The superposition of two Kerr solutions, *Phys. Lett.* **75A**, 259 (1980).
[62] Neugebauer, G. and Kramer, D. The soliton concept in general relativity, *General Relativity and Gravitation* **13**, 195 (1981).

9 *P. S. Lomdahl*

Soliton models of protein dynamics

9.1 How do proteins work?

The discipline of bioenergetics which is the study of how living cells generate and transfer their energy supply is primarily a study of how proteins work. It is widely accepted that proteins play a principal role in how the cell works. The operation of other common macromolecules like DNA and polysaccharides is always closely tied to that of proteins. From decades of careful chemical analysis and X-ray crystallography, we know the composition and three-dimensional structure of about two hundred proteins. Despite this massive amount of structural knowledge [43], however, there is no generally accepted model of how proteins operate dynamically. This presents one of the outstanding questions and challenges for bioenergetics and biophysics.

Proteins are constructed from individual building blocks called amino acids. About twenty different amino acids are commonly found in proteins. Each amino acid has an amino group (NH_2), a carboxyl group ($COOH$), and a side group, or radical (R) attached to the α carbon atom. The radical is what distinguishes one amino acid from another. Amino acids polymerise to form long chains of residues that constitute a protein. When two amino acids join together they release one molecule of water and form a *peptide bond* as shown in Figure 9.1. When the polypeptide chain has been formed it can fold into a variety of complex three-dimensional conformations. Of particular interest are three structural configurations that recur over and over in proteins: the α helix (Figure 9.2), the β sheet, and globular conformation. In the α helix the polypeptide chain is tightly coiled about its longitudinal axis. In the β sheet the chain can be visualised as pleated strands of protein. The globular conformation is the most complex; here the chain is folded irregularly into a compact near-spherical shape. Part of the chain can often be in the α helix or β sheet configuration, but in general the structure would be characterised as random.

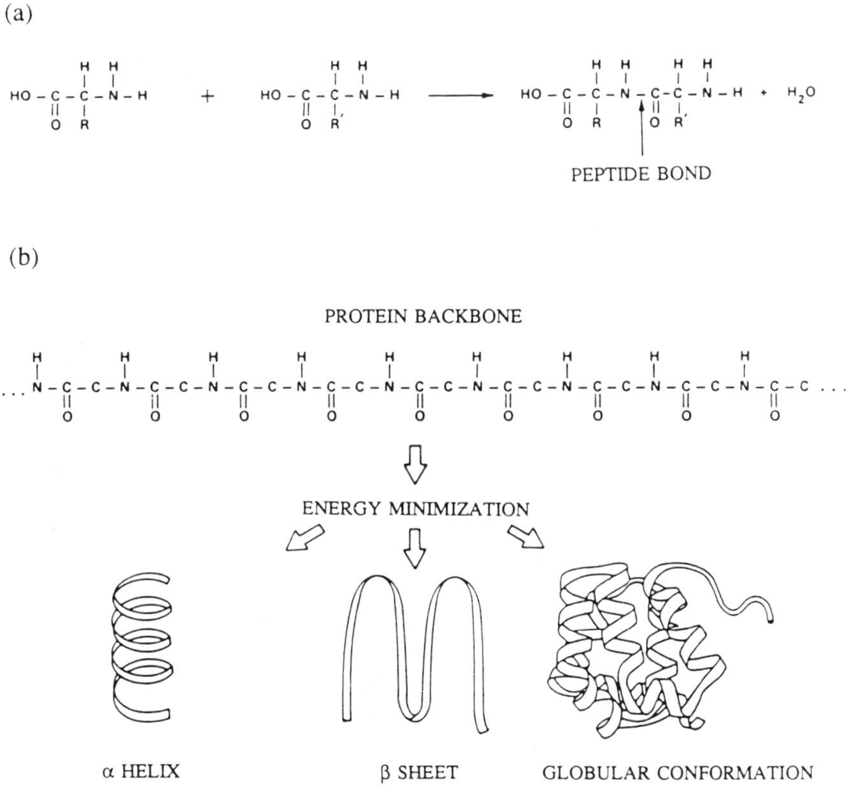

Figure 9.1 The formation and structure of proteins, from [37]. Creation of a polypeptide (a) and the folding of the polypeptide chain (b).

The energy supply for most protein activity is provided by the hydrolysis of ATP (adenosine triphosphate). An ATP molecule binds to a specific site on the protein, reacts with water, and under normal physiological conditions releases 0.42 eV of energy. The question for biophysics is 'what happens to this energy?' How does it perform useful work? Is the energy used through a non-equilibrium process or does it thermalise and then work through an equilibrium process? One hypothesis is that the energy is converted to a particular vibrational excitation within the protein. A likely recipient of such a resonant exchange is the *amide-I* vibration. This vibration is primarily stretch and contraction of the $C = O$ bond of the peptide groups. The energy of the amide-I vibration is about 0.21 eV or 1665 cm^{-1}. This energy is about half of the energy released during the ATP hydrolysis and almost equal to the H–O–H bending mode of water, suggesting that the excitation could proceed through an intermediate excitation of water. The amide-I vibration is also a promin-

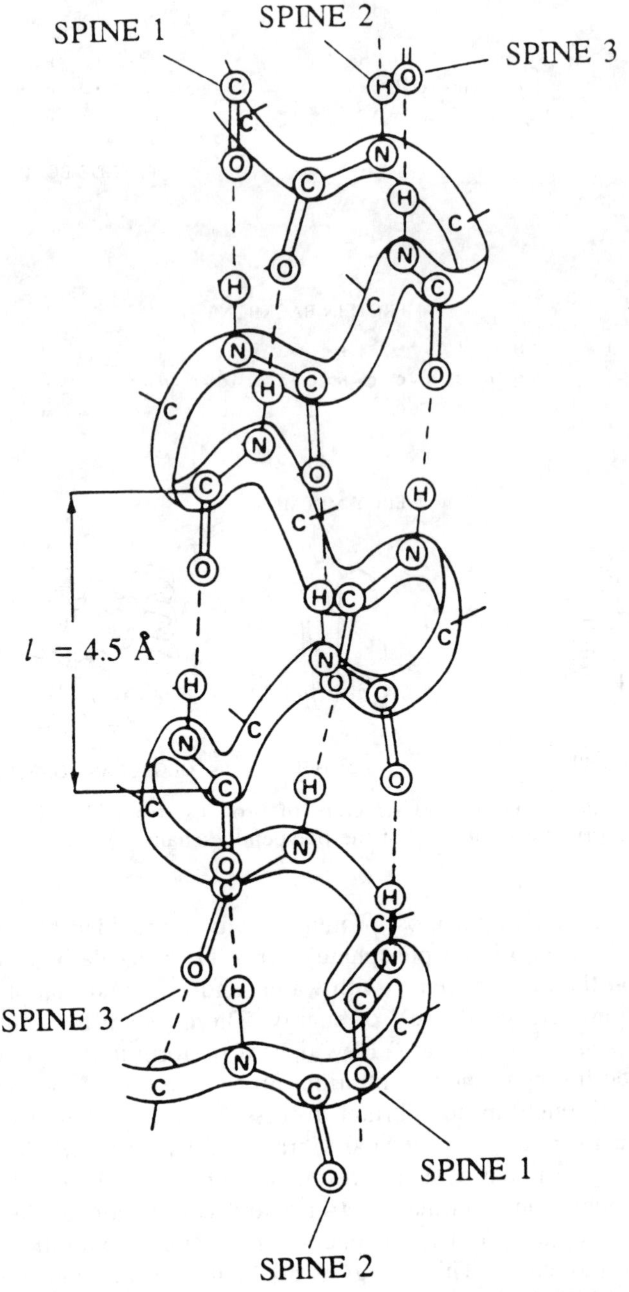

Figure 9.2 The a helix protein structure. There are three hydrogen-bonded spines that stabilise the structure, from [37].

ent feature in infrared and Raman spectra of proteins. Moreover, it stays nearly constant from protein to protein, indicating that it is rather weakly coupled to other degrees of freedom. All these factors lead to the conjecture that energy released by ATP might stay localised and stored in the amide-I vibration.

When an idea like this was discussed at the New York Academy of Sciences in 1973 [39], the objection raised was that the lifetimes of typical vibrations in proteins were too short (10^{-12} seconds) to be important for biological processes. In particular, the peptide group has a large electric dipole moment and therefore, dipole–dipole interactions would tend to spread the amide-I energy to neighbouring peptide groups. The energy would not stay localised and would be lost for biological processes. About the same time, the Soviet physicist A. S. Davydov [12, 20] formulated a soliton theory countering these objections. He suggested that the amide-I energy could stay localised through the nonlinear interactions of the vibrational excitation and the deformation in the protein structure caused by the presence of the excitation. The excitation and the deformation balance each other and form a soliton. Davydov worked out this idea for the α helix protein. His theory showed how a soliton could travel along the three hydrogen bonded spines of the α helix protein (see Figure 9.2). For more details on where this has biological relevance the reader is referred to [37].

Since then, a considerable amount of work has been done on the original Davydov model and related models. It is the purpose in this chapter to review this work and point out some of its limitations and possible extensions. We will also review some of the experimental evidence indicating the presence of soliton-like objects in polypeptides. The reference list is as complete as possible, so the interested reader can pursue more details in the original literature than can possibly be presented here.

9.2 The Davydov soliton model

Inspection of the α helix structure in Figure 9.2 reveals three channels of hydrogen-bonded peptide groups approximately in the longitudinal direction with the sequence:

$$\ldots\mathrm{H{-}N{-}C{=}O}\ldots\mathrm{H{-}N{-}C{=}O}\ldots\mathrm{H{-}N{-}C{=}O}\ldots$$

The dotted line indicates the hydrogen bond. For a detailed analysis [12, 18, 26, 44, 45] it is necessary to consider all three spines and their interaction. Here we will only consider one, since it suffices to convey the basic idea. The Hamiltonian Davydov used to describe the situation is:

$$\begin{aligned} H &= \sum_n [E_0 B_n^\dagger B_n - J(B_{n+1}^\dagger B_n + B_n^\dagger B_{n+1})] \\ &\quad + \sum_n \left[\frac{p_n^2}{2m} + \frac{1}{2}w(u_{n+1} - u_n)^2\right] + \chi \sum_n (u_{n+1} - u_{n-1}) B_n^\dagger B_n \\ &= H_\mathrm{v} + H_\mathrm{p} + H_\mathrm{int}. \end{aligned} \tag{9.2.1}$$

Here, $B_n^\dagger$ and B_n are boson creation and annihilation operators for quanta of intramolecular vibrations with energy $E_0 = 1665\,\mathrm{cm}^{-1}$ at site n (the C=O stretch mode or amide-I mode), u_n an p_n are the molecular displacement and momentum operators for the molecule at site n (the entire peptide group), m and w are the molecular mass and intermolecular force constant, and J is the intersite transfer energy produced by dipole–dipole interactions. The nonlinear coupling constant χ arises from modulation of the on-site energy by the molecular displacements. It is the derivative of the amide-I energy with respect to the length between peptide groups (l) of the adjacent hydrogen bond:

$$\chi \equiv \frac{\mathrm{d}E_0}{\mathrm{d}l}. \tag{9.2.2}$$

The vibrational part H_v, the phonon part H_p, and the interaction part H_int are defined to be the individual terms in (9.2.1). We note in passing that the Hamiltonian is identical to the Frohlich Hamiltonian for the 'polaron' problem with slightly different meaning of the symbols [24].

For later comparison we write here the equation of motion for the Heisenberg operator $B_n(t)$,

$$\mathrm{i}\hbar \dot{B}_n = E_0 B_n - J(B_{n+1} + B_{n-1}) + \chi B_n (u_{n+1} - u_{n-1}). \tag{9.2.3}$$

The form of this equation is such that a phase transformation

$$B_n(t) = \tilde{B}_n(t) \mathrm{e}^{-\mathrm{i}E_0 t/\hbar} \tag{9.2.4}$$

removes the energy of the amide-I quantum from the equation, that is, the equation for $\tilde{B}_n(t)$ is (9.2.3) but without the term proportional to E_0. Thus this simple transformation removes from the equations of motion any knowledge of the magnitude of E_0 relative to other energies in the problem, for example, the Debye energy of the acoustic phonon spectrum associated with H_p.

The phonon part of the Hamiltonian can be cast into familiar form in terms of phonon creation and annihilation operators by the use of the standard transformation

$$u_n = \sum_q \left(\frac{\hbar}{2Nm\omega_q}\right)^{1/2} \mathrm{e}^{\mathrm{i}qnl}(a_{-q}^\dagger + a_q), \tag{9.2.5a}$$

$$p_n = \sum_q \left(\frac{m\hbar\omega_q}{2N}\right)^{1/2} e^{iqnl} i(a_{-q}^{\dagger} + a_q). \tag{9.2.5b}$$

In these formulae

$$\omega_q = 2(w/m)^{1/2}|\sin(ql/2)| \tag{9.2.6}$$

is the dispersion relation for H_p and N is the number of groups in the polypeptide chain.

To understand the dynamics arising from the Hamiltonian (9.2.1), Davydov [12–16, 18, 19] makes the *Ansatz* for the state vector

$$|\Psi(t)\rangle = \sum_n a_n(t) B_n^{\dagger} \exp\left\{-\frac{i}{\hbar}\sum_j [\beta_j(t)p_j - \pi_j(t)u_j]\right\}|0\rangle, \tag{9.2.7}$$

where $|0\rangle$ is the ground state vector (i.e. it is annihilated both by B_n and by the phonon operators a_q). Assuming that the time evolution of this state vector is approximately the same as that of the (unknown) exact state vector, one can then understand the system behaviour by finding the time evolution of the three sets of unknown functions $a_n(t)$, $\beta_n(t)$, and $\pi_n(t)$.

Normalisation of this state vector requires

$$\langle\Psi(t)|\Psi(t)\rangle = \sum_n |a_n(t)|^2 = 1. \tag{9.2.8}$$

The form of the state vector restricts the system to have only a single amide-I quantum present,

$$\langle\Psi(t)|\sum_m B_m^{\dagger} B_m \,|\Psi(t)\rangle = 1, \tag{9.2.9}$$

and $|a_n(t)|^2$ is the probability that the single quantum is on the nth site. The interpretation of $\beta_n(t)$ and $\pi_n(t)$ is obtained as follows. Davydov [19] points out that the part of $|\Psi(t)\rangle$ depending on the displacement and momentum operators is a coherent state of the normal mode creation and annihilation operators. A coherent state for the mode with wave vector q is [25, 38]:

$$|\alpha_q\rangle = \exp(\alpha_q a_q^{\dagger} - \alpha_q^* a_q)|0\rangle. \tag{9.2.10}$$

To see that (9.2.7) is a coherent state of all the normal modes, we use (9.2.5) to show that

$$-\frac{i}{\hbar}\sum_n(\beta_n p_n - \pi_n u_n) = \sum_q(\alpha_q a_q^{\dagger} - \alpha_q^* a_q), \tag{9.2.11}$$

where

$$\alpha_q = \left(\frac{m\omega_q}{2\hbar}\right)^{1/2} \beta_q + i\left(\frac{1}{2m\hbar\omega_q}\right)^{1/2} \pi_q. \tag{9.2.12}$$

(Here β_q is the spatial Fourier transform of β_n,

$$\beta_q = \frac{1}{\sqrt{N}} \sum_n e^{-iqnl} \beta_n, \tag{9.2.13}$$

and similarly for π_q.) We substitute (9.2.11) into (9.2.7) and get a factor of the form (9.2.10) for every normal mode. With the property

$$<\alpha_q|a_q|\alpha_q> = \alpha_q, \tag{9.2.14}$$

and also using (9.2.5), (9.2.12), and (9.2.13), we straightforwardly obtain

$$<\Psi(t)|u_n|\Psi(t)> = \beta_n(t), \tag{9.2.15a}$$

$$<\Psi(t)|p_n|\Psi(t)> = \pi_n(t). \tag{9.2.15b}$$

The Davydov equations are now typically derived by first obtaining a formula for the average value of the energy $<\Psi(t)|H|\Psi(t)>$ in terms of a_n, β_n, π_n, and then using that result in Hamilton's equations of classical mechanics after identifying conjugate coordinates and momenta. We want to point out, though, that a complete quantum-mechanical derivation is possible by straightforward application of the Schrödinger equation and the properties of the coherent state (see [31, 36] for details). In both cases one obtains two equations for the average displacement of the nth peptide group and for the probability amplitude for excitation of the nth amide-I oscillator;

$$m\ddot{\beta}_n = w(\beta_{n+1} - 2\beta_n + \beta_{n-1}) + \chi(|a_{n+1}|^2 - |a_{n-1}|^2), \tag{9.2.16a}$$

$$i\hbar\dot{a}_n = -J(a_{n+1} + a_{n-1}) + \chi(\beta_{n+1} - \beta_{n-1})a_n. \tag{9.2.16b}$$

These equations are the main result of Davydov's analysis. The parameter values are all given in Table 9.1. It is interesting to note that there are no adjustable parameters in the equations. It is therefore important to decide whether the physical parameters are of such a magnitude that it is reasonable to expect soliton formation under the conditions of ATP hydrolysis. Extensive analytical and numerical work [26, 40, 44, 45, 47] has shown that χ is large enough to expect soliton formation. For $\chi \leq 0.45 \times 10^{-10}N$ solitons are not formed and for $\chi > 0.65 \times 10^{-10}N$ the soliton is so strongly pinned to the lattice that is does not move. There is a 'window' for soliton formation, with χ in the range $0.45 \times 10^{-10}N \leq \chi \leq 0.65 \times 10^{-10}N$ where the soliton seems capable of forming and being useful for transport. This range compares favourably with the value obtained by Careri [9] (see Table 9.1). Furthermore, one quantum of amide-I energy as imposed by (9.2.8) is enough energy to see self-focusing. In addition it seems that the resulting solitons travel rather slowly compared to the sound velocity.

To tie in with conventional soliton techniques, such as inverse-scattering-transform and perturbation methods it is often convenient to

Table 9.1 α helix parameters

Parameter	Value	Unit	Reference
E_0	1665	cm^{-1}	[41]
J	7.8	cm^{-1}	[41]
w	13	Nm^{-1}	[27]
m	114	m_p	[44]
l	4.5	10^{-10} m	[44]
χ	0.62	10^{-10} N	[9]

go to the continuum limit of (9.2.16). Considering only functions that vary slowly with the group number n, we can replace $a_n(t)$ and $\beta_n(t)$ with continuous functions $a(x, t)$ and $\beta(x, t)$; in this case (9.2.16) becomes

$$\frac{\partial^2 \beta}{\partial t^2} - v_s{}^2 \frac{\partial^2 \beta}{\partial x^2} = \frac{2\chi l}{m} \frac{\partial(|a|^2)}{\partial x}, \qquad (9.2.17a)$$

$$i\hbar \frac{\partial a}{\partial t} = -Jl^2 \frac{\partial^2 a}{\partial x^2} + \left[-2J + 2\chi l \frac{\partial \beta}{\partial x}\right] a, \qquad (9.2.17b)$$

where v_s is the sound velocity, $v_s = l(w/m)^{1/2}$. We note in passing that these equations are essentially identical to the Zakharov equations of plasma physics [59], emphasising again that soliton equations are abundant all over physics. We can now look for travelling-wave solutions of the form $\beta(x, t) = \beta(x - vt)$. Inserting into (9.2.17a) and integrating once we get

$$\frac{\partial \beta}{\partial x} = -\frac{2\chi}{wl(1 - (v/v_s)^2)} |a|^2. \qquad (9.2.18)$$

Substituting (9.2.18) into (9.2.17b) we get the nonlinear Schrödinger equation (NLS) for the probability amplitude $a(x, t)$:

$$i\hbar \frac{\partial a}{\partial t} + Jl^2 \frac{\partial^2 a}{\partial x^2} + 2Ja + \kappa |a|^2 a = 0, \qquad (9.2.19)$$

where $\kappa = 4\chi^2/w(1 - (v/v_s)^2)$. The self-trapped state of amide-I energy is described by the well-known one-soliton solution

$$a(x, t) \propto \text{sech}[(x - vt)/\xi], \qquad (9.2.20)$$

where ξ is a constant determined by the continuum version of (9.2.8). The situation described by (9.2.18) and (9.2.20) has a finite interval where the amide-I oscillators are excited, accompanied by a lattice displacement which pulls the peptide groups closer together in that region. It is easy to show [18] that this configuration has a lower energy than the spatially extended solution to (9.2.19) and thus is self-trapped.

The result of a computer simulation of (9.2.16) is seen in Figures 9.3 and 9.4 [35]. The results are presented with certain diagnostics: One is of 'waveform' graphs: that is, plots of $|a_n|^2$ and the discrete gradient $\beta_{n+1} - \beta_{n-1}$ as a function of n at a given time t. Also used are 'soliton detector' plots: on the (t, n)-plane, a mark was put at those times and positions where both $|a_n|^2$ exceeded a certain level (rather arbitrarily chosen to be 0.02) and $\beta_{n+1} - \beta_{n-1}$ is negative. The temporal extent of such a marked region shows the trajectory of a soliton. In Figure 9.3 we see how several solitons are nucleated from random initial conditions and how they move along the chain. In Figure 9.4 we see wave-forms at the end of the run in Figure 9.3. A clear correlation of the maximum in $|a_n|^2$ and the minimum in $\beta_{n+1} - \beta_{n-1}$ is seen, in accordance with the characteristics of a soliton (i.e. (9.2.18) and (9.2.20)).

Despite the elegance of the Davydov *Ansatz* and the way it leads to the NLS equation and its soliton solutions, one can with some justification ask whether this has anything to do with energy transport in a complex biological system. To answer this question, one would perhaps as a first step like to know the influence of a realistic temperature environment present under physiological conditions, that is, $T \approx 300\,\mathrm{K}$.

Davydov [17] also treated this situation. His analysis was again based on the Hamiltonian (9.2.1) with the *Ansatz* (9.2.7) for the state vector. After a number of approximations he again obtains an NLS equation, but now with a temperature-dependent coefficient for the nonlinear term. This coefficient goes to zero with increasing temperature and vanishes at $T \approx 400\,\mathrm{K}$ indicating that soliton solutions should be stable for lower temperatures. This result is in direct contradiction with recent computer simulations of (9.2.16) at finite temperatures [33, 35, 36]. The result of these simulations is summarised below.

To describe the interaction of the system with a thermal reservoir at temperature T, Lomdahl and Kerr [35, 36] added a damping force and noise force,

$$F_n = -m\Gamma\dot{\beta}_n + \eta_n(t), \tag{9.2.21}$$

to (9.2.16a) for the molecular displacements. The correlation function for the random noise was

$$<\eta_n(t)\eta_{n'}(t')> = 2m\Gamma k_B T\delta_{nn'}\delta(t - t') \tag{9.2.22}$$

(k_B is Boltzmann's constant and Γ is a phenomenological damping constant). This extension converts (9.2.16) to Langevin equations. The effect of the two terms in (9.2.22) is to bring the system to thermal equilibrium; it was verified numerically that over sufficiently long time intervals the mean kinetic energy satisfied

$$<\sum_n \tfrac{1}{2}m\dot{\beta}_n^2(t)> = \tfrac{1}{2}Nk_B T \tag{9.2.23}$$

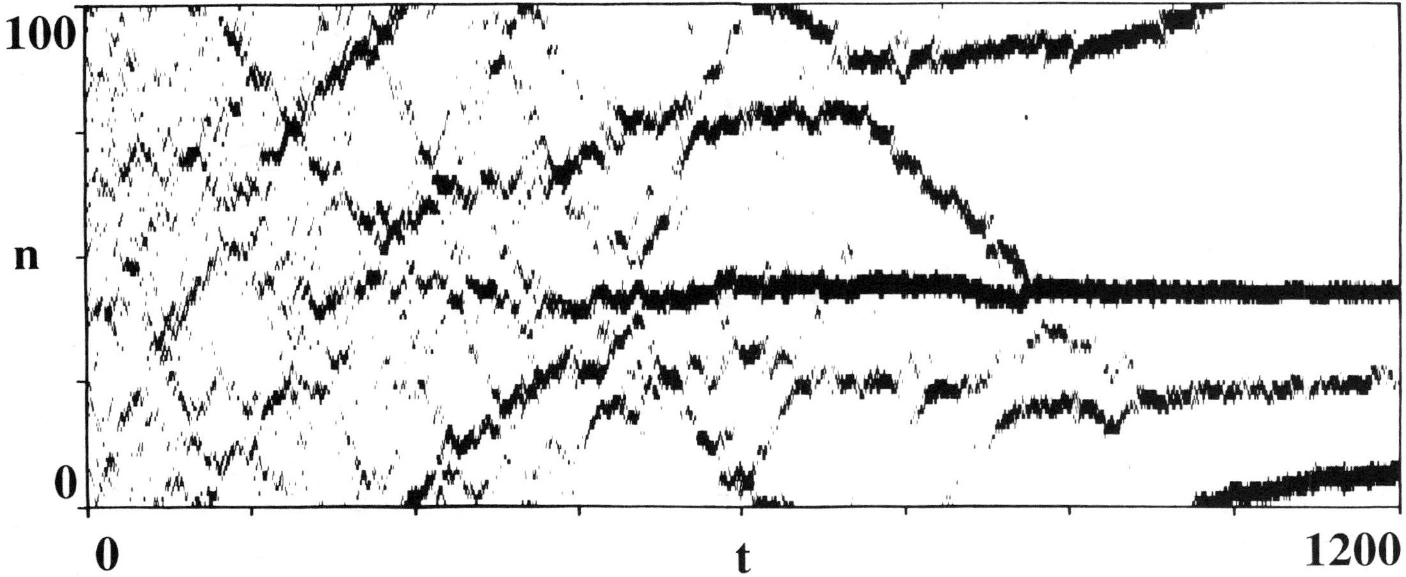

Figure 9.3 Soliton detector plot showing the formation of Davydov solitons from initial conditions consisting of one quantum of amide-I energy distributed randomly along the chain.

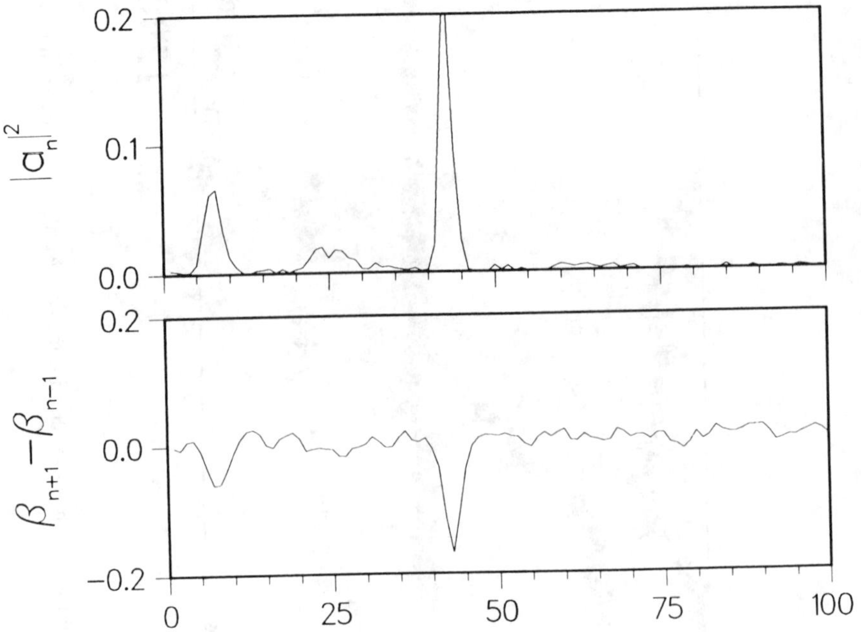

Figure 9.4 Waveform graph showing the spatial configuration at t = 1200 (end of Figure 9.3).

($<\ldots>$ denotes time average). Equations (9.2.16) with (9.2.21) included still imply the conservation of the norm (9.2.8).

With the same diagnostics as in Figures 9.3 and 9.4, we show the result of a simulation at T = 300 K in Figure 9.5. The initial conditions were constructed to mimic what might happen during ATP hydrolysis. A Davydov soliton (cf. (9.2.18) and (9.2.20)) has somehow been nucleated and now evolves under the influence of the random forces. The soliton is seen to disintegrate in a time corresponding to about 3 picoseconds. The filamentary black regions in the picture are seen to have a certain slope, which corresponds to the sound velocity in the units used in the calculation.

The calculation with (9.2.16) and (9.2.21) is a combination of the classical fluctuation–dissipation relation (9.2.23) with quantum-mechanical equations (9.2.16). The justification for this is that for the parameters relevant for α helix (Table 9.1) the highest acoustic frequency $\hbar\omega_{\max}$ is about 100 K. Since the equations were solved near 300 K, the occupation

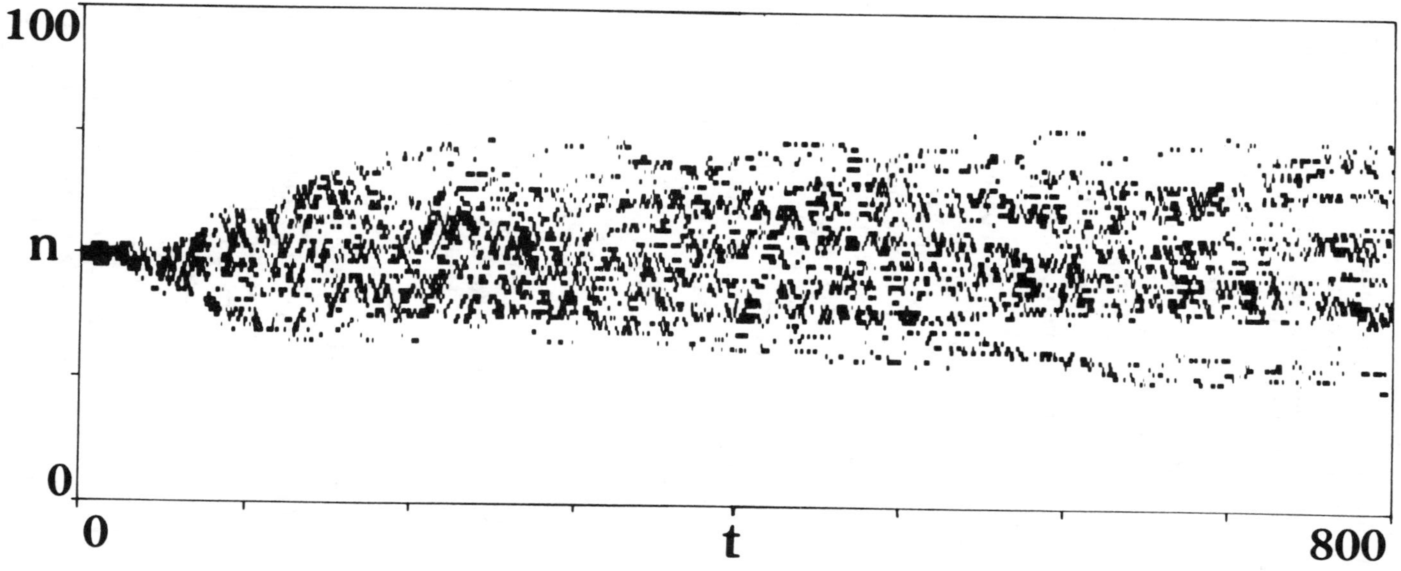

Figure 9.5 Soliton detector plot for a simulation of the Davydov equations at $T = 300\,K$. The initial condition was a Davydov soliton. It disappears in a few picoseconds, from [35].

numbers of all phonon modes are rather accurately given by the classical Boltzmann distribution and under those circumstances (9.2.23) is valid. At lower temperatures than say $T \approx 200\,\mathrm{K}$, the above approach would not be valid. The calculation with (9.2.16) and (9.2.21) is within the canonical ensemble, where the temperature is constant, but the energy allowed to vary. To check the consistency of the results, calculations were also done in the conventional microcanonical ensemble. The system was prepared with the use of (9.2.21) at $T = 300\,\mathrm{K}$; it was then allowed to evolve only under the influence of the deterministic equations (9.2.16). The results of these simulations were essentially the same as presented above. The soliton seems to disappear in a few picoseconds. Similar results were also obtained by Lawrence *et al.* [33].

These results can be interpreted as showing that the *Ansatz* used by Davydov, (9.2.7), is not a good approximation for description of soliton-like objects at biologically relevant temperatures. The assumption that the state vector is decomposable into a pure phonon part and pure exciton part seems broken. The *Ansatz* that the two components remain distinct at all times neglects the phase-mixing characteristic of the evolution of the coupled quantum-mechanical system. This deficiency was also criticised on more general grounds by Brown *et al.* [4, 6]. Since these results are somewhat negative, it is in order to state what *cannot* be concluded from this. The ability of the Hamiltonian (9.2.1) to support soliton-like objects at finite temperatures is still open. It is also not possibel based on the simulations to say what the lower temperature for soliton formation would be.

In addition to the observations made above about finite temperatures, it is also desirable to be able to treat more than one quantum of amide-I energy. Several attempts to circumvent this problem and to find alternative descriptions of the dynamics at finite temperatures are currently in progress. We will consider some of this work in the following sections.

9.3 The Takeno soliton model

Takeno [53–56] has recently proposed an alternative model for the propagation of biological energy in the α helix protein. He has argued that the dispersion term in the Davydov model (9.2.1), may not be appropriate for the migration of vibrational energy. This particular type of exchange interaction is more relevant for electrons or electronic excitons. His approach is basically classical and therefore does not have the constraint on the number of amide-I quanta (9.2.8). In this section we first introduce his Hamiltonian in its simplest form, with only acoustic phonons, in order to compare with the Davydov model. Takeno has also generalised his theory to deal with more complex systems where the amide-I energy is coupled to both acoustic and optic phonons (see Section 9.6). We consider an example of a simulation of such a system at $T = 300\,\text{K}$.

The Hamiltonian is

$$H_1 = \sum_n \left[\frac{\Pi_n^2}{2\mu} + \frac{1}{2}\mu\omega_0^2 q_n^2 - 2Lq_{n+1}q_n \right] + \sum_n \left[\frac{p_n^2}{2m} + \frac{1}{2}K_a(u_{n+1} - u_n)^2 \right]$$
$$+ \sum_n \frac{1}{2} A_a q_n^2 (u_{n+1} - u_{n-1})$$
$$= H_\text{v} + H_\text{p} + H'. \tag{9.3.1}$$

Here q_n and Π_n are the displacement and momentum coordinates for the high frequency intramolecular (amide-I) oscillator with mass μ and frequency ω_0; L is the coupling strength between neighbouring oscillators, which we have restricted to nearest neighbours. Also, u_n and p_n are the displacement and momentum coordinates for the molecule at site n; m and K_a are the molecular mass and intramolecular force constant (K_a is the same as w in Section 9.2). The last term couples these two oscillating fields with coupling constant A_a.

In order to make a comparison with the Davydov model, we now for a moment view (9.3.1) as a quantum Hamiltonian, with the displacement and momentum coordinates replaced by operators. We introduce creation and annihilation operators for the high-frequency oscillator at site n by the equations

$$q_n = \left(\frac{\hbar}{2\mu\omega_0} \right)^{1/2} (B_n^\dagger + B_n); \quad \Pi_n = \mathrm{i} \left(\frac{\hbar\mu\omega_0}{2} \right)^{1/2} (B_n^\dagger - B_n); \tag{9.3.2}$$

then the q_n-dependent parts of (9.3.1) can be written

$$H_\text{v} = \sum_n \hbar\omega_0 (B_n^\dagger B_n + \tfrac{1}{2})$$
$$- \frac{\hbar L}{\mu\omega_0} \sum_n (B_{n+1}^\dagger B_n^\dagger + B_n^\dagger B_{n+1} + \mathrm{B}_{n+1}^\dagger B_n + B_{n+1} B_n), \tag{9.3.3}$$

$$H' = \frac{\hbar A_a}{4\mu\omega_0} \sum_n (B_n^\dagger B_n^\dagger + 2B_n^\dagger B_n + B_n B_n)(u_{n+1} - u_{n-1}). \tag{9.3.4}$$

Comparing (9.3.4) with the Davydov Hamiltonian (9.2.1) it is clear that there are additional $B_n^\dagger B_n^\dagger$ and $B_n B_n$ terms both in the dispersive and interaction parts of the quantum version of the Takeno Hamiltonian. The equation of motion for the Heisenberg operator B_n obtained from (9.3.1) is

$$\mathrm{i}\hbar\dot{B}_n = \hbar\omega_0 B_n - \frac{\hbar L}{\mu\omega_0}(B_{n+1}^\dagger + B_{n+1} + B_{n-1}^\dagger + B_{n-1})$$
$$+ \frac{\hbar A_a}{2\mu\omega_0}(B_n^\dagger + B_n)(u_{n+1} - u_{n-1}). \tag{9.3.5}$$

This differs from (9.2.3), the corresponding equation using Hamiltonian (9.2.1), by the presence of the creation operators on the right-hand side. The presence of those terms means that a phase transformation of the form (9.2.4) cannot remove the energy of the amide-I quantum $\hbar\omega_0 = E_0$ from the equation. Carrying out that transformation on (9.3.5) produces factors oscillating at $2\omega_0$ in the creation operator terms. In this formulation the magnitude of E_0 relative to other energies in the problem remains important. The lack of $B_n^\dagger B_n^\dagger$ and $B_n B_n$ terms in the Davydov Hamiltonian has also been questioned by Fedyanin *et al.* [23].

We note that if we drop the creation operators from (9.3.4), then we can relate the parameters of the two theories by

$$L = (\mu\omega_0/\hbar)J, \quad A_a = (2\mu\omega_0/\hbar)\chi. \tag{9.3.6}$$

Using the parameters from Table 9.1 and the additional parameter $\mu = 1.145 \times 10^{-26}$ kg (reduced mass of the C=O unit) gives the values we have used in the calculations described below.

The equations of motion derived from the classical Hamiltonian (9.3.1) are

$$\mu\ddot{q}_n + \mu\omega_0^2 q_n - 2L(q_{n+1} + q_{n-1}) + A_a q_n(u_{n+1} - u_{n-1}) = 0, \tag{9.3.7a}$$

$$m\ddot{u}_n - K_a(u_{n+1} - 2u_n + u_{n-1}) - \tfrac{1}{2}A_a(q_{n+1}^2 - q_{n-1}^2) = 0. \tag{9.3.7b}$$

Takeno now proceeds by making a continuum approximation to (9.3.7) and obtains this way coupled nonlinear Klein–Gordon equations for the coordinates $q(x, t)$ and $u(x, t)$. A rotating-wave approximation then finally leads to an NLS equation (9.2.17), but now with a classical coordinate for the amplitude of the amide-I vibration compared to Davydov's NLS equation for the probability amplitude.

If (9.3.1) is augmented with the additional optic mode and interaction term:

$$H_{\text{op}} = \sum_n [\tfrac{1}{2}M_0\dot{y}_n^2 + \tfrac{1}{2}K_0 y_n^2] + \sum_n \tfrac{1}{2}A_0 q_n^2 y_n. \tag{9.3.8}$$

the equations of motion become

$$\mu\ddot{q}_n + \mu\omega_0^2 q_n - 2L(q_{n+1} + q_{n-1}) + A_a q_n(u_{n+1} - u_{n-1}) + A_0 q_n y_n = 0, \tag{9.3.9a}$$

$$m\ddot{u}_n - K_a(u_{n+1} - 2u_n + u_{n-1}) - \tfrac{1}{2}A_a(q_{n+1}^2 - q_{n-1}^2) = 0, \tag{9.3.9b}$$

$$M_0\ddot{y}_n + K_0 y_n + \tfrac{1}{2}A_0 q_n^2 = 0. \tag{9.3.9c}$$

Takeno [56] has used these equations to describe self-trapped states in crystalline acetanilide (see Section 9.6). In Figure 9.6 we show the result of a simulation [35] of (9.3.9) with damping and noise forces (9.2.21)

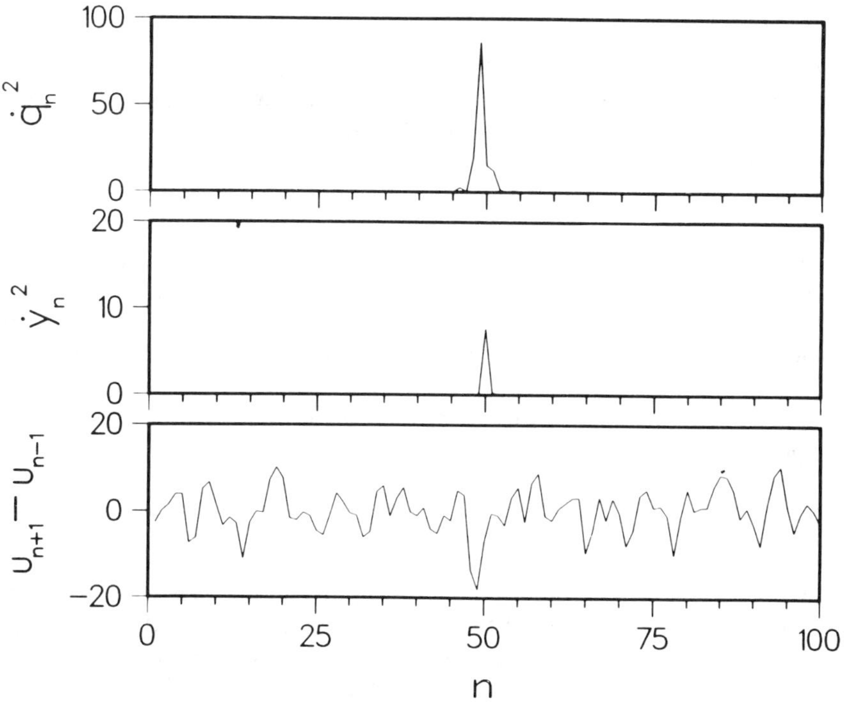

Figure 9.6 Simulation of the Takeno equations (9.3.9) at $T = 300\,\mathrm{K}$. The wave-form graphs show essentially the kinetic energy of the amide-I vibron field q_n and the lower frequency optic phonon y_n. The lower graph shows the 'discrete' gradient $(u_{n+1} - u_{n-1})$ of the acoustic field. A clear correlation between the fields is observed, indicating a stable self-trapped entity.

added in (9.3.9b) for the acoustic displacement field. The optic mode mass was $M_0 = 1.56 \times 10^{-27}\,\mathrm{kg}$ (the reduced mass of the N–H unit), and $A_a = A_0$ since both acoustic and optic mode couplings involve the hydrogen bond. The system was initially prepared in a state that had a large local displacement in the vibron (amide-I) field, no energy in the optic field, and kinetic energy and displacement of the acoustic field corresponding to 300 K. The wave-form graphs show essentially the kinetic energy in the vibron and optic fields and the discrete gradient $u_{n+1} - u_{n-1}$ approximately 60 ps into the simulation. The amide-I energy is clearly still localised and a significant correlation is seen to have developed in the acoustic and optic fields. Additional studies of the Takeno model are clearly needed, but it seems that the classical solitons described by (9.3.9) are more stable at biologically relevant temperatures than the Davydov soliton described by (9.2.16).

9.4 The Yomosa soliton model

Another classical soliton model for energy transport in the α helix has also recently been proposed by Yomosa [57, 58]. While both the Davydov and Takeno models are considering the interaction of the *intra*-molecular amide-I mode with lower frequency *inter*-molecular (acoustic) modes, the mechanism proposed by Yomosa focuses on large-amplitude motion of the peptide groups in the hydrogen-bonded spines of the α helix. His *lattice solitons* (see [11]) are established solely through the nonlinearity and asymmetry of the hydrogen bonds that stabilise the α helix structure. In this section we briefly outline Yomosa's theory and show results of some recent molecular dynamics simulations at 300 K in the spirit of his theory.

The potential of the nth hydrogen bond in the polypeptide chain is approximated as:

$$V_n(r_n) = Ar_n^2 - Br_n^3, \tag{9.4.1}$$

$$r_n = u_{n+1} - u_n, \tag{9.4.2}$$

where u_n is the displacement of the nth peptide group, r_n is the elongation of the nth peptide bond. The lattice constant is denoted l as in the previous sections. The values of the constants A and B can be determined from self-consistent-field molecular orbital calculations [58]. The Hamiltonian of the system is now

$$H_u = \tfrac{1}{2}m\sum_n \dot{u}_n^2 + \sum_n [A(u_n - u_{n-1})^2 - B(u_n - u_{n-1})^3], \tag{9.4.3}$$

where m is the mass of the peptide group. The equations of motion in terms of r_n are:

$$m\ddot{r}_n = 2A(r_{n+1} + r_{n-1} - 2r_n) - 3B(r_{n+1}^2 + r_{n-1}^2 - 2r_n^2). \tag{9.4.4}$$

Yomosa introduces the continuum limit of (9.4.4) ($nl \to x$, $r_n \to r(x, t)$) and looking only at right-going waves he obtains a KdV (Korteweg–de Vries) equation:

$$\phi_\tau - 12\phi\phi_\xi + \phi_{\xi\xi\xi} = 0, \tag{9.4.5}$$

where

$$\xi = x/l - (2A/m)^{1/2}t, \quad \tau = 1/24(2A/m)^{1/2}, \quad \phi = 3B/Ar. \tag{9.4.6}$$

In terms of the original elongation $r(x, t)$ the one-soliton solution is given as

$$r(x, t) = -\frac{A}{12B}\kappa^2 sech^2\left(\frac{\kappa}{2l}(x - vt) - \delta\right), \tag{9.4.7}$$

with

$$v = v_0\left(1 + \frac{\kappa^2}{24}\right). \qquad (9.4.8)$$

Here $v_0 = (2A/m)^{1/2}l$ is the sound velocity. It follows from (9.4.8) that Yomosa's solitons are supersonic. The parameter κ determines the amplitude of the soliton and δ its initial location. Yomosa estimates κ to be in the range 2–3 by equating the energy released during ATP hydrolysis (0.42 eV) to the energy of the KdV one-soliton (9.4.7) (using the continuum version of (9.4.3)). This gives an effective soliton width of about 4 Å., that is, approximatley one peptide group. Such a narrow pulse obviously puts some doubt on the validity of the continuum approximation.

The potential (9.4.1) was here taken to be cubic, reflecting the anharmonicity and asymmetry of the hydrogen bond. The exact form is probably not very important; Yomosa [57] has also considered a Toda-type potential, with essentially the same result. In fact, recent computer simulations with a Lenard–Jones potential [42] show effectively the same phenomena as predicted by Yomosa's continuum KdV theory, that is, the formation and propagation of supersonic solitons. These molecular dynamics simulations were also extended to biologically relevant temperatures by addition of noise and damping force terms (9.2.19) to the equations of motion. When 0.42 eV of energy was initialised on one bond, coherent pulses of energy were observed above the thermal noise for at least 25 ps at $T = 310$ K. A window of most efficient energy transport was found around 40–60 °C; at lower temperatures the viscosity of the solvent (modelled through Γ in (9.2.21)) inhibited transport, while at high temperatures the thermal noise is the limiting factor.

It seems that the supersonic lattice solitons proposed by Yomosa present a reasonable alternative to the Davydov and Takeno models of transport of biological energy. These lattice solitons may also be more efficient in doing mechanical work since they have no rest energy associated with them. More theoretical and numerical work is needed in this area.

9.5 The discrete self-trapping equation

The Hamiltonian (9.2.1) is by construction quantum-mechanical, Davydov's analysis is based on the coherent state *Ansatz* (9.2.7) and is therefore as 'classical' as a quantum-mechanical treatment can be. An alternative approach would be to start from purely classical equations, and then subsequently do semi-classical or Bohr–Sommerfeld quantisation. This has recently been done by Scott and collaborators [22, 48, 51, 52]. The advantage of this method is that it allows for an arbitrary number of amide-I quanta.

The starting point is to define classical amide-I displacement and momentum coordinates Q_n and P_n, for which the Hamiltonian is $\Sigma_n(P_n^2 + Q_n^2)$. The motion can then be described in terms of the complex mode amplitude

$$A_n \equiv \omega_0^{1/2}(P_n + \mathrm{i}Q_n). \tag{9.5.1}$$

It terms of A_n the classical analogue of H_v in (9.2.1) becomes

$$H_{\mathrm{v-clas}} = \sum_n [\omega_0|A_n|^2 - J(A_{n+1}^* A_n + A_n^* A_{n+1})], \tag{9.5.2}$$

where ω_0 is the amide-I frequency. It is then assumed that the amide-I mode interacts with some (unspecified) low-frequency mode q_n, with adiabatic energy

$$H_{\mathrm{p-clas}} = \tfrac{1}{2}w_{\mathrm{clas}} \sum_n q_n^2. \tag{9.5.3}$$

The interaction energy is taken to be

$$H_{\mathrm{int-clas}} = \chi_{\mathrm{cla}} \sum_n q_n |A_n|^2. \tag{9.5.4}$$

The total classical Hamiltonian is then

$$H_{\mathrm{clas}} = H_{\mathrm{v-clas}} + H_{\mathrm{p-clas}} + H_{\mathrm{int-clas}}. \tag{9.5.5}$$

Minimising (9.5.5) with respect to q_n yields

$$q_n = -\frac{\chi_{\mathrm{clas}}}{w_{\mathrm{clas}}}|A_n|^2, \tag{9.5.6}$$

whereupon (9.5.5) reduces to

$$H_{\mathrm{clas}} = \sum_n [\omega_0|A_n|^2 - J(A_{n+1}^* A_n + A_n^* A_{n+1}) - \tfrac{1}{2}\gamma|A_n|^4], \tag{9.5.7}$$

where $\gamma = \chi_{\mathrm{clas}}^2/w_{\mathrm{clas}}$. The equation of motion corresponding to (9.5.7) is

$$\left(\mathrm{i}\frac{\mathrm{d}}{\mathrm{d}t} - \omega_0\right)A_n + J(A_{n+1} + A_{n-1}) + \gamma|A_n|^2 A_n = 0. \tag{9.5.8}$$

In addition to the *energy* H_{clas}, (9.5.8) has another constant of motion, the *number*:

$$N = \sum_n |A_n|^2. \tag{9.5.9}$$

Equation (9.5.8) is a special case of the *discrete self-trapping equation*

$$\mathrm{i}\dot{\mathbf{A}} + M\mathbf{A} + \gamma D(|\mathbf{A}|^2)\mathbf{A} = 0, \tag{9.5.10}$$

where $\mathbf{A} = \{A_n\}$ *is a complex n*-vector, $D(|\mathbf{A}|^2)$ denotes the diagonal matrix $\text{diag}(|A_1|^2, |A_2|^2, \ldots, |A_n|^2)$, and M is a real symmetric matrix. This equation has recently been studied extensively [10, 22, 28, 49–52]. For small $n(\leq 2)$ it is completely integrable, for $n = 3$ and 4 the stability of stationary solutions has been mapped out. The DST equation has been used to describe self-trapping in a number of physical systems, including self-trapping on a dimer ($n = 2$) [30] and the dynamics of the CH stretch vibration in benzene ($n = 6$) ([50]; see also [11]). Note that M describes the influence of dispersion in very general terms. In (9.5.8) M is a tri-diagonal matrix with J on the two bidiagonals and $-\omega_0$ on the diagonal, appropriate for the description of the nearest-neighbour interaction of peptide groups on one spine of the α helix. If the problem above had been formulated for a globular protein [34], the matrix M would have elements reflecting the random structure.

Upon quantisation [48–52] the complex mode amplitudes are replaced by harmonic oscillator creation and annihilation operators $B_n^\dagger$ and B_n with the properties: $B_n^\dagger|n\rangle = \sqrt{(n+1)}|n+1\rangle$ and $B_n|n\rangle = \sqrt{n}|n-1\rangle$ and commutator $[B_n, B_m^\dagger] = \delta_{nm}$. *The number* (9.5.9) and the *energy* (9.5.7) then become operators

$$\hat{N} = \sum_n (B_n^\dagger B_n + \tfrac{1}{2}), \tag{9.5.11}$$

$$\hat{H} = (\omega_0 - \tfrac{1}{2}\gamma)\hat{N} - J\sum_n (B_{n+1}^\dagger B_n + B_n^\dagger B_{n+1}) - \tfrac{1}{2}\gamma\sum_n B_n^\dagger B_n B_n^\dagger B_n. \tag{9.5.12}$$

In the limit $J \ll \gamma$ (i.e. the zero dispersion limit) the eigenstates of $\hat{H}$ are those of a harmonic oscillator, $\hat{H}|n\rangle = E_n|n\rangle$ with energies given by

$$E_n = (\omega_0 - \tfrac{1}{2}\gamma)(n + \tfrac{1}{2}) - \tfrac{1}{2}\gamma_n{}^2. \tag{9.5.13}$$

As noted by Scott [48, 51] this defines an overtone spectrum $\nu(n) = E_n - E_0$, or

$$\nu(n) = (\omega_0 - \tfrac{1}{2}\gamma)n - \tfrac{1}{2}\gamma n^2, \tag{9.5.14}$$

which should be observable as absorption lines in a optical spectrum.

It is a major advantage of the DST approach that the method allows for treatment of more than one amide-I quantum; the validity (in the zero dispersion limit) is directly testable through (9.5.13). Any information about transport of self-trapped states is, however, precluded from (9.5.14). To obtain that kind of information a general treatment of (9.5.12) is required, which of course is as complicated as Davydov's original Hamiltonian (9.2.1). The influence of temperature is also not yet clear, since it enters only implicitly through the coupling to the low-frequency mode defined in (9.5.3).

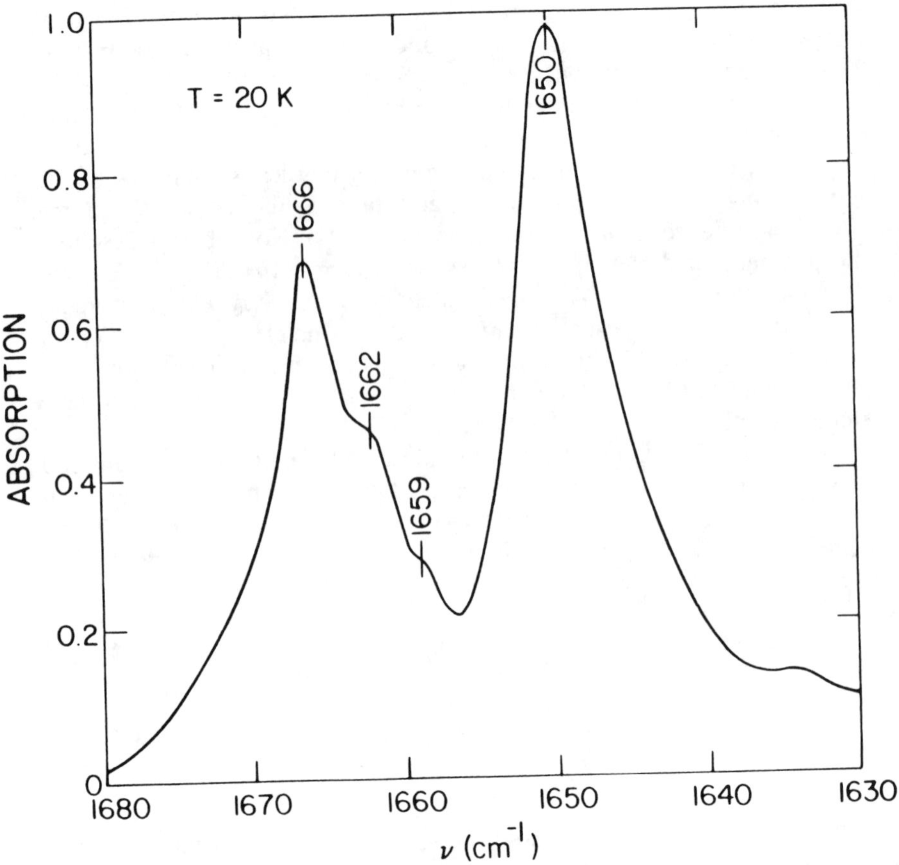

Figure 9.7 Low-temperature ($T = 20\,\mathrm{K}$) infrared spectrum of ACN showing the anomalous band at $1650\,\mathrm{cm}^{-1}$ (ν is the frequency). The shoulders on the amide-I band are due to the different normal modes of the amide-I excitation in the complicated unit cell of ACN. The unit cell has eight peptide groups. The splitting of the amide-I band can thus be explained by normal mode analysis based on group theory. The appearance of the band at $1650\,\mathrm{cm}^{-1}$ is interpreted as being due to the presence of a Davydov-like soliton (from [21], data courtesy E. Gratton).

9.6 Experimental evidence

There has been surprisingly little experimental work done in the search for solitons in proteins. The most convincing experimental evidence comes from spectra of the crystalline polymer acetanilide $(CH_3CONHC_6H_5)_x$, or ACN. ACN is organised into hydrogen-bonded chains of peptide groups and has a remarkable similarity to the chain structure of the α helix. Late in the 1960s Careri [7] noted that the peptide bond lengths and angels in ACN are very close to those in natural proteins, and he began an experimental programme to see whether ACN would show any unexpected physical properties that might be of biological interest. In 1973 he reported the observation of an anomalous line in the infrared absorption spectrum of ACN. This anomalous line is lower in wavenumber than the main amide-I peak by about $15\,\mathrm{cm}^{-1}$. Its intensity is low at room temperature but increases as the temperature is lowered. Figure 9.7 shows the anomalous and conventional amide-I bands in an infrared absorption spectrum at $T = 20\,\mathrm{K}$. Numerous attempts by Careri and co-workers to find a conventional assignment of this new line were unsuccessful throughout the 1970s.

In 1982 Scott became aware of Careri's data and he proposed that the anomalous line was due to a soliton-like excitation much like the Davydov soliton [8, 9, 48]. The amide-I vibration coupled to a low-frequency mode (9.5.4) gives rise to the self-trapping of amide-I energy. Scott's theory can be described essentially by the DST equation as outlined in Section 9.5. A major prediction of the soliton theory was the nonlinear overtone spectrum given by (9.5.14). Subsequent experimental findings compare very favourably with the predictions of (9.5.14) [51]. The temperature dependence of the intensity of the anomalous line was also predicted by Scott [9] based on the theory of coupling to the low-frequency optic mode. This approach has been disputed by Alexander and Krumhansl [1, 2] who argued that a better agreement is obtained considering the original Davydov model (Section 9.2), with coupling to acoustic phonons.

Other explanations of the anomalous amide-I line have been proposed. Blanchet and Fincher [3] suggested that the existence of a $3250\,\mathrm{cm}^{-1}$ band with the same temperature dependence as the $1650\,\mathrm{cm}^{-1}$ line, is evidence for a structural phase transition. Johnston and Swanson [29] have argued that the $1650\,\mathrm{cm}^{-1}$ can be explained as a Fermi resonance. The interested reader is referred to these references for more details.

The spectroscopic evidence for existence of a self-trapped state in ACN (Davydov-like or not) is quite exciting; however, it is not clear what its transport properties are. In fact, the theory indicates that the soliton is almost static [21]. At present, there is no experimental evidence for *transport* of energy via coherent soliton-like pulses in proteins. Experiments that directly deal with the question of how energy is taken from one point A to another point B are much needed. Experiments of this sort have recently been proposed [32] and are currently under way at the University of Rochester, NY.

9.7 **Concluding remarks**

The soliton theories of energy transport in proteins should attract the careful attention of the bioenergetics community. Clearly they cannot explain every aspect of protein dynamics, but they are motivating exciting questions and new experiments. We have reviewed the current theories; from Davydov's original quantum-mechanical approach to the classical theories put forward by the Japanese researchers and the recent semi-classical approach within the DST framework. There are clearly still many open problems and no single theory presently has answers to all questions. Continuing work on the extension and improvement of these theories makes it an exciting field of applied soliton research. The bulk of this work combined with carefully designed experiments will eventually succeed in establishing a better understanding of energy transport in biological systems.

Acknowledgements

I would like to thank my collaborators Irving Bigio, Chris Eilbeck, Bill Kerr, Scott Layne and Al Scott for their contributions to the work presented here and for many exciting discussions. The work was done under the auspices of the US Department of Energy.

References

[1] Alexander, D. M. Analog of small polaron in hydrogen-bonded amide systems, *Phys. Rev. Lett.* **54**, 138–41 (1985).

[2] Alexander, D. M. and Krumhansl J. A. Localized excitations in hydrogen-bonded molecular crystals, *Phys. Rev.* **33B**, 7172–85 (1986).

[3] Blanchet, G. B. and Fincher, C. R. Defects in a nonlinear pseudo one-dimensional solid, *Phys. Rev. Lett.* **54**, 1310–12 (1985).

[4] Brown, D. W., Lindenberg, K. and West, B. J. On the applicability of Hamilton's equations in the quantum soliton problem, *Phys. Rev.* **A33**, 4104–9 (1986).

[5] Brown, D. W., Lindenberg, K. and West, B. J. A nonlinear density matrix approach for the finite temperature soliton problem, *Phys. Rev. Lett.* **57**, 2341–3 (1986).

[6] Brown, D. W., West, B. J. and Lindenberg, K. Davydov solitons: new results at variance with standard derivation, *Phys. Rev.* **A33**, 4110–20 (1986).

[7] Careri, G. Search for cooperative phenomena in hydrogen-bonded amide structures. in *Cooperative Phenomena*, H. Haken and M. Wagner (eds), 391–4 Springer, Berlin, Heidelberg and New York. (1973).

[8] Careri, G., Buontempo, U., Carta, F., Gratton, E. and Scott, A. C., Infrared absorption in acetanilide by solitons, *Phys. Rev. Lett.* **51**, 304–7 (1983).

[9] Careri, G., Buontempo, U., Galluzzi, F., Scott, A. C., Gratton, E. and Shyamsunder, E. Spectroscopic evidence for Davydov-like solitons in acetanilide, *Phys. Rev.* **30B**, 4689–702 (1984).

[10] Carr, J. and Eilbeck, J. C. Stability of stationary solutions of the discrete self-trapping equation, *Phys. Lett.* **109A**, 201–4 (1985).

[11] Collins, M. A. Solitons in chemical physics, *Advan. Chem. Phys.* **53**, 225–339 (1983).

[12] Davydov, A. S. The theory of contraction proteins under their excitation, *J. Theor. Biol.* **38**, 559–69 (1973).

[13] Davydov, A. S. Quantum theory of muscular contraction, *Biofizika* **19**, 670–6 (1974).

[14] Davydov, A. S. Solitons and energy transfer along protein molecules, *J. Theor. Biol.* **66**, 379–87 (1977).

[15] Davydov, A. S. Solitons in molecular systems, *Physica Scr.* **20**, 387–94 (1979).

[16] Davydov, A. S. Solitons, bioenergetics and the mechanism of muscle contraction, *Int. J. Quantum Chem.* **16**, 5–17 (1979).

[17] Davydov, A. S. Soliton motion in a one-dimensional molecular lattice with account taken of thermal oscillations, *Sov. Phys. JETP* **51**, 397–400 (1980) (Transl. of *Zh. Exps. Teor. Fiz.* **78**, 789–96.)

[18] Davydov, A. S. *Biology and Quantum Mechanics*, Pergamon Press, New York (1982).

[19] Davydov, A. S. Solitons in quasi-one-dimensional molecular chains, *Usp. fiz. Nauk.* **138**, 603–43 (1982).

[20] Davydov, A. S. and Kislukha, N. I. Solitary excitations in one-dimensional molecular chains, *Physica Status Solidi* **B59**, 465–70 (1973).

[21] Eilbeck, J. C., Lomdahl, P. S. and Scott, A. S. Soliton structure in crystalline acetanilide, *Phys. Rev.* **B30**, 4703–12 (1984).

[22] Eilbeck, J. C., Lomdahl, P. S. and Scott, A. S. The discrete self-trapping equation, *Physica* **16D**, 318–38 (1985).

[23] Fedyanin, V. K., Makhankov, V. G. and Yakushevich, L. V. Exciton–phonon interaction in long-wave approximation, *Phys. Lett.* **61A**, 256–8 (1977).

[24] Frohlich, H. Interaction of electrons with lattice vibrations, *Proc. R. Soc. London Ser.* **A215**, 291–8 (1952).

[25] Glauber, R. J. Coherent and incoherent states of the radiation field, *Phys. Rev.* **131**, 2766–88 (1963).

[26] Hyman, J. M., McLaughlin, D. W. and Scott, A. C. On Davydov's alpha-helix solitons, *Physica* **3D**, 23–44 (1981).

[27] Itoh, K. and Shimanouchi, T. Vibrational spectra of crystalline formamide, *J. Molec. Spectrosc.* **42**, 86–99 (1972).

[28] Jensen, J. H., Christiansen, P. L., Elgin, J. N., Gibbon, J. D. and Skovgaard, O. Correlation exponents for trajectories in the low-dimensional discrete self-trapping equation, *Phys. Lett.* **110A**, 429–31 (1985).

[29] Johnston, C. T. and Swanson, B. I. Temperature dependence of the vibrational spectra of acetanilide: Davydov solitons or Fermi coupling, *Chem. Phys. Lett.* **114**, 547–52 (1985).

[30] Kenkre, V. M. and Campbell, D. K. Self-trapping on a dimer: time-dependent solutions of a discrete nonlinear Schroedinger equation, *Phys. Rev.* **34B**, 4959–61 (1986).

[31] Kerr, W. C. and Lomdahl, P. S. Quantum mechanical derivation of the equations of motion for Davydov solitons, *Phys. Rev.* **B35**, 3629–32 (1987).

[32] Knox, r. S. Hypothesis: detection and characterization of coherent vibrational energy transport in protein alpha-helix segments, *Biological Physics Group Technical Report* 58, University of Rochester (1985).

[33] Lawrence, A. F., McDaniel, J. C., Chang, D. B., Pierce, B. M. and Birge, R. R. Dynamics of the Davydov model in α-helix protein: effects of the coupling parameter and temperature, *Phys. Rev.* **A33**, 1188–2302 (1986).

[34] Lomdahl, P. S. Nonlinear dynamics of globular proteins. In *Nonlinear Elec-*

trodynamics in Biological Systems, W. Ross Adey and A. F. Lawrence (eds), Plenum Press, New York, 143–154 (1984).

[35] Lomdahl, P. S. and Kerr, W. C. Do solitons exist at 300 K? *Phys. Rev. Lett.* **55**, 1235–8 (1985).

[36] Lomdahl, P. S. and Kerr, W. C. Finite temperature effects on models of hydrogen-bonded polypeptides. In *Physics of Many Particle Systems*, A. S. Davydov (ed.), Kiev, USSR (1986).

[37] Lomdahl, P. S., Layne, S. P. and Bigio, I. J. Solitons in biology, *Los Alamos Science* **10**, 2–22 (1984). Available on request from the authors.

[38] Loudon, R. *The Quantum Theory of Light*, Oxford University Press, New York, p. 146 (1983).

[39] McClare, C. W. F. Resonance in bioenergetics, *Ann. N. Y. Acad. Sci.* **227**, 74–97 (1974).

[40] MacNeil, L. and Scott, A. C. Launching a Davydov soliton: II. Numerical studies, *Physica Scr.* **29**, 284–7 (1984).

[41] Nevskaya, N. A. and Chirgadze, Yu. N. Infrared spectra and resonance interactions of amide-I and amide-II vibrations of α-helix, *Biopolymers* **15**, 637–48 (1976).

[42] Perez, P. and Theodorakopoulos, N. Solitary excitations in the α-helix: viscous and thermal effects, *Phys. Lett.* **117A**, 405–8 (1986).

[43] Schulz, G. E. and Schirmer R. H. *Principles of Protein Structure*, Springer, New York (1979).

[44] Scott, A. C. Dynamics of Davydov solitons, *Phys. Rev.* **A26**, 578–95 (1982).

[45] Scott, A. C. The vibrational structure of Davydov solitons, *Physica Scr.* **25**, 651–8 (1982).

[46] Scott, A. C. Errata to Scott [44] *Phys. Rev.* **A27**, 2767 (1983).

[47] Scott, A. C. Launching a Davydov soliton. I. Soliton analysis, *Physica Scr.* **29**, 279–83 (1984).

[48] Scott, A. C. Davydov solitons in polypeptides, *Phil. Trans. R. Soc. Lond. A* **315**, 423–36 (1985).

[49] Scott, A. C. and Eilbeck, J. C. The quantized discrete self-trapping equation, *Phys. Lett.* **119A**, 60–4 (1986).

[50] Scott, A. C. and Eilbeck, J. C. On the CH stretch overtones of benzene, *Chem. Phys. Lett.* **132**, 23–8 (1986).

[51] Scott, A. C., Gratton, E. Shyamsunder, E. and Careri, G. IR overtone spectrum of the vibrational soliton in crystalline acetanilide, *Phys. Rev.* **B32**, 5551–3 (1985).

[52] Scott, A. C., Lomdahl, P. S. and Eilbeck, J. C. Between the local-mode and normal-model limits, *Chem. Phys. Lett.* **113**, 29–36 (1985).

[53] Takeno, S. Exciton soliton in one-dimensional molecular crystals, *Prog. Theor. Phys.* **69**, 1798–1801 (1983).

[54] Takeno, S. Vibron soliton in one-dimensional molecular crystal, *Prog. Theor. Phys.* **71**, 395–8 (1984).

[55] Takeno, S. Vibron solitons and coherent polarizatioon in an exactly tractable oscillator-lattice system, *Prog. Theor. Phys.* **73**, 853–73 (1985).

[56] Takeno, S. Vibron solitons and soliton-induced infrared spectra of crystalline acetanilide, *Prog. Theor. Phys.* **75**, 1–14 (1986).

[57] Yomosa, S. Toda-lattice solitons in α-helix proteins, *J. Phys. Soc. Jpn.* **53**, 3692–8 (1984).

[58] Yomosa, S. Solitary excitations in muscle protein, *Phys. Rev.* **32A**, 1752–8 (1985).

[59] Zakharov, V. E. Collapse of Langmuir waves, *Sov. Phys. JETP* **35**, 908–14 (1972). (Transl. of *Zh. Exps. Teor. Fiz.* **62**, 1745–59.)

Part IV

Hamiltonian theory

10 *S. Rauch-Wojciechowski*

Hamiltonian structures and complete integrability in analytical mechanics

10.1 Introduction

The basic problem of analytical mechanics is to solve Newton's equations of motion

$$\ddot{q}_k = F_k = -\frac{\partial}{\partial q_k} V(q_1, \ldots, q_n), \quad k = 1, \ldots, n \qquad (10.1.1)$$

where q_k, $k = 1, \ldots, n$ are position coordinates, F_k are forces and dot denotes time derivative. We shall assume that forces have a potential $V(q)$ which is time-independent.

Since the formulation of Newton's equations great efforts have been made to solve them explicitly. They were successful in solving the two-body problem $V(q) = \alpha/q$, $q^2 = q_1^2 + q_2^2 + q_3^2$, the harmonic oscillator $V(q) = \frac{1}{2}\sum_k \alpha_k q_k^2$, and one-dimensional potentials. However, the integration process usually leads to new functions and most equations are not amenable to any method. This stimulated development of different formulations of dynamics: the Lagrange equations and then the first-order Hamilton equations. The advantage of the Hamilton equations is that they are invariant with respect to a much larger (than the Lagrange equations) group of transformations: the canonical transformations. This invariance led to the discovery of the most powerful classical technique of integrating Hamilton equations – the method of separation of variables (MSV) in the Hamilton–Jacobi (HJ) equation. All examples in textbooks on classical mechanics are solvable by this technique.

Work on the exact solving of Hamilton equations was abandoned at the turn of this century. This was a consequence of the negative result of Poincaré which states that a Hamiltonian which is a small perturbation of an integrable one has usually no functionally independent integral which is analytic in the perturbation parameter and dynamical variables. This means that a generic, time-independent Hamiltonian has only one analytic integral, the Hamiltonian itself, and on the surface of constant energy motion can be extremely complicated. Integrable Hamiltonians are exceptional and only they can be expected to be analytically solvable. However, they still play an important role in understanding the dynamical behaviour of more general Hamiltonians. The famous KAM (Kolmogorov, Arnol'd and Moser) theorem states that for a small perturbation of an integrable Hamiltonian most of the invariant tori still exist, although deformed, and that the measure of destroyed tori goes to zero when the size of perturbation goes to zero. It has also been proved [40] that the deformed tori of the perturbed Hamiltonian can be completed to a foliation of the phase space. This foliation corresponds to another integrable perturbation of the initial Hamiltonian.

Resurgence of interest in integrable Hamiltonian systems is a consequence of the discovery of the isospectral deformation method (IDM) and of their importance for perturbation theory. The IDM was discovered (1967) in the context of an infinite-dimensional Hamiltonian system – the Korteweg–de Vries (KdV) equations [15] – and only later adapted for the finite-dimensional case. The Toda system and the Calogero–Moser (CM) system are the most prominent examples. It was the Moser paper on the CM system [32] which started and stimulated the main research on integrable natural Hamiltonians and led to the discovery of many new systems of this kind. Some of these will be described in the next sections.

Section 10.2 introduces definitions of standard and nonstandard Hamiltonian structures and of canonical transformations and explains the notion of complete integrability which is illustrated on the example of the Kepler two-body motion. Section 10.3 is concerned with the method of separation of variables for natural Hamiltonian system. The problem of a criterion of separability is discussed in detail. In particular there is formulated there the Stäckel characterisation of separable hamiltonians and an effective criterion of separability in two dimensions. Some new classes of separable potentials are constructed recursively. In Section 10.4 we introduce the isospectral deformation method on the example of the Toda system and then discuss general aspects of the method. The examples of the Calogero–Moser system and the Garnier system are chosen to each represent a large class of integrable natural Hamiltonians of similar nature. Flexibility of the isospectral deformation method and the use of nonstandard Hamiltonian structures are illustrated by the examples of Section 10.5. The examples are again chosen to represent families of integrable systems of similar type. The integrable Manakov case of the Euler equations of rigid-body motion is among the best known ones.

10.2 **Hamilton equations and complete integrability**

10.2.1 *Hamiltonian structure*

Let us start from the Newton equations (10.1.1) and write them as first-order equations by introducing $p_k = \dot{q}_k$. They take the form

$$\dot{q}_k = p_k = \frac{\partial H}{\partial p_k}, \tag{10.2.1}$$

$$\dot{p}_k = -\frac{\partial V}{\partial q_k} = -\frac{\partial H}{\partial q_k}, \tag{10.2.2}$$

where the function H is defined as

$$H(q, p) = \frac{1}{2}\sum_{k=1}^{n} p_k^2 + V(q_1, \ldots, q_n) = T + V. \tag{10.2.3}$$

The equations (10.2.1), (10.2.2) are *Hamilton equations*. They are generated in a skew-symmetric way by one function, the *Hamiltonian* $H(q, p)$. A Hamiltonian which is a sum of kinetic energy T and potential energy V is said to be *natural*.

The Hamilton equations can be given a more symmetric form by introducing the *Poisson bracket* (PB) of two functions:

$$\{f(q, p), g(q, p)\} = \sum_{k=1}^{n}\left(\frac{\partial f}{\partial q_k}\cdot\frac{\partial g}{\partial p_k} - \frac{\partial f}{\partial p_k}\cdot\frac{\partial g}{\partial q_k}\right). \tag{10.2.4}$$

Then equations (10.2.1), (10.2.2) become

$$\dot{q}_k = \{q_k, H\}, \tag{10.2.5}$$

$$\dot{p}_k = \{p_k, H\}. \tag{10.2.6}$$

The basic properties of the PB (10.2.4) are:

(i) linearity: $\{\alpha f + \beta g, h\} = \alpha\{f, h\} + \beta\{g, h\}$, for arbitrary functions f, g, h and constants α, β;
(ii) skew-symmetry: $\{f, g\} = -\{g, f\}$;
(iii) Jacobi identity: $\{\{f, g\}, h\} + \{\{h, f\}, g\} + \{\{g, h\}, f\} = 0$;
(iv) derivation property: $\{f, gh\} = \{f, g\}h + g\,\{f, h\}$.

There are many PB which satisfy properties (i)–(iv). Important examples are Lie–Poisson brackets (LPB) which for dynamical variables l_α are defined by structure constants of a Lie algebra:

$$\{l_\alpha, l_\beta\} = \sum_\gamma c^\gamma_{\alpha\beta}\, l_\gamma, \quad \alpha, \beta, \gamma = 1, \ldots, n.$$

The LP brackets can be *degenerate*: that is, there may exist non-constant functions, called *Casimir functions*, which have zero LPB with all dynamical variables. For example, if l_α, $\alpha = 1, 2, 3$ satisfy commutation relations of the $so(3)$ Lie algebra: $\{l_1, l_2\} = l_3$, $\{l_2, l_3\} = l_1$, $\{l_3, l_1\} = l_2$ then $l^2 = l_1^2 + l_2^2 + l_3^2$ commutes with all dynamical variables: $\{l^2, l_\alpha\} = 0$, $\alpha = 1, 2, 3$.

The PB (10.2.4) is *non-degenerate* – only the constant function commutes with all (q, p). The PB (10.2.4) and the hamiltonian (10.2.3) are called the *Hamiltonian structure* of the Newton equations (10.1.1). This is a *standard* Hamiltonian structure. Symmetric representation of the Hamilton equations (10.2.5), (10.2.6) is very convenient for generalisations. Replacing $\{.,\ .\}$ of (10.2.4) by a different PB and taking non-natural Hamiltonian $H(l)$ we can embody new types of equations into the framework of Hamiltonian formalism.

We shall say that dynamical equations

$$\dot{l}_\alpha = f_\alpha(l),\ \alpha = 1, \ldots, k, \tag{10.2.7}$$

where $f_\alpha(l)$ are vector fields depending on all variables l_α, have a *Hamiltonian structure* $\{.,\ .\}$, $H(l)$ whenever it is possible to find such PB $\{.,\ .\}$ and Hamiltonian $H(l)$ that

$$f_\alpha(l) \overset{l}{\equiv} \{l_\alpha, H(l)\}$$

identically with respect to all l_α. Sometimes it may be possible after change of independent variables.

10.2.2 *Canonical transformations*

In case of the standard Hamiltonian structure and a Hamiltonian $H(q, p)$ where H is an arbitrary differentiable function of (q, p) the Hamilton equations (10.2.5), (10.2.6) are invariant with respect to canonical transformations which we characterise by three basically equivalent conditions. An invertible transformation $(q_k, p_k) \to (Q_k(q, p), P_k(q, p))$, $k = 1, \ldots, n$ is called *canonical* if

(i) It preserves PB, that is, for any functions $f(Q, P)$, $g(Q, P)$

$$\{f(Q, P), g(Q, P)\}_{(Q,P)} = \{f(Q(q, p), P(q, p)), g(Q(q, p), P(q, p))\}_{(q,p)} \tag{10.2.8}$$

where subscripts indicate variables with respect to which differentiations in (10.2.4) are taken.

(ii) There exists a *generating function* $S(q, P)$ with $\det\left(\dfrac{\partial^2 S}{\partial q \partial P}\right) \neq 0$ such that the transformation is defined implicitly as:

$$p_k = \frac{\partial S}{\partial q_k}, \quad Q_k = \frac{\partial S}{\partial P_k}. \tag{10.2.9}$$

(iii) It carries the Hamilton equations (10.2.5), (10.2.6) into Hamilton equations, that is, in the new variables (Q, P) equations of motion have again the form $\dot{Q}_k = \{Q_k, \tilde{H}\}_{(Q,P)}$, $\dot{P}_k = \{P_k, \tilde{H}\}_{(Q,P)}$, $k = 1, \ldots, n$ with the new Hamiltonian $\tilde{H}(Q, P) = H(q(Q, P), p(Q, P))$. The existence of a generating function for canonical transformation is essential for the Hamilton–Jacobi method of solving Hamilton equations.

10.2.3 *Complete integrability*

Hamilton's equations are usually non-integrable, so that the motion is very complicated and, as the Poincaré theorem indicates, there are generically no additional analytic integrals of motion. We distinguish a class of Hamiltonians which can be called completely integrable.

A function $K(q(t), p(t))$ which is constant on trajectories of Hamilton equations is called an *integral of motion*. This condition is equivalent to

$$0 = \frac{\mathrm{d}}{\mathrm{d}t}K(q(t), p(t)) = \sum_k\left(\frac{\partial K}{\partial q_k}\dot{q}_k + \frac{\partial K}{\partial p_k}\dot{p}_k\right) = $$
$$= \sum_k\left(\frac{\partial K}{\partial q_k}\cdot\frac{\partial H}{\partial p_k} - \frac{\partial K}{\partial p_k}\cdot\frac{\partial H}{\partial q_k}\right) = \{K, H\}; \qquad (10.2.10)$$

that is, the function $K(q, p)$ commutes with $H(q, p)$ with respect to the PB (10.2.4). In particular, a time-independent Hamiltonian $H(q, p)$ is an integral of motion: $\dot{H} = \{H, H\} = 0$.

The geometric meaning of an integral is that the equation $K(q, p) =$ const. defines a foliation of the phase space into $2n$-1-dimensional leaves which are invariant with respect to flow generated by $H(q, p)$. Invariance of these leaves is obvious from (10.2.10) which is a condition of orthogonality of the gradient $(\partial K/\partial q; \partial K/\partial p)$ and of the vector field $(\partial H/\partial p; -\partial H/\partial q)$. Existence of an integral of motion usually allows for reduction of the order of ordinary differential equations (ODEs) by 1, but in the case of a Hamiltonian system it is possible to reduce the order of equations by 2. This is the reason why in the definition of integrability of a Hamiltonian system $H(q, p)$ of $2n$ variables we need only n integrals of motion.

Definition of complete integrability: A Hamiltonian system $H(q, p)$ is called *completely integrable* if it has n integrals of motion $I_k(q, p)$, $k = 1, \ldots, n$ which do not depend explicitly on time, are functionally independent, and are in involution.

We say that functions $I_k(q, p)$ are *functionally independent* if the vectors

$$\left(\frac{\partial I_k}{\partial q}; \frac{\partial I_k}{\partial p}\right)$$

are linearly independent; that is,

$$\text{rank}\left(\frac{\partial I_k}{\partial q_r}; \frac{\partial I_k}{\partial p_r}\right)_{k,r=1,\ldots,n} = n.$$

Equivalently: none of the integrals I_k is a function of any other; that is, $F(I_1(q, p), \ldots, I_n(q, p)) \overset{(q,p)}{\equiv} 0$ holds only for trivial function F. Integrals I_k are in *involution* if they *Poisson-commute*:

$$\{I_j, I_k\} = 0 \text{ for each pair } (j, k),\ j, k = 1, \ldots, n.$$

Geometrically this means that all Hamiltonian flows generated by I_k commute.

A set of integrals which satisfy all the conditions of the definition of complete integrability will be called a *complete set of integrals*, or in short a *complete set*.

Geometric content of the definition of integrability is explained by the Liouville–Arnold theorem [3] which asserts that if the surface of constant value of the Hamiltonian $H(q, p) = E$ is a compact connected manifold then

(i) the manifold $M_I = \{(q, p): I_k(q, p) = c_k = \text{const}, k = 1, \ldots, n\}$ is diffeomorphic to the n-torus T^n;

(ii) in the canonically conjugate coordinates (φ_k, I_k) Hamilton equations have a very simple form:

$$\dot{\varphi}_k = \frac{\partial \tilde{H}(I_1 \ldots I_n)}{\partial I_k} = \omega_k(I),\ \dot{I}_k = 0 \quad k = 1, \ldots, n \tag{10.2.11}$$

(where $\tilde{H}(I) = H(q(I, \varphi)p(I, \varphi)))$ and the motion is quasiperiodic on T^n;

(iii) the Hamilton equations can be solved in quadratures.

10.2.4 *Example: Kepler problem of two bodies*

The two-body problem in centre-of-mass coordinates is described by the Hamiltonian

$$H = \tfrac{1}{2}(p_1^2 + p_2^2 + p_3^2) + \frac{\alpha}{q}, \quad \alpha \text{ constant},$$
$$q = |\mathbf{q}|,\ \mathbf{q} = (q_1, q_2, q_3). \tag{10.2.12}$$

This Hamiltonian has four obvious integrals; the H and components of angular momenta $l_{jk} = q_j p_k - q_k p_j$, $j, k = 1, 2, 3$, which do not commute since l_{jk} satisfy commutation relations of the $so(3)$ Lie algebra. From these we can construct two commuting integrals of motion: $l^2 = l_{12}^2 + l_{23}^2 + l_{31}^2$ and l_{12}. Three integrals H, l^2, l_{12} are time-independent, functionally independent and in involution; thus the Hamiltonian (10.2.12) is completely integrable. The geometric meaning of the integral $\mathbf{l} = (l_{23}, l_{31}, l_{12})$ is that the vector $\mathbf{l}$ is orthogonal to trajectories: $\mathbf{l} \cdot \mathbf{p} = 0$ and, therefore, the plane orthogonal to $\mathbf{l}$ is an invariant manifold. Two-dimensional motion in this plane requires only two commuting integrals H, l^2 for complete integrability.

The Kepler problem has yet another integral of motion, the Lentz vector

$$\mathbf{A} = \mathbf{p} \times \mathbf{l} - \frac{\alpha \mathbf{q}}{q}. \tag{10.2.13}$$

This vector can be used for algebraic construction of the equation of trajectories. Scalar multiplication of both sides of (10.2.13) by $\bar{\mathbf{q}}$ yields

$$l^2 - \alpha q = Aq \cos \varphi \Rightarrow q = \frac{l^2/\alpha}{1 + \dfrac{A}{\alpha} \cos \varphi}, \qquad (10.2.14)$$

where A $|\mathbf{A}|$ and φ is the angle between the vectors $\mathbf{q}$ and $\mathbf{A}$. The vector $\mathbf{A}$ has direction to the nearest point on the trajectory, which is called the perihelion.

The manifold of constant energy $H(q, p) = E$ is compact and connected for $E < 0$. Then the trajectories (10.2.14) are closed curves and the motion is strictly periodic. The Kepler problem has seven integrals of motion H, $\mathbf{l}$, $\mathbf{A}$ altogether, but only five are independent. Each integral defines a foliation into 5-dimensional manifolds and the intersection of manifolds defined by five independent integrals yields the curve (10.2.14).

In general, a Hamiltonian system can have $2n$-1 time-independent integrals $K_j(q, p)$, $j = 1, \ldots, 2n$-1 at most; it is then called *completely degenerate*. Trajectories of such Hamiltonians are determined algebraically by the set of $2n$-1 equations $K_j(q, p) = c_j =$ const. and only time parametrisation of trajectories has to be found directly from the Hamilton equations. Note that $2n$-1 integrals cannot be in involution since at most n integrals can commute, as they can be taken as new canonical momenta in the Liouville–Arnold theorem. However, if there are more then n integrals then there exist inequivalent complete sets of integrals which define different invariant tori.

Occurrence of $2n$-1 integrals for a hamiltonian system is very exceptional. Two well-known classical examples are $V(q) = \alpha/q$ and $V(q) = \frac{1}{2}q^2$. Other, more recent, examples are the Calogero–Moser system and the Euler–Calogero–Moser (ECM) equations to be discussed in the next sections.

10.3 The method of separation of variables

The method of separation of variables (MSV) in the Hamilton–Jacobi (HJ) equation applies to natural Hamiltonians and to some other Hamiltonians which depend at most quadratically on generalised momenta. Such Hamiltonians often follow from the natural ones by change of variables.

The MSV was introduced by Jacobi in 1837 and successfully applied to many examples. The Kepler problem, the Jacobi problem of geodesic motion on 3-axial ellipsoid and the Neumann problem of motion on a sphere in anisotropic harmonic potential are among the best-known applications. In fact, nearly all mechanical problems solved in the 19th century are amenable to this method. The Kowalewsky integrable case of rigid-body motion is the only, remarkable, exception. A common feature of all separable Hamiltonians is that the complete set of integrals is at most quadratic in momenta.

10.3.1 *Description of the MSV*

The main idea of the MSV is to employ canonical transformations for reducing the Hamilton equations of $H(q, p)$ to a trivial form

$$\dot{\varphi}_k = \frac{\partial \tilde{H}(\eta_1, \ldots, \eta_n)}{\partial \eta_k} = \omega_k(\eta) = \text{const.}, \tag{10.3.1}$$

$$\dot{\eta}_k = -\frac{\partial \tilde{H}(\eta_1, \ldots, \eta_n)}{\partial \varphi_k} = 0 \tag{10.3.2}$$

in terms of the new variables (φ, η). Then

$$\eta_k = \text{const.}, \; \varphi_k(t) = \omega_k t + \varphi_k(0). \tag{10.3.3}$$

The generating function $S(q, \eta)$ of the transformation $(q, p) \to (\varphi, \eta)$ has to satisfy the Hamilton–Jacobi equation

$$E = H(q, p) = H\left(q_1, \ldots, q_n; \frac{\partial S}{\partial q_1}, \ldots, \frac{\partial S}{\partial q_n}\right) \tag{10.3.4}$$

since $p_k = \partial S(q, \eta)/\partial q_k$ by (10.2.9). The HJ equation is a first-order PDE which is solved when a complete integral $S(q, \eta)$ depending on n essential constants $\eta_1, \ldots, \eta_n$ is known. The form of the equation (10.3.4) and its solvability depends essentially on a Hamiltonian. For most Hamiltonians we are not able to solve the HJ equation.

Thus the main question is: for which Hamiltonians (potentials) are we able to solve effectively the HJ equation? Let us start from the natural Hamiltonian (10.2.3). The HJ equation reads

$$E = \tfrac{1}{2} \sum_k \left(\frac{\partial S}{\partial q_k}\right)^2 + V(q_1, \ldots, q_n) \tag{10.3.5}$$

and we can solve this equation if we take a potential of the particular form $V(q) = \Sigma_k w_k(q_k)$ where w_k are arbitrary differentiable functions of one variable. Then by assuming

$$S(q, \eta) = S_1(q_1; \eta) + \ldots + S_n(q_n; \eta)$$

we get

$$E = \sum_k \left(\frac{1}{2}\left(\frac{\partial S_k}{\partial q_k}\right)^2 + w_k(q_k)\right), \tag{10.3.6}$$

that is, a sum of n functions each depending on a different variable q_k.

Therefore the PDE (10.3.6) splits into n decoupled ODEs

$$\frac{1}{2}\left(\frac{\partial S_k}{\partial q_k}\right)^2 + w_k(q_k) = \eta_k = \text{const}, \sum_k \eta_k = E,$$

which can be immediately solved by quadratures

$$S(q, \eta) = \sum_k \int^{q_k} \sqrt{(2\eta_k - w_k(x))}\mathrm{d}x$$

The procedure of splitting the HJ equation into a set of decoupled ODEs is called *separation of variables* in the HJ equation; η_k are *constants of separation.*

The function $S(q, \eta)$ is a required complete integral of (10.3.5) which can be used for actual solving of Hamilton equations of motion. It defines the canonical transformation

$$p_k = \frac{\partial S(q, \eta)}{\partial q_k} \qquad (10.3.7)$$

$$k = 1, \ldots, n$$

$$\varphi_k = \frac{\partial S(q, \eta)}{\partial \eta_k} \qquad (10.3.8)$$

and $q_k(t)$ are implicitly given by (10.3.8) since $\varphi_k(t) = \omega_k(\eta)t + \varphi_k(0)$. The $p_k(t)$ follows from (10.3.7).

So far the result of the application of MSV is trivial because Hamilton equations of $H = \frac{1}{2} \Sigma_k p_k^2 + \Sigma_k w_k(q_k)$ are already decoupled:

$$\dot{q}_k = p_k, \dot{p}_k = -\frac{dw_k}{dq_k}$$

and it is simpler to solve them directly. An important observation is that natural Hamiltonians preserve quadratic dependence on momenta under point transformations

$$u_k = f_k(q_1, \ldots, q_n), p_k = \sum_r \frac{\partial f_r}{\partial q_k} v_r$$

generated by $\tilde{S}(q, v) = \Sigma_r f_r(q)v_r$; where v_k are canonical momenta conjugate to u_k. In new orthogonal coordinates $u_1, \ldots, u_n$ the Hamiltonian (10.2.3) takes the form

$$H = \frac{1}{2} \sum_k c_k(u)v_k^2 + \tilde{V}(u_1, \ldots, u_n).$$

Thus, in non-trivial cases the MSV consists of two steps: an orthogonal, point transformation of the Hamiltonian $H = T + V$ to new coordinates in which the HJ equation can be solved effectively by separation of variables.

10.3.2 *Example: the Garnier system*

The Garnier system [48] is characterised by the Hamiltonian

$$H = \frac{1}{2}\sum_k p_k^2 + \frac{1}{2}q^4 - \frac{1}{2}\sum_k \alpha_k q_k^2,\ q^2 = \sum_k q_k^2. \qquad (10.3.9)$$

We show that it is separable in generalised elliptic coordinates $u_1, \ldots, u_n$ defined by the rational equation

$$1 + \sum_{k=1}^{n} \frac{q_k^2}{z - \alpha_k} = \prod_{j=1}^{n} (z - u_j) / \prod_{k=1}^{n} (z - \alpha_k) = U(z)/A(z) \qquad (10.3.10)$$

where all α_k are positive and different $0 < \alpha_1 < \ldots < \alpha_n$. The second equality in (10.3.10) defines $U(z)$ and $A(z)$. In new coordinates the Hamiltonian (10.3.9) takes the form

$$H = \frac{1}{2}\sum_{j=1}^{n} \frac{-4A(u_j)}{U'(u_j)} v_j^2 + \sum_{j=1}^{n} \frac{1}{U'(u_j)}\left(\frac{1}{2}u_j^{n+1} - \frac{1}{2}b_1 u_j^n + \frac{1}{2}b_2 u_j^{n-1}\right)$$

where prime on U means differentiation with respect to z and $b_1 = \Sigma_k \alpha_k$, $b_2 = \frac{1}{2}(b_1^2 - \Sigma_k a_k^2)$. The HJ equation then reads

$$\sum_{j=1}^{n} \left[-4A(u_j)\left(\frac{\partial S}{\partial u_j}\right)^2 + u_j^{n+1} - b_1 u_j^n + b_2 u_j^{n-1}\right] \frac{1}{U'(u_j)} = 2\eta_1 \quad (10.3.11)$$

where η_1 denotes the constant value of energy. In order to separate this equation we make use of the identity

$$\eta_1 = \sum_{j=1}^{n} (P(u_j)/U'(u_j))$$

where $P(z) = \eta_1 z^{n-1} + \eta_2 z^{n-2} + \ldots + \eta_n$ is an arbitrary polynomial and we assume $S(u_1, \ldots, u_n) = \Sigma_j S_j(u_j)$. Then each S_j satisfies

$$-4A(u_j)\left(\frac{\partial S_j}{\partial u_j}\right)^2 + u_j^{n+1} - b_1 u_j^n + b_2 u_j^{n-1} - 2P(u_j) = 0.$$

The complete integral of the HJ equation (10.3.11) can be written as

$$S(u, \eta) = \frac{1}{2}\sum_{j=1}^{n} \int^{u_j} \sqrt{(Q(z)/A(z))}\mathrm{d}z$$

with

$$Q(z) = z^{n+1} - b_1 z^n + (2b_2 - 2\eta_1)z^{n-1} - 2\,\eta_2 z^{n-2} + \ldots -2\,\eta_n.$$

10.3.3 *Criterion of separability*

The basic problem of the MSV is characterisation of all Hamiltonians which can be solved by this method. For natural Hamiltonians the answer to this question has been given in terms of variables of separation by

Stäckel [41]. In terms of Cartesian coordinates this problem is not completely solved yet, as we shall discuss below.

***Theorem* – Stäckel** [41]
A necessary and sufficient condition for the Hamiltonian

$$H = \frac{1}{2}\sum_{k=1}^{n} c_k(u_1, \ldots, u_n)v_k^2 + V(u_1, \ldots, u_n) \qquad (10.3.12)$$

to be separable is the existence of a matrix Ψ and a vector ψ such that

$$(c_1, \ldots c_n)\begin{bmatrix} \Psi_{11}(u_1), & \ldots, & \Psi_{1n}(u_1) \\ \vdots & & \vdots \\ \Psi_{n1}(u_n), & \ldots, & \Psi_{nn}(u_n) \end{bmatrix}$$

$$= (1, 0, \ldots, 0), (c_1, \ldots, c_n)\begin{bmatrix} \psi(u_1) \\ \vdots \\ \psi_n(u_n) \end{bmatrix} = V(u_1, \ldots, u_n) \qquad (10.3.13)$$

where $\det \Psi \neq 0$ and each row of Ψ depends only on one variable.

We shall sketch the proof. Sufficiency follows immediately from the HJ equation, since

$$\frac{1}{2}\sum_r c_r\left(\frac{\partial S}{\partial u_r}\right)^2 + V = \eta_1 \qquad (10.3.14)$$

implies

$$\frac{1}{2}\sum_r c_r\left(\frac{\partial S}{\partial u_r}\right)^2 + \sum_r c_r\psi_r(u_r) = \eta_1\left(\sum_r c_r\Psi_{r1}(u_r)\right) + \ldots + \eta_n\left(\sum_r c_r\Psi_{rn}(u_r)\right)$$

due to (10.3.13). For necessity, note that if (10.3.14) is separable then it admits a solution $S(u, \eta) = \Sigma_r S_r(u_r; \eta_1, \ldots, \eta_n)$ and then the equation (10.3.14) is satisfied identically with respect to all u and η. By differentiating it with respect to each η_k we get

$$\sum_r c_r\left(\frac{\partial S_r}{\partial u_r}\right) \cdot \frac{\partial^2 S_r}{\partial u_r\, \partial \eta_1} = 1,$$

$$\sum_r c_r\left(\frac{\partial S_r}{\partial u_r}\right) \cdot \frac{\partial^2 S_r}{\partial u_r\, \partial \eta_k} = 0 \text{ for } 1 < k \leq n,$$

and the coefficient of c_r in each equation is a function of u_r only. Thus

$$\Psi_{rk} = \left(\frac{\partial S_r}{\partial u_r}\right)\frac{\partial^2 S_r}{\partial u_r\, \partial \eta_k}$$

and

$$\det \Psi = \left(\frac{\partial S_1}{\partial u_1}\right) \ldots \left(\frac{\partial S_n}{\partial u_n}\right) \det\left(\frac{\partial^2 S_r}{\partial u_r\, \partial \eta_n}\right) \neq 0$$

since S is a generating function of a canonical transformation. Furthermore, from (10.3.14) we get

$$V(u) = \eta_1 - \frac{1}{2}\sum_r c_r\left(\frac{\partial S}{\partial u_r}\right)^2 = \sum_r c_r\left(\eta_1\Psi_{r1} - \frac{1}{2}\left(\frac{\partial S_r}{\partial u_r}\right)^2\right) = \sum_r c_r\psi_r(u_r)$$

as required by (10.3.13).

For more general Hamiltonians,

$$H = \frac{1}{2}\sum_r\sum_s c_{rs}v_rv_s + V(u),$$

a necessary and sufficient condition has been found by Levi-Civita [25]. It is in the form of a set of nonlinear PDEs

$$\frac{\partial^2 H}{\partial v_i\partial v_j}\frac{\partial H}{\partial u_i}\frac{\partial H}{\partial u_j} - \frac{\partial^2 H}{\partial v_i\partial u_j}\frac{\partial H}{\partial v_j}\frac{\partial H}{\partial u_i} - \frac{\partial^2 H}{\partial u_i\partial v_j}\frac{\partial H}{\partial v_i} - \frac{\partial H}{\partial u_j} + \frac{\partial^2 H}{\partial u_i\partial u_j}\frac{\partial H}{\partial v_i}\frac{\partial H}{\partial v_j} = 0$$

for all pairs (i, j), $i \neq j$. Here, separation in non-orthogonal coordinates may occur and the structure of the basic types of $c_{rs}(u)$ have been worked out only recently through Benenti's theory of separability structures [4] (see also [22]).

In practical circumstances the Stäckel theorem (or the Levi-Civita conditions) is not very useful because it is not usually known which transformation has to be used in order to transform a natural Hamiltonian (10.2.3) to the separable form (10.3.13) or whether such a transformation exists at all.

There is a need for characterising separable Hamiltonians (separable potentials) in terms of Cartesian coordinates q_k in order to have an effective criterion of separability of a given potential.

In the case of two degrees of freedom, such a criterion is a direct consequence of the Bertrand–Darboux theorem [44].

Theorem

The following conditions on the Hamiltonian

$$H = \tfrac{1}{2}(p_1^2 + p_2^2) + V(q_1, q_2) \tag{10.3.15}$$

are equivalent.

(i) It has an integral of motion quadratic in momenta of the form

$$K = (Aq_2^2 + Bq_2 + C)p_1^2 + (Aq_1^2 + B_1q_1)p_2^2 + (-2Aq_1q_2 - B_1q_2 - Bq_1 + C_1)p_1p_2 + k(q_1, q_2). \tag{10.3.16}$$

(ii) Potential V satisfies the equation

$$(V_{22} - V_{11})(-2Aq_1q_2 - B_1q_2 - Bq_1 + C_1) \\ + 2V_{12}(Aq_2^2 - Aq_1^2 + Bq_2 - B_1q_1 + C) + \\ + V_1(6Aq_2 + 3B) - V_2(6Aq_1 + 3B_1) = 0, \tag{10.3.17}$$

where the subscripts 1, 2 on V denote partial derivatives with respect to q_1 and q_2.

(iii) Hamiltonian (10.3.15) is separable in one of the four orthogonal systems of coordinates on the plane: elliptic, parabolic, polar, or Cartesian. Separation coordinates are characteristic variables of the equation (10.3.16).

The equation (10.3.17) applied to a given potential $V(q_1, q_2)$ gives some functional expression to be satisfied identically with respect to q_1 and q_2. Equating to zero coefficients at linearly independent functions we get a set of linear homogeneous equations (usually overdetermined) on the, as yet unknown, parameters A, B, B_1, C, C_1.

Criterion of separability – for $n = 2$ [51]
If there exist non-zero solution on A, B, B_1, C, C_1 then there exist also an additional integral (10.3.16) and the Hamiltonian (10.3.15) is separable. If there is no non-trivial solution then (10.3.15) is not separable.

Example – separable cases of the Henon–Heiles system [46]
Application of the equation (10.3.17) to the Hamiltonian

$$H = \tfrac{1}{2}(p_1^2 + p_2^2) + \tfrac{1}{2}\omega_1q_1^2 + \tfrac{1}{2}\omega_2q_2^2 + aq_1^2q_2 - \tfrac{1}{3}bq_2^3 \tag{10.3.18}$$

gives eight equations on parameters

$$A(26a + 4b) = 0,\ 10aA = 0,\ 7aB_1 = 0,\ B_1(5b + 2a) = 0, \\ 2B(6a + b) - 8A(\omega_2 - \omega_1) = 0, \\ 2C_1(b + a) + B_1(4\omega_2 - \omega_1) = 0, \\ 4aC - B(\omega_2 - 4\omega_1) = 0, \\ C_1(\omega_2 - \omega_1) = 0.$$

The first four equations imply $A = B_1 = 0$ if either $a \neq 0$ or $b \neq 0$. The case $a = 0$ is separable in cartesian coordinates, so we shall assume $a \neq 0$. The remaining four equations then reduce to:

$$2B(6a + b) = 0, \quad 2C_1(a + b) = 0, \\ 4aC - B(\omega_2 - 4\omega_1) = 0, \quad C_1(\omega_2 - \omega_1) = 0.$$

If $B = 0$ then either $C = 0$ (for $a \neq 0$) and $C_1 \neq 0$ or $C \neq 0$ implies $a = 0$ and $C_1 = 0$ (since $b \neq 0$). The condition $B \neq 0$ gives $b = -6a$, $C_1 = 0$ and $4aC - B(\omega_2 - 4\omega_1) = 0$. Thus we obtain the following separable cases of the Henon–Heiles system:

(i) $a = 0$, ω_1 ω_2, b arbitrary: separable in Cartesian coordinates;
(ii) $b = -a$, $\omega_2 = \omega_1$: separable in $x = q_1 + q_2$, $y = q_1 - q_2$;
(iii) $b = -6a$, ω_1, ω_1 arbitrary: separable in translated parabolic coordinates.

Separation coordinates for the case (iii) can be determined from the equation (10.3.17) which here reads

$$Bq_1(V_{11} - V_{22}) + 2BV_{12}\left(q_2 + \frac{1}{4a}(\omega_2 - 4\omega_1)\right) + 3BV_1 = 0. \quad (10.3.19)$$

By the method of characteristics we find the variables ξ, η

$$q_1 = \sqrt{(\xi\eta)}, \quad q_2 = \tfrac{1}{2}(\xi - \eta) + (4\omega_1 - \omega_2)/a, \quad 0 < \xi, \eta < \infty$$

in which the highest derivative term of (10.3.19) reduces to $V_{\xi\eta}$. The potential V of the (10.3.19) has in these coordinates the form

$$V = \alpha(q_1^2 q_2 + 2q_2^3) + \tfrac{1}{2}(\omega_1 q_1^2 + \omega_2 q_2^2) =$$
$$= \frac{1}{\xi + \eta}\left(\alpha(\xi^4 - \eta^4) + \alpha(\xi^3 + \eta^3) + \tfrac{1}{2}\beta(\xi^2 - \eta^2) + \gamma(\xi + \eta)\right),$$

where

$\alpha = \frac{3}{2}\omega_1 - \frac{1}{4}\omega_2$, $\beta = \frac{1}{4}\,a(\omega_2 D + \frac{3}{2}D^2)$, $\gamma = \omega_1 D/8a^2$ and $D = 4\,\omega_1 - \omega_2$. For arbitrary n there is not, as yet, a similiar criterion of separability, but the structure of the problem is now understood [28] because of the following theorems.

Theorem – Iarov–Iarovoi [18]
A natural Hamiltonian system $H = T + V$ is separable only in orthogonal curvilinear systems of coordinates and the most general such system is the generalised elliptic coordinates (10.3.10) or their degenerations.

This theorem indicates that it is necessary to consider only the generalised elliptic coordinates or their degenerations. All degenerations have been classified in [23]. Here we shall consider only the case of generalised elliptic coordinates for which the n-dimensional analogue of the Bertrand–Darboux theorem has been formulated in the following way.

Theorem [28]
The following conditions on a natural hamiltonian $H = T + V$ are equivalent.

(i) H has n global, functionally independent, commuting, quadratic in momenta integrals of motion having in some Euclidean reference frame the form

$$K_i = \sum_{j\neq 1}^{n} (\alpha_i - \alpha_j)^{-1}\, l_{ij}{}^2 + p_i^2 + k_i(q) \qquad (10.3.20)$$

where $l_{ij} = q_i p_j - q_j p_i$, $\alpha_n > \alpha_{n-1} > \ldots > \alpha_1 > 0$ are n distinct constants and $k_i(q)$ are some functions.

(ii) Potential V satisfies a set of $n\binom{n-1}{2} + \binom{n}{2} = \frac{1}{2}n(n-1)^2$ equations

$$(\alpha_i - \alpha_r)^{-1}(q_i^2 V_{rs} - q_i q_r V_{is}) = (\alpha_i - \alpha_s)^{-1}(q_i^2 V_{sr} - q_i q_s V_{ir}) \qquad (10.3.21)$$

for all i, r, s different

$$(\alpha_i - \alpha_r)^{-1} q_i q_r (V_{ii} - V_{rr}) - \sum_{j\neq i,r} (\alpha_i - \alpha_j)^{-1} q_i q_j \cdot V_{jr}{}^{\cdot} +$$
$$+ V_{ir}\Big(\sum_{j\neq i} (\alpha_i - \alpha_j)^{-1} q_j^2 + (\alpha_r - \alpha_j{}^{-1}) q_i^2\Big) + V_{ir} +$$
$$+ 3(\alpha_i - \alpha_r)^{-1}(q_r V_i - q_i V_r) = 0 \text{ for } i \neq r \qquad (10.3.21)$$

where subscripts on V mean partial derivatives.

(iii) H is separable in generalised elliptic coordinates (10.3.10).

For a given potential $V(q)$ equations (10.3.21) have to be verified for separability of $V(q)$. However, there are several problems which have to be resolved in order to obtain an effective criterion as in the case $n = 2$.

(i) Integrals of motion K_j and the equations (10.3.21) have such a simple form only in some special reference frame. For testing a given potential $V(q)$ we have to allow for dependence on orthogonal transformations A and translations b, that is, to substitute $V(Aq + b)$ into (10.3.21). In general it is impossible to parametrise A in a linear way and after applying (10.3.21) to $V(Aq + b)$ we end up with an overdetermined nonlinear set of algebraic equations on components of A, b and, possibly, parameters of $V(q)$.

(ii) Additionally the parameters α_k of the elliptic coordinates are also unknown and they enter equations in a nonlinear way too. It may be difficult or even impossible to decide solvability of the, so obtained, set of algebraic equations and to determine separable values of the parameters involved in $V(q)$. In two dimensions the situation is exceptional because there is only one parameter $(\alpha_1 - \alpha_2)$ and the parameter describing rotations enters the final equation (10.3.17) in a linear way.

(iii) The generalisation of the Bertrand–Darboux theorem is formulated for the generalised elliptic coordinates while a given $V(q)$ may be separable in some of its degenerations – like the Henon–Heiles system for $n = 2$. In the case $n = 2$, equation (10.3.17) automatically embodies all degenerations. It may well happen here too that by proper choice of the parameters α_k, A, b in equations (10.3.21) we recover all degenerate cases. However, it is a conjecture, as yet. For the two most important degenerate cases, the generalised parabolic and the spherical-conical coordinates, equations analogous to (10.3.21) have been formulated in [28]. This makes it possible to investigate completely the case $n = 3$.

It is not clear how to overcome the difficulties indicated above, nor whether it is possible to find a satisfactory solution. Nevertheless the linear nature of the equations (10.3.21) is quite remarkable and can be used for the relatively simple *necessary* criterion of separability of a given $V(q)$. The idea is to treat all the coefficients entering the (10.3.21) (after substitution of $V(Aq + b)$) as independent quantities and to ask whether there exists a non-trivial solution for these quantities. Then the problem requires, like before, solving only a linear homogeneous system of equations which will still be heavily overdetermined. Linear nature of equations on coefficient follows from the linearity of (10.3.21). The idea of this necessary condition may serve as a cut-off procedure for *excluding* separability of a given potential of interest. For actual *proving* separability more detailed analysis would be necessary.

10.3.4 **Separable potentials**

The MSV makes it possible to determine the Cartesian form $V(q)$ of separable potentials only in a few cases: $n = 2$, $n = 3$ (except elliptic and parabolic coordinates), and of some symmetric potentials for arbitrary n. For $n = 2$ the relevant formulae can be obtained by the use of known [24] forms of separable potentials in terms of separation coordinates. A list of separable 3-dimensional potentials has been given by Eisenhart [12]. His potentials are written in some particular Euclidean reference frame, so that parameters of translations and orthogonal transformations are to be introduced there. The same applies to symmetric n-dimensional potentials derived below. Eisenhart's list is incomplete because in the elliptic and parabolic cases he gets an irreducible case of the third-order algebraic equation for which it is impossible to express a solution by the use of algebraic operations – inversion of trigonometric functions is necessary.

Symmetric potentials – separable in generalised elliptic coordinates [51]
By symmetric potentials we understand here those which in generalised elliptic coordinates take the form $V(u_1, \ldots, u_n) = \Sigma_{j=1}^{n} f(u_j)/U'(u_j)$

where the function of one variable $f(u_j)$ does not depend on j. Such potentials are symmetric with respect to permutations of the indices 1, ..., n and they remain symmetric in terms of the Cartesian coordinates $q_1, \ldots, q_n$ since the defining equation (10.3.10) has permutational symmetry.

It is not difficult to determine from (10.3.10) the first few symmetric potentials corresponding to $f(x) = x^{n+k}$, $k = 0, 1, 2, \ldots$ but calculations become increasingly involved with the growth of k. Here we construct such potentials recursively by the use of quadratic integrals (10.3.20).

Theorem [51]
If the functions

$$K_j^{(m)} = \sum_{k\neq j}^{n} (\alpha_j - \alpha_k)^{-1} l_{jk}^2 + p_j^2 + k_j^{(m)}(q) = P_j + k_j^{(m)}, \quad j = 1, \ldots, n \tag{10.3.22}$$

Poisson-commute for m and $m - 1$, and the functions $k_j^{(m)}$ are defined as

$$k_j^{(m+1)}(q) = q_j^2\Big(\sum_r k_r^{(m)}(q)\Big) + \alpha_j\, k_j^{(m)}(q), \tag{10.3.23}$$
$$k_j^{(1)}(q) = q_j^2$$

then the functions $K_j^{(m+1)}$ commute too.

The proof follows from a straightforward calculation that

$$\{K_j^{(m+1)}, K_k^{(m+1)}\} = \{P_j, k_k^{(m+1)}\} + \{k_j^{(m+1)}, P_k\} = 0. \tag{10.3.24}$$

The functions $K_j^{(m)}$ constitute a complete set of integrals for the Hamiltonian

$$H^{(m)} = \frac{1}{2}\sum_j K_j^{(m)} = \frac{1}{2}\sum_j p_j^2 + \frac{1}{2}\sum_j k_j^{(m)}.$$

The recursion starts from $k_j^{(1)} = q_j^2$ and provides the following sequence of separable potentials $V^{(m)} = \frac{1}{2}\sum_j k_j^{(m)}$:

$$q^2 = \sum_k q_k^2,\ q^4 - \sum_k \alpha_k q_k^2,\ q^6 - 2q^2\sum_k \alpha_k q_k^2 + \sum_k \alpha_k^2 q_k^2,$$
$$q^8 - 3q^4\sum_k \alpha_k q_k^2 + 2q^2\sum_k \alpha_k^2 q_k^2 + \Big(\sum_k \alpha_k q_k^2\Big)^2 - \sum_k \alpha_k^3 q_k^2, \ldots \tag{10.3.25}$$

which correspond to growing powers of $f(x) = x^{n+k}$. The recursion (10.3.23) can be inverted:

$$k_j^{(m)} = -\alpha_j^{-1}\rho^{-1}q_j^2\sum_r \alpha_r^{-1}k_r^{(m+1)} - \alpha_j^{-1}k_j^{(m+1)},$$
$$\rho = 1 - \sum_k \frac{q_k^2}{\alpha_k} \tag{10.3.26}$$

to define a downward family of integrals. Their commutativity can also be proved by two-step finite induction. The first few potentials are

$$\rho^{-1},\ \rho_1\rho^{-1},\ \rho_1^2\rho^{-3} + \rho_2\rho^{-2},\ \rho_1^3\rho^{-4} - 2\rho_1\rho_2\rho^{-3} + \rho_3\rho^{-3},\ \ldots \qquad (10.3.27)$$

where $\rho_k = \Sigma_r\, \alpha_r^{-(k+1)} q_r^2$. There is also a potential which is invariant with respect to the recursion formula (10.3.23): $V^{\mathrm{R}} = \Sigma_k\, c_k^2 q_k^{-2}$. Any linear combination of the potentials (10.3.25), (10.3.26) and V^{R} is integrable due to linearity of (10.3.23) and (10.3.24). This reflects the separable character of all these potentials.

Analogous symmetric families of potentials can be constructed for the generalised parabolic and spherical-conical coordinates [51] by taking the proper limits of the formulae found above.

10.4 **Isospectral deformation method (IDM)**

The discovery and development of the isospectral deformation method (IDM) signify great progress in constructing and solving integrable Hamiltonian systems. The particular feature of the MSV is that it applies only to quadratic-in-momenta Hamiltonians and that there exists a complete set of commuting integrals which are all quadratic in momenta. The IDM enables construction and treatment of large number of Hamiltonian systems with integrals of higher (than 2) order in momenta. The IDM was discovered in 1967 in connection with the KdV (Korteweg–de Vries) equation [15] but in the context of Hamiltonian mechanics it was applied first to the Toda lattice [13, 26, 33]. We shall consider this example and then discuss the more general features.

10.4.1 *Free-end Toda system*

The Toda system is defined by the Hamiltonian

$$H_{\mathrm{T}} = \frac{1}{2}\sum_{k=1}^{n} p_k^2 + \sum_{k=1}^{n-1} g_k^2\, e^{2(q_k - q_{k+1})} \qquad (10.4.1)$$

which describes n identical particles on a line each interacting with its nearest neighbours. Interaction potential depends on the mutual distances of particles. The Hamiltonian (10.4.1) is not separable even in the case of three particles when by the use of centre-of-mass coordinates we can reduce it to a two-dimensional problem and apply the separability criterion from Section 10.3.3.

The main discovery was that the Hamilton equations

$$\dot{q}_k = \frac{\partial H}{\partial p_k} = p_k,\ \dot{p}_k = -\frac{\partial H}{\partial q_k} = 2\left(g_{k-1}^2 \mathrm{e}^{2(q_{k-1} - q_k)} - g_k^2 \mathrm{e}^{2(q_k - q_{k+1})}\right)$$

are equivalent to a matrix equation of the form

$$\dot{L} = [A, L] \tag{10.4.2}$$

with the tridiagonal matrices L, A defined as:

$$L = \begin{bmatrix} p_1 & g_1 e^{2(q_1-q_2)} & & 0 \\ g_2 & p_2 & \ddots & \\ & \ddots & \ddots & g_{n-1}e^{2(q_{n-1}-q_n)} \\ 0 & & g_{n-1} & p_n \end{bmatrix}$$

$$A = \begin{bmatrix} 0 & g_1 e^{2(q_1-q_2)} & & 0 \\ 0 & 0 & \ddots & \\ & \ddots & \ddots & g_{n-1}e^{2(q_{n-1}-q_n)} \\ 0 & & 0 & 0 \end{bmatrix}$$

The equation (10.4.2) is usually called the Lax equation or Lax representation of equations of motion. An immediate consequence of the Lax equation is that the quantities $K_r = (1/r)\,\mathrm{Tr}L^r$ are integrals of motion since

$$\dot{K}_r = \mathrm{Tr}L^{r-1}\dot{L} = \mathrm{Tr}L^{r-1}(AL - LA) = 0.$$

The Hamiltonian (10.4.1) is $H_\mathrm{T} = 1/2\ \mathrm{Tr}L^2$. The first n integrals K_r are functionally independent because their leading terms are basic symmetric functions of momenta. They are also in involution $\{K_r, K_s\} = 0$, $r, s = 1, \ldots, n$ (as can be verified directly) and the Toda system is completely integrable in the sense of the Liouville–Arnold theorem.

The Toda system can be solved explicitly on the set of positively defined Hermitian matrices by the method due to Olshanetsky and Perelomov [36]. Every Hermitian, positively defined matrix Y has a unique decomposition $Y = ZXZ^+$ where Z is an invertible upper-diagonal matrix with unit diagonal elements, Z^+ is its Hermitian conjugate and X is a positively defined diagonal matrix.

If $X = \mathrm{diag}(e^{2q_1(t)}, \ldots, e^{2q_n(t)})$ and Z is defined as a solution of $\dot{Z} = -ZA$, $Z(0) = I$ = identity then

$$\begin{aligned}\dot{Y}Y^{-1} &= (\dot{Z}XZ^+ + Z\dot{X}Z^+ + ZX\dot{Z}^+)(Z^+)^{-1}X^{-1}Z^{-1} \\ &= Z(Z^{-1}\dot{Z} + \dot{X}X^{-1} + X\dot{Z}^+(Z^+)^{-1}X^{-1})Z^{-1} = 2ZLZ^{-1}\end{aligned} \tag{10.4.3}$$

since $\dot{Z}^+(Z^+)^{-1} = -A^+$. Next, differentiation yields

$$\begin{aligned}&\frac{d}{dt}(\dot{Y}Y^{-1}) \\ &\quad = 2Z(Z^{-1}\dot{Z}L + \dot{L} - L(Z^{-1}\dot{Z}))Z^{-1} = 2Z(\dot{L} - [A, L])Z^{-1} = 0\end{aligned} \tag{10.4.4}$$

since $(Z^{-1})^{\cdot} = -Z^{-1}\dot{Z}Z^{-1}$ as follows from differentiation of $ZZ^{-1} = I$.

A general solution of (10.4.4) in the class of positively defined Hermitian matrices can be represented in the form

$$Y(t) = B\, e^{2Dt} B^{+}$$

where the matrices B, D depend on the initial conditions
$Y(0) = \mathrm{diag}(e^{2q_1(0)}, \ldots, e^{2q_n(0)})$, $\dot{Y}(0)\, Y^{-1}(0) = 2L(0)$. Thus,
$B = \mathrm{diag}(e^{q_1(0)}, \ldots, e^{q_n(0)})$ and $D = B^{-1}L(0)B$, that is,
$D_{jk} = \delta_{jk}p_k(0) + g_{j-1}e^{(q_{j-1}(0) - q_j(0))}\delta_{j,k+1} + g_j e^{(q_j(0) - q_{j+1}(0))}\delta_{j,k-1}$.

Positions of particles are determined by the horospheric projection of $Y(t)$, that is, from $ZXZ^{+} = B\, e^{2Dt}B^{+}$ we get

$$X = Z^{-1}B\, e^{2Dt}B^{+}(Z^{+})^{-1} = BZ_1 e^{2Dt}Z_1{}^{+}B^{+}$$

because for every Z there exists such Z_1 (also upper triangular) that $Z^{-1}B = BZ_1$. Thus,

$$B^{-1}X(B^{+})^{-1} = Z_1 e^{2Dt}Z_1^{+}$$

and

$$e^{2(q_k(t) - q_0(t))} = \frac{\Delta_{n+1-j}}{\Delta_{n-j}}$$

where Δ_j are the lower principal minors of order j of the matrix e^{2Dt}. Momenta of particles follow in a similar way from the formula(10.4.3).

The example discussed above is one of the simplest representatives of a large class of integrable natural Hamiltonian systems with exponential-type potentials [5]. All these systems have been studied and proved to be integrable by the use of IDM.

10.4.2 *The Lax equation*

The essential content of the Lax equation (10.4.2) for the Toda system is that time evolution can be described by a similarity transformation, or in other words the system undergoes an isospectral deformation. Invariants of the similarity transformation are, therefore, integrals of motion.

It is easy to see that if a dynamical state of a system $(q(t), p(t))$ is represented by some matrix $L(q(t), p(t))$ and time evolution is given by

$$L(q(t), p(t)) = V(t)L(q(0), p(0))V^{-1}(t) \qquad (10.4.5)$$

with some invertible matrix $V(t)$, then $L(t) = L(q(t), p(t))$ satisfies the Lax equation. Indeed, by differentiation of

$$L(0) = V^{-1}L(t)V \quad \text{and} \quad VV^{-1} = I$$

we get

$$0 = \dot{V}^{-1}LV + V^{-1}\dot{L}V + V^{-1}L\dot{V}, \; \dot{V}V^{-1} + V(V^{-1})^{\cdot} = 0$$
$$0 = V^{-1}(-\dot{V}V^{-1}L + \dot{L} + L\dot{V}V^{-1})V, \; (V^{-1})^{\cdot} = -V^{-1}\dot{V}V^{-1}$$

and defining $A(t) = \dot{V}(t)V^{-1}(t)$ we have

$$\dot{L} = [A, L]$$

since V is invertible. However, in order to prove a converse statement we need some non-degeneracy assumptions about the matrix $L(t)$. There are known examples of systems having trivial or incomplete Lax representation. We shall discuss later an example of such a system: the multiwave interaction system (MTW). Here we present an example due to Calogero [10].

The matrix N:

$$N_{jk} = \delta_{jk}\, q_j \sum_{l \neq j}^{n} (q_j - q_l)^{-1} + (1 - \delta_{jk})q_j(q_j - q_k)^{-1}.$$

has the eigenvalues $0, 1, 2, \ldots, n - 1$ independently of the values of n numbers q_j. If the $q_j(t)$ are considered as functions of time then $N(t)$ undergoes an isospectral deformation

$$N(q(t)) = W(q(t), q(0))\; N(q(0))\; W^{-1}(q(t), q(0))$$

where

$$W_{jk}(q(t), q(0)) = [(q_j(t) - q_j(0))/(q_j(t) - q_k(0))] \cdot \prod_{l \neq j}^{n} [(q_j(t) - q_l(0))/(q_j(t) - q_l(t))]$$

and satisfies the Lax equation

$$\dot{N} = [A, N] \qquad (10.4.6)$$

with the matrix A defined as

$$A_{jk} = \delta_{jk} \sum_{l \neq j} \dot{q}_i(q_j - q_l)^{-1} + (1 - \delta_{jk})\dot{q}_j(q_j - q_k)^{-1}.$$

The Lax equation (10.4.6) is here an identity with respect to q_j and $\dot{q}_j$. For arbitrary dynamical equations $\dot{q}_j = f_j(q)$ we can substitute (in the definition of A) the $\dot{q}_j$ by $f_j(q)$, and then the Lax equation (10.4.6) reproduces these dynamical equations. Thus all dynamical equations (even non-integrable ones) have a Lax representation (10.4.6) which is, however, useless because constants of motion, the eigenvalues of N, are numbers, not functions of dynamical variables.

For studying integrability of a given Hamiltonian system we need a *complete* Lax representation with Lax matrix L which provides the number of integrals of motion required by the Liouville–Arnol'd theorem.

For a Hamiltonian system with complete Lax representation it is convenient to write the Lax equation in the form

$$\dot{L}(q, p) = \{L(q, p), H(q, p)\} \underset{(q,p)}{\equiv} [A(q, p), L(q, p)] \quad (10.4.7)$$

where the PB of the matrix L with H is defined componentwise. The first equality in (10.4.7) is satisfied for any L and H, being a consequence of Hamilton equations. The second equality is an identity which can be fulfilled for all q, p only for every exceptional H. Note that then the matrix $A(q, p)$ depends on time only through the variables $q(t)$ and $p(t)$.

The form (10.4.7) makes it obvious that (existence of) Lax representation is invariant under

(i) canonical transformations: $(q(x, y), p(x, y))$, $\{x_j, x_k\} = \{y_j, y_k\} = 0$, $\{x_j, y_k\} = \delta_{jk}$ since

$$\{L(q, p), H(q, p)\}_{(q,p)} = \{L(q(x, y), p(x, y)), H(q(x, y), p(x, y))\}_{(x,y)}$$

(ii) similarity transformations: $L = SLS^{-1}$, $A = SAS^{-1}$ where S is a constant matrix.

Using the invariance under canonical transformations we can easily construct a Lax representation (in terms of original variables (q, p)) for any integrable Hamiltonian system $H(q, p)$. By the Liouville–Arnold theorem there exist linearising canonical coordinates (φ_k, I_k), $k = 1, \ldots, n$ such that $H(q, p) = \tilde{H}(I_1, \ldots, I_n)$ and the Hamilton equations are

$$\dot{\varphi}_k = \frac{\partial \tilde{H}}{\partial I_k} = \omega_k(I),$$

$$\dot{I}_k = -\frac{\partial \tilde{H}}{\partial \varphi_k} = 0, \text{ hence } I_k = \text{const.}$$

These equations can be given Lax form (10.4.5) with $2n \times 2n$ Lax matrices defined

$$L = \text{diag}(L_1, \ldots, L_n),\ A = \text{diag}(A_1, \ldots, A_n), \text{ where}$$

$$L_k = \begin{bmatrix} 0 & \varphi_k \\ 0 & I_k \end{bmatrix}, \quad A_k = \begin{bmatrix} 0 & 0 \\ 0 & -\dfrac{\omega_k}{\varphi_k} \end{bmatrix}. \quad (10.4.8)$$

Expressing φ_k, I_k by (q, p) we get the required Lax representation. The eigenvalues of L, the $I_k(q, p)$, constitute a complete set of integrals for $H(q, p)$.

However not every Lax representation is satisfactory and the representation (10.4.8) determined for a system already known to be integrable is usually not. What we need is a simple, spectral parameter-dependent Lax representation with non-degenerate Lax matrix $L(\mu)$

which can be used for *proving* integrability of the system and for solving it either by some matrix method or by the method of spectral curve [11]. This means that we need L of such size that the matrix invariants provide sufficiently many integrals of motion. If the size of L is insufficient ($n \times n$ matrix has n independent invariants at most), we can still recover necessary integrals by the use of spectral parameter since $\mathrm{Tr}L^k(\mu)$ are polynomials in μ and all coefficients of these are integrals themselves. It often happens that we can trade the size of $L(\mu)$ for the order of polynomial dependence on μ. It will be illustrated in Section 10.5.1 on the example of multiwave interaction equations.

Invention of every particular Lax pair requires great ingenuity and discovery of a new algebraic type of Lax matrices usually leads to extensive study of a whole class of similar systems. For instance, the Toda system is one of the simplest examples of the large family of integrable natural Hamiltonian systems with exponential-type potential [5]. Similarly, two other examples, to be presented here, the Calogero–Moser (CM) system and the Garnier system, are respectively representative of the family of integrable many-body systems in one dimension [1, 37, 51] and of the family of integrable quartic potentials related to symmetric spaces [14],

10.4.3 *The Calogero–Moser system*

The Toda system was the first to be integrated by the IDM [33] but real progress in the application of IDM to Hamiltonian dynamics was stimulated by the discovery of Lax representation for the CM system [32].

The CM system is defined by the Hamiltonian

$$H_{\mathrm{CM}} = \frac{1}{2}\sum_{k=1}^{n} p_k^2 + \frac{1}{2}\sum_{\substack{j,k=1\\ j\neq k}}^{n} (q_j - q_k)^{-2}$$

which describes n identical particles on the line interacting with each other by inverse square potential. This system was first solved by Calogero [7] in the quantum case. In the classical case Moser [32] found the Lax pair L, A:

$$\begin{aligned} L_{jk} &= \delta_{jk}p_j + \mathrm{i}(1-\delta_{jk})(q_j - q_k)^{-1}, \\ A_{jk} &= \delta_{jk}\sum_{l\neq j} \mathrm{i}(q_j - q_l)^{-2} + \mathrm{i}(1-\delta_{jk})(q_j-q_k)^{-2},\ \mathrm{i}^2 = -1, \end{aligned} \qquad (10.4.9)$$

proved complete integrability, and studied scattering behaviour. He showed that asymptotically as $t \to \pm\infty$ particles are decoupled and move as free particles with their asymptotic momenta being the eigenvalues of L. These momenta are all different and ordered

$$p_1^- > p_2^- \ldots > p_n^-, \quad \text{for } t \to -\infty,$$
$$p_1^+ < p_2^+ < \ldots < p_n^+, \quad \text{for } t \to +\infty,$$

if the scattering is from the left to the right. The law of exchange of momenta is $p_k^+ = p_{n-k+1}^-$ and becomes obvious from the explicit solution derived below, since the off-diagonal elements $L_{jk} = \mathrm{i}(q_j - q_k)^{-1}$ go to zero as $t \to \pm\infty$. Calculation of phase shifts of the particles is more elaborate. The result is $q_k^+ = q_{n-k+1}^-$ [37].

Explicit solution is easiest to find by the method of Olshanetsky and Perelomov [34]. If we introduce a matrix $X = \mathrm{diag}(q_1(t), \ldots, q_n(t))$ then we have, incidentally, two matrix equations to solve:

$$\dot{L} = \{L, H\} \equiv [A, L], \tag{10.4.10}$$

$$\dot{X} = \{X, H\} \equiv [A, X] + L. \tag{10.4.11}$$

The matrix $Y(t) = V^{-1}XV$, with V a solution of $\dot{V} = AV$, $V(0) = I$ satisfies then

$$\dot{Y} = V^{-1}(-\dot{V}V^{-1}X + \dot{X} + X\dot{V}V^{-1})\,V = V^{-1}(\dot{X} - [A, X])V = V^{-1}LV,$$
$$\ddot{Y} = V^{-1}(\dot{L} - [A, L])V = 0,$$

and the solution is

$$Y(t) = \dot{Y}(0)t + Y(0) = L(0)t + X(0).$$

Positions of particles are eigenvalues of $Y(t)$ and asymptotically they grow linearly with time like $p_k^{\pm}t$. All $p_k^{\pm}$ are different (see [33]) and the off-diagonal elements $(q_j - q_k)^{-1} \to 0$ for $t \to \pm\infty$. The eigenvalues of L become the asymptotic momenta and the law of exchange $p_k^+ = p_{n-k+1}^-$ follows since the ordering of particles is fixed.

An exceptional property of the CM system is that it has $n-1$ additional integrals [45] functionally independent of $\mathrm{Tr}L^k$, $k = 1, \ldots, n$ and, therefore, trajectories are determined algebraically; that is, the CM system is completely degenerate. Instead of constructing these integrals directly we shall consider the Hamiltonian

$$H_\omega = H_{\mathrm{CM}} + \tfrac{1}{2}\omega^2 \sum_{k=1}^{n} q_k^2$$

which can also be solved explicitly by the Olshanetsky–Perelomov method. Equations (10.4.10), (10.4.11) are replaced by

$$\dot{L} = [A, L] - \omega^2 X,$$

$$\dot{X} = [A, X] + L,$$

and $Y = V^{-1}XV$, where $\dot{V} = AV$, $V(0) = I$, satisfies

$$\ddot{Y} = -\omega^2 Y$$

with the solution $Y(t) = 1/\omega L(0) \sin \omega t + X(0) \cos \omega t$. Thus positions of particles, given by eigenvalues of $Y(t)$, are periodic functions of time and the motion is completely degenerate on invariant tori of H_ω. This means that H_ω has $n - 1$ additional integrals of motion which in a proper limit $\omega \to 0$ give additional integrals of H_{CM}.

The CM system is one of the simplest in a class of integrable one-dimensional many-body systems characterised by

$$H = \frac{1}{2} \sum_{k=1}^{n} p_k^2 + \frac{1}{2} \sum_{\substack{j,k=1 \\ j \neq k}}^{n} V(q_j - q_k) + \sum_{k=1}^{n} W(q_k).$$

Admissible potentials were found by generalising the Lax matrices (10.4.9). Substitution $(q_j - q_k)^{-1} \to \phi(q_j - q_k)$, $(q_j - q_k)^{-2} \to \phi'(q_j - q_k)$ (prime denotes derivative) leads to the functional–differential equation [8, 9]

$$\phi'(x)\, \phi(y) - \phi(x)\, \phi'(y) = \phi(x + y)(\chi(x) - \chi(y))$$

on the function $\phi(x)$. Then an integrable potential, which follows from $H = \frac{1}{2} \operatorname{Tr} L^2$, is $V(x) = \phi(x)\, \phi(-x) = a\, \mathscr{P}(bx)$ where $\mathscr{P}$ is the Weierstrass elliptic function and a, b are constants. Extension of the dimension of L from n to $2n$ or $2n + 1$ leads to some non-trivial integrable external potentials $W(x)$ [19, 35, 47]. For instance, in the case of the CM potential $V(x) = x^{-2}$ an integrable external potential is $W(x) = \alpha x^4 + \beta x^3 + \gamma x^2 + \delta x$ where α, β, γ, δ are parameters.

10.4.4 *The Garnier system*

The Garnier Hamiltonian

$$H_{\mathrm{G}} = \frac{1}{2} \sum_{k=1}^{n} p_k^2 + \frac{1}{2} q^4 - \frac{1}{2} \sum_{k=1}^{n} \alpha_k q_k^2, \quad q^2 = \sum_{k=1}^{n} q_k^2 \qquad (10.4.12)$$

describes one particle in an external quartic anisotropic potential. In the case of two degrees of freedom and positive, different values of α_k this potential has a wine-bottle shape with two minima. By letting α_k become negative this potential can be continuously deformed to a potential with one minimum. The Garnier potential is an integrable example of a multi-dimensional symmetry breaking potential.

We have already solved the Garnier system by the MSV but H_{G} admits also a complete Lax representation [48] with

$$L = \left[\begin{array}{c|c} q^2 - 4\mu^2 & \cdots\ p_k - 2\,\mathrm{i}\mu q_k\ \cdots \\ \hline \vdots & \vdots \\ p_j + 2\mathrm{i}\,\mu q_j & \cdots\ -q_j q_k + \delta_{jk}\alpha_j\ \cdots \\ \vdots & \vdots \end{array}\right],$$

$$A = \left[\begin{array}{c|cccc} -\mathrm{i}\mu & \ldots & q_k & \ldots & \\ \hline \vdots & \mathrm{i}\mu & & & 0 \\ -q_k & & \ddots & & \\ \vdots & 0 & & & \mathrm{i}\mu \end{array}\right],$$

where μ is a spectral parameter and $\mathrm{i}^2 = -1$. The dimension of L is $(n + 1) \times (n + 1)$ but $\mathrm{Tr}L = \text{const.}$ and only n integrals $1/2k\ \mathrm{Tr}L^k$, $k = 2, \ldots, n + 1$ are functionally independent. The Hamiltonian is $H_G = 1/4\ \mathrm{Tr}L^2$. Integrals of motion following from L are simply related to the integrals K_k from the Section 10.3.1 by the Weinstein–Aronszajn formula [48]:

$$\frac{\det(L - Iz)}{\det(\Lambda - Iz)} = \frac{\prod_{j=0}^{n}(z - \lambda_j)}{\prod_{k=1}^{n}(z - \alpha_k)} = 1 - \frac{1}{z}\left(\sum_{k=1}^{n} \frac{K_k}{z - \alpha_k}\right)$$

where λ_j are eigenvalues of L and $\Lambda = \mathrm{diag}\ (\alpha_1, \ldots, \alpha_n)$.

The spectral parameter μ is not very important for this Lax representation but becomes indispensable for proving integrability of generalisations of the Garnier system: a family of quartic potentials related to symmetric spaces [14]. The simplest generalisation is represented by

$$L(\mu) = \left[\begin{array}{cc|c} q^2 + \delta_{-1} - n(n+2)\mu^2 & \rho^2 & \ldots p_k - \mathrm{i}(n+2)\mu q_k \ldots \\ \rho^2 & r^2 + \delta_0 - n(n+2)\mu^2 & \ldots s_k - \mathrm{i}(n+2)\mu r_k \ldots \\ \hline \vdots & \vdots & \vdots \\ p_j + \mathrm{i}(n+2)\,\mu q_j & s_j + \mathrm{i}(n+2)\,\mu r_j & -q_j q_k - r_j r_k + [2(n+2)\,\mu^2 + \alpha_j]\delta_{jk} \ldots \\ \vdots & \vdots & \vdots \end{array}\right]$$

and

$$\begin{aligned} H_r = \frac{1}{4}\ \mathrm{Tr}L^2|_{\mu=0} &= \frac{1}{2}\sum_{k=1}^{n}(p_k^2 + s_k^2) + \frac{1}{2}q^4 + \frac{1}{2}r^4 + \rho^4 \\ &- \frac{1}{2}\sum_{k=1}^{n}(\alpha_k - \delta_{-1})q_k^2 - \frac{1}{2}\sum_{k=1}^{n}(\alpha_k - \delta_0)r_k^2 \end{aligned}$$

where $r^2 = \Sigma_k r_k^2$, $\rho^2 = \Sigma_k q_k r_k$ and r_k are additional n position coordinates with canonically conjugate momenta s_k. Here the number of dynamical variables is $4n$, dimension of L is $(n + 2) \times (n + 2)$, $\mathrm{Tr}L = \text{const}$, and the spectral parameter μ becomes necessary to produce a sufficient number of integrals of motion. These integrals are coefficients of different powers of μ in $(1/2k)\ \mathrm{Tr}L^k(\mu)$. The number of independent integrals is $2n$ indeed and H_r is completely integrable as has been proved in [14]. More general systems can be obtained by adding more rows and columns in the off-diagonal blocks and (possibly) imposing some symmetry conditions on the elements of the matrix L.

10.5 Non-standard Hamiltonian structures

The great success of the IDM comes from the fact that it applies not only to natural (or quadratic-in-momenta) Hamiltonians like MSV, but to a large number of other Hamiltonians and to systems with non-standard Hamiltonian structures. The Euler equations on Lie algebras are the best-known example but we shall present below also some other examples which are representative of different classes of non-natural integrable systems.

In looking for Lax representation of some given equations we have a freedom of choice of several ingredients which we can play with: of a Hamiltonian structure (that is, a PB and a Hamiltonian), of dynamical variables, and of an algebraic form of Lax matrices. By algebraic form we mean here size of matrices and functional dependence of their matrix elements on spectral parameter and on dynamical variables. This indeterminacy is a source of flexibility of the IDM on the one hand and is responsible for non-existence of any reasonable rules of finding L, A matrices on the other hand. Every single problem requires great ingenuity in finding Lax representation but success of the exercise allows for more systematic investigations and extension of the range of applicability of the IDM.

10.5.1 *Multiwave interaction equations*

The multiple three-wave interaction equations (MTW) are

$$\begin{aligned} \dot{u} &= i \sum_{j=1}^{n} \alpha_j b_j c_j^*, \\ \dot{b}_j &= -\tfrac{1}{2}\,\mathrm{i}\varepsilon_j b_j + \mathrm{i}\alpha_j u c_j, \\ \dot{c}_j &= \tfrac{1}{2}\,\mathrm{i}\varepsilon_j c_j + \mathrm{i}\alpha_j u^* b_j, \end{aligned} \tag{10.5.1}$$

and their complex conjugates. The ε_j, α_j are real parameters, * denotes complex conjugation and $\mathrm{i}^2 = -1$. These equations model the growth of a low-frequency internal ocean wave by interaction with higher-frequency waves [42], but were also proposed [29] as a model of plasma turbulence. They have a Hamiltonian structure with the Hamiltonian

$$H = \frac{1}{2}\,\mathrm{i} \sum_{j=1}^{n} \varepsilon_j (c_j c_j^* - b_j b_j^*) + \mathrm{i} \sum_{j=1}^{n} \alpha_j (u b_j^* c_j + u^* b_j c_j^*) \tag{10.5.2}$$

and the PB

$$\{f, g\} = \left(\frac{\partial f}{\partial u}\frac{\partial g}{\partial u^*} - \frac{\partial f}{\partial u^*}\frac{\partial g}{\partial u}\right) + \sum_{j=1}^{n}\left[\left(\frac{\partial f}{\partial b_j}\frac{\partial g}{\partial b_j^*} - \frac{\partial f}{\partial b_j^*}\frac{\partial g}{\partial b_j}\right) + \left(\frac{\partial f}{\partial c_j}\frac{\partial g}{\partial c_j^*} - \frac{\partial f}{\partial c_j^*}\frac{\partial g}{\partial c_j}\right)\right]$$

but are not usually integrable unless all $\alpha_j = 1$. Then the equations (10.5.1) are called restricted multiple 3-wave interaction equations (rMTW). They have the Lax representation [50]:

$$\dot{L} = \{L, H\} \equiv [A, L] \tag{10.5.3}$$

in which L and A are $(2n + 2) \times (2n + 2)$ matrices:

$$L = \tfrac{1}{2}\begin{bmatrix} \pi & \sigma_1 & \ldots & \sigma_n \\ \tau_1 & \varepsilon_1 I & & 0 \\ \vdots & & \ddots & \\ \tau_n & 0 & & \varepsilon_n I \end{bmatrix}, \quad A = \tfrac{1}{2}\,\mathrm{i}\begin{bmatrix} 0 & \omega\sigma_1 & \ldots & \omega\sigma_n \\ \tau_1\omega & & & \\ \vdots & & \boxed{0} & \\ \tau_n\omega & & & \end{bmatrix}.$$

The π, σ_j and τ_j are 2×2 matrices

$$\pi = \begin{bmatrix} 0 & -2u \\ 2u^* & 0 \end{bmatrix}, \quad \tau_j = \begin{bmatrix} b_j^* & -c_j^* \\ -c_j & -b_j \end{bmatrix}, \quad \sigma_j = \begin{bmatrix} -b_j & -c_j^* \\ -c_j & b_j^* \end{bmatrix}$$

and $\omega = \mathrm{diag}(-1, 1)$, $I = \mathrm{diag}(1, 1)$ – the identity. In terms of L the Hamiltonian (10.5.2) is $\frac{2}{3}\cdot\mathrm{i}\cdot\mathrm{Tr}L^3 + \text{const}$. In the case of all $\varepsilon_j = 0$ this Lax representation has been found in [30] but it was incomplete and provided only three non-trivial integrals of motion: $\mathrm{Tr}L^2$, $\mathrm{Tr}L^3$, $\mathrm{Tr}L^4$.

For all ε_j different, the characteristic determinant $\det(2L - \lambda L)$ is easy to calculate and gives the simplest set of integrals which are quadratic (at most) in the canonical momenta u^*, b_j^*, c_j^*. We find

$$\det(I\lambda - 2L) = \det\begin{bmatrix} I & \dfrac{\sigma_1}{\lambda - \varepsilon_1} & \cdots & \dfrac{\sigma_n}{\lambda - \varepsilon_n} \\ 0 & I & & 0 \\ \vdots & & \ddots & \vdots \\ 0 & \cdots & \cdots & I \end{bmatrix}$$

$$\cdot \det\begin{bmatrix} \lambda I - \pi & -\sigma_1, & \ldots, & -\sigma_n \\ -\tau_1 & (\lambda - \varepsilon_1)I & & 0 \\ \vdots & & \ddots & \\ -\tau_n & 0 & & (\lambda - \varepsilon_n)I \end{bmatrix}$$

$$= \det\begin{bmatrix} \left(\lambda I - \pi - \sum_j \dfrac{\sigma_j\tau_j}{\lambda - \varepsilon_i}\right) & 0 & \ldots\ldots & 0 \\ -\tau_1 & (\lambda - \varepsilon_1)I & & 0 \\ \vdots & & \ddots & \\ -\tau_n & 0 & & (\lambda - \varepsilon_n)I \end{bmatrix}$$

$$= (\lambda - \varepsilon_1)^2 \ldots (\lambda - \varepsilon_n)^2 \det\left(I\lambda - \pi - \sum_j \frac{\sigma_j \tau_j}{\lambda - \varepsilon_j}\right)$$

$$= (\lambda - \varepsilon_1)^2 \ldots (\lambda - \varepsilon_n)^2 \left[\lambda^2 - 4\mathrm{i}J_0 + 4\mathrm{i} \sum_{j=1}^{n} \frac{K_j}{\lambda - \varepsilon_j} - \sum_{j=1}^{n} \frac{J_j^2}{(\lambda - \varepsilon_j)^2}\right]$$

where $J_0 = \mathrm{i} \cdot u \cdot u^* + \frac{1}{2}\,\mathrm{i} \cdot \sum_{j=1}^{n} (b_j b_j^* - c_j c_j^*)$, $\quad J_j = \mathrm{i}(b_j b_j^* + c_j c_j^*)$

and $\quad K_j = \mathrm{i}\left[u\, b_j^* c_j + u^* b_j c_j^* + \frac{1}{2}\,\varepsilon_j (c_j c_j^* - b_j b_j^*)\right]$

$$- \frac{1}{2}\,\mathrm{i} \sum_{m \neq j}^{n} [(b_j b_j^* - c_j c_j^*)(b_m b_m^* - c_m c_m^*) + 2b_j^* c_j b_m c_m^*$$

$$+ 2b_j c_j^* b_m c_m^*] \, \frac{1}{\varepsilon_j - \varepsilon_m}.$$

Proof of commutativity and of functional independence of these integrals is straightforward and the rMTW system is completely integrable.

It is interesting that this system has another complete Lax representation of dimension $n + 2$ which depends on a spectral parameter. The Lax matrices are

$$\tilde{L}(\mu) = \begin{bmatrix} \pi - \mu\omega & \frac{\sigma_1}{\sqrt{2}} & \cdots & \frac{\sigma_n}{\sqrt{2}} \\ \frac{\tau_1}{\sqrt{2}} & \frac{1}{2}\varepsilon_1 & & 0 \\ \vdots & & \ddots & \\ \frac{\tau_n}{\sqrt{2}} & 0 & & \frac{1}{2}\varepsilon_n \end{bmatrix}, \qquad \tilde{A}(\mu) = \begin{bmatrix} 0 & \omega\tau_1 & \cdots & \omega\sigma_n \\ \tau_1\omega & \frac{\mu}{\sqrt{2}} & & 0 \\ \vdots & & \ddots & \\ \tau_n\omega & 0 & & \frac{\mu}{\sqrt{2}} \end{bmatrix},$$

in which

$$\pi = \begin{bmatrix} 0 & -u \\ u^* & 0 \end{bmatrix}, \quad \omega = \begin{bmatrix} -1 & 0 \\ 0 & 1 \end{bmatrix}, \quad \sigma_j = \begin{bmatrix} -b_j \\ -c_j \end{bmatrix}, \; \tau_j = [b_j^*, -c_j^*].$$

Note that the spectral parameter μ isolates now $2n + 1$ different integrals of motion J_0, J_j, $\tilde{K}_j$ from

$\det(I\lambda - L(\mu))$

$$= \left[\lambda^2 - \mu^2 - \mathrm{i}J_0 - \sum_{j=1}^{n} \frac{\mathrm{i}\mu J_j}{2\lambda - \varepsilon_j} + \sum_{j=1}^{n} \frac{\mathrm{i}K_j}{2\lambda - \varepsilon_j}\right] \prod_{j=1}^{n} (\lambda - \varepsilon_j)$$

in which $\tilde{K}_j = K_j + \frac{1}{2}\,\mathrm{i} \cdot \sum_{m \neq j} J_j J_m (\varepsilon_j - \varepsilon_m)^{-1}$.

This second form of Lax matrices allows for generalisation of σ_j, τ_j and for construction of a large class of integrable MTW-type systems [21, 50].

10.5.2 *Euler equations on so(n) Lie algebra*

The usefulness of non-standard Hamiltonian structures is best illustrated by the example of Euler equations on the Lie algebra $so(n)$ which are a generalisation of ordinary Euler equations of rigid-body motion around a fixed point in an external potential.

Consider a set of dynamical variables $l_{jk} = -l_{kj}$ which satisfy LPB of the $so(n)$ Lie algebra

$$\{l_{rs}, l_{tw}\} = \delta_{rw} l_{ts} + \delta_{rt} l_{sw} + \delta_{sw} l_{rt} + \delta_{st} l_{wr}. \tag{10.5.4}$$

These commutation relations can be easily recovered if we think of l_{jk} as $l_{jk} = q_j p_k - q_k p_j$ and apply commutation relations of the standard PB. However, the l_{jk} are not $q_j p_k - q_k p_j$ because here they are considered as functionally independent variables: $\binom{n}{2} = \dfrac{n(n-1)}{2}$.

The LPB (10.5.4) is degenerate, that is, there are non-constant functions $C(l)$, called Casimir functions, which Poisson-commute with all dynamical variables. For the $so(n)$ Lie algebra there are $\left[\dfrac{n}{2}\right]$ of them ($[\cdot]$ stands for integer part). On manifolds of constant values of all Casimir functions the LPB (10.5.4) is non-degenerate and the Liouville theorem can be applied for the proving of complete integrability of some Hamiltonian $H(l)$. So the necessary number of non-trivial integrals, independent of Casimir functions, is $\dfrac{1}{2}\left(\dfrac{n}{2} - \left[\dfrac{n}{2}\right]\right)$.

The Euler equations on the $so(n)$ are generated by the diagonal Hamiltonian of the form

$$H = -\frac{1}{4} \sum_{i=1}^{n} \sum_{j=1}^{n} b_{ij} l_{ij}^2$$

where $b_{ij} = b_{ji}$ are parameters. Then the equations of motion are

$$\dot{l}_{rs} = \{l_{rs}, H\} = \sum_{j=1}^{n} (b_{rj} l_{rj} l_{js} - l_{rj} b_{js} l_{js}) \tag{10.5.5}$$

and they can be written in a Lax form [2]

$$\dot{L} = [A, L] \tag{10.5.6}$$

where $L_{ij} = l_{ij}$, $A_{ij} = b_{ij}\, l_{ij}$. Note that $H = -\frac{1}{2}\,\mathrm{Tr}LA$. This Lax form provides only trivial integrals $\mathrm{Tr}L^{2k}$, $k = 1, \ldots, \left[\dfrac{n}{2}\right]$ (only these are non-zero) which are equivalent to Casimir functions. The equations (10.5.5) are non-integrable for generic values of the parameters b_{ij}.

For constructing larger number of integrals we need a Lax representation with a spectral parameter. If we define

$$L(\mu) = L + \mu\alpha, \quad \alpha = \mathrm{diag}(\alpha_1, \ldots, \alpha_n),$$
$$A(\mu) = A + \mu\beta, \quad \beta = \mathrm{diag}(\beta_1, \ldots, \beta_n),$$

then the Lax equation

$$\dot{L}(\mu) = [A(\mu), L(\mu)] = [A, L] + \mu([\beta, L] + [A, \alpha]) + \mu^2[\beta, \alpha]$$

is satisfied if $[\beta, L] + [A, \alpha] = 0$, that is, $b_{jk} = (\beta_j - \beta_k)/(\alpha_j - \alpha_k)$. These b_{jk} define the $2n$-parameter family of integrable Euler equations selected by Manakov [27]. It has also been proved that only the Manakov case is algebraically integrable [17].

In order to calculate a number of additional integrals [27] we write

$$\mathrm{Tr}(L + \mu\alpha)^k = \mathrm{Tr}L^k + \mu\mathrm{Tr}(L^{k-1}, \alpha) + \mu^2\mathrm{Tr}(L^{k-2}, \alpha^2) + \ldots + \mu^{k-1}\mathrm{Tr}(L, \alpha^{k-1}) + \mu^k\mathrm{Tr}\alpha^k \tag{10.5.7}$$

where $\mathrm{Tr}(L^{k-r}, \alpha^r)$ stands for the sum of all symmetric expressions of order $k - r$ in L and of order r in α. Since L is a skew-symmetric matrix and α is a symmetric matrix, the only non-zero terms in (10.5.7) are those where L enters with even parity, that is, the k and the power of μ have the same parity. The first term in (10.5.7) is a Casimir function and the last is constant. Thus the number of non-trivial integrals following from (10.5.8) is

$$\frac{k-2}{2} \quad \text{if k is even,}$$
$$\frac{k-1}{2} \quad \text{if k is odd,}$$

that is, $\left[\frac{k-1}{2}\right]$. The total number of additional integrals is $\sum_{k=2}^{n}\left[\frac{k-1}{2}\right] = \frac{1}{2}\left(\binom{n}{2} - \left[\frac{n}{2}\right]\right)$ as necessary. Proof of functional independence of all these integrals [31] is the most difficult part and Manakov argued instead that his case of the Euler equations can be written in the form of the Dubrovin equation $[\alpha, \dot{V}] = [[\beta, V], [\alpha, V]]$ which is solvable in terms of Riemann θ-functions. The V is an arbitrary matrix with zero diagonal elements and α, β are arbitrary diagonal matrices. The correct choice here is $V_{ij} = l_{ij}(\alpha_i - \alpha_j)^{-1}$.

Most of the integrals of the Manakov case of Euler equations are complicated expressions of high order in dynamical variables but among them are n quadratic integrals of particularly simple form:

$$K_i = \sum_{j \neq i}^{n} \frac{l_{ij}^2}{\alpha_i - \alpha_j} \tag{10.5.8}$$

which mutually commute and generate the Manakov Hamiltonian

$$H_M = -\frac{1}{2}\sum_{i=1}^{n} \beta_i K_i = -\frac{1}{2}\sum_{i} \sum_{> i} \frac{\beta_i - \beta_j}{\alpha_i - \alpha_j} l_{ij}^2.$$

The integrals (10.5.8) have generalisation [39]:

$$\bar{K}_i = \sum_{j\neq i}^{n} \frac{(l_{ij} + m_{ij})^2}{\alpha_i - \alpha_j} + \sum_{j\neq i}^{n} \frac{(l_{ij} - m_{ij})^2}{\alpha_i + \alpha_j} \tag{10.5.9}$$

where the $2 \cdot \binom{n}{2}$ variables l_{ij}, m_{ij} each satisfy the commutation relations of $so(n)$ and commute themselves: $\{l_{ij}, m_{rs}\} = 0$. These integrals define an integrable Hamiltonian

$$\bar{H}(l, m) = \frac{1}{2}\sum_{i=1}^{n} \beta_i K_i = \frac{1}{2}\sum_{i} \sum_{> j} \frac{\beta_i - \beta_j}{\alpha_i - \alpha_j} (l_{ij} + m_{ij})^2 + \frac{1}{2}\sum_{i} \sum_{> j} \frac{\beta_i + \beta_j}{\alpha_i + \alpha_j} (l_{ij} - m_{ij})^2 \tag{10.5.10}$$

which describes two n-dimensional interacting rigid bodies. Equations of motion of (10.5.10):

$$\dot{l}_{ij} = \{l_{ij}, \bar{H}(l, m)\},$$
$$\dot{m}_{ij} = \{m_{ij}, \bar{H}(l, m)\},$$

have also a spectral parameter-dependent Lax representation [39]:

$$L(\mu) = \frac{1}{2}\left[\begin{array}{c|c} l+m & l-m \\ \hline l-m & l+m \end{array}\right] + i\mu \left[\begin{array}{c|c} \alpha & 0 \\ \hline 0 & -\alpha \end{array}\right], \tag{10.5.11}$$

$$A(\mu) = \frac{1}{2}\left[\begin{array}{c|c} \tilde{l}+\tilde{m} & \tilde{l}-\tilde{m} \\ \hline \tilde{l}-\tilde{m} & \tilde{l}+\tilde{m} \end{array}\right] + i\mu \left[\begin{array}{c|c} \beta & 0 \\ \hline 0 & -\beta \end{array}\right], \tag{10.5.12}$$

where $(\tilde{l} + \tilde{m})_{ij} = \dfrac{\beta_i - \beta_j}{\alpha_i - \alpha_j}(l_{ij} + m_{ij})$, $(\tilde{l} - \tilde{m})_{ij} = \dfrac{\beta_i + \beta_j}{\alpha_i + \alpha_j}(l_{ij} - m_{ij})$, and the meaning of $l + m$ and $l - m$ is obvious. The number of additional integrals following from (10.5.11) can be calculated in a similar way as for the Manakov case: it is $\frac{1}{2}\, n(n - 1)$ and suffices to prove the complete integrability of the Hamiltonian system generated by $\bar{H}$. The Lax representation (10.5.11), (10.5.12) is an example of an integrable reduction of the Manakov case for the $so(2n)$ rigid body. The problem of constructing all such reductions is non-trivial. Yet another example of such reduction which represents many interacting rigid bodies may be constructed [20] by the use of m-block matrices analogous to (10.5.11) and (10.5.12). Then, in the spectral parameter-dependent part, mth roots of unity are used instead of $\pm\alpha$. These equations are also equivalent to the Dubrovin equation.

10.5.3 *Euler–Calogero–Moser system*

As an illustration of the flexibility of the IDM and of the possibility of playing with different Hamiltonian structures in order to construct integrable systems we shall present the Euler–Calogero–Moser (ECM) equations [49].

Let us consider a space of dynamical variables (q_k, p_k), $l_{ij} = -l_{ji}$, $i, j, k = 1, \ldots, n$ with the PB defined by

$$\{f, g\} = \{f, g\}_p + \{f, g\}_l$$

a direct sum of the standard PB $\{.,\ .\}_p$ and of LPB of the $so(n)$ Lie algebra $\{.,\ .\}_l$. The variables (q_k, p_k) and l_{ij} are considered to be independent and to Poisson-commute with each other. On this phase space we shall consider the ECM Hamiltonian

$$H_{\mathrm{ECM}} = \frac{1}{2}\sum_{k=1}^{n} {p_k}^2 + \frac{1}{2}\sum_{\substack{j,k=1\\ j\neq k}}^{n} \frac{{l_{jk}}^2}{(q_j - q_k)^2} + \frac{1}{2}\omega^2 \sum_{k=1}^{n} {q_k}^2 \tag{10.5.13}$$

which generates the CM system:

$$\dot{q}_k = \{q_k, H_{\mathrm{ECM}}\} = \{q_k, H_{\mathrm{ECM}}\}_p = p_k, \tag{10.5.14}$$

$$\dot{p}_k = \{p_k, H_{\mathrm{ECM}}\} = \{p_k, H_{\mathrm{ECM}}\}_p = -\frac{\partial H_{\mathrm{ECM}}}{\partial q_k} = \sum_{j\neq k}^{n} \frac{2{l_{jk}}^2}{(q_k - q_j)^3} - \omega^2 q_k, \tag{10.5.15}$$

coupled to the Euler equations

$$\dot{l}_{ij} = \{l_{ij}, H_{\mathrm{ECM}}\} = \{l_{ij}, H_{\mathrm{ECM}}\}_l = \sum_{k\neq i,j}^{n} \left(-\frac{l_{ik}l_{kj}}{(q_i - q_k)^2} + \frac{l_{ik}l_{kj}}{(q_j - q_k)^2} \right). \tag{10.5.16}$$

The difference from the previously considered cases is that the coupling constant $l_{jk}^2(t)$ of the CM system depends on time according to the Euler equations and that the coefficients b_{jk} of the Euler equations follow the evolution of the CM system.

The equations (10.5.14)–(10.5.16) can be equivalently written in the Lax form:

$$\dot{L} = \{L, H_{\mathrm{ECM}}\} = \{L, H_{\mathrm{ECM}}\}_p + \{L, H_{\mathrm{ECM}}\}_l \equiv [A, L] - \omega^2 X, \tag{10.5.17}$$

$$\dot{X} = \{X, H_{\mathrm{ECM}}\} = \{X, H_{\mathrm{ECM}}\}_P \equiv [A, X] + L,$$

$$\dot{l} = \{l, H_{\mathrm{ECM}}\} = \{L, H_{\mathrm{ECM}}\}_l \equiv [A, l], \tag{10.5.18}$$

where the $n \times n$ matrices are defined:

$$L_{jk} = p_j\delta_{jk} - (1 - \delta_{jk})l_{jk}(q_j - q_k)^{-1},$$
$$A_{jk} = -(1 - \delta_{jk})l_{jk}(q_j - q_k)^{-2}, \ (l)_{jk} = l_{jk}. \qquad (10.5.19)$$

From these equations we get also a spectral parameter-dependent Lax representation

$$\frac{\mathrm{d}}{\mathrm{d}t}(L + \mu l) = [A, L + \mu l]$$

which provides a sufficient number of integrals of motion for proving complete integrability of the H_{ECM}. However, we can solve this system by the Olshanetsky–Perelomov method. If the invertible matrix V satisfies $\dot{V} = AV$, $V(0) = I$ then for $Y = V^{-1}XV$ we get $\dot{Y} = -\omega^2 Y$ with the solution $Y(t) = \omega^{-1}L(0)\sin\omega t + X(0)\cos\omega t$. Thus positions of particles, which are eigenvalues of $Y(t)$, are periodic in time. The maxtrix $V(t)$ which diagonalises $Y(t)$ determines also time dependence of $L(t) = V(t)L(0)V^{-1}(t)$ and of $l(t) = V(t)l(0)V^{-1}(t)$. Periodicity of the solution means that the motion of ECM is completely degenerate as in the CM system, while the Manakov case of the Euler equations has quasiperiodic solution on an invariant torus. That is, the physical variables (q_k, p_k) dominate the evolution of the angular momentum variables l_{jk}.

There is no difficulty in extending these results to other integrable potentials of the Calogero–Moser class. The other important examples of integrable systems where physical variables (q, p) are coupled to some other variables are the generalised Calogero–Moser system [16] and the rigid-body motion in an external quadratic potential [6].

10.5.4 *Integrable low-dimensional Riccati equations*

As a last example of the usefulness of non-standard Hamiltonian structures we present multi-parameter family of integrable Riccati equations, namely

$$\dot{e}_k = \gamma_k + \sum_{r=1}^{n} \beta_k{}^r e_r + \sum_{r=1}^{n}\sum_{s=1}^{n} \alpha_k{}^{rs} e_r\, e_s,$$

that is, dynamical systems with quadratic nonlinearities at most. Such systems are of great interest for applications since in many approximations the first nonlinear terms are quadratic. Moreover, they often arise as exact equations of chemical reactions or of ecological systems.

Construction of integrable equations is very simple and is based on the use of low-dimensional Lie algebras and their Casimir operators. As is known [43] with any Lie algebra defined by

$$[E_i, E_j] = \sum_{k=1}^{n} c_{ij}^k E_k$$

we can relate an LPB

$$\{e_i, e_j\} = \sum_{k=1}^{n} c_{ij}^k e_k \tag{10.5.20}$$

defined on a set of dynamical variables $e_1, \ldots, e_n$. Casimir operators of Lie algebra then become Casimir functions $C(e)$ which by definition commute with all dynamical variables $\{e_k, C(e)\} = 0$. Thus if we define dynamical equations by the formula

$$\dot{e}_k = \{e_k, H(e)\} = \sum_{r=1}^{n} \{e_k, e_r\} \frac{\partial H}{\partial e_r} \tag{10.5.21}$$

then the Casimir functions are automatically integrals of motion. For complete integrability of (10.5.21) we need m-2 Casimirs since the manifold defined by their constant values is 2-dimensional, the LPB is non-degenerate and by the Liouville theorem we need only one integral – the Hamiltonian. This Hamiltonian is functionally independent of the Casimirs $C_r(e)$ because equations of motion (10.5.21) are assumed to be non-trivial.

The above observations are rather trivial, but the equations of motion, following from (10.5.21), are not. It is easy to judge their integrability just by the construction but not if they are *a priori* given. Consider the simplest real, non-trivial Lie algebras of dimension 3. There are nine of these and they have one Casimir function each [38]. Thus any real 3-dimensional Lie algebra generates integrable Riccati equations.

As an example we shall take the Lie algebra $A_{3,7}^a$:

$$\{e_1, e_3\} = ae_1 - e_2, \{e_2, e_3\} = e_1 + ae_2 \quad (a > 1),$$

which depends on a parameter and has the Casimir function

$$C = (e_1^2 + e_2^2)\left[\frac{e_1 + \mathrm{i}e_2}{e_1 - \mathrm{i}e_2}\right]^{ia} = (e_1^2 + e_2^2)\exp\left(-2a \arctan \frac{e_2}{e_1}\right), \quad \mathrm{i}^2 = -1.$$

A most general Hamiltonian which generates Riccati equations is of the second order:

$$H = \alpha_1 e_1^2 + \alpha_2 e_2^2 + \alpha_3 e_3^2 + \beta_1 e_2 e_3 + \beta_2 e_3 e_1 + \beta_3 e_1 e_2 \\ + \gamma_1 e_1 + \gamma_2 e_2 + \gamma_3 e_3,$$

and then dynamical equations read

$$\dot{e}_1 = \{e_1, H\} = a\beta_2 e_1^2 - a\beta_1 e_2^2 - 2\alpha_3 e_2 e_3 + 2a\alpha_3 e_3 e_1 \\ + (a\beta_1 - \beta_2)e_1 e_2 + \gamma a e_1 - \gamma_3 e_2,$$

$$\dot{e}_2 = \{e_2, H\} = \beta_2 e_1^2 + a\beta_1 e_2^2 + 2a\alpha_3 e_2 e_3 + 2\alpha_3 e_3 e_1 \\ + (\beta_1 + a\beta_2)e_1 e_2 + \gamma_3 e_1 + a\gamma_3 e_2,$$

$$\begin{aligned}\dot{e}_3 = \{e_3, H\} &= -(2a\alpha_1 + \beta_3)e_1^2 - (2a\alpha_2 - \beta_3)e_2^2 \\ &- (a\beta_1 - \beta_2)e_2e_3 - (\beta_1 + a\beta_2)e_3e_1 + 2(\alpha_1 - \alpha_2 - a\beta_3)e_1e_2 \\ &- (a\gamma_1 + \gamma_2)e_1 - (a\gamma_2 - \gamma_1)e_2.\end{aligned}$$

They depend on ten parameters (!) and look quite non-trivial indeed. It would not be easy to find that they have the integrals $H(e)$ and $C(e)$.

There is no problem in using higher dimensional Lie algebras, listed in [38], for producing integrable equations. A more interesting question is how to study given Riccati equations and to recognise whether they have some integrals of Casimir type. The work on this is in progress [52] and it appears that it is necessary to make use of more general (than Lie) algebras and that the problem is basically reduced to the study of systems of compatible linear equations. The problem linearises but in a different way than in the Lax method.

Acknowledgements

The main part of this chapter has been given as lectures in the Department of Applied Mathematics of the University of Leeds and the Department of Mathematics at UMIST in the academic year 1984/5. It is my pleasure to thank Dr A. Fordy at Leeds for inviting me to give these lectures and for many useful discussions, and to thank Prof. R. Bullough for inviting me to UMIST and for encouragement. I would like also to thank Prof. Y. Kosmann-Schwarzbach for many interesting discussions during my stay at the Department of Mathematics of the University of Lille in June 1986 and for giving me an opportunity to lecture there on this topic. The friendly atmosphere of my own department helped me very much in the writing of this text, and my particular thanks go to Prof. L.-I. Hedberg. I thank also Nan Strömberg for excellent typing of the text. This piece of work has been written under the Swedish NFR contract No. F-FU 8677-102, which support I gratefully acknowledge here.

References

[1] Adler, M., Some finite dimensional integrable systems and their scattering behaviour, *Commun. Math. Phys.* **55**, 195–230 (1977).
[2] Arnol'd, V. I., Hamiltonian structure of Euler equations of rigid body and of ideal fluid, *Usp. Matem. Nauk* **26**, 225–6 (1969).
[3] Arnol'd, V. I., *Mathematical Methods of Classical Mechanics*, Springer, Berlin (1978).
[4] Benenti, S., *Separability Structures on Riemannian Manifolds*, Lecture Notes in Mathematics 836, Springer, Berlin (1980).
[5] Bogoyavlensky, O. I., On perturbations of the periodic Toda lattice, *Comm. Math. Phys.* **51**, 201–9 (1976).
[6] Bogoyavlensky, O. I., New integrable problem of classical mechanics, *Commun. Math. Phys.* **94**, 255–69 (1984).

[7] Calogero, F., Solution of the one-dimensional n-body problem with quadratic and/or inversely quadratic pair potentials, *J. Math. Phys.* **12**, 419–36 (1971).

[8] Calogero, F., Exactly solvable one-dimensional many-body problems, *Lett. Nuovo Cim.* **13**, 411 (1975).

[9] Calogero, F., On a functional equation connected with integrable many-body problems, *Lett. Nuovo Cim.* **16**, 77–80 (1976).

[10] Calogero, F., Finite transformations of certain isospectral matrices, *Lett. Nuovo Cim.* **28**, 502–4 (1980).

[11] Dubrovin, B. A., Matveev, V. B. and Novikov, S. P., Nonlinear equations of Korteweg de Vries types, finite zone linear operators and Abelian varieties, *Russian Math. Surveys* **31**, 59–146 (1976).

[12] Eisenhart, L. P., Enumeration of potentials for which one-particle Schrödinger equations are separable, *Phys. Rev.* **74**, 87–9 (1948).

[13] Flaschka, H., The Toda lattice, *Phys. Rev.* **B9**, 1924–5 (1974).

[14] Fordy, A., Wojciechowski, S. and Marshall, I., A family of integrable quartic potentials related to symmetric spaces, *Phys. Lett.* **113A**, 395–400 (1986).

[15] Gardner, C., Greene, J., Kruskal, M. and Miura, R., Method for solving the Korteweg de Vries equation, *Phys. Rev. Lett.* **19**, 1095–7 (1967).

[16] Gibbons, J. and Hermsen, T., A generalization of the Calogero–Moser system, *Physica* **11D**, 337–48 (1984).

[17] Haine, L., The algebraic complete integrability of geodesic flow on SO(N), *Commun. Math. Phys.* **94**, 271–87 (1984).

[18] Iarov-Iarovoi, M. S., Integration of the Hamilton–Jacobi equation by the method of separation of variables, *J. Appl. Math. Mech.* **27**, 1499–1520 (1964).

[19] Inozemtsev, V. I., On the motion of classical integrable systems of interacting particles in an external field, *Phys. Lett.* **98A**, 316–19 (1983).

[20] Jiang, Z. and Wojciechowski, S., Integrable systems of many interacting rigid bodies, *Il Nuovo Cim.* **101B**, 415–27 (1988).

[21] Jiang, Z. and Wojciechowski S., Integrable multiwave interaction systems related to Lie algebras (in preparation).

[22] Kalnins, E. G. and Miller, V. Jr., Intrinsic characterization of variable separation for the partial differential equations of mechanics, in *Modern Development in Analytical Mechanics*, Turin, S. Benenti, M. Francaviglio and A. Lichnerowicz, (eds), Academia delle Scienze di Torino (1982).

[23] Kalnins, E. G. and Miller, Jr. W., Separation of variables on n-dimensional Riemannian manifolds I. The n-sphere S_n and Euclidean n-space $\mathbb{R}_n$, *J. Math. Phys.* **27**, 1721–36 (1986).

[24] Landau, L. D. and Lifshitz E. M., *Mechanics*, Pergamon, London, ch. 7, § 48 (1960).

[25] Levi-Civita, T., Sulla integrazione della equazione de Hamilton–Jacobi per separazioni de variable, *Math. Ann.* **59**, 383–97 (1904).

[26] Manakov, S. V., Complete integrability and stochastization of discrete dynamical systems, *Sov. Phys. JETP* **40** 269–74 (1974).

[27] Manakov, S. V., Remarks on the integration of the Euler equations of the n-dimensional heavy top, *Funct. Anal. and its Appl.* **10**, 93–4 (1976).

[28] Marshal, I. and Wojciechowski, S., When is a natural hamiltonian system separable?, *J. Math. Phys.* **29**, 1338–46 (1988).

[29] Meiss, J. D., Integrability of multiple three-wave interactions, *Phys. Rev.* **A19**, 1780–9 (1979).

[30] Menyuk, C. R., Chen, H. H. and Lee, Y. C., Integrable hamiltonian systems and the Lax pair formalism, *Phys. Rev.* **A26**, 3731–3 (1982).

[31] Mishchenko, A. S. and Fomienko, A. T., Euler equations on finite dimensional Lie groups., *Math. USSR Izv.* **12**, 371–89 (1978).

[32] Moser, J., Three integrable hamiltonian systems connected with isospectral deformations, *Adv. Math.* **16**, 1–23 (1975).

[33] Moser, J., Finitely many mass points on the line under the influence of an exponential potential – an integrable system, *Lecture Notes in Physics* **38**, 97–101 (1976).

[34] Olshanetsky, M. A. and Perelomov. A. M., Explicit solution of the Calogero model in the classical case and geodesic flows of zero curvature, *Lett. Nuovo Cim.* **16**, 333–9 (1976).

[35] Olshanetsky, M. A. and Perelomov, A. M., Completely integrable hamiltonian systems connected with semisimple Lie algebras, *Invent. Math.* **37**, 93 (1976).

[36] Olshanetsky, M. A. and Perelomov, A. M., Explicit solution of classical generalized Toda models, *Inv. Math.* **54**, 261–9 (1979).

[37] Olshanetsky, M. A. and Perelomov, A. M., Classical integrable finite-dimensional systems related to Lie algebras, *Phys. Rep.* **71**, 313–400 (1981).

[38] Patera, J., Sharp, R. T., Winternitz, P. and Zassenhaus, M., Invariants of real low dimensional Lie algebras, *J. Math. Phys.* **17**, 986–94 (1976).

[39] Perelomov, A. M., Ragnisco, O. and Wojciechowski, S., Integrability of two interacting rigid bodies, *Comm. Math. Phys.* 573–83 (1986).

[40] Pöschel, J., Integrability of hamiltonian systems on Cantor sets, *Comm. Pure Appl. Math.* **35**, 653–96 (1982).

[41] Stäckel, P., Ueber die Integration der Hamilton–Jacobi'schen Differentialgleichung mittels Separation der Variablen, Habilitationsschrift, Halle (1891).

[42] Watson, K. M., West, B. J. and Cohen, B. I., Coupling of surface and internal gravity waves: a mode coupling model, *J. Fluid Mech.* **77**, 185–208 (1978).

[43] Weinstein, A., The local structure of Poisson manifolds, *J. Diff. Geom.* **18**, 523–57 (1983).

[44] Whittaker, E. T., *A treatise on the Analytical Dynamics of Particles and Rigid Bodies*, Dover New York, ch. 12 (1944).

[45] Wojciechowski, S., Superintegrability of the Calogero–Moser system, *Phys. Lett.* **95A**, 279–81 (1983).

[46] Wojciechowski, S., Separability of an integrable case of the Henon–Heiles system, *Phys. Lett.* **100A**, 277–8 (1984).

[47] Wojciechowski, S., On the integrability of the Calogero–Moser system in an external quartic potential and other many-body systems, *Phys. Lett.* **102A**, 85–8 (1984).

[48] Wojciechowski, S., Integrability of one particle in a perturbed central quartic potential, *Physica Scripta* **31**, 433–8 (1985).

[49] Wojciechowski, S., An integrable marriage of the Euler equations with the Calogero–Moser system, *Phys. Lett.* **111A**, 101–3 (1985).

[50] Wojciechowski, S., Zuhan Jiang and Bullough, R. K., Integrable multi-wave interaction systems of ODE's, *Phys. Lett.* **117A**, 399–404 (1986).

[51] Wojciechowski, S., Review of recent results on integrability of natural hamiltonian systems, *Proceedings of SMS Systèmes dynamiques non linéaires: intégrabilité et comportement qualitatif*, P. Winternitz (ed.), Presse Université de Montréal, Montreal (1986).

[52] Wojciechowski, S., A new test for integrability of Riccali systems of equations, in *Finite dimensional integrable nonlinear dynamical systems*, P.G.L. Leach and W.H. Steeb (eds), World Scientific, Singapore (1988).

11 *M. Antonowicz and A. P. Fordy*

Hamiltonian structure of nonlinear evolution equations

11.1 **Introduction**

In Chapter 10 Wojciechowski described the notion of complete integrability for *finite-dimensional* Hamiltonian systems (ODEs). In this chapter we give an elementary introduction to the extension of these notions to the *infinite-dimensional* case (PDEs). This is an extremely short introduction to a very large subject, so we have had to be highly selective. The specialist will be alarmed at our omissions but we have no choice if we are to limit ourselves to only a few pages. Most noticeably, we give no formulations of Lie-algebraic theory which, although essential for a deeper understanding of complete integrability [11, 29, 30] can be omitted at this elementary level. Sections 11.2 and 11.3 give a brief introduction to the basic theory, a more complete but equally elementary version being available in textbooks such as [28]. The remaining sections reflect our own research interests, but are still of mainstream interest. We first give a brief synopsis of the contents of the chapter.

Hamiltonian structures

In Section 11.2 we generalise the familiar notions of Hamiltonian particle dynamics to the continuum. We do this through the machinery of differential algebras, introduced below, but first present some of the basic ideas which distinguish the particle and continuum cases. We present the case relevant to equations in (1+1)-dimensions. For these introductory remarks we consider only a single dependent variable.

In this framework, Hamiltonians are functionals:

$$\mathbb{H}[u] = \int \mathscr{H}[u]\mathrm{d}x \qquad (11.1.1)$$

where $\mathscr{H}[u]$ is the corresponding Hamiltonian density. The limits of integration will depend upon boundary conditions. With appropriate boundary conditions (e.g. periodic or rapidly vanishing as $|x| \to \infty$) the integral (11.1.1) vanishes if $\mathcal{H}$ is an exact derivative. Thus, two densities which differ by an exact derivative will be considered equivalent.

The gradient operator on finite-dimensional manifolds is now replaced by the variational derivative:

$$\frac{\mathrm{d}}{\mathrm{d}t}\,\mathbb{H} = \int \mathcal{H}'[u]u_t\mathrm{d}x = \int \frac{\delta\mathcal{H}}{\delta u}\,u_t\mathrm{d}x \tag{11.1.2}$$

where the Fréchet derivative $\mathcal{H}'[u]$ is defined by $\mathrm{d}/\mathrm{d}\varepsilon\mathcal{H}[u + \varepsilon v]|_{\varepsilon=0} = \mathcal{H}'[u]v$. $\mathcal{H}'[u]$ is a differential operator acting on u_t whereas the variational derivative $\delta\mathcal{H}/\delta u \equiv \delta_u\mathcal{H}$ is just a function, defined by:

$$\frac{\delta\mathcal{H}}{\delta u} = \frac{\partial\mathcal{H}}{\partial u} - \partial\left(\frac{\partial\mathcal{H}}{\partial u_x}\right) + \ldots, \quad \partial \equiv \frac{\partial}{\partial x} \tag{11.1.3}$$

which multiplies u_t. We will write δ in place of δ_u wherever there is no ambiguity. The second step relies upon integration by parts so requires the above mentioned appropriate boundary conditions to be valid.

When passing from finite to infinite dimensions the Poisson bracket formulation is the most appropriate one to generalise. The Poisson matrix, which appears in Hamilton's equations, is replaced by a skew adjoint differential operator $\mathbf{B}$:

$$u_t = \mathbf{B}\delta_u\mathcal{H} \tag{11.1.4}$$

and this is used to define a Poisson bracket. If $\mathbb{K}[u]$ is another functional defined on our phase space:

$$\mathbb{K}[u] = \int \mathcal{K}[u]\mathrm{d}x \tag{11.1.5}$$

then the time derivative of $\mathbb{K}[u]$:

$$\frac{\mathrm{d}}{\mathrm{d}t}\,\mathbb{K} = \int \frac{\delta\mathcal{K}}{\delta u}\,u_t\mathrm{d}x = \int \frac{\delta\mathcal{K}}{\delta u}\,\mathbf{B}\,\frac{\delta\mathcal{H}}{\delta u}\,\mathrm{d}x \tag{11.1.6a}$$

gives us a candidate for the Poisson bracket:

$$\{\mathbb{K}, \mathbb{H}\} = \int \frac{\delta\mathcal{K}}{\delta u}\,\mathbf{B}\,\frac{\delta\mathcal{H}}{\delta u}\,\mathrm{d}x \tag{11.1.6b}$$

Skew symmetry follows from the same property of $\mathbf{B}$, but the Jacobi identities are not satisfied except for very special examples of operator $\mathbf{B}$. This imposes fairly technical conditions upon $\mathbf{B}$ which are discussed later, so we just present the three most common examples of Hamiltonian operator here:

$$\begin{aligned} &\text{(i)} && \mathbf{B} = \partial, \\ &\text{(ii)} && \mathbf{B} = u\partial + \partial u, \\ &\text{(iii)} && \mathbf{B} = \partial^3 + 2(u\partial + \partial u). \end{aligned} \tag{11.1.7}$$

We can think of a Hamiltonian operator as a map from function(al)s to vector fields (flows): $\mathbf{B}: \mathcal{H} \to \mathbf{B}\delta\mathcal{H}$. It is for this reason that we insist that $\mathbf{B}$ is purely differential, thus guaranteeing that the vector field $\mathbf{B}\delta\mathcal{H}$ is locally defined for *arbitrary* $\mathcal{H}$. Some people drop this condition and allow integro-differential Hamiltonian operators.

It can be seen from (11.1.6) that the functional $\mathbb{K}[u]$ is a constant of motion if and only if $\mathbb{K}$ and $\mathbb{H}$ Poisson-commute. Thus, using these notions it is possible to extend much of Hamiltonian mechanics to the infinite-dimensional case.

Bi-Hamiltonian systems

A particularly interesting and important class of Hamiltonian systems is the class of those which are bi-Hamiltonian (discussed in Section 11.3). In this case we can write our equation in terms of two different Hamiltonian structures, $\mathbf{B}_0$ and $\mathbf{B}_1$, and two different Hamiltonians $\mathcal{H}$, $\mathcal{G}$ such that:

$$u_t = \mathbf{B}_0\delta\mathcal{G} = \mathbf{B}_1\delta\mathcal{H}. \tag{11.1.8}$$

In particular, if $\mathbf{B}_0$ and $\mathbf{B}_1$ satisfy a compatibility condition (given in Appendix A), equivalent to asking for $\mathbf{B}_0 + \mathbf{B}_1$ to also be Hamiltonian, then (11.1.8) belongs to an infinite hierarchy of bi-Hamiltonian systems defined recursively by:

$$u_{t_n} = \mathbf{B}_0\delta\mathcal{H}_{n+1} = \mathbf{B}_1\delta\mathcal{H}_n. \tag{11.1.9}$$

It is assumed here that the condition $\mathbf{B}_1\delta\mathcal{H}_n \in$ *Range* $\mathbf{B}_0$ is satisfied for any $\mathcal{H}_n$. The hierarchy starts with $\mathcal{H}_0 \in \mathrm{Ker}\mathbf{B}_0\delta$, that is, $\mathbf{B}_0\delta\mathcal{H}_0 = 0$, $\mathcal{H}_{-1} = 0$. Formula (11.1.9) suggests a way of climbing the hierarchy of *equations* without having to calculate variational and inverse variational derivatives. Namely, the (integro-differential) *recursion operator* (formally inverting $\mathbf{B}_0$):

$$\mathbf{R} = \mathbf{B}_1\mathbf{B}_0^{-1} \tag{11.1.10}$$

maps flows into flows as shown in Figure 11.1. Thus, we can generate an infinite hierarchy of flows from any given member, purely by use of the recursion operator $\mathbf{R}$, without using the Hamiltonian formalism. If $\mathbf{R}$ satisfies the hereditary property (see [18]) then these flows mutually commute. In fact, recursion operators exist in a more general setting (see [18, 27, 28]) and need not be related to Hamiltonian structures.

Hamiltonian operators from Lax equations

Any flow in a hierarchy generated by a recursion operator (whether or not derived from a Hamiltonian pair) has a Lax representation defined in terms of the recursion operator and the Fréchet derivative of the right-

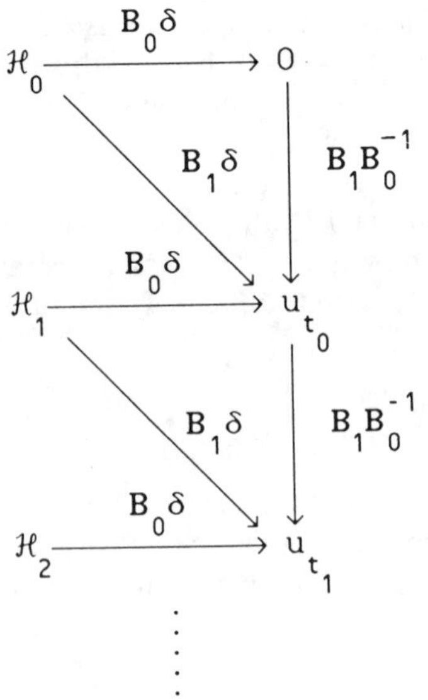

Figure 11.1

hand side of the flow (see equation (11.3.6)). On the other hand, given any spectral problem it is possible to construct (using the adjoint representation of the corresponding Lie algebra [17]) an associated recursion operator. However, it is not so easy to factorise this into the product (11.1.10) (assuming this to be possible), so as to derive the associated Hamiltonian structure(s). This would be particularly difficult for those systems with only *one* locally defined Hamiltonian structure.

In Section 11.4 we consider the problem of deriving the Hamiltonian structure of the isospectral flows of a given Lax operator. The results we present are, in fact, much stronger than this. We use an unusual form of the Lax approach [3, 4], which enables us to obtain, in *one construction*, the time evolutions of the eigenfunctions, the associated Hamiltonian operators and the corresponding Hamiltonians.

Miura maps

The remarkable explicit nonlinear transformation of Miura [25] was the first example of what is now called a Miura map. These are generally of the form:

$$u = M[v] \tag{11.1.11}$$

where M is a function of v and its x-derivatives. Such maps are sometimes referred to as many-to-one Bäcklund transformations, since the target function u is isolated. The variable v is called the 'modified variable' by analogy with the KdV–MKdV example. For us, the important feature of such maps is that they are Hamiltonian: if v satisfies an equation of Hamiltonian type:

$$v_t = \widetilde{\mathbf{B}}\delta_v \widetilde{\mathscr{H}} \tag{11.1.12a}$$

then so does u:

$$u_t = \mathbf{B}\delta_u \mathscr{H} \tag{11.1.12b}$$

where $\mathbf{B} = M'\widetilde{\mathbf{B}}(M')^\dagger|_{u=\mathrm{M}[v]}$ and $\widetilde{\mathscr{H}}[v] = \mathscr{H} \circ M[v]$, with M' the Fréchet derivative of M. Two pieces of magic occur here that distinguish Miura maps from arbitrary ones. It should be noted that since M contains derivatives of v (being not just a point transformation) the map cannot be inverted without solving a differential equation. Thus it would not be generally possible to write $\mathbf{B}$ and $\mathscr{H}$ *locally* in terms of u and its derivatives. However, maps with such properties do exist in abundance [4, 11, 15, 17, 23]. We refer to M as Hamiltonian rather than canonical since $\mathbf{B}$ and $\widetilde{\mathbf{B}}$ are not usually canonical structures.

An important aspect here is that a complicated Hamiltonian operator (such as (11.1.7iii)) is often the image, under a Miura map, of a much simpler operator (in this case (11.1.7i)). Thus, in infinite dimensions, Miura maps (and invertible transformation) can be used to simplify the structure of a given Hamiltonian operator (a role similiar to Darboux's theorem in finite dimensions).

We discuss these in Section 11.5.

Stationary flows

Section 11.6 discusses an explicit connection between the Hamiltonian framework of this chapter and that of Chapter 10. Starting with a PDE in (1+1)-dimensions there are several ways of generating an associated ODE (similarity solutions, for example). We consider solutions which are independent of a time variable. Such stationary solutions satisfy a system of ODEs which, often, can be put into Hamiltonian form (in the classical sense). Starting with an integrable PDE the resulting Hamiltonian system can often be shown to be completely integrable [7]. It is particulary interesting when looking at the stationary solutions of Miura-related PDEs. In this case a Miura map between the corresponding finite-dimensional phase spaces is induced [6].

11.2 **Hamiltonian structures**

We now present the elements of infinite-dimensional Hamiltonian theory in a more general context. We start with some basic definitions.

Phase space

Let $\mathbf{u} = (u_0, \ldots, u_{N-1})$ be a collection of smooth, real-valued functions of x. Let successive x-derivatives of u_j be denoted $u_{j,i} = \partial^i u_j / \partial x^i$. The infinitely many coordinates $(u_{j,i})$ locally define our phase space U.

We are interested in functions on this phase space. Specifically, consider smooth functions $K[\mathbf{u}]$ of a finite number of the above coordinates (i.e. differential functions of $\mathbf{u}$). Under pointwise multiplication, such functions form a commutative algebra, which we denote by A_u. With the addition of the total derivative operator ∂, A_u becomes a differential algebra:

$$\partial\colon A_u \to A_u,\ \partial K = \sum_{i,j} \frac{\partial K}{\partial u_{j,i}} u_{j,i+1}. \tag{11.2.1}$$

In what follows we are mainly interested in functionals (11.1.1) which are conveniently described (in an algebraic set-up) as densities, being equivalence classes of differential functions with respect to the total derivative. We denote the quotient space $A_u/\partial A_u$ by $\mathscr{A}_u$ and its typical element $K + \partial A_u$ by $\mathscr{K}$.

Variational derivative

This is defined purely algebraically. Let the Cartesian product of N copies of A_u be denoted by $A_u^N = A_u \times \ldots \times A_u$. With

$$\delta_j = \sum_{i=0}^{\infty} (-\partial)^i \frac{\partial}{\partial u_{j,i}} \tag{11.2.2}$$

we define $\delta_u = (\delta_0, \ldots, \delta_{N-1})^{\mathrm{T}}$. This variational derivative defines a mapping $\delta_u\colon A_u \to A_u^N$. Since elements of A_u are only functions of a finite number of coordinates, no problems of convergence arise. When there is no ambiguity we will use the simple notation δ in place of δ_u. It is only when discussing Miura maps that we need to be more careful.

It is easy to see that $\delta\partial = 0$. In fact, it is possible to prove (see [28], for example) that:

$$\operatorname{Ker} \delta = \operatorname{Im} \partial. \tag{11.2.3}$$

This enables us to define the quotient space $\mathscr{A}_u$ purely algebraically, without any reference to boundary conditions.

Differential operators

We denote by $M_N(A_u)$ the algebra of $N \times N$ matrices with entries in A_u. Let $A_u^N[\partial]$ denote the algebra of differential operators on A_u^N with coefficients in $M_N(A_u)$. Each element $\mathbf{B} \in A_u^N[\partial]$ can be uniquely represented as

$$\mathbf{B} = \sum_0^m \mathbf{b}_r \partial^r \tag{11.2.4a}$$

where $\mathbf{b}_r \in M_N(A_u)$. The adjoint of $\mathbf{B}$ is defined by:

$$\mathbf{B}^\dagger = \sum_0^m (-\partial)^r \mathbf{b}_r^{\mathrm{T}}, \tag{11.2.4b}$$

and satisfies the usual properties:

$$\partial^\dagger = -\partial, \quad \mathbf{b}^\dagger = \mathbf{b}^{\mathrm{T}}, \quad (\mathbf{B}_1\mathbf{B}_2)^\dagger = \mathbf{B}_2^\dagger \mathbf{B}_1^\dagger \tag{11.2.4c}$$

for $\mathbf{b} \in M_N(A_u)$, $\mathbf{B}_i \in A_u^N[\partial]$.

Hamiltonian operators and Poisson brackets

Let us define the bracket $\{,\}: A_u \times A_u \to A_u$ of two differential functions $H, G \in A_u$ with respect to the differential operator $\mathbf{B} \in A_u^N[\partial]$ as:

$$\{H, G\} = (\delta H)^{\mathrm{T}} \mathbf{B}\, \delta G. \tag{11.2.5}$$

It is clear that (11.2.5) induces a bracket on the space of densities $\{,\}$: $\mathscr{A}_u \times \mathscr{A}_u \to \mathscr{A}_u$. For special operators $\mathbf{B}$, which are called Hamiltonian, this is a Poisson bracket on the phase space.

Definition
$\{,\}: \mathscr{A}_u \times \mathscr{A}_u \to \mathscr{A}_u$ defines a Poisson bracket if:

$$\text{(i)} \quad \{\mathscr{G}, \mathscr{H}\} + \{\mathscr{H}, \mathscr{G}\} \cong 0, \tag{11.2.6a}$$

$$\text{(ii)} \quad \{\{\mathscr{G}, \mathscr{H}\}, \mathscr{K}\} + \{\{\mathscr{H}, \mathscr{K}\}, \mathscr{G}\} + \{\{\mathscr{K}, \mathscr{G}\}, \mathscr{H}\} \cong 0. \tag{11.2.6b}$$

Here we use the symbol '$\cong$' to indicate that the equality holds in $\mathscr{A}$, and therefore mod Im∂ only.

Remark
In terms of functions $G, H, K \in A_u$ the right-hand side of (11.2.6a, b) would be exact derivatives.

Definition
An operator $\mathbf{B} \in A_u^N[\partial]$ is called Hamiltonian if the bracket on the space of densities $\mathscr{A}_u$, induced by (11.2.5) satisfies the conditions (11.2.6).

Remark
The first of these conditions is very easy to characterise, since it follows from the definition (11.2.5) that $\{,\}$ satisfies (11.2.6a) if $\mathbf{B}$ is skew-adjoint. However, (11.2.6b) is more difficult to check but, fortunately, does not need to be checked directly. The Jacobi identity is conveniently characte-

rised in terms of the Schouten bracket [19]. We relegate the technical details to Appendix A.

Remark
Any constant coefficient, skew-symmetric operator is Hamiltonian.

Remark
The above construction of a Poisson bracket is in direct analogy with the finite-dimensional case, replacing δ_u by ∇_u and $\mathbf{B}$ by J we have $\{\mathscr{h}, \mathscr{f}\} = (\nabla \mathscr{h})^{\mathrm{T}} J \nabla \mathscr{f}$.

Hamiltonian flows

A Hamiltonian operator $\mathbf{B}$ serves to map differential functions into vector fields:

$$\mathbf{B}\colon H \to \mathbf{B}\delta H. \tag{11.2.7a}$$

The corresponding Hamiltonian flow depends on the equivalence class $\mathscr{H}$ of H only and defines an evolutionary derivation ∂_H (meaning that $[\partial_H, \partial] = 0$) on A_u:

$$\partial_H u_j = (\mathbf{B}\delta H)_j \tag{11.2.7b}$$

where $(\mathbf{B}\delta H)_j$ denotes the jth component. This is extended to any $P \in A_u$ by the chain rule:

$$\partial_H P[\mathbf{u}] = \sum_{i,j} \frac{\partial P}{\partial u_{j,i}} \partial^i(\partial_H u^j). \tag{11.2.8a}$$

Integrating by parts it is easy to see that:

$$\partial_H P \cong \{P, H\}, \tag{11.2.8b}$$

in direct analogy with the finite-dimensional case. As a consequence of the Jacobi identities (11.2.6b) we have the usual homomorphism between the Poisson bracket of densities and the commutator of their corresponding vector fields (evolutionary derivations):

$$\partial_{\{P,Q\}} = -[\partial_P, \partial_Q]. \tag{11.2.9}$$

Definition
A system of evolution equations;

$$\mathbf{u}_t = \mathbf{K}[\mathbf{u}] \tag{11.2.10a}$$

is said to be Hamiltonian if there exists a Hamiltonian operator $\mathbf{B}$ and Hamiltonian density $\mathscr{H}$, such that $\mathbf{K}[\mathbf{u}] = \mathbf{B}\delta\mathscr{H}$:

$$\mathbf{u}_t = \mathbf{B}\delta\mathscr{H}. \tag{11.2.10b}$$

Example
The KdV (Korteweg–de Vries) equation

$$u_t = u_{xxx} + 6uu_x \tag{11.2.11a}$$

is Hamiltonian with $\mathbf{B} = \partial$, $\mathscr{H} = u^3 - \frac{1}{2}u_x^2$, whereas the Burgers equation

$$u_t = u_{xx} + 2uu_x \tag{11.2.11b}$$

is not.

Conservation laws

Given any Hamiltonian equation (11.2.10) we can consider the time evolution of any element $\mathscr{P} \in \mathscr{A}_u$ and use (11.2.8) to write this in terms of the Poisson bracket:

$$\mathscr{P}_t \cong \{\mathscr{P}, \mathscr{H}\}. \tag{11.2.12}$$

Definition
$\mathscr{P} \in \mathscr{A}_u$ is a conserved density of (11.2.10) if $\mathscr{P}_t \cong 0$, or equivalently $\{\mathscr{P}, \mathscr{H}\} \cong 0$.

If $P \in A_u$ is a representative of a conserved density $\mathscr{P}$ then $P_t \in \operatorname{Im} \partial$, so that

$$P_t = Q_x \tag{11.2.13}$$

for some $Q \in A_u$.

Equation (11.2.13) is called a local conservation law which, when integrated over a suitable domain (depending upon the boundary conditions), gives rise to an integral of the motion:

$$\mathbb{P}[\mathbf{u}] = \int \mathscr{P}[\mathbf{u}] \, \mathrm{d}x, \tag{11.2.14}$$

satisfying $\mathbb{P}_t = 0$. We then say that $\mathbb{P}[\mathbf{u}]$ Poisson-commutes with $\mathbb{H}$:

$$\{\mathbb{P}, \mathbb{H}\} = \int \{\mathscr{P}, \mathscr{H}\} \, \mathrm{d}x = 0. \tag{11.2.15}$$

Liouville integrability

Liouville's theorem for finite-dimensional Hamiltonian systems is discussed by Wojciechowski in Chapter 10. For N degrees of freedom we only require N (functionally independent) integrals of motion in involution to prove complete integrability. In infinite dimensions the situation is slightly more complicated. The existence of infinitely many conserved densities in involution, though necessary, is not thought to be sufficient for the complete integrability. We need the additional requirement of the completeness of this infinite set of conserved densities. In some cases (such as the KdV equation) we can use a spectral representation of the dynamical variables to introduce the so-called action-angle variables, which seems to be the closest possible imitation of the Liouville–Arnold description of finite-dimensional completely integrable systems [13].

11.3 Bi-Hamiltonian systems

This section is concerned with a particularly interesting and important class of Hamiltonian systems: those equations which can be represented as Hamiltonian flows in two distinct ways. Under certain circumstances such equations can be embedded into an infinite family of bi-Hamiltonian systems and the whole family shown to be completely integrable.

Let $\mathbf{B}_0$ and $\mathbf{B}_1$ be Hamiltonian operators and $\{,\}_0$, $\{,\}_1$ the corresponding Poisson brackets. An evolution equation is bi-Hamiltonian if we can write:

$$\mathbf{u}_{t_0} = \mathbf{K}_0[\mathbf{u}] = \mathbf{B}_0\delta\mathcal{H}_1 = \mathbf{B}_1\delta\mathcal{H}_0 \tag{11.3.1a}$$

for two Hamiltonian operators and two densities $\mathcal{H}_0$, $\mathcal{H}_1$. Under such circumstances we can write down at least two related flows:

$$\mathbf{u}_{t_{-1}} = \mathbf{B}_0\delta\mathcal{H}_0, \quad \mathbf{u}_{t_1} = \mathbf{B}_1\delta\mathcal{H}_1. \tag{11.3.1b}$$

Using such arguments as:

$$\{\mathcal{H}_0, \mathcal{H}_1\}_0 = (\delta\mathcal{H}_0)^{\mathrm{T}}\mathbf{B}_0\delta\mathcal{H}_1 = (\delta\mathcal{H}_0)^{\mathrm{T}}\mathbf{B}_1\delta\mathcal{H}_0 = \{\mathcal{H}_0, \mathcal{H}_0\}_1 \cong 0$$

it is easy to see that $\mathcal{H}_0$ and $\mathcal{H}_1$ commute with respect to both Poisson brackets so that, from (11.2.9), both flows (11.3.1b) commute with (11.3.1a). In general the flows (11.3.1b) will not themselves be bi-Hamiltonian and will not mutually commute. To achieve these properties it is necessary to impose some additional conditions on $\mathbf{B}_0$, $\mathbf{B}_1$.

Definition
Hamiltonian operators $\mathbf{B}_0$ and $\mathbf{B}_1$ are said to be compatible if $\lambda\mathbf{B}_0 + \mu\mathbf{B}_1$ is also Hamiltonian for any numbers λ, μ. In this case they are called a Hamiltonian pair.

Remark
This combination is automatically skew-symmetric but would generally fail to satisfy the Jacobi identities. It is enough to show that $\mathbf{B}_0 + \mathbf{B}_1$ is Hamiltonian.

To directly check the Jacobi identities is rather messy, but there are several equivalent simpler ways of testing a pair of Hamiltonian structures for compatibility. One of these [28] is given in Appendix A.

Definition
A differential operator $\mathscr{D}: A_u^N \to A_u^N$ is degenerate if there is a non-zero differential operator $\widetilde{\mathscr{D}}: A_u^N \to A_u^N$ such that $\widetilde{\mathscr{D}} \cdot \mathscr{D} = 0$.

Remark
Scalar differential operators are always non-degenerate so we only need to check this condition in the case of systems of differential equations. This will be necessary when constructing the recursion operator (11.1.10).

We can now state the basic theorem, largely attributable to Magri [24], relating bi-Hamiltonian structures to complete integrability. We present the formulation given in [28].

Theorem
Let

$$\mathbf{u}_{t_0} = \mathbf{K}_0[\mathbf{u}] = \mathbf{B}_0\delta\mathscr{H}_1 = \mathbf{B}_1\delta\mathscr{H}_0 \qquad (11.3.2a)$$

be a bi-Hamiltonian system of evolution equations. Assume that $\mathbf{B}_0$, $\mathbf{B}_1$ are compatible and the operator $\mathbf{B}_0$ is non-degenerate. Let $\mathbf{R} = \mathbf{B}_1\mathbf{B}_0^{-1}$ be the corresponding recursion operator. Assume that for each $n = 0, 1, 2, \ldots$ we can recursively define:

$$\mathbf{K}_{n+1} = \mathbf{R}\mathbf{K}_n, \qquad (11.3.2b)$$

thus assuming that for each n, $\mathbf{K}_n \in \operatorname{Im} \mathbf{B}_0$. Then, there exists a sequence of Hamiltonian densities $\mathscr{H}_0, \mathscr{H}_1, \mathscr{H}_2, \ldots$ such that:

(i) for each $n \geq 0$, the evolution equation

$$\mathbf{u}_{t_n} = \mathbf{K}_n[\mathbf{u}] = \mathbf{B}_0\delta\mathscr{H}_{n+1} = \mathbf{B}_1\delta\mathscr{H}_n \qquad (11.3.2c)$$

is a bi-Hamiltonian system;

(ii) these Hamiltonian densities are all in involution with respect to either Poisson bracket:

$$\{\mathscr{H}_n, \mathscr{H}_m\}_0 \cong \{\mathscr{H}_n, \mathscr{H}_m\}_1 \cong 0, \quad n, m \geq 0, \qquad (11.3.2d)$$

and hence provide an infinite collection of conservation laws for each of the above bi-Hamiltonian systems;

(iii) the corresponding evolutionary derivations commute:

$$[\partial_{\mathscr{H}_n}, \partial_{\mathscr{H}_m}] = 0. \qquad (11.3.2e)$$

Proof
We only outline part of the proof, referring to [28] for details. Proof of (i) relies on the following lemma.

Lemma
If $\mathbf{B}_0$, $\mathbf{B}_1$ are compatible Hamiltonian operators with $\mathbf{B}_0$ non-degenerate and

$$\mathbf{B}_1\delta\mathscr{H}_{n-1} = \mathbf{B}_0\delta\mathscr{H}_n, \quad \mathbf{B}_1\delta\mathscr{H}_n = \mathbf{B}_0\mathbf{P}_{n+1} \tag{11.3.3a}$$

then

$$\mathbf{P}_{n+1} = \delta\mathscr{H}_{n+1} \tag{11.3.3b}$$

for some density $\mathscr{H}_{n+1} \in \mathscr{A}$.

Starting with $n = 1$ we have an inductive proof of (i).
Proof of (ii). From the above construction it follows that $\mathbf{B}_0\delta\mathscr{H}_n = \mathbf{B}_1\delta\mathscr{H}_{n-1}$, so that:

$$\{\mathscr{H}_m, \mathscr{H}_n\}_0 = \{\mathscr{H}_m, \mathscr{H}_{n-1}\}_1. \tag{11.3.4a}$$

Combining this with the skew symmetry of a Poisson bracket we get

$$\{\mathscr{H}_m, \mathscr{H}_n\}_0 \cong \{\mathscr{H}_{m+1}, \mathscr{H}_{n-1}\}_0. \tag{11.3.4b}$$

If $n - k = m + k$, applying this k times gives:

$$\{\mathscr{H}_m, \mathscr{H}_n\}_0 \cong \{\mathscr{H}_{m+k}, \mathscr{H}_{n-k}\}_0 \cong 0. \tag{11.3.4c}$$

If $n - 1 - k = m + k$ we obtain analogously

$$\{\mathscr{H}_m, \mathscr{H}_n\}_0 \cong \{\mathscr{H}_m, \mathscr{H}_{n-1}\}_1 \cong \{\mathscr{H}_{m+k}, \mathscr{H}_{n-1-k}\}_1 \cong 0. \tag{11.3.4d}$$

(iii) follows from (ii) by virtue of (11.2.9).

Remark
When constructing a bi-Hamiltonian hierarchy there is a basic assumption made: that $\mathbf{B}_1\delta\mathscr{H}_n$ is in the range of $\mathbf{B}_0$. It is possible to prove this for simple examples, such as the KdV equation [28] or multi-Hamiltonian coupled KdV systems [4] discussed in the next section, but not in general. In fact as Kupershmidt's counter example [22] indicates it need not be the case.

Remark
A recursion operator $\mathbf{R} = \mathbf{B}_1\mathbf{B}_0^{-1}$ obtained from a compatible pair of Hamiltonian operators $\mathbf{B}_0$, $\mathbf{B}_1$ possesses the properly known as hereditarity [18] or regularity [19]. This property implies that the (usually integro-differential) operator $\mathbf{B}_n = \mathbf{R}^n\mathbf{B}_0$ is automatically Hamiltonian and compatible with $\mathbf{B}_0$. In particular, if $\mathbf{R} = \mathbf{B}_1\mathbf{B}_0^{-1}$ with $\mathbf{B}_0$, $\mathbf{B}_1$ both Hamiltonian and $\mathbf{R}$ is hereditary then $\mathbf{B}_0$ and $\mathbf{B}_1$ are compatible.

Remark
In this chapter our main interest in $\mathbf{R}$ is in its role within Hamiltonian theory. However, recursion operators are more general than this and sometimes cannot be written as a quotient $\mathbf{B}_1\mathbf{B}_0^{-1}$, but *always* satisfy a relation such as (11.3.2b).

Linear spectral problem

In the next section we show how to derive the Hamiltonian structure(s) associated with the isospectral flows of a given spectral problem (of Lax type). However, to end the present section we recall that any bi-Hamiltonian system can be written as the compatibility conditions of a Lax pair:

$$\mathbf{R}_t^{\dagger}\Phi = \lambda^2\Phi \quad \text{or} \quad \mathbf{B}_1\Phi = \lambda^2\mathbf{B}_0\Phi. \tag{11.3.5}$$

The recursion operator satisfies the following Lax-type equations [10, 27, 28], for each specific flow $\mathbf{u}_t = \mathbf{K}[\mathbf{u}]$ in the hierarchy:

$$\mathbf{R}_t^{\dagger} = [-(\mathbf{K}')^{\dagger}, \mathbf{R}^{\dagger}] \tag{11.3.6}$$

where $\mathbf{K}'$ is the Fréchet derivative of $\mathbf{K}[\mathbf{u}]$.

Expanding Φ asymptotically in λ (or certain projective coordinates derived from the components of Φ) gives an alternative method of generating the infinite sequence of conserved densities.

Remark

In terms of the associated Lie algebra a linear spectral problem takes the simplest form when expressed in terms of the fundamental representation. The recursion operator spectral problem (11.3.5) corresponds to the adjoint (or squared eigenfunction) representation and is rather difficult to handle in its original form.

Example – KdV

$$\mathbf{B}_0 = \partial, \quad \mathbf{B}_1 = \partial^3 + 2(u\partial + \partial u),$$
$$\mathcal{H}_0 = \tfrac{1}{2}u, \quad \mathcal{H}_1 = \tfrac{1}{2}u^2, \quad \mathcal{H}_2 = -\tfrac{1}{2}u_x^2 + u^3. \tag{11.3.7a}$$

The hierarchy $u_{t_{2n-1}} = \mathbf{B}_0\delta\mathcal{H}_n = \mathbf{B}_1\delta\mathcal{H}_{n-1}$ starts with

$$u_{t_1} = u_x, \quad u_{t_3} = (u_{xx} + 3u^2)_x,$$
$$u_{t_5} = (u_{4x} + 10uu_{xx} + 5u_x^2 + 10u^3)_x. \tag{11.3.7b}$$

In this case it is easy to prove that $\mathbf{B}_1\delta\mathcal{H}_n$ is in the range of $\mathbf{B}_0$ [4, 28].

11.4 Hamiltonian operators from Lax equations

In this section we pose the problem of deriving the Hamiltonian structure of the isospectral flows of a given linear problem. We use an unusual form of the Lax approach which enables us to obtain, in *one construction*, the time evolutions of the eigenfunctions, the associated Hamiltonian operators and the corresponding Hamiltonians.

The method we give for constructing the time evolutions is applicable to a much broader class of spectral problems than the usual Gel'fand–

Dikii approach. For instance, the latter cannot easily be applied to energy-dependent operators. However, even in the most standard cases, such as the KdV hierarchy, our approach gives a very simple and elegant construction of the associated Hamiltonian operators. Applied to the usual third-order Lax operator our method immediately gives the two Hamiltonian structures for the Boussinesq hierarchy. The method can also be applied to Kupershmidt's non-standard Lax operators [21], but when written in purely differential form, and to super-Lax equations [5]. Many of these systems can be extended to multi-component versions, associated with energy-dependent extensions of the corresponding spectral problem, as shown in [4], [5] and below.

Let us consider a fairly general second-order scalar spectral problem:

$$\mathbb{L}\psi \equiv (\varepsilon\partial^2 + u)\psi = 0 \tag{11.4.1a}$$

where ε and u depend polynomially upon the spectral parameter λ:

$$\varepsilon = \sum_0^N \varepsilon_i\lambda^i, \quad u = \sum_0^N u_i\lambda^i, \tag{11.4.1b}$$

with ε_i being constant and u_i functions of x.

We look for time evolutions of the wave function ψ of the form:

$$\psi_t = \mathbb{P}\psi \equiv \tfrac{1}{2}(P\partial + Q)\psi \tag{11.4.1c}$$

where P and Q are functions of u_i and the spectral parameter λ. A simple calculation leads to

$$\mathbb{L}_t - [\mathbb{P}, \mathbb{L}] = u_t + \tfrac{1}{2}\varepsilon Q_{xx} - \tfrac{1}{2}Pu_x + \tfrac{1}{2}\varepsilon(P_{xx} + 2Q_x)\partial + \varepsilon P_x\partial^2. \tag{11.4.2a}$$

Evidently, we cannot expect the usual Lax equation to hold. However, the integrability conditions of (11.4.1a, c) imply that $(\mathbb{L}_t - [\mathbb{P}, \mathbb{L}])\psi = 0$ for eigenfunctions of (11.4.1a). To match the coefficient of ∂^2 we must take:

$$\mathbb{L}_t - [\mathbb{P}, \mathbb{L}] = P_x\mathbb{L}. \tag{11.4.2b}$$

This further implies that $P_{xx} + 2Q_x = 0$, so that (11.4.2b) takes the remarkably simple form:

$$u_t = (\tfrac{1}{4}\varepsilon\partial^3 + \tfrac{1}{2}(u\partial + \partial u))P \equiv JP \tag{11.4.2c}$$

Remark

On the phase space defined by just one function u, the operator J, defined by (11.4.2c) is Hamiltonian, being (when $\varepsilon = 1$) just the second Hamiltonian structure of the KdV equation. The operator J is the basic unit out of which all our Hamiltonian operators are built.

Example – KdV: $\varepsilon = 1, \quad u = u_0 - \lambda$

$$u_{0t} = (\tfrac{1}{4}\partial^3 + \tfrac{1}{2}(u_0\partial + \partial u_0) - \lambda\partial)P \equiv (J_0 - \lambda J_1)P. \qquad (11.4.3a)$$

Example – Harry Dym: $\varepsilon = 1, \quad u = u_1\lambda$

$$u_{1t} = (\tfrac{1}{4}\partial^3 + \tfrac{1}{2}\lambda(u_1\partial + \partial u_1))P \equiv (J_0 + \lambda J_1)P. \qquad (11.4.3b)$$

In [4] we developed general results concerning (11.4.2c), including both the generalised KdV and generalised Harry Dym cases. However, to save space, we only present the multi-component KdV case here.

With ε and u defined by (11.4.1b) (but with $\varepsilon_N = 0$, $u_N = -1$) the operator J takes the form:

$$J = \sum_{k=0}^{N} J_k\lambda^k \qquad (11.4.4a)$$

where

$$J_k = \tfrac{1}{4}\varepsilon_k\partial^3 + \tfrac{1}{2}(u_k\partial + \partial u_k). \qquad (11.4.4b)$$

The operator $J_N = -\partial$ is constant coefficient. Equation (11.4.2c) then takes the form

$$\sum_{k=0}^{N-1} \lambda^k u_{kt} = \left(\sum_{k=0}^{N} J_k\lambda^k\right)P. \qquad (11.4.4c)$$

Substituting a polynomial expansion for P:

$$P = \sum_{k=0}^{m} P_{m-k}\lambda^k \qquad (11.4.5a)$$

we find the recursion relations for the coefficients P_{m-k} and the equations of motion for u_i:

$$J_0P_{k-N} + J_1P_{k-N+1} + \ldots + J_NP_k = 0, \quad k = 0, \ldots, m, \qquad (11.4.5b)$$

$$\begin{pmatrix} u_0 \\ \cdot \\ \cdot \\ \cdot \\ u_{N-1} \end{pmatrix}_t = \begin{pmatrix} & & & J_0 \\ & 0 & \cdot^{\cdot^{\cdot}} & \vdots \\ & \cdot^{\cdot^{\cdot}} & & \vdots \\ J_0 & \cdots & \cdots & J_{N-1} \end{pmatrix} \begin{pmatrix} P_{m-N+1} \\ \cdot \\ \cdot \\ \cdot \\ P_m \end{pmatrix}. \qquad (11.4.5c)$$

It is a remarkable fact that the scalar recursion relation (11.4.5b) can be written as an $N \times N$ matrix equation in exactly $(N + 1)$ different ways:

$$\mathbf{B}_n\mathbf{P}^{(k-1)} = \mathbf{B}_{n-1}\mathbf{P}^{(k)}, \quad n = 1, \ldots, N, \qquad (11.4.6a)$$

where $\mathbf{P}^{(k)} = (P_{k-N+1}, \ldots, P_k)^{\mathrm{T}}$ and the matrix differential operators $\mathbf{B}_n$ are determined by the following requirement: $\mathbf{B}_n$ is skew-adjoint and the

nth row of each matrix equation (11.4.6a) is just (11.4.5b), the remaining ones being identities. Explicitly, $\mathbf{B}_n$ are:

$$\mathbf{B}_n = \left(\begin{array}{ccc|ccc} 0 & & J_0 & & & \\ & \cdot^{\cdot^{\cdot}} & \vdots & & 0 & \\ J_0 & \cdots & J_{n-1} & & & \\ \hline & & & -J_{n+1} & \cdots & -J_N \\ & 0 & & \vdots & \cdot^{\cdot^{\cdot}} & \vdots \\ & & & -J_N & \cdots & 0 \end{array}\right) \tag{11.4.6b}$$

and satisfy the formal relation $\mathbf{B}_n = \mathbf{R}\mathbf{B}_{n-1}$ where:

$$\mathbf{R} = \mathbf{B}_1\mathbf{B}_0^{-1} = \left(\begin{array}{ccc|c} 0 & \cdots & 0 & -J_0J_N^{-1} \\ \hline 1 & & & -J_1J_N^{-1} \\ \vdots & \ddots & 0 & \vdots \\ 0 & \cdots & 1 & -J_{N-1}J_N^{-1} \end{array}\right) \tag{11.4.7}$$

One can prove three basic facts [4]:

(i) The operators $\mathbf{B}_n$ are each Hamiltonian and, furthermore, are mutually compatible.
(ii) For each $m \geq 0$, the recursion relation (11.4.5b) can be successively solved for all polynomial expansions (11.4.5a).
(iii) The vectors $\mathbf{P}^{(k)}$ given by (11.4.6a) are variational derivatives of a sequence of function(al)s $\mathcal{H}_n$ (the Hamiltonians).

Then it follows from (11.4.6a) that the equations of motion (11.4.5c) can be written in Hamiltonian form in $(N + 1)$ distinct ways:

$$\mathbf{u}_{t_m} = \mathbf{B}_N\delta\mathcal{H}_m = \ldots = \mathbf{B}_0\delta\mathcal{H}_{m+N}. \tag{11.4.8a}$$

We summarise these results as follows.

Theorem
There exists an infinite sequence of isospectral flows of (11.4.1a) which can be represented as the integrability conditions (11.4.2) of (11.4.1a) and (11.4.1c) with P given by (11.4.5a). These equations are Hamiltonian with respect to the $(N + 1)$ mutually compatible Hamiltonian operators (11.4.6b):

$$\mathbf{u}_{t_m} = \mathbf{B}_{N-l}\,\delta\mathcal{H}_{m+l}, \quad l = 0, \ldots, N, \quad m = 0, 1, \ldots \tag{11.4.8b}$$

All the flows (11.4.8) commute.

Inverse hierarchy

In [2] we discussed the inverse hierarchy which exists whenever $\varepsilon_0 = 0$. This corresponds to having P polynomial in λ^{-1} rather than in λ. The inverse hierarchy is again $(N + 1)$-Hamiltonian with the same Hamiltonian structures (11.4.6b) and the Hamiltonians $\mathcal{H}_{-1}, \mathcal{H}_{-2}, \ldots$, generated by the inverse of the multi-Hamiltonian ladder (11.4.6a).

We illustrate the above construction with the simplest non-trivial example, $N = 2$.

Example – DWW(S)
Taking $\varepsilon_0 = 1$, $\varepsilon_1 = 0$ we obtain a tri-Hamiltonian dispersive water waves (DWW) hierarchy [21] written in rather unusual coordinates. We will refer to it as DWW(S) so as to distinguish it from its more familiar form (DWW(K)) due to Kaup and Kupershmidt (see below). The Hamiltonian operators are given by

$$\mathbf{B}_0 = \begin{pmatrix} -\frac{1}{2}u_1\partial - \frac{1}{2}\partial u_1 & \partial \\ \partial & 0 \end{pmatrix}, \quad \mathbf{B}_1 = \begin{pmatrix} \frac{1}{4}\partial^3 + \frac{1}{2}u_0\partial + \frac{1}{2}\partial u_0 & 0 \\ 0 & \partial \end{pmatrix},$$

$$\mathbf{B}_2 = \begin{pmatrix} 0 & \frac{1}{4}\partial^3 + \frac{1}{2}u_0\partial + \frac{1}{2}\partial u_0 \\ \frac{1}{4}\partial^3 + \frac{1}{2}u_0\partial + \frac{1}{2}\partial u_0 & \frac{1}{2}u_1\partial + \frac{1}{2}\partial u_1 \end{pmatrix}. \tag{11.4.9}$$

The first non-trivial flow $\mathbf{u}_{t_1} = \mathbf{B}_2\delta\mathcal{H}_1$ $(\mathcal{H}_1 = 2u_0 + \frac{1}{2}u_1^2)$ is

$$\begin{aligned} u_{0t_1} &= \tfrac{1}{4}u_{1xxx} + \tfrac{1}{2}u_1u_{0x} + u_0u_{1x}, \\ u_{1t_1} &= u_{0x} + \tfrac{3}{2}u_1u_{1x}. \end{aligned} \tag{11.4.10}$$

Example – DWW(K)
An invertible change of variables:

$$\begin{aligned} q &= u_0 + \tfrac{1}{4}u_1^2 - \tfrac{1}{2}u_{1x} = F_1[u_0, u_1], \\ r &= u_1 \qquad\qquad\quad\;\; = F_2[u_0, u_1]. \end{aligned} \tag{11.4.11}$$

transforms the equation (11.4.10) into a more symmetric form [2, 21]:

$$\begin{aligned} q_{t_1} &= \tfrac{1}{2}(-q_x + 2qr)_x, \\ r_{t_1} &= \tfrac{1}{2}(r_x + 2q + r^2)_x, \end{aligned} \tag{11.4.12}$$

which is just the standard dispersive water waves equation.

The change of variables (11.4.11) corresponds to a simple gauge transformation relating the linear problem (11.4.1) to the corresponding Kupershmidt version [21]:

$$\tilde{\mathbb{L}}\varphi = (\partial^2 - (r - 2\lambda)\partial + q)\varphi = 0. \tag{11.4.13a}$$

We can generalise this linear problem to its energy-dependent form:

$$\tilde{\mathbb{L}}\varphi = (\varepsilon\partial^2 - r\partial + q)\varphi = 0, \tag{11.4.13b}$$

where ε, r, and q are now polynomials in λ and construct the associated Hamiltonian operators and Hamiltonians in exactly the same way as was done for the energy-dependent Schrödinger problem (11.4.1). Assuming a time evolution of the wave function φ of the form (11.4.1c) we obtain the generalised Lax equation:

$$\mathbb{L}_t - [\mathbb{P}, \mathbb{L}] - P_x\mathbb{L} = (-r_t + \tfrac{1}{2}\varepsilon P_{xx} + \varepsilon Q_x + \tfrac{1}{2}(Pr)_x)\partial + q_t + \tfrac{1}{2}\varepsilon Q_{xx} - \tfrac{1}{2}rQ_x - P_xq - \tfrac{1}{2}Pq_x = 0, \tag{11.4.14}$$

which gives us the Hamiltonian form of the isospectral flows:

$$\begin{pmatrix} q \\ r \end{pmatrix}_t = \frac{1}{2}\begin{pmatrix} q\partial + \partial q & -\partial^2 + r\partial \\ \partial^2 + \partial r & 2\partial \end{pmatrix}\begin{pmatrix} P \\ Q \end{pmatrix}. \tag{11.4.15}$$

In the DWW(K) case (11.4.15) leads to the following two Hamiltonian operators:

$$\bar{\mathbf{B}}_0 = \begin{pmatrix} 0 & \partial \\ \partial & 0 \end{pmatrix}, \quad \bar{\mathbf{B}}_1 = \frac{1}{2}\begin{pmatrix} q\partial + \partial q & -\partial^2 + r\partial \\ \partial^2 + \partial r & 2\partial \end{pmatrix}, \tag{11.4.16}$$

which are, of course, the images of $\mathbf{B}_0$, $\mathbf{B}_1$ of (11.4.9) under the map (11.4.11). However, the third Hamiltonian structure of DWW(K), given by

$$\bar{\mathbf{B}}_2 = \frac{1}{4}\begin{pmatrix} (r-\partial)(q\partial + \partial q) + (q\partial + \partial q)(r+\partial) & (r-\partial)^2\partial + 2(q\partial + \partial q) \\ \partial(r+\partial)^2 + 2(q\partial + \partial q) & 2(r\partial + \partial r) \end{pmatrix}, \tag{11.4.17}$$

does not occur naturally in the above picture and to explain its existence one must revert to the DWW(S) formulation.

11.5 Miura maps

The discovery of the (original) Miura map [25] was a key element in the foundations of soliton theory. As described in Chapter 1 (see also [26]) Miura discovered this 'remarkable, explicit nonlinear transformation' while comparing the hierarchies of conserved densities connected with the KdV and MKdV equations. At that time, the Hamiltonian nature of the KdV equation was not known. However, the Miura map is fundamental in the explanation of the second Hamiltonian structure of the KdV equation.

If $M \in A_v^N$ is a differential function, we define the Fréchet derivative $M'\colon A_v^N \to A_v^N$ as a matrix differential operator acting on $\mathbf{P} \in A_v^N$ according to

$$M'[\mathbf{v}]\mathbf{P} = \frac{\mathrm{d}}{\mathrm{d}\varepsilon} M[\mathbf{v} + \varepsilon\mathbf{P}]\big|_{\varepsilon=0}. \tag{11.5.1a}$$

In components we have $(M'\mathbf{P})_j = (M')_j^i \mathrm{P}_i$ where:

$$(M')_j^i = \sum_k \frac{\partial M_j}{\partial v_{i,k}} \partial^k. \qquad (11.5.1b)$$

Now consider a differential algebra A_u of differential functions of $\mathbf{u}$ and use M to define a mapping $M{:}V \to U$ given by:

$$u_i = M_i[\mathbf{v}] \qquad (11.5.2)$$

and their derivatives. This gives a homomorphism between the differential algebras A_u and A_v. An element $P \in A_u$ is 'pulled back' to $\tilde{P} = M^*P \in A_v$:

$$\tilde{P}[\mathbf{v}] = P \circ M[\mathbf{v}] = P[M[\mathbf{v}]]. \qquad (11.5.3)$$

Whenever $M_i[\mathbf{v}]$ are independent in A_v (so that there is no non-trivial differential function $P \in A_u$ such that $P \circ M = 0$) we can identify A_u with a subalgebra $\tilde{A}_u$ of A_v. Since $\tilde{P} \in A_v$ is a total derivative whenever $P \in A_u$ is, (11.5.3) induces a quotient mapping to act on densities in $\mathscr{A}_u$. It is easily checked that:

$$\partial_t \tilde{\mathbf{u}} = M' \partial_t \mathbf{v} \qquad (11.5.4a)$$

for any evolutionary derivation ∂_t of A_v and:

$$\delta_v \tilde{\mathscr{H}} = (M')^\dagger \widetilde{\delta_u \mathscr{H}} \qquad (11.5.4b)$$

where δ_v and δ_u are the variational derivatives on $\mathscr{A}_v$ and $\mathscr{A}_u$ respectively. Thus, if $\mathbf{v}$ satisfies:

$$\mathbf{v}_t = \mathbf{B}_v \delta_v \tilde{\mathscr{H}} \qquad (11.5.5a)$$

then $\tilde{\mathbf{u}}$ satisfies:

$$\tilde{\mathbf{u}}_t = M' \mathbf{B}_v (M')^\dagger \widetilde{\delta_u \mathscr{H}}. \qquad (11.5.5b)$$

This motivates the following definition of a (Hamiltonian) Miura map.

Definition
The mapping (11.5.2) is called a Miura map for the Hamiltonian structure $\mathbf{B}_v \in A_v^N[\partial]$ if:

(i) $M_i[\mathbf{v}]$, $i = 0, \ldots, N-1$ are independent in A_v;
(ii) $\mathbf{B}_u = M' \mathbf{B}_v (M')^\dagger$ is a differential operator on $\tilde{A}_u^N \subset A_v^N$;
(iii) the homomorphism M^*: $A_u \to A_v$ is not invertible.

Remark
Part (ii) of this definition means that $M'\mathbf{B}_v(M')^\dagger$ depends upon $\mathbf{v}$ only in the combinations $M_i[\mathbf{v}]$ so that we can consider this as an operator on $\tilde{A}_u^N \cong A_u^N$. This operator is Hamiltonian whenever B_v is, so (11.5.5b)

represents a Hamiltonian system on A_u. If M were invertible this would be trivial. However, since we have (iii), property (ii) is quite remarkable.

The following theorem holds.

Theorem [23]
If the map (11.5.2) defined by $M \in A_v^N$ is a Miura map for the Hamiltonian structure $\mathbf{B}_v$ on A_v^N then

(i) $\mathbf{B}_u = M'\mathbf{B}_v(M')^\dagger$ is a Hamiltonian structure on $A_u^N \cong \tilde{A}_u^N$,
(ii) for any $\mathcal{H} \in \mathcal{A}_u$ the evolutionary derivation $\partial_{\mathcal{H}}$ of A_u determined by $\mathbf{B}_u$, is a restriction to $A_u \cong \tilde{A}_u$ of the evolutionary derivation $\partial_{\tilde{\mathcal{H}}}$ of A_v determined by $\mathbf{B}_v$.

Proof
We have:

$$\partial_{\tilde{\mathcal{H}}}\tilde{\mathbf{u}} = M'\mathbf{B}_v\delta_v\tilde{\mathcal{H}} = M'\mathbf{B}_v(M')^\dagger\widetilde{\delta_u\mathcal{H}} = \mathbf{B}_u\delta_u\mathcal{H} = \partial_{\mathcal{H}}\,\mathbf{u}, \quad (11.5.6a)$$

which proves (ii).

To prove (i) we notice that $\mathbf{B}_u$ is skew-adjoint and for any $\mathcal{H}, \mathcal{G} \in \mathcal{A}_u$ we have:

$$\begin{aligned}\{\tilde{\mathcal{H}}, \tilde{\mathcal{G}}\}_{B_v} &= (\delta_v\tilde{\mathcal{H}})^{\mathrm{T}}\mathbf{B}_v\delta_v\tilde{\mathcal{G}} = ((M')^\dagger(\widetilde{\delta_u\mathcal{H}}))^{\mathrm{T}}\mathbf{B}_v(M')^\dagger\widetilde{\delta_u\mathcal{G}} \\ &= (\widetilde{\delta_u\mathcal{H}})^{\mathrm{T}}M'\mathbf{B}_v(M')^\dagger\widetilde{\delta_u\mathcal{G}} = (\delta_u\mathcal{H})^{\mathrm{T}}\mathbf{B}_u\delta_u\mathcal{G} = \{\mathcal{H}, \mathcal{G}\}_{B_u}.\end{aligned} \quad (11.5.6b)$$

Thus the Jacobi identity for $\{,\}_{B_u}$ on A_u follows from the one for $\{,\}_{B_v}$ on A_v.

Because of the identification $A_u = \tilde{A}_u \subset A_v$ we can drop the tildes above elements of $\tilde{A}_u$. From now on we will also denote $\mathbf{B}_u$ simply by $\mathbf{B}$ and $\mathbf{B}_v$ by $\tilde{\mathbf{B}}$.

Example – DWW(K)–MDWW(K) [21]
These represent a two-component generalisation of the KdV–MKdV hierarchies. Factorisation of the Lax operator (11.4.13a), described in Appendix B, provides us with a non-invertible map

$$\begin{aligned} q &= M^1[v, w] = -v_x - v^2 + vw, \\ r &= M^2[v, w] = w, \end{aligned} \quad (11.5.7)$$

which is a Miura map for the Hamiltonian operator $\tilde{\mathbf{B}}_1 \in A_v^2$:

$$\tilde{\mathbf{B}}_1 = \begin{pmatrix} 0 & \frac{1}{2}\partial \\ \frac{1}{2}\partial & \partial \end{pmatrix}. \quad (11.5.8)$$

The image of $\tilde{\mathbf{B}}_1$ is:

$$\mathbf{B}_1 = M'\tilde{\mathbf{B}}_1(M')^\dagger = \frac{1}{2}\begin{pmatrix} q\partial + \partial q & -\partial^2 + r\partial \\ \partial^2 + \partial r & 2\partial \end{pmatrix} \quad (11.5.9a)$$

which is the second of the Hamiltonian structures of DWW(K) hierarchy. The third Hamiltonian structure (11.4.17) of this hierarchy is the image (under the map (11.5.7)) of

$$\tilde{\mathbf{B}}_2 = \frac{1}{4}\begin{pmatrix} v\partial + \partial v & -\partial^2 + w\partial + 2\partial v \\ \partial^2 + \partial w + 2v\partial & 2w\partial + 2\partial w \end{pmatrix} \qquad (11.5.9b)$$

which is the second Hamiltonian structure of the modified DWW(K) hierarchy (MDWW(K)). The recursion operators $\mathbf{R} = \mathbf{B}_1 B_0^{-1}$ and $\widetilde{\mathbf{R}} = \tilde{\mathbf{B}}_2\tilde{\mathbf{B}}_1^{-1}$ are 'intertwined' by M. We thus have the following diagram:

$$\begin{array}{ccc} q_{t_n} & \xleftarrow{\;M'\;} & v_{t_n} \\ \mathbf{R}\Big\downarrow & & \Big\downarrow\widetilde{\mathbf{R}} \\ \mathbf{R}q_{t_n} & \xleftarrow{\;M'\;} & \widetilde{\mathbf{R}}v_{t_n} \end{array} \qquad (11.5.10)$$

Remark

The existence of a first-order Hamiltonian structure $\tilde{\mathbf{B}}_0$ given by (11.4.16) for the DWW(K) is not guaranteed by the above theorem and is, indeed, independent of it.

We list the first few Hamiltonian of these two hierarchies:

$$\mathcal{H}_0 = 2r, \quad \mathcal{H}_1 = 2q, \quad \mathcal{H}_2 = qr,$$
$$\widetilde{\mathcal{H}}_0 = 2w, \; \widetilde{\mathcal{H}}_1 = 2vw - 2v^2, \; \widetilde{\mathcal{H}}_2 = vw^2 - v^2w - v_xw, \qquad (11.5.11)$$

with $\widetilde{\mathcal{H}}_n = \mathcal{H}_n \circ M$. The first non-trivial flows of both hierarchies are given by:

$$q_{t_1} = \tfrac{1}{2}(-q_x + 2qr)_x,$$
$$r_{t_1} = \tfrac{1}{2}(r_x + r^2 + 2q)_x, \qquad (11.5.12a)$$

and

$$v_{t_1} = \tfrac{1}{2}(-v_x + 2vw - v^2)_x,$$
$$w_{t_1} = \tfrac{1}{2}(w_x - 2v_x - 2v^2 + 2vw + w^2)_x. \qquad (11.5.12b)$$

Miura maps for energy-dependent Schrödinger operators

We present now a large class of Miura maps which relate isospectral flows of (11.4.1) to their modifications. They were constructed in [4] using the generalised factorisation approach described in Appendix B. We present our results in the context of the generalised KdV reduction, referring the reader to [4] for the general case.

Proposition 1
Under the change of variables $\mathbf{u} = M_N[\mathbf{v}]$ defined by

$$u_k = -v_{kx} - \frac{1}{2}\sum_{i=0}^{k} v_{k+i}v_{N-i}, \quad k = 0, \ldots, N-1 \quad (11.5.13)$$

the Hamiltonian operator $\mathbf{B}_N \in A_u^N[\partial]$, given by (11.4.6b), is the image:

$$\mathbf{B}_N = M'_N \tilde{\mathbf{B}}_N (M'_N)^\dagger|_{u=M_N[v]} \quad (11.5.14a)$$

of the constant, first-order Hamiltonian operator $\tilde{\mathbf{B}}_N \in A_v^N[\partial]$:

$$\tilde{\mathbf{B}}_N = \frac{1}{4}\begin{pmatrix} 0 & & -\partial \\ & ⋰ & \\ -\partial & & 0 \end{pmatrix}. \quad (11.5.14b)$$

The proof of this proposition relies on the fact that the Fréchet derivative of the mapping M_N is given by

$$M'_N = \begin{pmatrix} m_0 & & 0 \\ \vdots & \ddots & \\ m_{N-1} & \cdots & m_0 \end{pmatrix} \quad (11.5.15a)$$

where:

$$m_0 = -\partial - 2v_0, \quad m_k = -2v_k, \quad k = 1, \ldots, N-1. \quad (11.4.15b)$$

The Miura map $\mathbf{u} = M_N[\mathbf{v}]$ can be decomposed into N primitive ones. Define a sequence of non-invertible maps $\mathbf{u}^{(k)} = M_N^{k+1}[\mathbf{u}^{(k+1)}]$ by:

$$\begin{aligned} u_k^{(k)} &= -u_{kx}^{(k+1)} - \frac{1}{2}\sum_{i=1}^{k} u_i^{(k+1)}u_{k-i}^{(k+1)}, \\ u_i^{(k)} &= u_i^{(k+1)} \quad \text{for } i \neq k \end{aligned} \quad (11.5.16a)$$

where $\mathbf{u}^{(0)} \equiv \mathbf{u}$, $\mathbf{u}^{(N)} = \mathbf{v}$.

We can write $\mathbf{u} \equiv \mathbf{u}^{(0)} = M_N[\mathbf{u}^{(N)}] \equiv M_N[\mathbf{v}]$ as the composition of these maps:

$$\mathbf{u}^{(0)} = M_N[\mathbf{u}^{(N)}] = M_N^1 \circ M_N^2 \circ \ldots \circ M_N^N[\mathbf{u}^{(N)}]. \quad (11.5.16b)$$

Starting with $\mathbf{B}_N^N = \tilde{\mathbf{B}}_N$ of (11.5.14b) define $\mathbf{B}_N^k \in A_{u^{(k)}}^N[\partial]$ inductively by:

$$\mathbf{B}_N^{k-1} = (M_N^k)'\mathbf{B}_N^k((M_N^k)')^\dagger|_{u^{(k-1)} = M_N^k[\mathbf{u}^{(k)}]}, \quad k = N, \ldots, 1. \quad (11.5.17)$$

B_N^0 is just our original $\mathbf{B}_N$ of (11.4.6b).

Direct calculation shows that, as indicated by the notation, each $\mathbf{B}_N^k$ is *locally* defined in terms of the variables $\mathbf{u}^{(k)}$. Thus, each of the maps M_N^k is a genuine Hamiltonian Miura map. Let $M^{(r)} = M_N^1 \circ M_N^2 \circ \ldots \circ M_N^r$. We can summarise our results as follows.

Proposition 2
There exist local Hamiltonian operators $\mathbf{B}_k^r$ such that $(M^{(r)})'\mathbf{B}_k^r((M^{(r)})')^\dagger = \mathbf{B}_k^0 \equiv \mathbf{B}_k$ for $k = r, \ldots, N$. These constitute $(N - r + 1)$ compatible Hamiltonian structures for the rth modification. The sequence of modified Hamiltonians is defined by $\mathscr{H}_n^r = \mathscr{H}_n \circ M^{(r)}$ and the rth modified hierarchy is written as:

$$\mathbf{u}_{t_n}^{(r)} = \mathbf{B}_{N-k}^r \delta \mathscr{H}_{n+k}^r, \quad k = 0, \ldots, N - r, \quad n = 0, 1 \ldots \tag{11.5.18}$$

We can represent these modifications and their Hamiltonian structures schematically as in Figure 11.2.

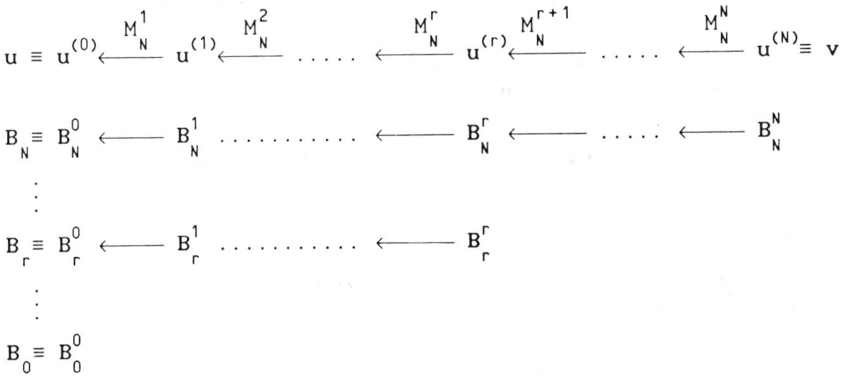

Figure 11.2

Example – DWW(S)
The Miura maps $\mathbf{u} = M_2^1[\mathbf{w}]$, $\mathbf{w} = M_2^2[\mathbf{v}]$ (we denote $\mathbf{u}^{(1)}$ by $\mathbf{w}$ here) are given by:

$$u_0 = -w_{0x} - w_0^2, \quad u_1 = w_1, \tag{11.5.19a}$$

$$w_0 = v_0, \quad w_1 = -v_{1x} - 2v_0v_1. \tag{11.5.19b}$$

Their composition $\mathbf{u} = M_2[\mathbf{v}] = M_2^1 \circ M_2^2[\mathbf{v}]$ is

$$u_0 = -v_{0x} - v_0^2, \quad u_1 = -v_{1x} - 2v_0v_1. \tag{11.5.19c}$$

The Miura map (11.5.19a) is easily seen to be equivalent to Kupershmidt's first modification of DWW(K) hierarchy (11.5.7). The second Miura map (11.5.19b), however, is not equivalent to either of Kupershmidt's second modifications [21]. The corresponding modified DWW(S) hierarchy is given by:

$$\mathbf{v}_{t_i} = \mathbf{B}_2^2 \delta \mathscr{H}_i^2, \tag{11.5.20}$$

where $\mathcal{H}_i^2[\mathbf{v}] = \mathcal{H}_i \circ M_2[\mathbf{v}]$. The first non-trivial flow is

$$v_{0t_1} = (\tfrac{1}{4}v_{1xx} + \tfrac{1}{2}v_1 v_{0x} - v_0^2 v_1)_x,$$
$$v_{1t_1} = (v_0 - v_0 v_1^2 - \tfrac{1}{2}v_1 v_{1x})_x. \qquad (11.5.21)$$

Normal forms

In finite dimensions, Darboux's theorem guarantees the existence of canonical coordinates with respect to which the Poisson bracket takes canonical form $\begin{pmatrix} 0 & I \\ -I & 0 \end{pmatrix}$. In infinite dimensions such a theorem does not exist. Indeed, this theorem *cannot* be true in such a simple form since the Hamiltonian operator ∂ *cannot* be transformed into canonical form by any local change of coordinates. We thus have at least two simple forms to which we might transform a given Hamiltonian operator: canonical or $\mathbf{b}\partial$, where $\mathbf{b}^{\mathrm{T}} = \mathbf{b}$ is a constant symmetric matrix.

As an example of such a construction one could mention the Dubrovin and Novikov results on Hamiltonian operators of hydrodynamical type [12].

We see from the earlier part of this section that it is possible to use Miura maps to relate the operators (11.4.6b) to (11.5.14b). In this section we have only given this result in the case $\varepsilon \equiv 1$, but it is possible to show this is general [4]. There are, however, important differences between the case $\varepsilon_0 = 1$ (Miura maps) and the case $\varepsilon_0 = 0$ (invertible transformations). We propose here a method of determining whether a given Hamiltonian operator can be transformed to a simpler form using an invertible change of variables rather than a Miura map. Our approach is based on studying kernels of Hamiltonian operators.

Kernels of Hamiltonian operators play an important part in the theory of integrable bi-Hamiltonian systems. For example, any bi-Hamiltonian hierachy starts from the Hamiltonian $\mathcal{H}_0$ such that its variational derivative $\delta\mathcal{H}_0$ belongs to the kernel of the first Hamiltonian operator $\mathbf{B}_0$. The explicit form of Ker$\mathbf{B}$ is also necessary when studying stationary flows of integrable Hamiltonian evolution equations (see Section 11.6). Thus the explicit description of kernels of Hamiltonian operators is important for the theory of integrable evolution equations. In what follows we will consider only that part of the kernel Ker$\mathbf{B}$ of a Hamiltonian operator $\mathbf{B} \in A_u^N[\partial]$ which is a subspace of a vector space A_u^N. Thus kernels of the Hamiltonian operators ∂ and ∂^3 are considered to be identical, both consisting of just constant vectors: $\mathrm{Ker}\partial = \mathrm{Ker}\partial^3 = \mathbb{R}^1 \subset A_u^1$, while the kernel of the operator $\mathbf{B} = \partial^3 + 2u\partial + 2\partial u$ has trivial intersection with A_u^1.

We introduce here a simple notion of regularity for a Hamiltonian operator.

Definition
A Hamiltonian operator $\mathbf{B} \in A_u^N[\partial]$ is said to be regular if it is possible to bring it, through a suitable *invertible* change of variables $\mathbf{w} = F[\mathbf{u}]$, into a form $\bar{\mathbf{B}} \in A_w^N[\partial]$ such that

$$\text{Ker}\bar{\mathbf{B}} = \mathbb{R}^N \subset A_w^N. \tag{11.5.22}$$

The Dubrovin and Novikov results [12] show that any non-degenerate (see Section 11.3) Hamiltonian operator of hydrodynamical type is regular. Such operators are, in fact, transformed into constant-coefficient, first-order differential form. It is not difficult to see that any (non-degenerate) constant-coefficient Hamiltonian operator is regular.

Definition
The operator $\bar{\mathbf{B}}$ of the above definition will be referred to as a *normal form* of the operator $\mathbf{B}$. The corresponding variables $\mathbf{w}$ will be called normal.

Remark
Neither normal coordinates nor normal forms of a given regular operator $\mathbf{B}$ are defined uniquely. The property of normality is certainly preserved under linear changes of variables. However, it will usually be possible to introduce a much larger group of transformations which preserve normality.

To explain the notion of regular operators we note that the kernel $\text{Ker}\mathbf{B}$ of any regular Hamiltonian operator $\mathbf{B} \in A_u^N[\partial]$ can be represented as:

$$\text{Ker}\mathbf{B} = c^0\delta F_0 + c^1\delta F_1 + \ldots + c^{N-1}\delta F_{N-1} \tag{11.5.23}$$

where $c^k \in \mathrm{R}$ and $F_k \in A_u^N$ are independent differential functions of $u_0, \ldots, u_{N-1}$. (11.5.23) follows easily from the fact that for a regular operator $\mathbf{B} \in A_u^N[\partial]$ (11.5.22) can be written as:

$$\text{Ker}\bar{\mathbf{B}} = c^0\delta w_0 + \ldots + c^{N-1}\delta w_{N-1} \tag{11.5.24}$$

and, since $w_k = F_k(u_0, \ldots, u_{N-1})$, (11.5.23) is simply (11.5.24) written in the original coordinates $\mathbf{u}$.

Conversely, the following proposition holds.

Proposition
If the kernel of a Hamiltonian operator $\mathbf{B} \in A_u^N[\partial]$ can be represented in the form (11.5.23) and the formulae

$$w_k = F_k(u_0, \ldots, u_{N-1}), \quad k = 0, \ldots, N-1, \tag{11.5.25}$$

define an (invertible) change of variables then $\mathbf{B}$ is regular and $w_0, \ldots, w_{N-1}$ are normal coordinates.

The proposition provides us with a method of constructing normal coordinates and normal forms of regular Hamiltonian operators.

Example – normal forms for the operators (11.4.6b)
To obtain normal coordinates and a normal form for any regular Hamiltonian operator $\mathbf{B}_n$ of the energy-dependent Schrödinger operator (11.4.1) we use the proposition. We will concentrate on the coupled KdV case again. If $\varepsilon_0 \neq 0$, all $\mathbf{B}_r$, $r = 1, \ldots, N$, require genuine Miura (and therefore non-invertible) maps to relate them to the simpler Hamiltonian operators, so are not regular. If $\varepsilon_0 = 0$, then there exist both direct and inverse hierarchies and the kernel of the Hamiltonian operator $\mathbf{B}_r$, $r = 0, \ldots, N$, contains the set $\{\delta\mathcal{H}_{-r}, \delta\mathcal{H}_{-r+1}, \ldots, \delta\mathcal{H}_{-r+N-1}\}$. It is easy to see that the formulae

$$\begin{aligned} w_k &= \mathcal{H}_{-1-k}, \quad k = 0, \ldots, r-1, \\ w_k &= \mathcal{H}_{N-1-k}, \quad k = r, \ldots, N-1, \end{aligned} \tag{11.5.26}$$

define an invertible change of variables $\mathbf{w} = F[\mathbf{u}]$. The Fréchet derivative of F has a block-diagonal triangle form:

$$F' = \left[\begin{array}{ccc|ccc} u_0^{-1/2} & & 0 & & & \\ & \ddots & & & 0 & \\ * & & u_0^{-1/2} & & & \\ \hline & & & 1 & & * \\ & 0 & & & \ddots & \\ & & & 0 & & 1 \end{array}\right], \tag{11.5.27}$$

where * = 'stuff', which should be compared with that of $\mathbf{B}_r$ (11.4.6b):

$$\mathbf{B}_r = \left[\begin{array}{ccc|ccc} 0 & & \sqrt{u_0}\partial\sqrt{u_0} & & & \\ & ⋰ & & & 0 & \\ \sqrt{u_0}\partial\sqrt{u_0} & & * & & & \\ \hline & & & * & & \partial \\ & 0 & & & ⋰ & \\ & & & \partial & & \end{array}\right]. \tag{11.5.28}$$

From (11.5.27) and (11.5.28) it immediately follows that $\bar{\mathbf{B}}_r = F'\mathbf{B}_r(F')^\dagger|_{w=F[u]}$ is of the form:

$$\bar{\mathbf{B}}_r = \left[\begin{array}{ccc|ccc} 0 & & \partial & & & \\ & ⋰ & & & 0 & \\ \partial & & * & & & \\ \hline & & & * & & \partial \\ & 0 & & & & \\ & & & \partial & & 0 \end{array}\right]. \tag{11.5.29}$$

From (11.5.29) (and the fact that $R^N \in \mathrm{Ker}\bar{\mathbf{B}}_r$) it follows that $\bar{\mathbf{B}}_r$ is a normal form of $\mathbf{B}_r$ and w_r are normal coordinates.

As we have seen previously, if $\varepsilon_0 \neq 0$ all the $\mathbf{B}_r$, $r = 1, \ldots, N$ are Miura-related to the constant Hamiltonian operators $\tilde{\mathbf{B}}_r$. It is not difficult to see that, in this case, the dimension of the kernel of $\mathbf{B}_r$ is less then N. If the rank of a Hamiltonian operator $\mathbf{B}$ is defined as $\mathrm{rank}\mathbf{B} = N$-$\dim(\mathrm{Ker}\mathbf{B})$ one easily finds that $\mathrm{rank}\mathbf{B}_r = r$ (if $\varepsilon_0 \neq 0$). This suggests the existence of a close connection between the rank of a Hamiltonian operator $\mathbf{B}$ and the number of (primitive) Miura maps relating it to a regular operator $\tilde{\mathbf{B}}$ (with $\mathrm{rank}\tilde{\mathbf{B}} = 0$). Such a relation requires, however, that the (primitive) Miura map increases the rank of a Hamiltonian operator by one. Since this is false in general (example: $\mathbf{B} = \partial^3$ and $\tilde{\mathbf{B}} = \partial$ are Miura-related and both of rank zero), one should consider the above remarks as a rather crude approximation of the real picture which is yet to be established.

11.6 **Hamiltonian structures of stationary flows**

Let us start with an infinite hierarchy of commuting Hamiltonian flows (in the PDE context):

$$\mathbf{u}_{t_{n-1}} = \mathbf{B}\delta\mathscr{H}_n, \quad n = 1, 2, \ldots, \tag{11.6.1}$$

where $\mathbf{B} \in A_u^N[\partial]$ is some Hamiltonian structure, and $\delta \equiv \delta_u$ the variational derivative with respect to $\mathbf{u}$. We restrict attention to those solutions independent of 'time' t_{m-1} for some *fixed* m:

$$\mathbf{B}\delta\mathscr{H}_m = 0, \tag{11.6.2a}$$

so that:

$$\delta\mathscr{H}_m \in \mathrm{Ker}\mathbf{B}. \tag{11.6.2b}$$

For the discussion of this chapter we make the following assumption [31].

Assumption

The form of elements of Ker**B** can be explicitly written down and, furthermore, represented as the variational derivative of some functional $\mathscr{K}_B$.

Thus, we assume

$$\delta\mathscr{H}_m = \delta\mathscr{K}_B, \tag{11.6.3a}$$

so that the stationary solutions satisfy the Lagrange equation:

$$\delta\mathscr{L} = 0 \tag{11.6.3b}$$

with the Lagrangian $\mathscr{L} = \mathscr{H}_m - \mathscr{K}_B$. Here $\mathscr{K}_B$ will depend upon some parameters C_j, $j = 1, \ldots, l$, which are constants of integration.

Remark

The x-variable is now playing the role of time; $\mathrm{d}f/\mathrm{d}x$ will sometimes be denoted by $\dot{f}$.

Since $\mathcal{L}$ will generally depend not just upon $\mathbf{u}$ and $\dot{\mathbf{u}}$, but also on higher derivatives, we need to use a generalised Legendre transformation [32] to change (11.5.3b) into a canonical Hamiltonian system.

Generalised Legendre transformation

Consider a Lagrangian $\mathcal{L}(u_j, u_{j,1}, u_{j,2}, \ldots, u_{j,k})$ where $u_{j,k} = \mathrm{d}^k u_j/\mathrm{d}x^k$. We assume that $\mathcal{L}$ is non-degenerate, meaning that $\det[\partial^2\mathcal{L}/\partial u_{j,n}\partial u_{i,n}] \neq 0$. Generalised coordinates $q_{j,k}$ and their conjugate momenta $p_{j,k}$ are defined by the formulae:

$$q_{j,k} = u_{j,k-1}, \quad p_{j,k} = \frac{\partial\mathcal{L}}{\partial u_{j,k}} + \sum_{i=1}^{n-k}(-\partial)^i\,\frac{\partial\mathcal{L}}{\partial u_{j,i}}, \qquad (11.6.4a)$$

$k = 1, \ldots, n$; $j = 0, \ldots, N-1$. Lagrange's equation (11.6.3b) is then equivalent to Hamilton's canonical equations with the Hamiltonian:

$$h_{\mathcal{L}}(q, p) = \sum_{j=0}^{N-1}\left(p_{j,n}\dot{q}_{j,n} + \sum_{k=1}^{n-1} p_{j,k}q_{j,k+1}\right) - \mathcal{L}. \qquad (11.6.4b)$$

Example

To motivate the above definition, consider the simple example with $\mathcal{L}(u, u_{,1}, u_{,2}, u_{,3})$. Here we have only one function $u_0(x) = u(x)$ but the Lagrangian depends upon second and third derivatives. The time derivative is given by the chain rule:

$$\frac{\mathrm{d}\mathcal{L}}{\mathrm{d}x} = \frac{\partial\mathcal{L}}{\partial u}u_{,1} + \frac{\partial\mathcal{L}}{\partial u_{,1}}u_{,2} + \frac{\partial\mathcal{L}}{\partial u_{,2}}u_{,3} + \frac{\partial\mathcal{L}}{\partial u_{,3}}u_{,4} \qquad (11.6.5a)$$

Substituting $\partial\mathcal{L}/\partial u$ from Lagrange's equation leads to a conserved quantity:

$$\frac{\mathrm{d}}{\mathrm{d}x}\left[u_{,3}\frac{\partial\mathcal{L}}{\partial u_{,3}} + u_{,2}\left(\frac{\partial\mathcal{L}}{\partial u_{,2}} - \frac{\mathrm{d}}{\mathrm{d}x}\left(\frac{\partial\mathcal{L}}{\partial u_{,3}}\right)\right) + u_{,1}\left(\frac{\partial\mathcal{L}}{\partial u_{,1}} - \frac{\mathrm{d}}{\mathrm{d}x}\left(\frac{\partial\mathcal{L}}{\partial u_{,2}}\right) + \frac{\mathrm{d}^2}{\mathrm{d}x^2}\left(\frac{\partial\mathcal{L}}{\partial u_{,3}}\right)\right) - \mathcal{L}\right] = 0 \qquad (11.6.5b)$$

In [7] Bogoyavlenskii and Novikov proved (for the stationary KdV case) that the canonical Hamiltonian system resulting from $\mathcal{L} = \mathcal{H}_m - C_1 u$ is completely integrable in the Liouville sense. A complete set of commuting integrals h_r is given by the formula:

$$(h_r)_x = -\{\mathcal{L}, \mathcal{H}_r\} = \{\mathcal{H}_m - \mathcal{K}_B, \mathcal{H}_r\}, \quad r = 1, \ldots, m-1. \qquad (11.6.6a)$$

The fact that the right-hand side of (11.6.6a) is an exact derivative follows from $\{\mathcal{H}_m, \mathcal{H}_r\} = 0$ (mod Im∂) and $\{\mathcal{K}_B, \mathcal{H}_r\} = -\{\mathcal{H}_r, \mathcal{K}_B\}$ (mod Im∂), the last bracket being zero by definition: $\mathbf{B}\delta\mathcal{K}_B = 0$. The fact that h_r is a first integral of (11.6.3b) is a direct consequence of the definition (11.2.5) of the Poisson bracket:

$$(h_r)_x = -\{\mathcal{L}, \mathcal{H}_r\} = (\delta\mathcal{L})^{\mathrm{T}}\mathbf{B}\delta\mathcal{H}_r = 0$$

on solutions of $\delta\mathcal{L} = 0$.

In [7] it was proved that the Hamiltonian flow $\phi(h_r)$ on the phase space (11.6.4a) generated by the integral h_r is a restriction (to a finite dimensional space of stationary solutions) of the flow $\mathbf{u}_{t_{r-1}} = \mathbf{B}\delta\mathcal{H}_r$. In particular, whenever $\mathbf{u}_{t_0} = \mathbf{u}_x$, $\phi(h_1)$ is an x shift (or time-translational flow, since x is the '*time*' for stationary problems) and thus h_1 must be identical to the original Hamiltonian $h_{\mathcal{L}}$.

It is not difficult to see that the formula (11.6.6a) is equivalent to the following prescription: consider the conservation law $(\mathcal{H}_r)_{t_{m-1}} = (\mathcal{F}_{m-1})_x$ (with respect to the flow $\mathbf{u}_{t_{m-1}} = \mathbf{B}\delta\mathcal{H}_m$) associated with the conserved functional $\mathcal{H}_r$. Use the stationary equation (11.6.3b) and the formulae (11.6.4a) to express $\mathcal{F}_{m-1}$ in terms of canonical variables (q, p) and define h_r as

$$h_r(q, p) = \mathcal{F}_{m-1}(q, p). \tag{11.6.6b}$$

In fact, as we see from (11.6.6a) all h_r depend not only on (q, p) but on the parameters $C_1, \ldots, C_l$ as well. The Legendre transformation (11.6.4) is to be performed for any set of fixed values of C_r independently. Thus we end up with an l-parameter family of Hamiltonian flows on an l-parameter family of $2nN$-dimensional symplectic manifolds with the canonical Poisson brackets resulting from a Lagrangian nature of stationary flows. From the very construction it follows that we can glue all these manifolds together and consider the $(2nN + l)$-dimensional extended phase space Q of a given stationary flow. This is no longer a symplectic manifold of course. However, it carries a natural (degenerate) Poisson bracket π_0 which, in local coordinates (q, p, C), is given by:

$$\pi_0 = \left(\begin{array}{cc|c} 0 & I & \\ -I & 0 & \\ \hline & & 0 \end{array}\right) \tag{11.6.7}$$

where I is a $(nN) \times (nN)$ identity matrix. Now, C_r, $r = 1, \ldots, l$, are Casimirs of this Poisson structure and the Hamiltonian form of the stationary flow (11.6.3) is:

$$\dot{\mathbf{q}} = \pi_0 \nabla h \tag{11.6.8a}$$

where $$\mathbf{q} = (q, p, C)^{\mathrm{T}}, \ \nabla = \left(\frac{\partial}{\partial q}, \frac{\partial}{\partial p}, \frac{\partial}{\partial C}\right)^T.$$

Equation (11.6.8a) gives Hamilton's canonical equations for q, p:

$$\dot{q} = \frac{\partial h}{\partial p}, \quad \dot{p} = -\frac{\partial h}{\partial q}, \tag{11.6.8b}$$

as well as the trivial evolution of the Casimirs:

$$\dot{C} = 0. \tag{11.6.8c}$$

The original $2nN$-dimensional symplectic manifolds Q_c are leaves of a symplectic foliation of the Poisson manifold (Q, π_0) given by:

$$Q_c = \{(q, p, C) \in Q: C = c = \text{const.}\} \tag{11.6.9}$$

The above extended phase space picture seems to be of rather marginal importance for studying the intrinsic nature of a given stationary flow. Its advantages can be easily seen, however, when trying to describe a stationary flow analogue of a Miura map [6].

Miura maps

Let us consider now the stationary flows of two hierarchies of completely integrable evolution equations connected by a Miura map as in Section 11.5. The Miura map $\mathbf{u} = M[\mathbf{v}]$ transforms the solutions of $\mathbf{v}_{t_m} = \tilde{\mathbf{B}}_1\delta_v\tilde{\mathcal{H}}_m$ onto the solutions of $\mathbf{u}_{t_m} = \tilde{\mathrm{B}}_1\delta_u\mathcal{H}_m$.

However, for the purposes of this section we assume that there exists a lower-order Hamiltonian structure $\mathbf{B}_0 \in A_q^N$ such that $\mathbf{u}_{t_m} = \mathbf{B}_0\delta_u\mathcal{H}_{m+1}$. It is usually more convenient and sometimes even necessary to use $\mathbf{B}_0$ in order to satisfy the assumption (11.6.3a). Such a $\mathbf{B}_0$ usually exists and certainly does in our example. Since M is time-independent, it maps the solutions of the corresponding stationary equation $\tilde{\mathbf{B}}_1\delta_v\tilde{\mathcal{H}}_m = 0$ onto solutions of $\mathbf{B}_0\delta_u\mathcal{H}_{m+1} = 0$. If both these stationary equations can be cast into canonical form by (11.6.4) the stationary version of the Miura transformation $\mathbf{u} = M[\mathbf{v}]$ is a map M from the extended phase space $\tilde{\mathbf{Q}}$ of the stationary flow $\mathbf{v}_{t_m} = 0$ into the extended phase space Q of the stationary flow $\mathbf{u}_{t_m} = 0$ which transforms solutions of $\dot{\tilde{\mathbf{q}}} = \tilde{\pi}_1\nabla\tilde{h}$ into solutions of $\dot{\mathbf{q}} = \pi_0\nabla h$ where π_0 and $\tilde{\pi}_1$ are the canonical brackets (11.6.7) on spaces Q and $\tilde{Q}$ respectively.

Using definitions (11.6.4a) of canonical variables and the stationary equations $\mathbf{B}_0\delta_u\mathcal{H}_{m+1} = 0$, $\tilde{\mathbf{B}}_1\delta_v\tilde{\mathcal{H}}_m = 0$ one can explicitly calculate M from the Miura map $\mathbf{u} = M[\mathbf{v}]$ and its differential consequences $\mathbf{u}_{,r} = \partial^r M[\mathbf{v}]$. As will be seen in our example, this map M: $\tilde{Q} \to Q$ does not preserve the symplectic foliation; that is, the image $M[\tilde{Q}_{\tilde{c}}]$ of a leaf $\tilde{Q}_{\tilde{c}}$ of the symplectic foliation of $(\tilde{Q}, \tilde{\pi}_1)$ is not contained in any Q_c. This feature of M makes the use of the extended phase space indispensable.

The map M: $\tilde{Q} \to Q$ need not be a (local) diffeomorphism in general.

If it is, however, we can transport the Poisson bracket $\tilde{\pi}_1$ from $\tilde{Q}$ to Q, resulting in:

$$\pi_1 = M_*\tilde{\pi}_1 = J\tilde{\pi}_1 J^{\mathrm{T}}\big|_{q=M(\tilde{q})}. \tag{11.6.10}$$

π_0 and π_1 are *different* Poisson brackets on Q and constitute a Hamiltonian pair for the stationary flows.

Theorem
The two Poisson structures π_0 and $\pi_1 = M_*\tilde{\pi}_1$ and the Hamiltonians $\mathscr{h}_r$ generated by the fluxes $\mathscr{F}_{mr}$ constitute a bi-Hamiltonian ladder for the stationary flow $\mathbf{B}_0\delta_u\mathscr{H}_{m+1} = 0$:

$$\begin{aligned} \pi_0\nabla\mathscr{h}_0 &= 0, \\ \pi_0\nabla\mathscr{h}_r &= \pi_1\nabla\mathscr{h}_{r-1}, \quad r = 1, 2, \ldots \end{aligned} \tag{11.6.11a}$$

Proof
Since $\widetilde{\mathscr{H}}_r = M^*\mathscr{H}_r$, $\widetilde{\mathscr{F}}_{mr} = M^*\mathscr{F}_{mr}$ we have $\tilde{\mathscr{h}}_r = M^*\mathscr{h}_r$ and

$$\pi_0\nabla\mathscr{h}_r = M_*(\tilde{\pi}_1\tilde{\nabla}\tilde{\mathscr{h}}_{r-1}) = (M_*\tilde{\pi}_1)(M^{-1})^*(\widetilde{\nabla\mathscr{h}}_{r-1}) = \pi_1\nabla\mathscr{h}_{r-1}$$

because the Miura transformation maps flows on $\tilde{Q}$ into corresponding flows on Q.

The existence of the bi-Hamiltonian ladder (11.6.11a) supplies another proof of commutativity of the integrals $\mathscr{h}_r$. Of course, $\mathscr{h}_r$, $r = 1, \ldots$ are not independent since our system is finite-dimensional only. However, it is usually not difficult to prove that $\mathscr{h}_r$, $r = 1, 2, \ldots, m$, are functionally independent. In fact one can concentrate on a finite interval of the ladder (11.6.11a) only:

$$\pi_0\nabla\mathscr{h}_0 = 0, \quad \pi_0\nabla\mathscr{h}_r = \pi_1\nabla\mathscr{h}_{r-1}, \quad \pi_1\nabla\mathscr{h}_m = 0 \tag{11.6.11b}$$

with $r = 1, \ldots, m$; $\mathscr{h}_0$, $\mathscr{h}_m$ are Casimirs of respectively π_0, π_1. The latter follows from the theorem by Bogoyavlenskii and Novikov [7] which states that the Hamiltonian flows $\phi(\mathscr{h}_r)$ are restrictions of the infinite-dimensional flows $\mathbf{u}_{t_{r-1}} = \mathbf{B}_0\delta_u\mathscr{H}_r$ to an invariant (stationary) set of $\mathbf{u}_{t_m} = \mathbf{B}_0\delta_u\mathscr{H}_{m+1}$. Thus the flow $\phi(\mathscr{h}_{m+1})$ generated by $\pi_0\nabla\mathscr{h}_{m+1} = \pi_1\nabla\mathscr{h}_m$, being the restriction of $\mathbf{u}_{t_m}$ to its stationary set, must be trivial.

The complete integrability of a stationary flow (11.6.8) relies on the completeness of the set of integrals $\mathscr{h}_1, \ldots, \mathscr{h}_m$. This is often easy to check for specific examples but difficult to determine in general.

To illustrate the above construction we consider the KdV hierarchy and its modification [6].

Example – KdV, $u_{t_3} = 0$ and MKdV, $v_{t_3} = 0$
The stationary third-order flow of the KdV hierarchy is

$$u_{xx} = -3u^2 + C, \tag{11.6.12a}$$

with Lagrangian $\mathcal{L}_2 = -\frac{1}{2}u_x^2 + u^3 - uC$. This has one degree of freedom with coordinates $q = u$, $p = -u_x$. The Casimir h_0 and non-trivial Hamiltonian h_1 are given by:

$$h_0 = \tfrac{1}{2}C, \quad h_1 = -\tfrac{1}{2}p^2 - q^3 + qC. \tag{11.6.13a}$$

The corresponding stationary flow of the MKdV hierarchy:

$$v_{xx} = 2v^3 + \tilde{C} \tag{11.6.12b}$$

has Lagrangian $\tilde{\mathcal{L}}_2 = \frac{1}{2}(v_x^2 + v^4) + v\tilde{C}$. This has one degree of freedom with coordinates $\tilde{q} = v$, $\tilde{p} = v_x$. The Casimir $\tilde{h}_{-1}$ and non-trivial Hamiltonian $\tilde{h}_0$ are

$$\tilde{h}_{-1} = \tilde{C}, \tilde{h}_0 = \tfrac{1}{2}(\tilde{p}^2 - \tilde{q}^4) - \tilde{q}\tilde{C}, \quad \tilde{h}_1 = -\tfrac{1}{2}\tilde{C}^2, \tag{11.6.13b}$$

with $\tilde{h}_0 = \tilde{q}_x\tilde{p} - \tilde{\mathcal{L}}_2$.

The Miura map (11.B.2b) leads to

$$u_{xx} + 3u^2 - C = -(v_{3x} - 6v^2v_x) - 2v(v_{xx} - 2v^3) + v_x^2 - v^4 - C. \tag{11.6.14a}$$

Thus when u and v satisfy respectively (11.6.12a) and (11.6.12b):

$$C = v_x^2 - v^4 - 2v\tilde{C}, \tag{11.6.14b}$$

with the expression on the right-hand side being a conserved quantity of (11.6.12b), thus *not* contradicting the fact that C is a constant!

Remark

It is evident that we cannot just set $C = 0$ since this corresponds to choosing the 'zero energy' solution of (11.6.12b).

We must include (11.6.14b) in our definition of the Miura map. The phase spaces must thus be extended to include the parameters C and $\tilde{C}$. From the definitions of q, p, $\tilde{q}$ and $\tilde{p}$ it is easy to construct M: $(\tilde{q}, \tilde{p}, \tilde{C}) \to (q, q, C)$ as:

$$q = -\tilde{p} - \tilde{q}^2, \quad p = 2\tilde{q}\tilde{p} + 2\tilde{q}^3 + \tilde{C}, \quad C = \tilde{p}^2 - \tilde{q}^4 - 2\tilde{q}\tilde{C} = 2\tilde{h}_0. \tag{11.6.15}$$

We must similarly extend the canonical Poisson bracket to these spaces. Since C and $\tilde{C}$ are Casimirs, this is a standard procedure, reflecting the trivial equations of motion $\dot{C} = \dot{\tilde{C}} = 0$. We label the corresponding skew-symmetric matrices π_0 and $\tilde{\pi}_1$:

$$\pi_0 = \tilde{\pi}_1 = \begin{pmatrix} 0 & 1 & 0 \\ -1 & 0 & 0 \\ 0 & 0 & 0 \end{pmatrix}. \tag{11.6.16a}$$

The stationary flows (11.6.12a, b) are written:

$$\dot{q} = \pi_0 \nabla h_1, \quad \dot{\tilde{q}} = \tilde{\pi}_1 \nabla \tilde{h}_0. \tag{11.6.16b}$$

We can use the Jacobian J of the Miura map M to construct the *second* Hamiltonian structure π_1 in the standard way,

$$\pi_1 = J\tilde{\pi}_1 J^{\mathrm{T}}|_{q=M(\tilde{q})}, \tag{11.6.17a}$$

where J^{T} denotes the transpose of J, so that $\dot{\tilde{q}} = \tilde{\pi}_1 \nabla\tilde{h}_0$ is mapped onto $\dot{q} = \pi_1 \nabla h_0$. This second Hamiltonian structure is given explicitly by:

$$\pi_1 = \begin{pmatrix} 0 & -2q & -2p \\ 2q & 0 & 6q^2 - 2C \\ 2p & -6q^2 + 2C & 0 \end{pmatrix} \tag{11.6.17b}$$

Remark

The Jacobian $J = \partial(q, p, C)/\partial(\tilde{q}, \tilde{p}, \tilde{C})$ is a complicated matrix function of the variables $\tilde{q}$, $\tilde{p}$ and $\tilde{C}$. The matrix $J\tilde{\pi}_1 J^{\mathrm{T}}$ is an even more complicated function of these variables. However, these variables have combined in such a way that we can write the relatively simple matrix (11.6.17b) involving only the (q, p, C) coordinates! This remarkable fact is part of the 'magic' which makes the Miura maps very special.

Remark

As a consequence of formula (11.6.17a), π_1 is guaranteed to be skew-symmetric and to satisfy the Jacobi identities. Furthermore, since $\tilde{\pi}_1$ is of co-rank 1, π_1 must also be of co-rank 1. And, it is easily checked that π_0 and π_1 are compatible.

In the usual way, the bi-Hamiltonian property provides a ladder for generating a sequence of Hamiltonians from h_0, the Casimir of π_0:

$$0 = \pi_0 \nabla h_0, \quad \pi_1 \nabla h_0 = \pi_0 \nabla h_1, \quad \pi_1 \nabla h_1 = 0. \tag{11.6.18}$$

This ladder is only finite with h_1 the Casimir of π_1.

Acknowledgement

We thank SERC for financial support. MA was on leave of absence at the Department of Applied Mathematical Studies and Centre for Nonlinear Studies, University of Leeds.

References

[1] Adler, M. and Moser, J. On a class of polynomials connected with KdV equation, *Commun. Math. Phys.* **61**, 1–30 (1978).
[2] Antonowicz, M. and Fordy, A. P. Coupled KdV equations with multi-Hamiltonian structures, *Physica* **28D**, 345–57 (1987).
[3] Antonowicz, M. and Fordy, A. P. Coupled KdV equations associated with a novel Schrödinger spectral problem, in *Nonlinear Evolutions*, 145–60, J. Leon (ed.), World Scientific, Singapore (1988).

[4] Antonowicz, M. and Fordy, A. P. Factorisation of energy dependent Schrödinger operators: Miura maps and modified systems, *Commun. Math. Phys.* **124**, 465–86 (1989).
[5] Antonowicz, M. and Fordy, A. P. Super-extensions of energy dependent Schrödinger operators, *Commun. Math. Phys.* **124**, 487–500 (1989).
[6] Antonowicz, M. Fordy, A. P. and Wojciechowski, S. Integrable stationary flows: Miura maps and bi-Hamiltonian structures, *Phys. Lett. A* **124**, 143–50 (1987).
[7] Bogoyavlenskii, O. I. and Novikov, S. P. The relationship between Hamiltonian formalisms of stationary and nonstationary problems, *Func. Anal. and Apps.* **10**, 8–11 (1976).
[8] Calogero, F. and Degasperis, A. *Spectral Transform and Solitons I*, North-Holland, Amsterdam (1982).
[9] Chen, H. H. General derivation of Bäcklund transformations from inverse scattering problems, *Phys. Rev. Letts.* **33**, 925–8 (1974).
[10] Chen, H. H. Lee, Y. C. and Lin, C. S. Integrability of nonlinear Hamiltonian systems by inverse scattering method, *Phys. Scr.* **20**, 490–2 (1979).
[11] Drinfel'd, V. G. and Sokolov, V. V. Lie algebras and equations of Korteweg–de Vries type, *J. Sov. Math.* **30**, 1975–2036 (1985).
[12] Dubrovin, B. A. and Novikov, S. P. Hamiltonian formalism of one-dimensional systems of hydrodynamic type, and Bogolyubov–Whitam averaging method, *Sov. Math. Dokl.* **27**, 665–9 (1983).
[13] Faddeev, L. D. and Takhtajan, L. A. *Hamiltonian Methods in the Theory of Solitons*, Springer, Berlin (1987).
[14] Fordy, A. P. Projective representations and deformations of integrable systems, *Proc. R. Ir. Acad.* **83A**, 75–94 (1983).
[15] Fordy, A. P. and Gibbons, J. Factorisation of operators I. Miura transformations, *J. Math. Phys.* **21**, 2508–10 (1980).
[16] Fordy, A. P. and Gibbons, J. Integrable nonlinear Klein–Gordon equations and Toda lattices, *Commun. Math. Phys.* **77**, 21–30 (1980).
[17] Fordy, A. P. and Gibbons, J. Factorisation of operators II, *J. Math. Phys.* **22**, 1170–5 (1981).
[18] Fuchssteiner, B. and Fokas, A. S. Symplectic structures, their Bäcklund transformations and hereditary symmetries, *Physica* **4D**, 47–66 (1981).
[19] Gel'fand, I. M. and Dorfman, I. Ya. Hamiltonian operators and algebraic structures related to them, *Func. Anal. & Apps.* **13**, 248–62 (1979).
[20] Kupershmidt, B. A. Discrete Lax equations and differential difference calculus, *Revue Astérisque*, **123**, Paris (1985).
[21] Kupershmidt, B. A. Mathematics of dispersive water waves, *Commun. Math. Phys.* **99**, 51–73 (1985).
[22] Kupershmidt, B. A. Is a bi-Hamiltonian system necessarily integrable? *Phys. Letts. A.* **123**, 55–9 (1987).
[23] Kupershmidt, B. A. and Wilson, G. Modifying Lax equation and the second Hamiltonian structure, *Invent. Math.* **62**, 403–36 (1981).
[24] Magri, F. A simple model of the integrable Hamiltonian equation, *J. Math. Phys.* **19**, 1156–62 (1978).
[25] Miura, R. M. Korteweg–de Vries equation and generalizations I. A remarkable explicit nonlinear transformation, *J. Math. Phys.* **9**, 1202–4 (1968).
[26] Miura, R. M. The Korteweg–de Vries equation: a survey of results, *SIAM Review* **18**, 412–59 (1976).
[27] Olver, P. J. Evolution equations possessing infinitely many symmetries, *J. Math. Phys.* **18**, 1212–5 (1977).

[28] Olver, P. J. *Application of Lie Groups to Differential Equations*, Springer, Berlin (1986).
[29] Reyman, A. G. and Semenov-Tian-Shansky, M. A. Reduction of Hamiltonian systems, affine Lie algebras and Lax equations, *Invent. math.* **54**, 81–100 (1979).
[30] Reyman, A. G. and Semenov-Tian-Shansky, M. A. Reduction of Hamiltonian systems, affine Lie algebras and Lax equations II, *Invent. math.* **63**, 423–32 (1981).
[31] Veselov, A. P. *Fun. Anal. Apps.* **13**, 1–6 (1979).
[32] Whittaker, E. T. *A Treatise on the Analytic Dynamics of Particles and Rigid Bodies*, Cambridge University Press, Cambridge (1988).

Appendix A

We present here a simple method of checking the property of being Hamiltonian for any skew-adjoint differential operator $\mathbf{B} \in A_u^N[\partial]$. A more detailed description of this method (due to Olver) can be found in his book [28].

Let X be a space of vector functions $\theta(x) = (\theta_0(x), \ldots, \theta_{N-1}(x))^{\mathrm{T}}$ of x. We define a functional bi-vector:

$$\Theta \cong \theta^{\mathrm{T}} \wedge \mathbf{B}\theta \tag{11.A.1}$$

where $\wedge$ is an alternating symbol (i.e. $\alpha\wedge\beta = -\beta\wedge\alpha$, $\alpha\wedge(f\beta + g\gamma) = f\alpha\wedge\beta + g\alpha\wedge\gamma$, for α, β, $\gamma \in X$, and f, $g \in A_u$) and Θ is defined modulo the total derivative (we adopt the definition $\partial(\alpha\wedge\beta) = (\partial\alpha)\wedge\beta + \alpha\wedge(\partial\beta)$).

Proposition (Olver)
A skew-adjoint operator $\mathbf{B} \in A_u^N[\partial]$ is Hamiltonian if and only if:

$$\sum_k \left\{\left(\frac{\partial}{\partial u_{0,k}}, \ldots, \frac{\partial}{\partial u_{N-1,k}}\right)\Theta\right\}\wedge\partial^k(\mathbf{B}\theta) \cong 0 \tag{11.A.2}$$

for some representative Θ of (11.A.1) and any vector function θ.

Remark
The equality (11.A.2) holds modulo total derivatives only. It follows from (11.A.2) that any constant-coefficient skew-adjoint operator $\mathbf{B} \in A_u^N[\partial]$ is Hamiltonian.

Example
Let $N = 1$ and $\mathbf{B} = f\partial f$ where $f \in A_u$ is a function of u only. Then

$$\Theta \cong \theta\wedge f(f\theta)_x \cong f^2\theta\wedge\theta_x \tag{11.A.3a}$$

and the left-hand side of (11.A.2) is:

$$2f\frac{\partial u}{\partial u}\,\theta\wedge\theta_x\wedge f(f\theta)_x = 2f\frac{\partial u}{\partial u}\theta\wedge\theta_x\wedge(ff_x\theta + f^2\theta_x) = 0. \tag{11.A.3b}$$

Thus any operator of the form $f\partial f$ (or equivalently $\frac{1}{2}(f^2\partial + \partial f^2)$) with $f = f(u)$ is Hamiltonian.

Remark
This follows more simply, of course, from the fact that $f\partial f$ is a special case of $M'\mathbf{B}(M')^\dagger$ with $\mathbf{B} = \partial$; any constant-coefficient, skew-symmetric operator trivially satisfies the criterion (11.A.2).

Substituting $\mathbf{B} = \mathbf{B}_0 + \mathbf{B}_1$ into (11.A.2) and assuming that $\mathbf{B}_0$, $\mathbf{B}_1$ are Hamiltonian we obtain the condition:

$$0 \cong \sum_k \left\{\left(\frac{\partial}{\partial u_{0,k}}, \ldots, \frac{\partial}{\partial u_{N-1,k}}\right)\Theta_0\right\} \wedge \partial^k(\mathbf{B}_1\theta)$$
$$+ \sum_k \left\{\left(\frac{\partial}{\partial u_{0,k}}, \ldots, \frac{\partial}{\partial u_{N-1,k}}\right)\Theta_1\right\} \wedge \partial^k(\mathbf{B}_0\theta) \qquad (11.A.4a)$$

for their compatibility. Condition (11.A.4a) simplifies to:

$$\sum_k \left\{\left(\frac{\partial}{\partial u_{0,k}}, \ldots, \frac{\partial}{\partial u_{N-1,k}}\right)\Theta_1\right\} \wedge \partial^k(\mathbf{B}_0\theta) \cong 0 \qquad (11.A.4b)$$

when $\mathbf{B}_0$ is constant-coefficient.

Example

We now show that any constant-coefficient Hamiltonian operator $\mathbf{B}_0$ is compatible with:

$$\mathbf{B}_1 = \begin{pmatrix} -q\partial - \partial q & \partial^2 - r\partial \\ -\partial^2 - \partial r & -2\partial \end{pmatrix}. \qquad (11.A.5a)$$

If

$$\mathbf{B}_0 = \begin{pmatrix} a & b \\ b & d \end{pmatrix}\partial \qquad (11.A.5b)$$

then

$$\Theta_1 \cong -2q\theta_1 \wedge \theta_{1x} - 2r\theta_1 \wedge \theta_{2x} - 2\theta_1 \wedge \theta_{2x} \qquad (11.A.6a)$$

and the left-hand side of (11.A.4b) is equal to:

$$-2\theta_1 \wedge \theta_{1x} \wedge (a\theta_{1x} + b\theta_{2x}) - 2\theta_1 \wedge \theta_{2x}(b\theta_{1x} + d\theta_{2x}) =$$
$$-2b\theta_1 \wedge \theta_{1x} \wedge \theta_{2x} - 2b\theta_1 \wedge \theta_{2x} - 2\theta_1 \wedge \theta_{2x} \wedge \theta_{1x} = 0. \qquad (11.A.6b)$$

(11.A.6b) shows that $\mathbf{B}_0$ and $\mathbf{B}_1$ are compatible.

Appendix B

In Sections 11.5 and 11.6 we assumed that we were given two systems of nonlinear evolution equations, related through a Miura map. We gave no indication of how we were to know that the two systems were Miura-related or how to find the Miura map. We briefly discuss this question here. There is in fact, no systematic way of *explicity* constructing a Miura map and modification for an arbitrary, given (integrable) system. However, there are some methods which work for large classes of equations. Here we present the methods of factorisation (first presented in [1] and later developed in [15, 17, 23]) and of projective coordinates [9, 14], and also exhibit the role of gauge transformations.

Factorisation of differential operators

Consider the linear spectral problem of the KdV hierarchy

$$(\partial^2 + u)\psi = \lambda^2\psi \qquad (11.B.1a)$$

together with the time evolution of ψ:

$$\psi_{t_n} = P_n\psi. \qquad (11.B.1b)$$

This can be written as:

$$(\partial + v)(\partial - v)\psi = \lambda^2\psi \tag{11.B.2a}$$

if and only if u and v are related by:

$$u = -v_x - v^2 \tag{11.B.2b}$$

which is the original Miura map. Equation (11.B.2a) can now be written as a system:

$$\begin{pmatrix}\psi_1\\ \psi_2\end{pmatrix}_x = \begin{pmatrix} v & \lambda \\ \lambda & -v\end{pmatrix}\begin{pmatrix}\psi_1\\ \psi_2\end{pmatrix} \tag{11.B.3a}$$

with $\psi_1 \equiv \psi$. We can then consider (polynomial in λ) time evolutions of ψ

$$\psi_{t_n} = T_n(u, u_x, \ldots; \lambda)\psi, \quad T_n = \sum_0^n T_n^{(i)}\lambda^i. \tag{11.B.3b}$$

With $n = 3$ in (11.B.1b) and (11.B.3b) we obtain the KdV and MKdV equations:

$$u_{t_3} = u_{xxx} + 6uu_x, \tag{11.B.4a}$$

$$v_{t_3} = v_{xxx} - 6v^2v_x. \tag{11.B.4b}$$

The Miura map (11.B.2) relates corresponding elements of these hierarchies and, in particular, the KdV and MKdV equations.

This approach was extended to third-order Lax operators in [15, 17]:

$$(\partial^3 + q\partial + \tfrac{1}{2}q_x + r)\psi = \lambda^3\psi \tag{11.B.5a}$$

which can be factorised as:

$$(\partial + w - v)(\partial + 2v)(\partial - w - v)\psi = \lambda^3\psi \tag{11.B.5b}$$

to give the two-component Miura map:

$$\begin{aligned} q &= -(2w_x + w^2 + 3v^2), \\ r &= -(v_{xx} + 3wv_x + vw_x + 2v(w^2 - v^2)). \end{aligned} \tag{11.B.5c}$$

The second-order flows isospectral to (11.B.5a) and (11.B.5b) are respectively:

$$\begin{aligned} q_{t_2} &= 2r_x, \\ r_{t_2} &= -\tfrac{1}{6}\,(q_{xxx} + 4qq_x), \end{aligned} \tag{11.B.6a}$$

$$\begin{aligned} w_{t_2} &= v_{xx} + 2(wv)_x, \\ v_{t_2} &= -\tfrac{1}{3}\,(w_x + 3w^2 - v^2)_x, \end{aligned} \tag{11.B.6b}$$

referred to as the Boussinesq and modified Boussinesq equations. These, and indeed the whole hierarchy, are related through the Miura map (11.B.5c).

This 'factorisation' method is easily extended to the higher-order Lax operators [16, 23]. However, this method is restricted to Lax equations and cannot be applied to first-order matrix equations such as (11.B.3a).

Factorisation of energy-dependent Lax operators

The above factorisation method can be modified slightly so as to cover the energy-dependent Schrödinger linear problem discussed in Section 11.4. The main idea is to replace the standard factorisation $\mathbb{L} = A(-A^\dagger)$ with a more general one $\mathbb{L} = l\Lambda(-l^\dagger)$ where Λ is a constant (λ-dependent) symmetric matrix [4]. We will illustrate the construction on the simple example of the DWW hierarchy. We have two gauge-equivalent linear problems here:

$$\widetilde{\mathbb{L}}\phi \equiv (\partial^2 - (r - 2\lambda)\partial + q)\phi = 0, \tag{11.B.7a}$$

$$\mathbb{L}\psi \equiv (\partial^2 + u_0 + \lambda u_1 - \lambda^2)\psi = 0, \tag{11.B.7b}$$

the corresponding invertible change of variables being

$$\begin{aligned} u_0 &= q + \tfrac{1}{2}r_x - \tfrac{1}{4}r^2, \\ u_1 &= r. \end{aligned} \tag{11.B.8}$$

The standard factorisation procedure applied to $\widetilde{\mathbb{L}} - 2\lambda\partial$ leads to

$$(\partial^2 - r\partial + q) = (\partial + v - w)(\partial - v) \tag{11.B.9}$$

which gives us the Miura map [21]:

$$\begin{aligned} q &= -v_x - v^2 + vw, \\ r &= w, \end{aligned} \tag{11.B.10}$$

reducing to the original Miura map when $r = w = 0$. To factorise $\mathbb{L} + \lambda^2$ we write:

$$\mathbb{L} + \lambda^2 = l\Lambda(-l^\dagger) \tag{11.B.11}$$

where $l = (\partial + v_0, v_1)$,

$$\Lambda = \begin{pmatrix} 1 & \lambda \\ \lambda & 0 \end{pmatrix}.$$

This gives us the Miura map

$$\begin{aligned} u_0 &= -v_{0x} - v_0^2, \\ u_1 &= -v_{1x} - 2v_0v_1. \end{aligned} \tag{11.B.12}$$

In Section 11.5 we show that the Miura transformations (11.B.10) and (11.B.12) are in fact Hamiltonian Miura maps. More about this non-standard factorisation can be found in [4].

Projective coordinates

This approach just reverses the process of linearizing Riccati equations. Defining the variable $y = \psi_x/\psi$ and using (11.B.1a) we obtain:

$$y_x + y^2 = \lambda^2 - u \tag{11.B.13}$$

which defines a one-parameter family of Miura maps. By considering the time evolution of ψ (for a given flow) we can obtain an expression for y_t. By eliminating u and ψ we obtain an equation which is autonomous in y. The third-order time evolution:

$$\psi_{t_3} = 4\psi_{xxx} + 6u\psi_x + 3u_x\psi, \tag{11.B.14a}$$

corresponding to the KdV flow, gives rise to the one-parameter family of MKdV equations:

$$y_{t_3} = y_{xxx} - 6y^2y_x + 6\lambda^2y_x \tag{11.B.14b}$$

related to the KdV equation (11.B.4a) by the parameter-dependent Miura map (11.B.13).

Remark
A disadvantage of this method is that it does not include a direct way of obtaining the modified linear spectral problem (11.B.3a).

The main advantage of the projective coordinate method over factorisation is its applicability to first-order matrix spectral problems such as (11.B.3a). In this case we introduce the projective coordinate $z = \psi_1/\psi_2$, which satisfies:

$$z_x = \lambda(1 - z^2) + 2vz. \qquad (11.B.15a)$$

After a change of variable $z = e^{2\theta}$, this becomes:

$$v = \theta_x + \lambda \sinh^2 2\theta. \qquad (11.B.15b)$$

The third-order time evolution of θ is the one-parameter family of equations:

$$\theta_{t_3} = \theta_{xxx} - 2\theta_x^3 - 6\lambda^2 \sinh^2 2\theta \qquad (11.B.16)$$

which is related to (11.B.4b) through (11.B.15b).

This method has been applied to several examples of 2×2 and 3×3 matrix spectral problems in [9, 14].

Gauge transformations

From the point of view of theoretical physics, the gauge transformation approach is perhaps the most natural. Suppose we start with a linear spectral problem:

$$\phi_x = \mathbb{F}\phi \qquad (11.B.17a)$$

where matrix $\mathbb{F}$ depends in some way upon the potential functions v_i and the spectral parameter λ. Now consider the change of basis:

$$\psi = \mathbf{T}\phi. \qquad (11.B.18)$$

This ψ satisfies the equation:

$$\psi_x = \hat{\mathbb{F}}\psi = (\mathbf{T}\mathbb{F}\mathbf{T}^{-1} + \mathbf{T}_x\mathbf{T}^{-1})\psi. \qquad (11.B.17b)$$

Remark
$\mathbf{T}$ will depend upon the potential functions v_i, so that $\hat{\mathbb{F}}$ depends upon certain combinations of v_i and their x-derivatives, used to define the Miura map. When $\mathbf{T}$ depends upon the spectral parameter λ, this is often called a Darboux transformation and can be used to construct Bäcklund transformations [8].

Example
After a simple scale transformation on the wave functions, (11.B.3a) becomes:

$$\phi_x = \mathbb{F}\phi, \ \mathbb{F} = \begin{pmatrix} v & 1 \\ \lambda^2 & -v \end{pmatrix}. \qquad (11.B.19a)$$

With

$$T = \begin{pmatrix} 1 & 0 \\ v & 1 \end{pmatrix} \qquad (11.B.19b)$$

we find:

$$\hat{\mathbb{F}} = \begin{pmatrix} 0 & 1 \\ \lambda^2 + v_x + v^2 & 0 \end{pmatrix} = \begin{pmatrix} 0 & 1 \\ \lambda^2 - u & 0 \end{pmatrix} \qquad (11.B.19c)$$

when $u = -v_x - v^2$.

Remark

An interesting feature of this approach is that it directly transforms wave functions and therefore, through a fractional linear transformation, the projective variables previously discussed. In the 2×2 case, $y = \psi_2/\psi_1$, $z = \phi_2/\phi_1$, $\psi = \mathbf{T}\phi$ gives:

$$y = \frac{T_{21} + T_{12}z}{T_{11} + T_{12}z}. \tag{11.B.20}$$

This transformation enables us to directly relate the Hamiltonian densities associated with the two hierarchies. It is easily seen that as a consequence of (11.B.19a, c) y and z satisfy:

$$y_x + y^2 + u = \lambda^2, \tag{11.B.21a}$$

$$z_x + z^2 + 2vz = \lambda^2. \tag{11.B.21b}$$

The Riccati equations (11.B.21) admit formal asymptotic series solutions of the form:

$$y = \lambda + \sum_0^\infty y_i\lambda^{-i}, \quad z = \lambda + v + \sum_0^\infty z_i\lambda^{-i} \tag{11.B.22}$$

with (11.B.20) taking the simple form $y = v + z$, which gives us a direct transformation between asymptotic expansions (11.B.22) and therefore the conserved densities of the two hierarchies.

Part V

Algebraic and geometric structures

12 *A. P. Fordy*

Integrable equations associated with simple Lie algebras and symmetric spaces

12.1 Introduction

In Chapter 1 we considered several of the standard (and simpler) linear spectral problems. The simplest of these are second-order:

$$\mathbb{L}\psi \equiv (\partial^2 + u)\psi = \lambda^2\psi, \tag{12.1.1a}$$

$$\begin{pmatrix}\psi_1\\ \psi_2\end{pmatrix}_x = \begin{pmatrix}\lambda & q\\ r & -\lambda\end{pmatrix}\begin{pmatrix}\psi_1\\ \psi_2\end{pmatrix}, \tag{12.1.1b}$$

where $\partial \equiv \partial/\partial x$, q, r, u are 'potential functions' and λ the spectral parameter. The Schrödinger operator (12.1.1a), which is the simplest of all scalar Lax operators, and the ZSAKNS system of two first-order equations (12.1.1b) are both associated with the smallest simple Lie algebra, $sl(2, \mathbb{C})$. The first of these can, when written as a system, be embedded in (12.1.1b) with $r = 1$, $q = -u$.

To derive integrable systems of equations with many components we need to use other higher-order Lax operators [15, 35] or to enlarge the matrix in (12.1.1b) [28]. While higher-order Lax operators can easily be written as first-order systems, it is worthwhile treating these as a separate case in their Lax form. The simplest generalisations of (12.1.1) are of the form:

$$\mathbb{L}\psi = (\partial^n + u_{n-2}\partial^{n-2} + \ldots + u_1\partial + u_0)\psi = \lambda^n\psi, \tag{12.1.2a}$$

where u_k are scalar functions, and the system of n equations:

$$\psi_x = (\lambda\mathbf{A} + \mathbf{Q})\psi \tag{12.1.2b}$$

where $\mathbf{A}$ is a constant matrix and $\mathbf{Q}$ a matrix of potential functions. The number of potential functions increases linearly with n in the first and quadratically in the second. It is of practical importance to find reduced spectral problems where the number of potential functions is considerably less than n^2. For instance, when (12.1.2a) is written as a system, it is an example of (12.1.2b) with only $(n - 1)$ of them.

A number of such reduced systems will be presented in this chapter. The approach will be to use Lie-algebraic and symmetric-space characterisations to isolate particularly interesting classes of spectral problems (and associated isospectral flows). Such Lie-algebraic characterisations are very interesting in themselves, since they give an 'explanation' of the existence of the spectral problems. Furthermore, they lend themselves to immediate generalisations to Kač–Moody, super-algebraic and other extensions. Some basic mathematical preliminaries are given in the Appendix.

There are a number of interesting generalisations which will not be considered in this chapter. We shall not be concerned with *matrix* Lax equations [31, 32] or with 'energy-dependent' spectral problems [1, 2, 19, 21]. The latter are briefly discussed in Chapter 11.

12.2 Zero curvature representations

In this section some basic results concerning general zero curvature representations are presented, leaving specific examples to later sections.

Consider a pair of linear, first-order differential equations:

$$\psi_x = \mathbf{F}\psi \tag{12.2.1a}$$

$$\psi_t = \mathbf{G}\psi \tag{12.2.1b}$$

where $\psi(x, t)$ is a vector-valued function, $\mathbf{F}$ and $\mathbf{G}$ are matrices depending upon a number of potential functions q_i and their partial derivatives, together with a spectral parameter λ.

The integrability conditions of (12.2.1a,b) are:

$$\mathbf{F}_t - \mathbf{G}_x + [\mathbf{F}, \mathbf{G}] = 0. \tag{12.2.2}$$

With particular choices of $\mathbf{F}$ and $\mathbf{G}$, (12.2.2) gives rise to a (system of) nonlinear partial differential equation(s). For specific classes of $\mathbf{F}$ it is possible to construct infinite hierarchies of $\mathbf{G}$ corresponding to infinite hierarchies of isospectral flows. This will be carried out in this chapter for the case in which $\mathbf{F}$ is linear in λ:

$$\mathbf{F} = \lambda\mathbf{A} + \mathbf{Q} \tag{12.2.3}$$

where $\mathbf{A}$ is a constant diagonalisable matrix and $\mathbf{Q} \in \operatorname{Im} \operatorname{ad}\mathbf{A}$ is the matrix of potential functions.

Remark

A will not generally be regular (distinct eigenvalues) and not necessarily diagonalised.

Remark

Case (12.2.3) is the simplest case to work with but is not essential. Zakharov and Mikhailov [34] use a pole expansion: $\mathbf{F} = \sum \mathbf{F}_i(q)(\lambda - \lambda_i)^{-1}$, while others [19, 21] favour polynomial expansions.

Substitute (12.2.3) into (12.2.2), so that:

$$\mathbf{Q}_t = \mathbf{G}_x - [\mathbf{Q}, \mathbf{G}] - \lambda[\mathbf{A}, \mathbf{G}]. \tag{12.2.4a}$$

We seek solutions of $\mathbf{G}$ which are polynomial in λ:

$$\mathbf{G}^{(m)} = \sum_{i=0}^{m} G_{m-i}\lambda^i. \tag{12.2.4b}$$

By equating coefficients of λ^k, $\mathbf{G}_0, \ldots, \mathbf{G}_m$ are found recursively, leading to the equations of motion:

$$\mathbf{Q}_{t_m} = \mathbf{G}_{mx} - [\mathbf{Q}, \mathbf{G}_m]. \tag{12.2.4c}$$

To achieve this for all $m \geq 0$ we seek an asymptotic solution:

$$\bar{\mathbf{G}} = \sum_{i=0}^{\infty} \mathbf{G}_i\lambda^{-i} \tag{12.2.5a}$$

of the equation:

$$\bar{\mathbf{G}}_x - [\mathbf{Q}, \bar{\mathbf{G}}] = \lambda[\mathbf{A}, \bar{\mathbf{G}}]. \tag{12.2.5b}$$

This corresponds to the infinite recursion:

$$\mathbf{G}_{kx} - [\mathbf{Q}, \mathbf{G}_k] = [\mathbf{A}, \mathbf{G}_{k+1}]. \tag{12.2.5c}$$

The solution (12.2.4b) is then given by:

$$\mathbf{G}^{(m)} = (\lambda^m\bar{\mathbf{G}})_+ \tag{12.2.6a}$$

where $(\ \ )_+$ is the truncation containing only non-negative powers of λ. Using (12.2.5c) with $k = m$, the equations of motion (12.2.4c) can be written:

$$\mathbf{Q}_{t_m} = \mathbf{G}_{mx} - [\mathbf{Q}, \mathbf{G}_m] = [\mathbf{A}, \mathbf{G}_{m+1}] \tag{12.2.6b}$$

thus confirming that $\mathbf{Q}$ does indeed lie in Im ad$\mathbf{A}$.

In general, $\mathbf{A} \in g$ is a semi-simple element of Lie algebra g, which admits the following vector-space decomposition:

$$g = \text{Ker ad}A \oplus Im\ ad\mathbf{A}. \tag{12.2.7a}$$

Furthermore, Im ad$\mathbf{A}$ can be written as the vector space direct sum of eigenspaces g_i (with eigenvalue a_i) of ad$\mathbf{A}$. Thus:

$$g = g_0 \oplus \sum_i g_i, \quad [\mathbf{A}, g_0] = 0, \quad [\mathbf{A}, g_i] = a_i g_i. \tag{12.2.7b}$$

Thus we can write $\mathbf{G}_k = \mathbf{G}_k^0 + \sum \mathbf{G}_k^i$. The $\mathbf{Q}$ of (12.2.3) is chosen to be a general element of the vector space complement of g_0 in g.

Remark

Any non-zero component of $\mathbf{Q}$ in g_0 can be gauged away.

The element $\mathbf{Q}$ can thus be written:

$$\mathbf{Q} = \sum_{\alpha \in \theta^+} (q^\alpha E_\alpha + r^{-\alpha} E_{-\alpha}) \tag{12.2.8}$$

where we have used the notation of reductive homogeneous spaces of the Appendix. However, since we have not necessarily diagonalised the element $\mathbf{A}$, the notation $\mathbf{E}_\alpha$ has been used in place of e_α.

The recursion

The recursion (12.2.5c) starts with the condition that $\mathbf{G}_0$ be an arbitarary *constant* element of Ker ad$\mathbf{A}$. To obtain the component $\mathbf{G}_1^0$ it is necessary to perform an integration and this gives rise to non-local expressions unless $\mathbf{G}_0$ is chosen to be in the *centre* of Ker ad$\mathbf{A}$. In the case of the modified Lax systems of Section 12.3, where $\mathbf{A}$ is regular, the whole of Ker ad$\mathbf{A}$ is commutative. In the highly degenerate Hermitian symmetric space case of Section 12.4, the centre of Ker ad$\mathbf{A}$ is just the single element $\mathbf{A}$.

The element $\mathbf{G}_1^0$ is, in fact, the only hurdle to cross. Wilson [33] proves the following lemma.

Lemma

For each element $\mathbf{G}_0$ of the *centre* of Ker ad$\mathbf{A}$, there is a formal series $\bar{\mathbf{G}}$, satisfying (12.2.5c), with each term $\mathbf{G}_k$ given as a differential polynomial in the components of $\mathbf{Q}$.

Remark

At each step of the recursion we can add a constant element of g_0. If we agree not to do so then the above formal series is unique.

Constants of motion and Hamiltonian structure

The conserved densities of the hierarchy (12.2.6) are given by:

$$\mathcal{H}_m = \frac{1}{m+1} \operatorname{tr} \mathbf{A}\mathbf{G}_{m+2} \tag{12.2.9a}$$

while the 'off-diagonal' components of $\mathbf{G}_{m+1}$ are given as variational derivaties [33]:

$$G_{m+1}^{\alpha} = \frac{\delta \mathcal{H}_m}{\delta r^{-\alpha}}, \quad G_{m+1}^{-\alpha} = \frac{\delta \mathcal{H}_m}{\delta q^{\alpha}} \tag{12.2.9b}$$

so that the second part of (12.2.6b) gives:

$$q^{\alpha}_{t_m} = \alpha(\mathbf{A}) \frac{\delta \mathcal{H}_m}{\delta r^{-\alpha}}, \quad r^{-\alpha}_{t_m} = -\alpha(\mathbf{A}) \frac{\delta \mathcal{H}_m}{\delta q^{\alpha}}. \tag{12.2.9c}$$

Thus, the hierarchy (12.2.6) is Hamiltonian with the densities (12.2.9a) being constant *wrt* each of its flows. The flows (12.2.9c) mutually commute and (equivalently) the densities (12.2.9a) are in involution *wrt* the corresponding Poisson bracket.

The next two sections are concerned with special examples of the above construction. The symmetric-space case of Section 12.4 just corresponds to a highly degenerate choice of **A**. However, the modified Lax systems of Section 12.3 have a regular element **A** but a highly restricted **Q**, no longer in general position. In this case the Hamiltonian structure (12.2.9c) is no longer appropriate.

12.3 Modified Lax hierachies and 2D Toda systems

The results of this section can be found in many guises in [9, 11, 12, 22, 24, 26, 27, 33]. I shall present the version found in [11, 12].

Scalar Lax equations and their modifications

The scalar Lax operator of order $(N + 1)$:

$$\mathbb{L} = \partial^{N+1} + u_{N-1}\partial^{N-1} + \ldots + u_1\partial + u_0 \tag{12.3.1a}$$

can be written as the product of $N + 1$ first-order operators:

$$\mathbb{L} = (\partial - v_{N+1}) \ldots (\partial - v_1) \tag{12.3.1b}$$

$\left(\text{with } \sum_{i=1}^{N+1} v_i = 0\right)$ giving rise to some identifications:

$$u_i = F_i[v_1, \ldots, v_{N+1}] \tag{12.3.1c}$$

where F_i depend upon the functions v_j and their x-derivative (of order $\leq N$). The spectral problem:

$$\mathbb{L}\psi_1 = \lambda^{N+1}\psi_1 \tag{12.3.2a}$$

may then be written in first-order differential, trace-free matrix form:

$$\begin{bmatrix} \psi_1 \\ \cdot \\ \cdot \\ \cdot \\ \cdot \\ \psi_{N+1} \end{bmatrix}_x = \begin{bmatrix} v_1 & \lambda & & 0 \\ & \ddots & \ddots & \\ & & \ddots & \lambda \\ 0 & & & \\ \lambda & & & v_{N+1} \end{bmatrix} \begin{bmatrix} \psi_1 \\ \cdot \\ \cdot \\ \cdot \\ \cdot \\ \psi_{N+1} \end{bmatrix} \equiv (\lambda\mathbf{A} + \mathbf{v})\psi. \tag{12.3.2b}$$

The identifications (12.3.1c) then play the role of a Miura map between the isospectral flows of the two spectral problems (12.3.2a, b).

The spectral problem (12.3.2b) is an example of (12.2.1a) with $\mathbf{F}$ given by (12.2.3). In this case the element $\mathbf{A}$ is regular so that its centraliser g_0 (lying within $sl(N+1)$ is Abelian and of dimension N, the rank of $sl(N+1)$. However, since $\mathbf{v}$ is not in general position within Im ad$\mathbf{A}$, not every flow given by the construction of Section 12.2 will exist. The sequence of matrices in $\mathbf{G}_i$ always exists, but now we demand that $\mathbf{G}_m \in h$, so that:

$$\mathbf{v}_t = \mathbf{G}_{mx} - [\mathbf{v}, \mathbf{G}_m] = \mathbf{G}_{mx} \tag{12.3.2c}$$

is consistent. For a given m this uniquely fixes $\mathbf{G}_0$ (up to a constant factor). With our choice of $\mathbf{A}$, $C_g(\mathbf{A}) = \text{span}\{\mathbf{A}, \mathbf{A}^2, \ldots, \mathbf{A}^N\}$ and $\mathbf{A}^{N+1} = I_{N+1}$ (the $(N+1) \times (N+1)$ identity matrix). For each $\bar{m}$, $1 \leq \bar{m} \leq N$, there exists an infinite sequence of flows running through the orders m such that $m \equiv \bar{m} (\text{mod } N+1)$ and for each m there exists exactly one flow and this corresponds to choosing $\mathbf{G}_0 = \mathbf{A}^{\bar{m}}$.

Since $\mathbf{v}$ is not in general position within Im ad$\mathbf{A}$, the Hamiltonian structure (12.2.9c) is no longer appropriate. However, the second part of (12.3.2c) shows that the equations are Hamiltonian (with structure ∂) since $\mathbf{G}_m$ can be written [33] as a variational derivative of Hamiltonian $\mathcal{H}_{m-1} = 1/m' \text{tr} \mathbf{A}\mathbf{G}_{m+1}$. Since $\mathbf{A}$ is of height 1 this trace is only non-zero when $\mathbf{G}_{m+1}$ has height -1. Thus the only Hamiltonians which remain non-trivial after $\mathbf{Q}$ has been reduced to $\mathbf{v}$ are those corresponding to the non-trivial flows of (12.3.2c) (as expected). All these flows commute.

In this section, however, we are interested in the generalised sinh-Gordon flow. Consider a time evolution of the form

$$\begin{bmatrix} \psi_1 \\ \vdots \\ \psi_{N+1} \end{bmatrix}_t = \frac{1}{\lambda} \begin{bmatrix} 0 & \cdots & \cdots & a_{1N+1} \\ a_{21} & \ddots & & \vdots \\ & \ddots & \ddots & \vdots \\ 0 & & a_{N+1N} & 0 \end{bmatrix} \begin{bmatrix} \psi_1 \\ \vdots \\ \psi_{N+1} \end{bmatrix} \tag{12.3.2d}$$

where a_{nn-1} and a_{1N+1} are the only non-zero elements.

The integrability conditions of (12.3.2b, d) are:

$$a_{n+1nx} = a_{n+1n}(v_{n+1} - v_n), \tag{12.3.3a}$$

$$v_{nt} = a_{nn-1} - a_{n+1n}, \tag{12.3.3b}$$

where $n = 1, \ldots, N+1$ and indices are considered modulo $N+1$. To solve (12.3.3a), define θ_i by $v_i = \theta_{ix}$, so that:

$$a_{n+1n} = A_{n+1n} \exp(\theta_{n+1} - \theta_n) \tag{12.3.4a}$$

where A_{n+1n} are the constants of integration. The equations of motion (12.3.3b) then take the form of a 2D Toda lattice:

$$\theta_{nxt} = A_{nn-1}\exp(\theta_n - \theta_{n-1}) - A_{n+1n}\exp(\theta_{n+1} - \theta_n) \quad (12.3.4b)$$

The two simplest examples (taking $A_{n+1n} = 1$, for all n) are:

$N = 1$:
$$-\theta_2 = \theta_1 = \theta, \quad \theta_{xt} = 2\sinh 2\theta \quad (12.3.5)$$

which is just the sinh-Gordon equation.

$N = 2$:
$$\theta_1 = \theta + \phi, \quad \theta_2 = -2\phi, \quad \theta_3 = -\theta + \phi,$$
$$\theta_{xt} = e^{2\theta} - e^{-\theta}\cosh 3\phi, \quad \phi_{xt} = e^{-\theta}\sinh 3\phi \quad (12.3.6a)$$

A reduction of this system is:

$$\phi \equiv 0, \quad \theta_{xt} = e^{2\theta} - e^{-\theta} \quad (12.3.6b)$$

which is the Dodd–Bullough [8] equation.

More examples can be found in [11].

Lie-algebraic generalisation

The form of the exponents in (12.3.4b) are the principal clue towards placing these equations in a Lie-algebraic framework, since they are just the root forms of the algebra $sl(N+1)$ (root space A_N).

As a basis for the Cartan subalgebra h of $sl(N+1)$ choose the diagonal elements $\{h_i = e_{ii}\}_{i=1}^{N+1}$ of $gl(N+1)$ (where e_{ij} is the matrix with 1 in the i-j position and 0 elsewhere). Any element $h \in h$ is then written:

$$h = \sum_{i=1}^{N+1} \lambda_i h_i, \quad \sum_{i=1}^{N+1} \lambda_i = 0. \quad (12.3.7a)$$

Remark

This basis satisfies convenient orthogonality relations *wrt* the trace form: $\mathrm{tr} h_i h_j = \delta_{ij}$. This 'trick' will not be necessary for the other simple Lie algebras, since the standard basis for h is already orthogonal (but not normalised).

A set of simple roots are $\{\alpha_i = \lambda_i - \lambda_{i+1}\}_{i=1}^{N}$ and $\alpha_{\max} = \lambda_1 - \lambda_{N+1}$. The corresponding eigenvectors are:

$$e_{\alpha_i} = e_{i\,i+1}, \quad i = 1, \ldots, N+1 \quad (12.3.7b)$$

with $e_{\alpha_{N+1}} = e_{N+11} = e_{-\alpha_{\max}}$. In this framework the elements $\mathbf{A}$ and $\mathbf{v}$ of (12.3.2b) are just

$$\mathbf{A} = \sum_{i=1}^{N+1} e_{\alpha_i}, \quad \mathbf{v} = \sum_{i=1}^{N+1} v_i h_i, \quad \sum_{i=1}^{N+1} v_i = 0. \quad (12.3.7c)$$

The spectral problem (12.3.2b) with $v_i = \theta_{ix}$ and time evolution (12.3.2d) are then:

$$\psi_x = \left(\lambda \sum_{i=1}^{N+1} e_{\alpha_i} + \sum_{i=1}^{N+1} \theta_{ix} h_i\right)\psi, \tag{12.3.8a}$$

$$\psi_t = \frac{1}{\lambda} \sum_{i=1}^{N+1} a_i e_{-\alpha_i} \psi. \tag{12.3.8b}$$

The integrability conditions of (12.3.8a, b) are then:

$$\sum_{i=1}^{N+1} \theta_{ixt} h_i + \sum_{i=1}^{N+1} a_i [e_{\alpha_i}, e_{-\alpha_i}] + \frac{1}{\lambda}\left(\sum_{i,k=1}^{N+1} \theta_{kx} a_i [h_k, e_{-\alpha_i}] - \sum_{i=1}^{N+1} a_{ix} e_{-\alpha_i}\right) = 0 \tag{12.3.9}$$

so that, using the commutation relations (12.A.1):

$$a_{kx} = -a_k \sum_{i=1}^{N+1} \alpha_k(h_i)\theta_{ix}, \tag{12.3.10a}$$

$$\theta_{kxt} = -\sum_{i=1}^{N+1} a_j d_{jk}. \tag{12.3.10b}$$

For $sl(N+1)$ the solution of (12.3.10a) is given by (12.3.4a) and equations (12.3.10b) reduce to (12.3.4b). However, having placed our linear equations in such an abstract context the subsequent calculations can be carried out *wrt* an arbitrary simple Lie algebra. The Cartan elements, h_i, root forms α_i and root vectors e_{α_i} must now be considered as belonging to an arbitrary simple Lie algebra. However, we must now remember that although we still have $(N + 1)$ roots $\{\alpha_1, \ldots, \alpha_N, \alpha_{N+1} = -\alpha_{\max}\}$, where N is the rank of the algebra we have only N Cartan elements h_i. Thus, in the above formulae, summation over h_i is now only taken to N.

The solution of (12.3.10a) and consequent equations of motion (12.3.10b) are then:

$$a_k = A_k \exp\left(-\sum_{i=1}^{N} \alpha_k(h_i)\theta_i\right), \tag{12.3.11a}$$

$$\theta_{kxt} = -\sum_{j=1}^{N+1} d_{jk} A_j \exp\left(-\sum_{i=1}^{N} \alpha_k(h_i)\theta_i\right). \tag{12.3.11b}$$

Remark

The coefficients A_k are, of course, arbitrary, but the other coefficients are determined by the specific root space and chosen basis. In particular, $\alpha_k(h_i)$ defines the Cartan matrix [18] associated with the particular root space.

Remark

This generalisation is modelled directly on Bogoyavlenskii's extension of the (finite-dimensional) periodic Toda lattices to those associated with simple Lie algebras [5]. Such extensions had previously been introduced by Olshanetsky and Perelomov in the context of the Calogero–Moser and similar systems (see [29] for a review).

Remark

The 2D Toda lattice equations (12.3.11b) have appeared in several places [12, 24, 27]. The appearance of the extended root system $\{\alpha_1, \ldots, \alpha_N, -\alpha_{\max}\}$ indicates that these equations are actually associated with the infinite-dimensional affine Lie algebras [9, 23].

Classification

The Cartan classification of simple Lie algebras can be used directly to classify the spectral problems (12.3.8a) and field equations (12.3.11b). In [12] a number of lower-dimensional examples are explicitly given. Below we give the form of the potentials which give rise to equations (12.3.11b):

$$\theta_{ixt} = -\frac{\partial \mathcal{V}}{\partial \theta_i} \tag{12.3.12a}$$

with

$$\begin{aligned}
\mathcal{V}_{A_n} &= \mathcal{V}_n + \exp(\theta_1 - \theta_{n+1}), \quad n \geq 2,\\
\mathcal{V}_{B_n} &= \mathcal{V}_{n-1} + \exp(-\theta_n) + \exp(\theta_1 + \theta_2), \quad n \geq 2,\\
\mathcal{V}_{C_n} &= \mathcal{V}_{n-1} + \exp(-2\theta_1) + \exp(2\theta_n), \quad n \geq 3,\\
\mathcal{V}_{D_n} &= \mathcal{V}_{n-1} + \exp(-\theta_{n-1} - \theta_n) + \exp(\theta_1 + \theta_2), \quad n \geq 4,\\
\mathcal{V}_{E_6} &= \mathcal{V}_5 + \exp(\tfrac{1}{2}(\theta_1 + \theta_2 + \theta_3 - \theta_4 - \theta_5 - \theta_6) - \theta_7/\sqrt{2}) + \exp(\sqrt{2}\theta_7),\\
\mathcal{V}_{E_7} &= \mathcal{V}_5 + \exp(\tfrac{1}{2}(\theta_1 - \theta_2 - \cdots - \theta_7 + \theta_8)) + \exp(\theta_1 + \theta_2)\\
&\quad + \exp(-\theta_7 - \theta_8),\\
\mathcal{V}_{E_8} &= \mathcal{V}_6 + \exp(\tfrac{1}{2}(\theta_1 - \theta_2 - \cdots - \theta_7 + \theta_8)) + \exp(\theta_1 + \theta_2)\\
&\quad + \exp(-\theta_7 - \theta_8),\\
\mathcal{V}_{F_4} &= \mathcal{V}_2 + \exp(-\theta_3) + \exp(\tfrac{1}{2}(\theta_1 + \theta_2 + \theta_3 - \theta_4)) + \exp(\theta_1 + \theta_4),\\
\mathcal{V}_{G_2} &= \exp(\theta_2 - \theta_1) + \exp(2\theta_1 - \theta_2 - \theta_3) + \exp(-\theta_1 - \theta_2 + 2\theta_3),
\end{aligned} \tag{12.3.12b}$$

where the coefficients $A_j = 1$ and $\mathcal{V}_m = \sum_{i=1}^{m} \exp(\theta_{i+1} - \theta_i)$.

Generalised symmetries

The polynomial flows of the spectral problem (12.3.8a) mutually commute and indeed, commute with the generalised sinh-Gordon flow (12.3.11b), generated by the evolution (12.3.8b). Once again $\mathbf{v} = \Sigma_{i=1}^{N} \theta_{ix} h_i$ is not in general position within Im ad$\mathbf{A}$ and not every flow survives this reduction. With our choice of $\mathbf{A}$, $C_g(\mathbf{A}) = \text{span}\{\mathbf{A}^{m_i}\}$ where m_i runs through the exponents of g (see Appendix). For each such m_i there exists an infinite sequence of flows whose orders are now given by $m \equiv m_i (\text{mod } h)$, where h is the Coxeter number of g.

Remark

This is proved in [33], but roughly follows from the following argument: since e_{α_i}, $i = 1, \ldots, 1 + \text{rank}\, g$ are each of height 1 (mod h), formula (12.2.5c) gives $\text{ht}\mathbf{G}_{k+1} = \text{ht}\mathbf{G}_k - 1$. Thus if $\text{ht}\mathbf{G}_0 = m_i$, then $\text{ht}\mathbf{G}_m = m_i - m$ and for $\mathbf{G}_m \in h$, we require $m_i - m \equiv 0 \ (\text{mod } h)$.

Thus, for each $m \equiv m_i (\text{mod } h)$ there exists exactly one isospectral flow of (12.3.8a) (in the case of D_{2k} there exist two flows when $m_i = 2k - 1$).

Scalar Lax operators associated with simple Lie algebras

Recall that at the beginning of this section the 'modified' spectral problem (12.3.2b) was derived from the scalar Lax equation (12.3.2a) through the factorisation (12.3.1b). It is interesting that for the spectral problem (12.3.8a) we can reverse this process and obtain Lax operators associated with simple Lie algebras. Some examples are given in [12] and below. A complete treatment is given in [9]. One example is given below.

Example – B_3

This rank-3 root space corresponds to the Lie algebra $so(7)$. For the purposes of this section it is convenient to use the (less usual) representation given in [18]. The spectral problem is:

$$\begin{bmatrix} \psi_1 \\ \cdot \\ \cdot \\ \cdot \\ \cdot \\ \cdot \\ \psi_7 \end{bmatrix}_x = \begin{bmatrix} 0 & 0 & 0 & 0 & 0 & 0 & -\lambda \\ 0 & v_1 & \lambda & 0 & 0 & 0 & 0 \\ 0 & 0 & v_2 & \lambda & 0 & 0 & 0 \\ \lambda & 0 & 0 & v_3 & 0 & 0 & 0 \\ 0 & 0 & -\lambda & 0 & -v_1 & 0 & 0 \\ 0 & \lambda & 0 & 0 & -\lambda & -v_2 & 0 \\ 0 & 0 & 0 & 0 & 0 & -\lambda & -v_3 \end{bmatrix} \begin{bmatrix} \psi_1 \\ \cdot \\ \cdot \\ \cdot \\ \cdot \\ \cdot \\ \psi_7 \end{bmatrix} \qquad (12.3.13a)$$

The coefficients in the time evolution (12.3.8b) are defined by $a_1 = \exp(\theta_2 - \theta_1)$, $a_2 = \exp(\theta_3 - \theta_2)$, $a_3 = \exp(-\theta_3)$ and $a_4 = \exp(\theta_1 + \theta_2)$, where $v_i = \theta_{ix}$. The equations of motion (12.3.11b) are:

$$\begin{aligned} \theta_{1xt} &= \exp(\theta_1 + \theta_2) - \exp(\theta_2 - \theta_1), \\ \theta_{2xt} &= \exp(\theta_2 - \theta_1) - \exp(\theta_3 - \theta_2) + \exp(\theta_1 + \theta_2), \\ \theta_{3xt} &= \exp(\theta_3 - \theta_2) - \exp(\theta_3). \end{aligned} \qquad (12.3.13b)$$

If we define $\mathbb{L} = (\partial - v_3)(\partial - v_2)(\partial - v_1)$ then it follows form (12.3.13a) that:

$$\mathbb{L}\psi_2 = \lambda^3\psi_1, \quad \partial\psi_1 = -\lambda\psi_7, \quad \mathbb{L}^\dagger\psi_7 = 2\lambda^2\partial\psi_2 \qquad (12.3.14a)$$

so that

$$\mathbb{L}^\dagger\partial\mathbb{L}\psi_2 = -2\lambda^b\partial\psi_2. \qquad (12.3.14b)$$

This an unusual form of Lax equation giving rise to isospectral flows for u_i, defined by $\mathbb{L} = \partial^3 + u_2\partial^2 + u_1\partial + u_0$, which defines a Miura map between v_i and u_i.

Remark
The exceptional root space G_2 can be embedded in B_3. The scalar Lax equation is still (12.3.14b) but with the reduced $\mathbb{L}$ given by $u_2 = 0$.

12.4 **Hierarchies associated with Hermitian symmetric spaces**

In this section we give another class of spectral problem (12.2.1) with **F** of the form (12.2.3). Given a simple Lie algebra g, let **A** be an element of the Cartan subalgebra: $\mathbf{A} \in h \subset g$. If **A** is regular $C_g(\mathbf{A}) = h$. Otherwise $C_g(\mathbf{A}) \supset h$. Let $k = C_g(\mathbf{A})$ and m be the complementary vector subspace of k in g:$g = k \oplus m$. As in (12.A.3), m is identified with the tangent space $T_{p_0}(G/K)$ of the homogeneous space G/K. In this chapter we use representations in which all elements of h are diagonalised. Because of the construction, this homogeneous space is automatically reductive.

In [13] the more general reductive homogeneous spaces were considered, but for the purposes of this chapter it is enough to discuss the symmetric-space case. The relations (12.A.3) play an important role in decoupling the integrability conditions (12.4.2) below.

Consider the linear equations:

$$\psi_x = (\lambda\mathbf{A} + \mathbf{Q}(x, t))\psi \qquad (12.4.1a)$$

$$\psi_t = \mathbf{G}(x, t; \lambda)\psi \qquad (12.4.1b)$$

where $\mathbf{Q} \in m$ and $\mathbf{G} \in g$. In terms of the symmetric space coordinates we can write:

$$\mathbf{Q} = \sum_{\alpha\in\theta^+} (q^\alpha e_\alpha + r^{-\alpha} e_{-\alpha}). \qquad (12.4.1c)$$

The integrability conditions of (12.4.1a, b) are:

$$\mathbf{Q}_t = \mathbf{G}_x - [\mathbf{Q}, \mathbf{G}] - \lambda[\mathbf{A}, \mathbf{G}]. \qquad (12.4.2)$$

The relations (12.A.3) enable us to project into the three eigenspaces of ad**A** and thus decouple (12.4.2):

$$\mathbf{Q}_t^{\pm} = \mathbf{G}_x^{\pm} - [\mathbf{Q}^{\pm}, \mathbf{G}^0] \mp \lambda a \mathbf{G}^{\pm}, \tag{12.4.3a}$$

$$\mathbf{G}_x^0 = [\mathbf{Q}^+, \mathbf{G}^-] + [\mathbf{Q}^-, \mathbf{G}^+]. \tag{12.4.3b}$$

Following the procedure of Section 12.2, we seek an (infinite) asymptotic series

$$\bar{\mathbf{G}} = \sum_{i=0}^{\infty} \mathbf{G}_i \lambda^{-i} \tag{12.4.4}$$

satisfying (12.4.3b) and the 'stationary' version of (12.4.3a). The coefficients $\mathbf{G}_i$ thus satisfy the infinite recursion:

$$\mathbf{G}_{i+1}^{\pm} = \pm \frac{1}{a}(\mathbf{G}_{ix}^{\pm} - [\mathbf{Q}^{\pm}, \mathbf{G}_i^0]), \tag{12.4.5a}$$

$$\mathbf{G}_{ix}^0 = [\mathbf{Q}^+, \mathbf{G}_i^-] + [\mathbf{Q}^-, \mathbf{G}_i^+]. \tag{12.4.5b}$$

The recursion starts with $\mathbf{G}_0$ satisfying:

$$\mathbf{G}_0^{\pm} = 0, \quad \mathbf{G}_{0x}^0 = 0, \tag{12.4.6a}$$

so that $\mathbf{G}_0$ is a constant element of $C_g(\mathbf{A})$: $\mathbf{G}_0 = \mathbf{B} \in C_g(\mathbf{A})$. The next step in the recursion leads to:

$$\mathbf{G}_1^{\pm} = \pm \frac{1}{a}[\mathbf{B}, \mathbf{Q}^{\pm}], \quad \mathbf{G}_{ix}^0 = \frac{1}{a}[[\mathbf{Q}^+, \mathbf{Q}^-], \mathbf{B}]. \tag{12.4.6b}$$

It is seen that unless **B** lies in the centre of $C_g(\mathbf{A})$, then $\mathbf{G}_1^0$ is non-local. For Hermitian symmetric spaces the centre of $C_g(\mathbf{A})$ is 1-dimensional, namely **A**. Thus we have (up to a constant factor):

$$\mathbf{G}_0 = \mathbf{A}, \quad \mathbf{G}_1 = \mathbf{Q}. \tag{12.4.7a}$$

The next two terms in this recursion are given by:

$$\mathbf{G}_2 = \frac{1}{a}(\mathbf{Q}_x^+ - \mathbf{Q}_x^- - [\mathbf{Q}^+, \mathbf{Q}^-]), \tag{12.4.7b}$$

$$\mathbf{G}_3 = \frac{1}{a^2}([\mathbf{Q}, \mathbf{Q}_x] + \mathbf{Q}_{xx} + [\mathbf{Q}^+, [\mathbf{Q}^+, \mathbf{Q}^-]] + [\mathbf{Q}^-, [\mathbf{Q}^-, \mathbf{Q}^+]]). \tag{12.4.7c}$$

The mth-order flow is then given by (12.2.6b), the first two non-trivial ones being:

$$\pm a \mathbf{Q}_{t_2}^{\pm} = \mathbf{Q}_{xx}^{\pm} + [\mathbf{Q}^{\pm}, [\mathbf{Q}^{\pm}, \mathbf{Q}^{\mp}]], \tag{12.4.8a}$$

$$a^2 \mathbf{Q}_{t_3} = \mathbf{Q}_{xxx}^{\pm} + 3[\mathbf{Q}_x^{\pm}, [\mathbf{Q}^{\pm}, \mathbf{Q}^{\mp}]]. \tag{12.4.8b}$$

In terms of the symmetric space coordinates q^α, $r^{-\alpha}$ these equations take the form:

$$aq^\alpha_{t_2} = q^\alpha_{xx} + R^\alpha{}_{\beta\gamma-\delta}q^\beta q^\gamma r^{-\delta}, \qquad (12.4.9a)$$

$$a^2q^\alpha_{t_3} = q^\alpha_{xxx} + 3R^\alpha{}_{\beta\gamma-\delta}q^\beta_x q^\gamma r^{-\delta}, \qquad (12.4.9b)$$

where $R^\alpha{}_{\beta\gamma-\delta}$ are the components of the curvature tensor (12.A.8) *wrt* the basis $e_{\pm\alpha}$ and where $r^{-\alpha}$ satisfies analogous equations.

Remark
The second-order flow (12.4.9a), written in the real form $r^{-\alpha} = \pm(q^\alpha)^*$, is a generalised NLS system and was considered in detail in [13]. The third-order flow (12.4.9b) possesses KdV- and MKdV-like reductions, considered in [3].

Classifications

The Hermitian symmetric spaces are listed in Helgason ([17] p. 518), thus providing a classification of all such equations. There are four infinite families: AIII, CI, DIII and BDI, and two further examples associated with exceptional algebras of the E-series: EIII and EVII. Examples with explicit representations are given in [3, 13]. One such example is given below.

Hamiltonian structure and conserved densities

The Hamiltonian structure of the above hierarchy of equations was given by Fordy and Kulish [13]:

$$q^\alpha_{t_m} = g^{\alpha-\beta}\frac{\delta\mathcal{H}_m}{\delta r^{-\beta}}, \quad r^{-\beta}_{t_m} = -g^{-\alpha\beta}\frac{\delta\mathcal{H}_m}{\delta q^\beta} \qquad (12.4.10a)$$

with conserved densities $\mathcal{H}_m$ defined by:

$$a^{m-1}\mathcal{H}_m = \frac{1}{m+1}\mathrm{tr}\mathbf{AG}_{m+2} \qquad (12.4.10b)$$

and where $g^{\alpha-\beta}$ is the metric defined by (12.A.9) (see also Section 12.2 and [33]). The conserved densities can be written explicitly in terms of geometric quantities. The first five are given below:

$$\mathcal{H}_0 = \sum_{\alpha,\beta} g_{\alpha-\beta}q^\alpha r^{-\beta}, \qquad (12.4.11a)$$

$$\mathcal{H}_1 = \frac{1}{2}\sum_{\alpha,\beta} g_{\alpha-\beta}(q^\alpha_x r^{-\beta} - q^\alpha r^{-\beta}_x), \qquad (12.4.11b)$$

$$a\mathscr{H}_2 = -\sum_{\alpha,\beta} g_{\alpha-\beta} q_x^{\alpha} r_x^{-\beta} + \frac{1}{2} \sum_{\alpha,\beta,\gamma,\delta} R_{-\alpha\beta\gamma-\delta} r^{-\alpha} q^{\beta} q^{\gamma} r^{-\delta}, \quad (12.4.11c)$$

$$a^2\mathscr{H}_3 = \frac{1}{2}\sum_{\alpha,\beta} g_{\alpha-\beta}(q_{xxx}^{\alpha} r^{-\beta} - q^{\alpha} r_{xxx}^{-\beta})$$
$$- \frac{3}{4} \sum_{\alpha,\beta,\gamma,\delta} \mathrm{R}_{-\alpha\beta\gamma-\delta}(r_x^{-\alpha} q_x^{\beta} q^{\gamma} r^{-\delta} - r^{-\alpha} q_x^{\beta} q^{\gamma} r^{-\delta}), \quad (12.4.11d)$$

$$a^3\mathscr{H}_4 = \sum_{\alpha,\beta} g_{\alpha-\beta} q_{xx}^{\alpha} r_{xx}^{-\beta} + \frac{1}{2} \sum_{\alpha,\beta,\gamma,\delta,\varepsilon,\rho,\sigma} \mathrm{R}_{-\alpha\beta\gamma-\delta} \mathrm{R}^{\gamma}{}_{\varepsilon\rho-\sigma} r^{-\alpha} q^{\beta} r^{-\delta} q^{\varepsilon} q^{\rho} r^{-\sigma}$$
$$- \frac{1}{2} \sum_{\alpha,\beta,\gamma,\delta} \mathrm{R}_{-\alpha\beta\gamma-\delta}(r^{-\alpha} q^{\beta})_x (q^{\gamma} r^{-\delta})_x - 3 \sum_{\alpha,\beta,\gamma,\delta} \mathrm{R}_{-\alpha\beta\gamma-\delta} r_x^{-\alpha} q_x^{\beta} q^{\gamma} r^{-\delta}. \quad (12.4.11e)$$

Example – class AIII and CI NLS equations
Class AIII symmetric spaces are given by $\mathrm{SU}(p + q)/\mathrm{S}(\mathrm{U}(p) \times \mathrm{U}(q))$. When $p = 1$ the second-order flow is the familiar vector NLS equation. The linear problem is an equation in the Lie algebra $su\,(p + q)$, which is the compact real form associated with the root space $\mathscr{A}_{p+q-1}$. When $p = q = 2$ we have (with $r^{-\alpha} = -(q^{\alpha})^*$):

$$\begin{bmatrix} \psi_1 \\ \psi_2 \\ \psi_3 \\ \psi_4 \end{bmatrix}_x = \left[\begin{array}{cc|cc} \frac{1}{2}\mathrm{i}\lambda & 0 & q_1 & q_2 \\ 0 & \frac{1}{2}\mathrm{i}\lambda & q_4 & q_3 \\ \hline -q_1^* & -q_4^* & -\frac{1}{2}\mathrm{i}\lambda & 0 \\ -q_2^* & -q_3^* & 0 & -\frac{1}{2}\mathrm{i}\lambda \end{array}\right] \begin{bmatrix} \psi_1 \\ \psi_2 \\ \psi_3 \\ \psi_4 \end{bmatrix} \quad (12.4.12)$$

The choice of $\mathbf{A}$ makes $\alpha(\mathbf{A}) = \mathrm{i}$ in the top right-hand block. The first two components of the second-order flows are:

$$\mathrm{i}q_{1t_2} = q_{1xx} + 2q_1 \sum_{i=1}^{4} q_i q_i^* - 2q_3^*(q_1 q_3 - q_2 q_4), \quad (12.4.13a)$$

$$\mathrm{i}q_{2t_2} = q_{2xx} + 2q_2 \sum_{i=1}^{4} q_i q_i^* + 2q_4^*(q_1 q_3 - q_2 q_4). \quad (12.4.13b)$$

The second two are generated from (12.4.13) by the interchange $1 \leftrightarrow 3$ $2 \leftrightarrow 4$. There are then four complex-conjugate equations.

Class CI symmetric spaces are given by $\mathrm{Sp}(n)/\mathrm{U}(n)$, which is a submanifold of $\mathrm{SU}(2n)/\mathrm{S}(\mathrm{U}(n) \times \mathrm{U}(n))$:

$$\frac{\mathrm{Sp}(n)}{\mathrm{U}(n)} \subset \frac{\mathrm{SU}(2n)}{\mathrm{S}(\mathrm{U}(n) \times \mathrm{U}(n))}$$

corresponding to each of the off-diagonal (square) blocks being symmetric. The NLS equations corresponding to $\mathrm{Sp}(2)/\mathrm{U}(2)$ are thus derived from (12.4.13) by setting $q_4 \equiv q_2$.

Remark

A similar reduction occurs in the case of DIII symmetric spaces. Here we use the property:

$$\frac{SO(2n)}{U(n)} \subset \frac{SU(2n)}{S(U(n) \times U(n))}$$

corresponding to the off-diagonal blocks being skew-symmetric.

Generalised KdV reduction

Equation (12.4.9b) admits both a KdV- and MKdV-like reduction [3]. The KdV reduction corresponds to $r^{-\alpha}$ being constant so that, by use of the Bianchi identities, (12.4.9b) can be written:

$$a^2 q^{\alpha}_{t_3} = (q^{\alpha}_{xx} + \tfrac{3}{2} R^{\alpha}{}_{\beta\gamma-\delta} q^{\beta} q^{\gamma} r^{-\delta})_x. \tag{12.4.14}$$

This is a generalisation of the KdV system which can, in all but class BDI symmetric spaces, be embedded in a matrix KdV equation:

$$a^2 \mathbf{u}_{t_3} = \mathbf{u}_{xxx} - 3(\mathbf{u}^2)_x \tag{12.4.15}$$

with $\mathbf{u}$ an $n \times n$ matrix defined below. Equation (12.4.15) is isospectral to a matrix Schrödinger equation which can be derived from the spectral problem (12.4.1a). In all but class BDI symmetric spaces the spectral problem (12.4.1a) has block-diagonal form:

$$\begin{bmatrix} \psi_1 \\ \psi_2 \end{bmatrix} = \left[\begin{array}{c|c} b\lambda I_m & \mathbf{q} \\ \hline \mathbf{r} & c\lambda I_n \end{array}\right] \begin{bmatrix} \psi_1 \\ \psi_2 \end{bmatrix} \tag{12.4.16a}$$

where $\mathbf{q}$ and $\mathbf{r}$ are $m \times n$ and $n \times m$ matrices respectively, I_m is the $m \times m$ identity matrix and ψ_1, ψ_2 are m- and n-dimensional column vectors. Since tr$\mathbf{A} = 0$ we have $mb + nc = 0$. The matrix $\mathbf{r}$ is constant and $\mathbf{q}$ contains at most mn functions. The $n \times n$ matrix $\mathbf{u}$ of (12.4.15) is defined by $\mathbf{u} = \mathbf{rq}$. The vector ϕ, defined by $\phi = \psi_2 \exp[\frac{1}{2}(b + c)\lambda x]$, satisfies the matrix Schrödinger equation:

$$\phi_{xx} - \mathbf{u}\phi = \tfrac{1}{4}(b - c)^2 \lambda^2 \phi. \tag{12.4.16b}$$

When $m = n$, $b + c = 0$ so that this is the identity transformation. This condition can hold in class AIII and always holds in classes CI (in which case $\mathbf{q}$ and $\mathbf{r}$ are square and symmetric) and DIII (in which case $\mathbf{q}$ and $\mathbf{r}$ are square and skew-symmetric). In the class AIII it is usual that $m \neq n$, so that without loss of generality $m < n$, in which case (12.4.15) consists of n^2 scalar equations whereas (12.4.14) represents $mn(<n^2)$ equations. Indeed, in both cases there are only mn independent functions q^{α}, making the system (12.4.15) degenerate. This is a reduction of the n^2 independent equations derived from (12.4.16a) when $m = n$.

Furthermore, the same manipulation can be carried out for the $m \times m$ matrix $\mathbf{w} = \mathbf{qr}$. Here we have only $m^2 < mn$ independent equations, this time for certain linear combinations of the (at most) mn independent functions in $\mathbf{q}$.

Hamiltonian formulation of matrix KdV equations

The hierarchy of equations isospectral to (12.4.16b) have Hamiltonian form:

$$u_{ijt_k} = \partial \frac{\delta}{\delta u_{ji}} H_k \tag{12.4.17a}$$

where the conserved densities H_k are derived in a standard way [3] from the matrix Riccati equation associated with (12.4.16b). The matrix KdV equation (12.4.15) is derived from the density H_3:

$$H_3 = -\frac{1}{a^2}\,\mathrm{tr}(\tfrac{1}{2}\mathbf{u}_x^2 + \mathbf{u}^3). \tag{12.4.17b}$$

Remark

In the reduction $r^{-\alpha}$ = const., the odd densities of (12.4.11) are exact derivatives whereas the even ones correspond to those of (12.4.17). In particular, under this reduction $\mathcal{H}_4$ corresponds to H_3 of (12.4.17).

We can similarly write the equations for the $m \times m$ matrix $\mathbf{w} = \mathbf{qr}$ in Hamiltonian form. If $m \le n$ then the equation for $\mathbf{w}$ is a system of m^2 scalar equations while that for $\mathbf{u}$ contains n^2 scalar equations. Thus, the original mn equations for variables q^α are squeezed between two Hamiltonian systems: $m^2 \le mn \le n^2$. They constitute an interesting reduction of the matrix system for $\mathbf{u}$ while the m^2 variables w_{ij} are constructed out of linear combinations of the variables q^α and thus satisfy a further reduction of (12.4.14). This will be seen more clearly in the following examples in class AIII spaces.

Class AIII symmetric spaces

In this case the spectral problem is of the form (12.4.16a) with $\mathbf{q}$ an arbitary $m \times n$ matrix, giving rise to equations in mn components. The constants of motion (12.4.11) are written in terms of the coordinates q^α and can only be considered as Hamiltonians (*wrt* Hamiltonian structure (12.4.17)) when written in terms of the components of matrices $\mathbf{u}$ and $\mathbf{w}$. In the case of the third-order, KdV-like flow the Hamiltonian H_3 (using the summation convention) is:

$$a^2 H_3 = -\tfrac{1}{2}u_{ijx}u_{jix} - u_{ij}u_{jk}u_{ki} \equiv -\tfrac{1}{2}w_{ijx}w_{jix} - w_{ij}w_{jk}w_{ki} \tag{12.4.18a}$$

with equations (12.4.17) being written:

$$a^2 u_{ijt_3} = (u_{ijxx} - 3u_{ik}u_{kj})_x \qquad (12.4.18b)$$

and similarly for w_{ij}. Since there are (generally) only mn independent functions in $\mathbf{q}$ the system of equations for u_{ij} is apparently over determined. However, $\mathbf{u}$ has only m independent row vectors so that there are precisely mn independent equations (12.4.18b), which can always be rearranged as equations (12.4.14) for q^α.

Remark

If rank $\mathbf{q} = m$ but rank $\mathbf{r} < m$ then rank $\mathbf{u} < m$ and the system (12.4.18b) is underdetermined for q^α.

The corresponding m^2 equations for w_{ij} are always underdetermined for q^α unless $m = n$.

The apparently degenerate (when $m < n$) Hamiltonian system (12.4.18b) for $\mathbf{u}$ is a consistent reduction of the corresponding system for the case $m = n$. Thus, when $m < n$ the system (12.4.14) is squeezed between two Hamiltonian systems: one for $\mathbf{u}$, too large but reduced, the other for $\mathbf{w}$, too small, being equations for certain linear combinations of the q^α. However, when $m < n$ the equations for q^α are not themselves Hamiltonian *wrt* a simple Hamiltonian structure, even though they possess an infinite number of conserved densities and an infinite number of commuting flows. When $m = n$ the system (12.4.14) is Hamiltonian when r is chosen to be symmetric. In particular, when $\mathbf{r}$ is the identity matrix, $\mathbf{u} = \mathbf{w} = \mathbf{q}$.

Example $1 = m < n$: vector KdV.

Here $u_{ij} = r_i q_j$ and without loss of generality $r_1 \neq 0$. Equation (12.4.18b) for u_{ij} now takes the form (after dividing by r_1):

$$a^2 q_{jt_3} = [q_{jxx} - 3(\mathbf{q}\cdot\mathbf{r})q_j]_x. \qquad (12.4.19)$$

Since $w = \mathbf{q}\cdot\mathbf{r}$ is a scalar it satisfies the scalar KdV equation, which is also easily derived from (12.4.19).

Example $2 = m \leq n$.

Here:

$$\mathbf{q} = \begin{pmatrix} q_1 \dots q_n \\ p_1 \dots p_n \end{pmatrix} \quad \text{and} \quad \mathbf{r} = \begin{pmatrix} r_1 \dots r_n \\ s_1 \dots s_n \end{pmatrix}^{\mathrm{T}}$$

so that $u_{ij} = r_i q_j + s_i p_j$ and

$$\mathbf{w} = \begin{pmatrix} \mathbf{q}\cdot\mathbf{r} & \mathbf{q}\cdot\mathbf{s} \\ \mathbf{p}\cdot\mathbf{r} & \mathbf{p}\cdot\mathbf{s} \end{pmatrix}.$$

Since both matrices are of rank 2 we need only consider equations (12.4.18b) for $\mathbf{u}_{1j}$ and $\mathbf{u}_{2j}$. Multiplying these equations by $\begin{pmatrix} r_1 & s_1 \\ r_2 & s_2 \end{pmatrix}^{-1}$ leads to:

$$\begin{aligned} a^2 q_{jt_3} &= q_{jxxx} - 3[(\mathbf{r}\cdot\mathbf{q})q_j + (\mathbf{s}\cdot\mathbf{q})p_j]_x, \\ a^2 p_{jt_3} &= p_{jxxx} - 3[(\mathbf{r}\cdot\mathbf{p})q_j + (\mathbf{s}\cdot\mathbf{p})p_j]_x. \end{aligned} \tag{12.4.20}$$

The four equations for w_{ij} can be obtained by taking scalar products of the vector equations (12.4.20) with $\mathbf{r}$ and $\mathbf{s}$.

Other examples can be found in [3].

Non-local conserved quantities

The choice (12.4.7a) for $\mathbf{G}_0$ guarantees that the hierarchy of equations and conservation laws are locally defined. Recently, Crumey [6, 7] has considered the hierarchies corresponding to the general element $\mathbf{B} \in k$. Equations and constants of motion are generally non-local and we must label the latter both by the index n and the element ℓ of k. Crumey derived the commutation relations for these:

$$\{\mathcal{H}^{k}_{m}, \mathcal{H}^{l}_{n}\} = \mathcal{H}^{[k,l]}_{m+n}. \tag{12.4.21}$$

We must first notice that if $\ell = \mathbf{A}$, so that $[\ell, \ell] = 0$ then the right-hand side of (12.4.21) is zero. Thus *all* $\mathcal{H}^{l}_{n}$ are indeed constants of motion for the local hierarchy associated with $\mathbf{A}$. Thus the local hierarchy possesses dim k infinite sequences of constants of motion while each of the non-local hierarchies possesses exactly one infinite sequence of constants (namely, the local sequence).

More details of this can be found in Crumey's papers.

12.5 **Conclusions**

This chapter has been concerned with hierarchies of equations isospectral to matrix eigenvalue problems of the form (12.1.2b). The case when $\mathbf{A}$ is a regular, diagonal, matrix, $\mathbf{Q}$ general but off-diagonal, was discussed by Newell [28]. This case was studied from a purely algebraic point of view (and in the context of simple Lie algebras) by Wilson [33], who also discussed the reduction to modified Lax equations.

The present chapter has been particularly concerned with spectral problems (12.1.2b) with special choices of $\mathbf{A}$ and $\mathbf{Q}$ giving rise to highly reduced systems associated with Lie algebras and symmetric spaces.

The modified Lax systems of Section 12.3 form a particularly important class since they include the 2D Toda lattice equations. A similar construction generalises the Cologero–Moser system and its close relatives [29].

The generalised NLS hierarchies of Section 12.4 can be extended in several ways. With a less degenerate choice of **A** we construct hierarchies associated with reductive homogeneous spaces [13]. These are more complicated since the torsion tensor (12.A.8) is no longer zero. In [13] we also discussed the 'Heisenberg ferro-magnets' associated with the generalised NLS equations of Section 12.4. By considering spectral problems which are quadratic in λ it is possible to construct generalised DNLS (derivative nonlinear Schrödinger) equations [10]. The inverse spectral transform relevant to all these equations is discussed by Gerdjikov [16]. The stationary flows of the generalised NLS equations form interesting completely integrable classical mechanical systems with quartic potential [14, 25, 30]. Equations in (2 + 1) dimensions such as generalised Davey–Stewartson and KP like equations are given in [4].

Acknowledgement

I thank the SERC for their financial support

References

[1] Antonowicz, M. and Fordy, A. P Factorisation of energy dependent Schrödinger operators: Miura maps and modified systems. *Commun. Math. Phys.* **124**, 465–86 (1989).
[2] Antonowicz, M. and Fordy, A. P. Super-extensions of energy dependent Scrhödinger operators. *Commun. Math. Phys.* **124**, 487–500 (1989).
[3] Athorne, C. and Fordy, A. P. Generalised KdV and MKdV equations associated with symmetric spaces, *J. Phys.* **A20**, 1377–86 (1987).
[4] Athorne, C. and Fordy, A. P. Integrable equations in (2+1)-dimensions associated with Hermitian symmetric and homogeneous spaces, *J. Math. Phys.* **28**, 2018–24 (1987).
[5] Bogoyavlenskii, O. I. On perturbations of the periodic Toda lattice, *Commun. Math. Phys.* **51**, 201–9 (1976).
[6] Crumey, A. D. W. B. Local and non-local conserved quantities for generalised non-linear Schrödinger equations, *Commun. Math. Phys.* **108**, 631–49 (1987).
[7] Crumey, A. D. W. B. Kač–Moody symmetry of generalized non-linear Schrödinger equations, *Commun. Math. Phys.* **111**, 167–79 (1987).
[8] Dodd, R. K. and Bullough, R. K. Polynomial conserved densities for the sine-Gordon equations, *Proc. R. Soc. Lond.* **A352**, 481–503 (1977).
[9] Drinfel'd, V. G. and Sokolov, V. V. Lie algebras and equation of Korteweg–de Vries type, *J. Sov. Math.* **30**, 1975–2036 (1985).
[10] Fordy, A. P. Generalised derivative nonlinear Schrödinger equations and Hermitian symmetric spaces, *J. Phys.* **A17**, 1235–45 (1984).
[11] Fordy, A. P. and Gibbons, J. Integrable nonlinear Klein–Gordon equations and Toda lattices, *Commun. Math. Phys.* **77**, 21–30 (1980).
[12] Fordy, A. P. and Gibbons, J. Integrable nonlinear Klein–Gordon equations and simple Lie algebras, *Proc. R. Ir. Acad.* **83A**, 33–45 (1983).
[13] Fordy, A. P. and Kulish, P. P. Nonlinear Schrödinger equations and simple Lie algebras, *Commun. Math. Phys.* **89**, 427–43 (1983).

[14] Fordy, A. P., Wojciechowski, S. and Marshall, I. A family of integrable quartic potentials related to symmetric spaces, *Phys. Letts. A* **113**, 395–400 (1986).
[15] Gel'fand, I. M. and Dikii, L. Fractional powers of operators and Hamiltonian systems, *Funct. Anal. and Apps.* **10**, 259–73 (1976).
[16] Gerdjikov, V. S. Generalised Fourier transforms for the soliton equations. Gauge-covariant formulation, *Inverse Problems* **2**, 51–74 (1986).
[17] Helgason, S. *Differential Geometry, Lie Groups and Symmetric Spaces* (2nd edn), Academic Press, New York (1978).
[18] Humphreys, J. E. *Introduction to Lie Algebras and Representation Theory*, Springer, Berlin (1972).
[19] Kaup, D. J. and Newell, A. C. An exact solution for a derivative nonlinear Schrödinger equation, *J. Math. Phys.* **19**, 798–801 (1978).
[20] Kobayashi, S. and Nomizu, K. *Foundations of Differential Geometry*, Vol. 2, Wiley, New York (1969).
[21] Konopelchenko, B. G. *Nonlinear Integrable Equations*, Springer, Berlin (1987).
[22] Kupershmidt, B. A. and Wilson, G. Conservation laws and symmetries of generalized sine-Gordon equations, *Commun. Math. Phys.* **81**, 189–202 (1981).
[23] Leznov, A. N. On the complete integrability of a nonlinear system of partial differential equations in two dimensional space, *Theor. Math. Phys.* **42**, 343–9 (1980).
[24] Leznov, A. N. and Saveliev, M. V. Representation of zero curvature for the system of nonlinear partial differential equations $x_{\alpha,zz} = \exp(kx)_\alpha$ and its integrability, *Lett. Math. Phys.* **3**, 489–94 (1979).
[25] Marshall, I. Some integrable systems related to affine Lie algebras and homogeneous spaces, *Phys. Letts. A* **127**, 19–26 (1988).
[26] Mikhailov, A. V. Integrability of a two-dimensional generalisation of the Toda chain, *JETP Letts.* **30**, 414–18 (1979).
[27] Mikhailov, A. V., Olshanetsky, M. A. and Perelomov, A. M. Two-dimensional generalised Toda lattices, *Commun. Math. Phys.* **79**, 473–88 (1981).
[28] Newell, A. C. The general structure of integrable evolution equations, *Proc. R. Soc.Lond.* **A365**, 283–311 (1979).
[29] Olshanetsky, M. A. and Perelomov, A. M. Classical integrable finite dimensional systems related to Lie algebras, *Phys. Rep.* **71**, 313–400 (1981).
[30] Reyman, A. G. An orbit interpretation of Hamiltonian systems of harmonic oscillator type, *Proc. Steklov. Inst. Leningrad*, **155**, 187–90 (1986) (in Russian).
[31] Wadati, M. and Kamijo, T. On the extension of inverse scattering method, *Prog. Theor. Phys.* **52**, 397–414 (1974).
[32] Wilson, G. Commuting flows and conservation laws for Lax equations, *Math. Proc. Camb. Phil. Soc.* **86**, 131–43 (1979).
[33] Wilson, G. The modified Lax and two-dimensional Toda lattice equations associated with simple Lie algebras, *Ergod. Th. and Dynam. Sys.* **1**, 361–80 (1981).
[34] Zakharov, V. E. and Mikhailov, A. V. Relativistically invariant two-dimensional models of field theory which are integrable by means of the inverse scattering problem method, *Sov. Phys. JETP* **47**,1017–27 (1978).
[35] Zakharov, V. E. and Shabat, A. B. A scheme for integrating the nonlinear equations of mathematical physics by the method of the inverse scattering problem I, *Func. Anal. and Apps.* **8**, 226–35 (1974).

Appendix: Mathematical preliminaries

In this appendix I state a number of relevant facts concerning simple Lie algebras homogeneous and symmetric spaces, giving the barest details. The full theory can be found in [17, 18, 20].

Simple Lie algebras; Cartan–Weyl basis

In terms of the Cartan–Weyl basis a complex, simple Lie algebra $\mathfrak{g}$ has the following commutation relations [17, 18].

$$\text{(i)} \quad [h_i, h_j] = 0, \quad \forall h_i, h_j \in h,$$

$$\text{(ii)} \quad [h, e_\alpha] = \alpha(h)e_\alpha, \quad \forall h \in h, \quad \alpha \in \phi,$$

$$\text{(iii)} \quad [e_\gamma, e_{-\gamma}] = h_\gamma = \sum_{i=1}^{l} d_{\gamma i} h_i,$$

$$\text{(iv)} \quad [e_\gamma, e_\beta] = \begin{cases} N_{\gamma,\beta} e_{\gamma+\beta}, & 0 \neq \gamma + \beta \in \phi \\ 0 & \gamma + \beta \notin \phi. \end{cases} \tag{12.A.1}$$

It will be necessary to explain some of the terms:

(a) h is the Cartan subalgebra, which is any maximal Abelian subalgebra of diagonalisable elements of g. The subalgebra h has basis $\{h_i\}_1^l$ and $d_{\gamma i}$ are the components of $[e_\gamma, e_{-\gamma}] \in h$ with respect to this basis. The number l is the rank of the algebra.

(b) α: $h \rightarrow \mathbb{C}$ are linear functionals, called roots, on h and their values on given $h \in h$ are the eigenvalues of the matrix ad$h$. The corresponding eigenvectors, e_α, are called root vectors.

(c) The coefficients $N_{\alpha,\beta}$ are the most complicated part of these commutation relations. They satisfy various identities which are used in this chapter:

$$N_{-\alpha,-\beta} = -N_{\alpha,\beta}, \quad N_{\alpha,\beta} = N_{\beta,-\alpha-\beta} = N_{-\alpha-\beta,\alpha}. \tag{12.A.2}$$

(d) We choose a basis $\{\alpha_i\}_{i=1}^l$ for the root space and these are referred to as positive simple roots.

(e) We define the 'height' of a root vector as follows: ht$e_{\alpha_i} = 1$, ht$e_{-\alpha_i} = -1$, $\forall$ positive simple roots α_i and ht$[\mathbf{x}, \mathbf{y}] =$ ht$\mathbf{x}$ + ht$\mathbf{y}$. The highest root is called the maximal root $\alpha_{\max}$.

The discussion of Section 12.3 uses the exponents and Coxeter number of a Lie algebra. Table 12.1 below (which I extracted from [33]) gives the values of the exponents m_i for the various simple Lie algebras. The Coxeter number h is given by $m_i + 1$ for the largest m_i.

Lie algebra	$m_1, \ldots, m_l$
$A_{l-1}(l \geq 2)$	$1, 2, 3, \ldots, l-1$
$B_l, C_l(l \geq 2)$	$1, 3, 5, \ldots, 2l-1$
$D_l(l \geq 3)$	$1, 3, 5, \ldots, 2l-3, l-1$
E_6	1, 4, 5, 7, 8, 11
E_7	1, 5, 7, 9, 11, 13, 17
E_8	1, 7, 11, 13, 17, 19, 23, 29
F_4	1, 5, 7, 11
G_2	1, 5

Table 12.1

Homogeneous and symmetric spaces

A homogeneous space of a Lie group G is any differentiable manifold M on which G acts transitively ($\forall p_1, p_2 \in M \, \exists \, g \in G \mid g \cdot p_1 = p_2$). The subgroup of G which leaves a given point $p_0 \in M$ fixed is called the isotropy group at p_0 and is defined by:

$$K \equiv K_{p_0} = \{g \in G\colon g \cdot p_0 = p_0\}.$$

It is a theorem that each such M can be identified with a coset space G/K for some subgroup K and that this K plays the role of isotropy group of some point. There are many topological and differential geometric subtleties, but we have no need of them in this chapter. We are only interested in the decompositions of the corresponding Lie algebras.

Let g and k be the Lie algebras of G and K respectively, and let m be the vector space complement of k in g. Then

$$\mathit{g} = \mathit{k} \oplus \mathit{m}, \quad [\mathit{k}, \mathit{k}] \subset \mathit{k}, \tag{12.A.3}$$

and m is identified with the tangent space $T_{p_0}M$ of $M = G/K$ at point p_0. At the moment we have $[\mathit{k}, \mathit{k}] \subset \mathit{k}$, but know nothing of $[\mathit{k}, \mathit{m}]$ and $[\mathit{m}, \mathit{m}]$.

When g satisfies the more stringent conditions:

$$\mathit{g} = \mathit{k} \oplus \mathit{m}, \quad [\mathit{k}, \mathit{k}] \subset \mathit{k}, \quad [\mathit{k}, \mathit{m}] \subset \mathit{m}, \tag{12.A.4}$$

then G/K is called a reductive homogeneous space. These spaces possess canonically defined connections with curvature and torsion. Evaluated at fixed point p_0, the curvature and torsion tensors are given purely in terms of the Lie bracket operation on m:

$$\begin{aligned} (\mathrm{R}(\mathbf{X}, \mathbf{Y})\mathbf{Z})_{p_0} &= -[[\mathbf{X}, \mathbf{Y}]_{\mathit{k}}, \mathbf{Z}], \quad \mathbf{X}, \mathbf{Y}, \mathbf{Z} \in \mathit{m}, \\ \mathrm{T}(\mathbf{X}, \mathbf{Y})_{p_0} &= -[\mathbf{X}, \mathbf{Y}]_{\mathit{m}}, \quad \mathbf{X}, \mathbf{Y} \in \mathit{m} \end{aligned} \tag{12.A.5}$$

where subscript k and m refer to the components of $[\mathbf{X}, \mathbf{Y}]$ in those vector subspaces.

When g satisfies the conditions

$$\mathit{g} = \mathit{k} \oplus \mathit{m}, \quad [\mathit{k}, \mathit{k}] \subset \mathit{k}, \quad [\mathit{k}, \mathit{m}] \subset \mathit{m}, \quad [\mathit{m}, \mathit{m}] \subset \mathit{k}, \tag{12.A.6}$$

then g is called a symmetric algebra and G/K is a symmetric space. For these spaces the above mentioned canonical connection is derived from a metric, which is itself given by the restriction of the Killing form to m. This connection is torsion-free. Evaluated at fixed point p_0, the curvature tensor is given as in (12.A.5):

$$(R(\mathbf{X}, \mathbf{Y})\mathbf{Z})_{p_0} = -[[\mathbf{X}, \mathbf{Y}], \mathbf{Z}], \quad \mathbf{X}, \mathbf{Y}, \mathbf{Z} \in \mathit{m}, \tag{12.A.7}$$

where we now automatically have $[\mathbf{X}, \mathbf{Y}] \in \mathit{k}$.

The components R^i_{jkl} and T^i_{jk} of the curvature and torsion tensors with respect to a basis $\mathbf{X}_i$ of $T_{p_0}M$ are defined by:

$$\mathrm{R}(\mathbf{X}_k, \mathbf{X}_l)X_j = \mathrm{R}^i{}_{jkl}\mathbf{X}_i, \quad \mathrm{T}(\mathbf{X}_j, \mathbf{X}_k) = \mathrm{T}^i{}_{jk}\mathbf{X}_i. \tag{12.A.8}$$

For a symmetric space, the corresponding metric is given by the Killing form:

$$\mathit{g}(\mathbf{X}, \mathbf{Y}) = \mathrm{tr}(\mathrm{ad}\mathbf{X}\mathrm{ad}\mathbf{Y}), \quad \mathit{g}_{ij} = \mathit{g}(\mathbf{X}_i, \mathbf{X}_j) \tag{12.A.9}$$

where ad$\mathbf{X}$ is defined by $(\mathrm{ad}\mathbf{X})\mathbf{Y} = [\mathbf{X}, \mathbf{Y}]$. Tensorial indices are lowered and raised in the usual way by means of the metric tensor and its inverse. Since, for a semi-simple Lie algebra, all trace forms are proportional, we will often use that of the fundamental representation, without change of notation.

For spaces of constant curvature, the Riemann curvature tensor is related to the metric tensor in a simple way:

$$R^i_{jkl} = K(\delta^i_k \, g_{jl} - \delta^i_l \, g_{jk}) \tag{12.A.10}$$

where K is the constant Gaussian curvature.

We are particulary interested in those homogeneous and symmetric spaces which have a complex structure. This is a linear endomorhism $J: m \to m$ satisfying $J^2 = -1$. The vector subspace m must have even real dimension. Hermitian symmetric spaces are very special. For this chapter, the most useful properties are algebraic:

(i) $\exists \mathbf{A} \in h$ such that $k = C_g(\mathbf{A}) = \{\mathbf{B} \in g: [\mathbf{B}, \mathbf{A}] = 0\}$.
(ii) For a particular scaling of $\mathbf{A}$, $J = \mathrm{ad}\mathbf{A}$.
(iii) $\exists$ a subset $\theta^+ \subset \phi^+$ of the positive root system such that $m =$ span $\{e_{\pm\alpha}\}_{\alpha\in\theta^+}$ and $\alpha(\mathbf{A})$ is constant on θ^+.)
(iv) Following from (iii) $[e_\alpha, e_\beta] = 0$ if $\alpha, \beta \in \theta^+$ or $\alpha, \beta \in \theta^-$.

Thus m is decomposed into two eigenspaces: $m = m^+ \oplus m^-$. The components of an element $\mathbf{X} \in g$ will be labelled $\mathbf{X}^0 \in k$, $\mathbf{x}^\pm \in m^\pm$, satisfying $[\mathbf{A}, \mathbf{X}^0] = 0$, $[\mathbf{A}, \mathbf{X}^\pm] = \pm a\mathbf{X}^\pm$.

In the case of reductive homogeneous spaces a similar, but more complicated, decomposition occurs. Here $\mathrm{ad}\mathbf{A}$ has more than just two eigenvalues so that $\alpha(\mathbf{A})$ is no longer constant on θ^+. However, it is constant on blocks within the representation, depending upon the degeneracy of $\mathbf{A}$. The θ^+ can be written as the union of a number of subsets θ_j^+ on each of which $\alpha(\mathbf{A})$ takes a constant value $\alpha(\mathbf{A}) = a_j$, $\forall\ \alpha \in \theta_j^+$. Furthermore, the property (iv) above no longer holds.

13 *T. Miwa*

Infinite-dimensional Lie algebras as hidden symmetries of soliton equations

13.1 Introduction

Recently the role of infinite-dimensional Lie algebras in the theory of integrable systems has attracted much attention. The first appearance of such a connection was in the 2-dimensional Ising model solved by Onsager [9]. The algebra of free fermions was used to transform the Hamiltonian into the quadratic form [6]. This was a great advantage which led to the characterisation of the correlation functions [11–13, 15]. Unfortunately this is not a common feature of the integrable systems in statistical mechanics or in quantum field theory.

A general link was first found in the classical theory so that the hidden symmetries of the soliton equations were identified with the affine Lie algebras [2, 4, 14]. It is the aim of this chapter to describe this link in the case of the Korteweg–de Vries (KdV) and the Kadomtsev–Petviashvili (KP) equations [3]. It is now not too extreme to say that the algebraic structure of soliton equations is well understood.

For other categories of integrable systems the situation is less advanced. In the case of higher-dimensional equations such as the self-dual Yang–Mills equation or the Einstein equation, the geometric approach is dominant. In statistical mechanics and quantum field theory in two dimensions, conformal field theory is finding another remarkable link between infinite-dimensional Lie algebras and integrable systems [1]. This is concerned with the massless theory in the continuum. As for the massive theory on a lattice, it is still a challenging problem to find any such connection beyond the Ising model.

13.2 The KdV hierarchy

13.2.1 *The N-soliton solution*

The KdV (Korteweg–de Vries) equation

$$\frac{\partial u}{\partial t} = \frac{3}{2} u \frac{\partial u}{\partial x} + \frac{1}{4} \frac{\partial^3 u}{\partial x^3} \tag{13.2.1}$$

stands representative among soliton equations. It has been playing the role of a test runner for several new approaches to soliton theory. Let us first recall the N-soliton solution for (13.2.1). It is simple and explicit, and at the same time it is directly connected to the theme of this chapter.

Choose $2N$ parameters c_j, k_j ($j = 1, \ldots, N$) and set

$$\tau = 1 + \sum_{n=1}^{N} \sum_{1 \leqq j_1 < \ldots < j_n \leqq N} \prod_{j=j_1,\ldots,j_n} c_j \prod_{\substack{j,j'=j_1,\ldots,j_n \\ j<j'}} c_{jj'} \exp\Big(\sum_{j=j_1,\ldots,j_n} \xi_j\Big) \tag{13.2.2}$$

where

$$\xi_j = 2(k_j x + k_j^3 t), \tag{13.2.3}$$

$$c_{jj'} = \frac{(k_j - k_{j'})^2}{(k_j + k_{j'})^2}. \tag{13.2.4}$$

Then

$$u = 2 \frac{\partial^2}{\partial x^2} \ln \tau \tag{13.2.5}$$

solves the KdV equation (13.2.1). The solution of this form is called the N-soliton solution. If $N = 0$ then (13.2.5) merely gives $u \equiv 0$. If $N = 1$ then the solution represents the propagation of a solitary wave with the velocity $-k_1^2$. The solution with $N = 2$ is more complicated. It represents coexistence of two solitary waves. If c_{12} were 1, the solution $u(x, t)$ would be just a sum of two 1-solitons with the different velocities k_1 and k_2. Therefore the term $c_{12}(\neq 1)$ is responsible for the interaction of the two solitons. The quantity $\ln c_{12}$ is called the phase shift. The crucial fact concerning the formula (13.2.2) is that the phase shift of the N-soliton is the sum of the phase shifts of two soliton pairs.

Although the N-soliton solution (13.2.2) has a rather simple structure, it is not easy to prove that (13.2.5) actually solves (13.2.1). Until the end of the chapter we shall consider a proof based on the representation theory of the affine Lie algebra $A_1^{(1)}$. In fact, $A_1^{(1)}$ is the algebra of hidden symmetries of the KdV equation, which acts infinitesimally on the solution space.

Before proceeding to the representation theory we recall briefly two mathematical schemes for solving the KdV equation: the linearisation in the sense of Lax and the bilinearisation in the sense of Hirota.

13.2.2 *The Lax pair*

One of the common features of the soliton equations is that they are obtained as compatibility conditions for linear systems. The KdV equation is equivalent to the compatibility of the following system:

$$\left(\frac{\partial^2}{\partial x^2} + u(x, t)\right)w(x, t) = k^2 w(x, t), \tag{13.2.6}$$

$$\frac{\partial}{\partial t} w(x, t) = \left(\frac{\partial^3}{\partial x^3} + v_1(x, t)\frac{\partial}{\partial x} + v_2(x, t)\right)w(x, t). \tag{13.2.7}$$

These two linear differential equations are called the Lax pair.

Lax showed that the KdV equation (13.2.1) is just one of infinitely many evolution equations which preserve the spectre k^2 of the linear problem (13.2.6) [7]. He found that instead of (13.2.7) we can impose

$$\frac{\partial}{\partial t}w(x, t) = \left(\frac{\partial^{2n+1}}{\partial x^{2n+1}} + v_1(x, t)\,\frac{\partial^{2n-1}}{\partial x^{2n-1}} + \ldots + v_{2n}(x, t)\right)w(x, t)$$

and obtain a higher-order nonlinear equation for $u(x, t)$. It is called the $(2n + 1)$th-order KdV equation.

The N-soliton solution to the $(2n + 1)$th-order KdV equation is given by the same formulae with k_j^3 in (13.2.3) replaced by k_j^{2n+1}.

A giant step was made by Sato when he introduced different names for the time variables of the higher-order equations [10]. Let x denote infinitely many variables $(x_1, x_3, x_5, \ldots)$, and replace (13.2.3) by

$$\xi_j = 2 \sum_{l:\text{odd}} k_j^l\, x_l.$$

Then $u(x)$ given by (13.2.5) (we replace $\partial/\partial x$ by $\partial/\partial x_1$) satisfies simultaneously all the higher-order KdV equations – the KdV hierarchy. The reason for calling it a giant step will be clear later.

13.2.3 *The Hirota equation*

Rewriting the KdV equation in terms of τ, Hirota [5] found the following equation (see also Chapter 4):

$$(\mathrm{D}_1^4 - 4\mathrm{D}_1\mathrm{D}_3)\tau(x) \cdot \tau(x) = 0, \tag{13.2.8}$$

where, in general, Hirota's bilinear operator $P(\mathrm{D})$ is defined by

$$P(\mathrm{D})f(x)\cdot g(x) = P(\partial_y)(f(x + y)g(x - y))|_{y=0}.$$

For example,

$$\mathrm{D}_1^2 f(x)\cdot g(x) = \frac{\partial^2 f(x)}{\partial x_1^2}g(x) - 2\frac{\partial f(x)}{\partial x_1}\frac{\partial g(x)}{\partial x_1} + f(x)\frac{\partial^2 g(x)}{\partial x_1^2}.$$

The form of the N-soliton (13.2.2) is easily deduced from (13.2.8). (The proof is still not straightforward.)

The KdV hierarchy as a whole can be transformed into Hirota's bilinear equations. The explicit form will be given later. We remark here that the simultaneous introduction of higher-order time variables is crucial. It is not possible to rewrite the $(2n + 1)$th-order KdV equation into Hirota's bilinear form by using only D_1 and D_{2n+1}.

13.3 The vertex representation of $A_1^{(1)}$

13.3.1 *The basic $A_1^{(1)}$ module*

The affine Lie algebra $A_1^{(1)}$ is the one-dimensional central extension of $sl(2, C) \otimes C[t, t^{-1}]$:

$$A_1^{(1)} = sl(2, \mathbb{C}) \otimes \mathbb{C}[t, t^{-1}] \oplus \mathbb{C}c,$$

and the Lie bracket is given by

$$[C \otimes t^m, Y \otimes t^n] = [X, Y] \otimes t^{m+n} + m\delta_{m,-n}\mathrm{tr}XYc,$$

where $X, Y \in sl(2, \mathbb{C})$ and $m, n \in \mathbb{Z}$. The basic module is an irreducible $A_1^{(1)}$ module with a vector $|\mathrm{vac}>$ satisfying

$$\begin{aligned} X \otimes t^m \mid \mathrm{vac}> &= 0, \qquad (m > 0), \\ \begin{pmatrix} & 1 \\ & \end{pmatrix} \mid \mathrm{vac}> &= 0, \\ \begin{pmatrix} 1 & \\ & -1 \end{pmatrix} \mid \mathrm{vac}> &= 0, \\ \mathrm{c} \mid \mathrm{vac}> &= |\mathrm{vac}>. \end{aligned}$$

We call $|\mathrm{vac}>$ the highest weight vector.

Lepowsky and Wilson [8] constructed the basic representation on the polynomial ring $V = C[x_1, x_3, x_5, \ldots]$ in such a way that the highest weight vector is represented by 1 in V. The fact is that a polynomial $\tau(x) \in V$ solves the KdV hierarchy in Hirota's form if and only if it is on the group orbit of 1. In Section 13.2.1 we presented the N-soliton solution which is an exponential function in the time variables. It is closely related to the Lepowsky–Wilson construction as we shall see in Section 13.3.3.

13.3.2 *The Heisenberg subalgebra*

Let s be the principal Heisenberg subalgebra of $A_1^{(1)}$:

$$s = \bigoplus_{n:\mathrm{odd}} \mathbb{C}H_n \oplus \mathbb{C}c,$$

$$H_{2n+1} = \begin{pmatrix} & t^n \\ t^{n+1} & \end{pmatrix}. \tag{13.3.1}$$

The term 'Heisenberg subalgebra' refers to the commutation rule

$$[H_m, H_n] = m\delta_{m,-n}c. \tag{13.3.2}$$

If $c = 1$ (this is true for the basic module), an irreducible s module is isomorphic to V and the representation goes as follows $(n > 0)$:

$$H_n = \frac{\partial}{\partial x_n},$$

$$H_{-n} = n x_n.$$

The basic $A_1^{(1)}$ module has the special property that it is irreducible as an s module. Therefore V itself is the representation space of the whole algebra $A_1^{(1)}$. Those elements in $A_1^{(1)}$ that are not in s are represented on V in terms of the so-called vertex operators as we shall see next.

13.3.3 *The vertex operator*

Consider an operator

$$X(k) = \exp\left(2\sum_{n=1}^{\infty} k^{2n+1} x_{2n+1}\right)\exp\left(-2\sum_{n=1}^{\infty} \frac{k^{-2n-1}}{2n+1}\frac{\partial}{\partial x_{2n+1}}\right). \tag{13.3.3}$$

This is called a vertex operator. If we expand it as $X(k) = \sum_{n \in Z} k^n X_n$, then each X_n is a well-defined linear operator on V. They actually realise the basic representation in such a way that

$$X_{2n} = 2\begin{pmatrix} -t^{-n} & \\ & t^{-n} \end{pmatrix},$$

$$X_{2n+1} = 2\begin{pmatrix} & t^{-n-1} \\ -t^{-n} & \end{pmatrix}.$$

The proof of this fact is also given in Section 13.4. Here we concentrate on the link between the N-soliton formula (13.2.2) and the vertex operator (13.2.3).

Noting that (see (13.2.4))

$$X(k_1)X(k_2) = c_{12}\exp\left[2\sum_{n=1}^{\infty}(k_1^{2n+1} + k_2^{2n+1})x_{2n+1}\right]$$
$$\times \exp\left[-2\sum_{n=1}^{\infty}\frac{1}{2n+1}\left(\frac{1}{k_1^{2n+1}} + \frac{1}{k_2^{2n+1}}\right)\frac{\partial}{\partial x_{2n+1}}\right],$$

one can show that

$$\tau = \exp\left[\sum_{j=1}^{N} c_j X(k_j)\right]\cdot 1.$$

To put it differently the action of $e^{cX(k)}$ produces the $(N + 1)$-soliton out of the N-soliton.

The following two requirements are to be answered in the later sections:

(i) Show that the vertex operator represents the Lie algebra $A_1^{(1)}$.
(ii) Show that the τ function (13.2.2) solves the KdV hierarchy in Hirota's bilinear form.

We answer them by dealing with a more generic equation – the KP equation. We stress again that the introduction of infinitely many variables $x_1, x_3, \ldots$ is absolutely necessary in order to establish the above link between the KdV hierarchy and the vertex representation of $A_1^{(1)}$.

13.4 The boson fermion correspondence

13.4.1 *The Lie algebra* $gl(\infty)$

Consider an associative algebra A generated by fermions ψ_n, ψ_n^* ($n \in \mathbb{Z}$) satisfying the canonical anti-commutation relation:

$$[\psi_m, \psi_n]_+ = [\psi_m^*, \psi_n^*]_+ = 0,$$
$$[\psi_m^*, \psi_n]_+ = \delta_{m,n}. \tag{13.4.1}$$

Let g be the Lie algebra of elements in A of the form $\Sigma c_{mn}\psi_m\psi_n^*$. Since

$$[\psi_m\psi_n^*, \psi_p\psi_q^*] = \delta_{np}\psi_m\psi_q^* - \delta_{mq}\psi_p\psi_n^*,$$

$\psi_m\psi_n^*$ can be identified with the matrix E_{mn} of infinite rank given by

$$(E_{mn})_{pq} = \delta_{mp}\delta_{nq}.$$

We need a larger Lie algebra than g so that it contains a Heisenberg subalgebra (bosons). Since an element in g is a finite sum of $\psi_m\psi_n^*$'s the canonical commutation relation characteristic to bosons cannot be realised in g.

Consider an expression of the form

$$X = \Sigma a_{mn}:\psi_m\psi_n^*: + ac \tag{13.4.2}$$

where c denotes the central element and $a_{m,n}$, a $\in \mathbb{C}$ with the restriction

$$a_{m,n} = 0 \text{ if } |m - n| >> 0.$$

We define a Lie bracket among such elements extending the following:

$$[:\psi_m\psi_n^*:, :\psi_p\psi_q^*:] = \delta_{np}:\psi_m\psi_q^*: - \delta_{mq}:\psi_p\psi_n^*: + \delta_{np}\delta_{mq}(Y_-(q) - Y_-(n))c,$$

where

$$Y_-(m) = 0 \quad m \geqq 0,$$
$$= 1 \quad m < 0.$$

We call the Lie algebra obtained in this way $gl(\infty)$. The Lie algebra g is contained in $gl(\infty)$ by the identification

$$\psi_m\psi_n^* = :\psi_m\psi_n^*: + \delta_{mn}Y_-(n)c.$$

The affine algebra $A_1^{(1)}$ is also contained in $gl(\infty)$:

$$A_1^{(1)} = \{X \in gl(\infty) \mid a_{mn} = a_{m+2\,n+2}\}.$$

The identification goes as

$$\sum \begin{pmatrix} a_{0\,2k} & a_{0\,2k+1} \\ a_{1\,2k} & a_{1\,2k+1} \end{pmatrix} \otimes t^k \in sl(2, C) \otimes C[t, t^{-1}].$$

We set

$$H_l = \sum_{n \in Z} :\psi_n\psi_{n+l}^*: .$$

If l is odd, this is consistent with the previous definition (13.3.1). Then H_l satisfy the canonical commutation relation (13.3.2). We set

$$\mathrm{H} = \bigoplus_{n \neq 0} CH_n \oplus Cc$$

and call it the Heisenberg subalgebra.

In the following sections we consider the Fock representation of A, which induces a representation of $gl(\infty)$. Then H serves as a dictionary between the soliton theory and the representation theory.

13.4.2 *The vertex representation I*

The Fock representation of the algebra A is the irreducible representation of A generated by a vector $|0>$ satisfying

$$\begin{aligned} \psi_n|0> &= 0 \quad n < 0, \\ \psi_n^*|0> &= 0 \quad n \geqq 0. \end{aligned} \tag{13.4.3}$$

The vector $|0>$ is called the vacuum vector. The representation space is called the Fock space and is denoted by $\mathscr{F}$. We also note that

$$H_n|0> = 0 \quad (n > 0). \tag{13.4.4}$$

We denote by $\mathscr{F}^*$ the dual space of $\mathscr{F}$. It is generated by a vector $<0|$ satisfying

$$\begin{aligned} <0 \mid 0> &= 1, \\ <0 \mid \psi_n &= 0 \quad n \geqq 0, \\ <0 \mid \psi_n^* &= 0 \quad n < 0. \end{aligned}$$

Here we denote by $<0|0>$ the bilinear form evaluated on the vectors $<0|$ and $|0>$. In general, the bilinear form evaluated on $<0|\psi$ and $\varphi|0>$ (ψ, φ

$\in A$) is denoted by $<0|\psi\varphi|0>$. To put it in a different way we consider $\mathscr{F}^*$ as a right A module.

Since g is contained in A the Fock space $\mathscr{F}$ is considered as a g module. The action of $\mathsf{gl}(\infty)$ over $\mathscr{F}$ is uniquely defined by the conditions that it is an extension of the above action of g and that $c = 1$. In fact, because of (13.4.1) and (13.4.3) all but finite of the summand in (13.4.2) do not contribute.

The element H_0 belongs to the centre of $\mathsf{gl}(\infty)$. As a $\mathsf{gl}(\infty)$ module $\mathscr{F}$ decomposes into

$$\mathscr{F} = \bigoplus_{l \in Z} \mathscr{F}_l,$$

$$\mathscr{F}_l = \{|v> \in \mathscr{F}|\ H_0|v> = l|v>\} = \bigoplus_{s-r=l} \mathbb{C}\psi^*_{m_1} \ldots \psi^*_{m_r}\psi_{n_s} \ldots \psi_{n_1}|0>.$$

$$m_1 < \ldots < m_r < 0 \leqq n_s < \ldots < n_1$$

One can show that each $\mathscr{F}_l$ is an irreducible H module. This implies $\mathscr{F}_l$ is isomorphic to $V = \mathrm{C}[x_1, x_2, x_3, \ldots]$ and the action of H is given by $H_n = \partial/\partial x_n$, $H_{-n} = nx_n$ $(n > 0)$. The isomorphism is constructed explicitly as follows.

We set

$$\begin{aligned} <l| &= <0|\psi^*_0 \ldots \psi^*_{l-1} \quad l > 0, \\ &= <0|\psi_{-1} \ldots \psi_{-l} \quad l < 0, \end{aligned} \tag{13.4.5}$$

and

$$H(x) = \sum_{l=1}^{\infty} x_l H_l.$$

The latter is not an element of $\mathsf{gl}(\infty)$, but its action on $\mathscr{F}_l$ is well-defined. The isomorphism ρ_l is given by

$$\begin{array}{ccc} \mathscr{F}_l & \overset{\rho_l}{\rightarrow} & V \\ \cup & & \cup \\ |v> & \rightarrow & <l|e^{H(x)}|v>. \end{array}$$

In 13.6.1 we give an explicit formula for $<l|e^{H(x)}|v>$ by using the Schur polynomials.

Our goal is to obtain an explicit formula for the action of $\rho_0(X)$ ($X \in \mathsf{gl}(\infty)$):

$$\begin{array}{ccc} \mathscr{F}_0 & \overset{\rho_0}{\rightarrow} & V \\ \downarrow X & & \downarrow \rho_0(X) \\ \mathscr{F}_0 & \overset{\rho_0}{\rightarrow} & \mathrm{V} \end{array},$$

$$\begin{array}{ccc} |v> & \rightarrow & <0|e^{H(x)}|v> = f(x) \\ \downarrow & & \downarrow \\ X|v> & \rightarrow & <0|e^{H(x)}X|v> = \rho_0(X)f(x). \end{array}$$

In fact, $\rho_0(X)$ is expressed in terms of a vertex operator similar to (13.3.3).

13.4.3 *The vertex representation II*

We set

$$\psi(k) = \Sigma k^n \psi_n, \quad \psi^*(k) = \Sigma k^{-n} \psi_n^*.$$

These are common eigenvectors of the adjoint action of H_n ($n \in \mathbb{Z}$):

$$[H_n, \psi(k)] = k^n \psi(k),$$
$$[H_n, \psi^*(k)] = -k^n \psi^*(k). \tag{13.4.6}$$

In particular, we have

$$e^{H(x)}\psi(k)e^{-H(x)} = e^{\xi(x,k)}\psi(k),$$
$$e^{H(x)}\psi^*(k)e^{-H(x)} = e^{-\xi(x,k)}\psi^*(k). \tag{13.4.7}$$

Here $\xi(x, k) = \sum_{l=1}^{\infty} k^l x_l.$

We define the vertex operators $\Psi(k)$ and $\Psi^*(k)$ by

$$\Psi(k) = e^{\xi(x,k)} \, e^{-\xi(\tilde{\partial},k^{-1})},$$
$$\Psi^*(k) = e^{-\xi(x,k)} e^{\xi(\tilde{\partial},k^{-1})}, \tag{13.4.8}$$

where $\tilde{\partial} = (\partial/\partial x_1, \frac{1}{2}\partial/\partial x_2, \frac{1}{3}\partial/\partial x_3, \ldots)$. Every coefficient in k of $\Psi(k)$ or $\Psi^*(k)$ acts on $V = \mathrm{C}[x_1, x_2, x_3, \ldots]$.

The commutativity of the following diagram is the key lemma. We give a proof in 13.6.2.

$$\begin{array}{ccc}
\mathcal{F}_l & \xrightarrow{\rho_l} & V \\
\downarrow \psi^*(k) & & \downarrow k^{1-l}\Psi^*(k) \\
\mathcal{F}_{l-1} & \xrightarrow{\rho_{l-1}} & V \\
\downarrow \psi(k) & & \downarrow k^{l-1}\Psi(k)V \\
\mathcal{F}_l & \xrightarrow{\rho_l} & V
\end{array} \tag{13.4.9}$$

Setting l = 0 obtain

$$\rho_0(\psi(p)\psi^*(q)) = \frac{q}{p}\Psi(p)\Psi^*(q)$$
$$= \frac{q}{p-q} :\Psi(p)\Psi^*(q):$$

where

$$:\Psi(p)\Psi^*(q): = \exp^{[\xi(x,p)-\xi(x,q)]}\exp^{[-\xi(\tilde{\partial},p^{-1})+\xi(\tilde{\partial},q^{-1})]}. \tag{13.4.10}$$

This formula gives the vertex representation of $gl(\infty)$ on V. The Lepowsky–Wilson result (see (13.3.3)) is recovered by specialising $p = -q = k$,

$$X(k) = -2\rho_0(\psi(k)\psi^*(-k)).$$

In conclusion, we have answered the question (i) in 13.3.3. Moreover, we have shown that the vertex operator (13.4.10) represents the Lie algebra $gl(\infty)$. In the following we shall answer the question (ii) in 13.3.3. In fact, the τ function

$$\begin{aligned}\tau(x) &= \langle 0|e^{H(x)}\exp^{[\Sigma c_i\psi(p_i)\psi^*(q_i)]}|0\rangle \\ &= \exp\left[\Sigma a_i\frac{q_i}{p_i - q_i}:\Psi(p_i)\Psi^*(q_i):\right]\cdot 1. \qquad (13.4.11)\end{aligned}$$

satisfies the KP hierarchy in Hirota's bilinear form, and if $p_i = -q_i$ it is independent of even variables $x_2, x_4, \ldots$ and satisfies the KdV hierarchy.

13.5 **The bilinear identity**

13.5.1 *The KP hierarchy*

The τ function (13.4.11) belongs to the orbit of the highest weight vector 1 in the vertex representation of $gl(\infty)$. The question is: What is the equation characterising these τ functions?

This section is organised as follows. First we introduce the KP hierarchy by generalising the argument in Section 13.1. Secondly we give a characterisation of the orbit of $|0\rangle$ in $\mathcal{F}_0$ by using the fermion language. Thirdly we translate the characterisation into the boson language, and find that it gives rise to the KP hierarchy.

We denote by x the variables $(x_1, x_2, \ldots)$. Consider the following system of linear differential equations for w,

$$\frac{\partial}{\partial x_n} w(x, n) = B_n\left(x, \frac{\partial}{\partial x_1}\right)w(x, n), \qquad (13.5.1)$$

assuming that

$$w(x, k) = \left(1 + \sum_{j=1}^{\infty} \frac{w_j(x)}{k^j}\right)e^{\xi(x,k)} \qquad (13.5.2)$$

and

$$B_k\left(x, \frac{\partial}{\partial x_1}\right) = \left(\frac{\partial}{\partial x_1}\right)^k + \sum_{j=0}^{k-2} b_{k,j}(x)\left(\frac{\partial}{\partial x_1}\right)^j. \qquad (13.5.3)$$

We call $w(x, k)$ the wave function. The compatibility condition leads to a system of nonlinear evolution equations for $u(x) = b_{2,0}(x)$ called the KP hierarchy. The simplest equation reads

$$\frac{3}{4}\frac{\partial^2 u}{\partial x_2^2} = \frac{\partial}{\partial x_1}\left(\frac{\partial u}{\partial x_3} - \frac{3}{2}\frac{\partial u}{\partial x_1} - \frac{1}{4}\frac{\partial^3 u}{\partial x_1^3}\right).$$

If we restrict to the case that w_j's are independent of even variables x_2, x_4, ..., then the KP hierarchy reduces to the KdV hierarchy. In fact, the linear equation (13.5.1) with $n = 2$ reduces to (13.2.6).

As in the KdV case one can introduce τ and rewrite the equations in Hirota's bilinear form as we shall see in the next section.

13.5.2 The bilinear identity

The vacuum vector $|0>$ is uniquely (up to a constant multiple) characterised by (13.4.3). From (13.4.3) follows the identity

$$\sum_{n\in Z} \psi_n|0> \otimes \psi_n^*|0> = 0.$$

In general, a vector $|v> \in \mathscr{F}_0$ belongs to the orbit of $|0>$, that is,

$$|v> = e^{X_1} \dots e^{X_n}|0>$$

with some $X_1, \dots, X_n \in g$, if and only if it satisfies the following bilinear identity:

$$\sum_{n\in Z} \psi_n|v> \otimes \psi_n^*|v> = 0.$$

The bilinear identity can be written in terms of $\psi(k)$ and $\psi^*(k)$ as follows:

$$\oint_{k=\infty} \frac{dk}{2\pi ik}\, \psi(k)|v> \otimes \psi^*(k)|v> = 0.$$

Here we mean $\oint_{k=\infty} \frac{dk}{2\pi ik} F(k) = F_0$ if $F(k) = \sum_{j\in Z} F_j k^j$. We further rewrite it by taking the bilinear product with $<1|e^{H(x)} \otimes < -1|e^{H(y)}$:

$$\oint_{k=\infty} \frac{dk}{2\pi ik} <1|e^{H(x)}\psi(k)|v><-1|e^{H(y)}\psi^*(k)|v> = 0.$$

Let us define $\tau(x)$, $w(x, k)$ and the adjoint wave function $w^*(x, k)$ as follows:

$$\tau(x) <0|e^{H(x)}|v>,$$

$$w(x, k) = <1|e^{H(x)}\psi(x)|v>/\tau(x),$$

$$w^*(x, k) = <-1|e^{H(x)}\psi^*(x)|v>/(k\tau(x)).$$

Then, w and w* are related to τ by the vertex operators (13.4.8) as follows. (This is a corollary of (13.4.9).)

$$\begin{aligned} w(x, k) &= \Psi(k)\tau(x)/\tau(x), \\ &= \frac{\tau\left(x_1 - \dfrac{1}{k}, x_2 - \dfrac{1}{2k^2}, \ldots\right)}{\tau(x)} \mathrm{e}^{\xi(x,k)}. \\ w^*(x, k) &= \Psi^*(k)\tau(x)/\tau(x), \\ &= \frac{\tau\left(x_1 + \dfrac{1}{k}, x_2 + \dfrac{1}{2k^2}, \ldots\right)}{\tau(x)} \mathrm{e}^{-\xi(x,k)}. \end{aligned}$$

In particular, $w(x, k)$ is of the form (13.5.2) and $w^*(y, k)$ is of the form

$$w^*(y, k) = \left(1 + \sum_{j=1}^{\infty} \frac{w^*{}_j(y)}{k^j}\right) \mathrm{e}^{-\xi(y,k)}. \tag{13.5.4}$$

The bilinear identity reads

$$\oint_{k=\infty} \frac{\mathrm{d}k}{2\pi \mathrm{i}} w(x, k) w^*(y, k) = 0. \tag{13.5.5}$$

Instead of (13.5.2) assume that $w(x, k)$ is of the form

$$w(x, k) = \left(\sum_{j=1}^{\infty} \frac{w_j(x)}{k^j}\right) \mathrm{e}^{\xi(x,k)}. \tag{13.5.6}$$

One can show that if w of the form (13.5.6) and w^* of the form (13.5.4) satisfy the bilinear identity (13.5.5) then $w(x, k) \equiv 0$. Take $B_k(x, \partial/\partial x_1)$ of the form (13.5.3) so that

$$\left(\frac{\partial}{\partial x_k} - B_k\left(x, \frac{\partial}{\partial x_1}\right)\right) w(x, k) = \left(\sum_{j=1}^{\infty} \frac{a_j(x)}{k^j}\right) \mathrm{e}^{\xi(x,k)}.$$

This is always possible. From (13.5.5) follows

$$\oint_{k=\infty} \frac{\mathrm{d}k}{2\pi \mathrm{i}} \left(\frac{\partial}{\partial x_k} - B_k\left(x, \frac{\partial}{\partial x_1}\right)\right) w(x, k) \cdot w^*(y, k) = 0.$$

Hence we obtain (13.5.1). Note that (see (13.2.5))

$$u(x) = b_{2,0}(x) = 2\frac{\partial^2}{\partial x_1{}^2} \ln\tau(x).$$

In this way the single identity (13.5.5) implies the linear system (13.5.1) for $w(x, k)$ and the KP hierarchy for $u(x)$.

If we write (13.5.5) in terms of $\tau(x)$, we have

$$\oint_{k=\infty} \frac{\mathrm{d}k}{2\pi \mathrm{i}} \tau\left(x_1 - \frac{1}{k}, \ldots\right) \tau\left(y_1 + \frac{1}{k}, \ldots\right) \mathrm{e}^{\xi(x-y,k)} = 0. \tag{13.5.7}$$

For any polynomial $P(x)$ we have

$$\oint_{k=\infty} \frac{dk}{2\pi i} P(\partial_x)\left(\tau\left(x_1 - \frac{1}{k}, \ldots\right) e^{\xi(x,k)}\right)\tau\left(x_1 + \frac{1}{k}, \ldots\right) e^{-\xi(x,k)} = 0.$$

This is rewritten as an equation for τ in Hirota's bilinear form. In fact, we can rewrite (13.5.7) as

$$\sum_{j=0}^{\infty} p_j(-2y)p_{j+1}(\tilde{\mathrm{D}}_x)\exp\left[\sum_{l=1}^{\infty} y_l \mathrm{D}_{x_l}\right] \tau(x)\cdot\tau(x) = 0$$

where $\tilde{\mathrm{D}}_x = (\mathrm{D}_{x_1}, \frac{1}{2}\mathrm{D}_{x_2}, \ldots)$ denote Hirota's bilinear differential operators.

13.6 **Appendices**

13.6.1 *The Schur polynomials*

In this appendix we give an explicit formula for the polynomial

$$\langle s - r|e^{H(x)}\psi^*_{m_1} \ldots \psi^*_{m_r}\psi_{n_s} \ldots \psi_{n_1}|0\rangle$$

where $m_1 < \ldots < m_r < 0 \leqq n_s < \ldots < n_1$.

First consider the case $s = 0$ and $r = 1$:

$$\langle -1|e^{H(x)}\psi^*_m|0\rangle \quad (m < 0).$$

We proceed as follows.

$$\begin{aligned}
&\sum_{m<0} \langle -1|e^{H(x)}\psi^*_m|0\rangle k^{-m}\\
&= \langle -1|e^{H(x)}\psi^*(k)|0\rangle; \quad \text{by (13.4.3)}\\
&= \langle -1|e^{H(x)}\psi^*(k)e^{-H(x)}|0\rangle; \quad \text{by (13.4.4)}\\
&= e^{-\xi(x,k)}\langle 0|\psi_{-1}\psi^*(k)|0\rangle; \quad \text{by (13.4.5) and (13.4.7)}\\
&= ke^{-\xi(x,k)}.
\end{aligned}$$

In conclusion, if we define $q_l(x)$ $(l \geqq 0)$ by

$$e^{-\xi(x,k)} = \sum_{l=0}^{\infty} q_l(x)(-k)^l,$$

then we have

$$\langle -1|e^{H(x)}\psi^*_m|0\rangle = (-1)^{m+1}q_{-1-m}(x) \quad (m < 0).$$

Wick's theorem states that if $\Psi_j = \Sigma c_{jk}\psi_k$ and $\Psi^*_j = \Sigma c^*_{jk}\psi^*_k$ then

$$\langle 0|\Psi_1 \ldots \Psi_n\Psi^*_n \ldots \Psi^*_1|0\rangle = \det(\langle 0|\Psi_j\Psi^*_k|0\rangle)_{j,k=1,\ldots,n}.$$

From this follows that

$$
\begin{aligned}
&<-r|e^{H(x)}\psi^*_{m_1} \ldots \psi^*_{m_r}|0> \times (-1)^{m_1+\ldots+m_r+r(r+1)/2} \\
&= \det \begin{bmatrix} q_{k_1}(x) & q_{k_1+1}(x) & \ldots & q_{k_1+r-1}(x) \\ q_{k_2-1}(x) & q_{k_2}(x) & & q_{k_2+r-2}(x) \\ \vdots & \vdots & & \vdots \\ q_{k_r-r+1}(x) & q_{k_r-r+2}(x) & \ldots & q_{k_r}(x) \end{bmatrix},
\end{aligned}
$$

where $m_1 < \ldots < m_r < 0$ and $k_1 = -m_1 - r, \ldots, k_r = -m_r - 1$.

This polynomial, denoted by $\chi_Y(x)$, is called the Schur polynomial corresponding to the Young diagram Y of the form:

$$Y = \boxed{k_1} \ \ldots \ \boxed{k_r}\,.$$

Consider the irreducible representation ρ_Y of GL(N, C) corresponding to Y. If N is large enough we have

$$\chi_Y(x) = \mathrm{tr}\ \rho_Y(e^h)$$

where

$$h = \begin{pmatrix} \varepsilon_1 & & \\ & \ddots & \\ & & \varepsilon_N \end{pmatrix}, \quad x_j = \frac{1}{j}\ tr\ e^{jh}.$$

The general case of s, r is given by

$$
\begin{aligned}
\chi_Y(x) &= (-1)^{m_1+\ldots+m_r+(s-r)(s-r-1)/2} \\
&\quad \times <s - r|e^{H(x)}\psi^*_{m_1} \ldots \psi^*_{m_r}\psi_{n_s} \ldots \psi_{n_1}|0>
\end{aligned}
$$

(see Figure 13.1).

13.6.2 *The vertex operator $\Psi(k)$*

In this appendix we show that

$$<l|e^{H(x)}\psi(x)|v> = k^{l-1}\Psi(k)<l - 1|e^{H(x)}|v>$$

for $|v> \in \mathscr{F}_{l-1}$, which gives rise to the commutativity of (13.4.9) (the lower half). The identity for $\Psi^*(k)$ is similarly derived. Because of (13.4.7) it is sufficient to show that

$$<l|\psi(x) = k^{l-1}<l - 1|e^{\tilde{H}}$$

where

$$\tilde{H} = -\sum_{j=1}^{\infty} \frac{H_j}{jk^j}.$$

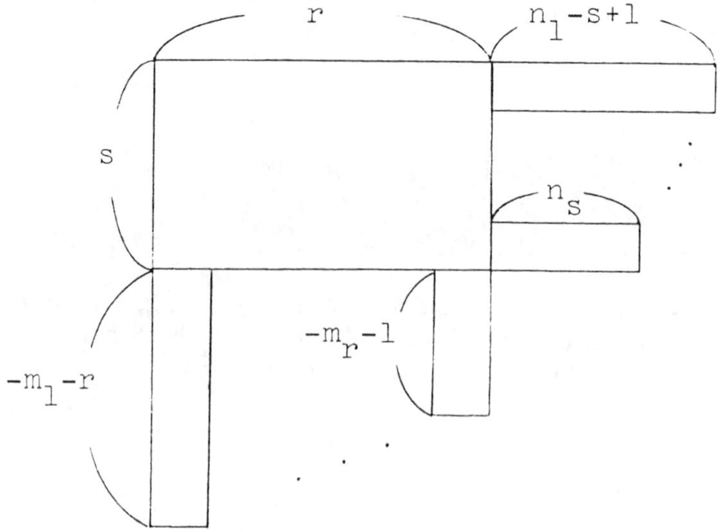

Figure 13.1

We define

$$|l\rangle = \psi_{l-1} \ldots \psi_0|0\rangle \quad l > 0,$$
$$= \psi^*_{-l} \ldots \psi^*_{-1}|0\rangle \quad l < 0.$$

Then $|l\rangle \in \mathcal{F}_l$ and $H_j|l\rangle = 0$ $(j > 0)$. Since $\mathcal{F}_l$ is H irreducible, it is spanned by those elements of the form

$$H_{j_1} \ldots H_{j_k}|l\rangle \quad (j_1, \ldots, j_k < 0).$$

Therefore it is enough to show

$$\langle l|\psi(k)H_{j_1} \ldots H_{j_k}|l-1\rangle = k^{l-1}\langle l-1|e^{\tilde{H}}H_{j_1} \ldots H_{j_k}|l-1\rangle.$$

for $j_1, \ldots, j_k < 0$.

The case $k = 0$ follows immediately. Noting that $\langle l|H_j = 0$ $(j < 0)$ and using (13.4.6) and (13.3.2) we have (for $j < 0$)

$$\langle l|\psi(k)H_j = \langle l|[\psi(k), H_j] = -k^j\langle l|\psi(k),$$
$$\langle l-1|e^{\tilde{H}}H_j = \langle l-1|(H_j - k^j)e^{\tilde{H}} = -k^j\langle l-1|e^{\tilde{H}}.$$

Therefore the case $k > 0$ reduces to the case $k = 0$.

References

[1] Belavin, A. A., Polyakov, A. M. and Zamolodchikov, A. B. *Nuclear Physics* **B241**, 333–80 (1984).
[2] Date, E., Kashiwara, M. and Miwa, T. *Proceedings of Japan Academy* **57**, 387–92 (1981).

[3] Date, E., Jimbo, M., Kashiwara, M. and Miwa, T. in *Non-linear Integrable Systems – Classical Theory and Quantum Theory*, M. Jimbo and T. Miwa (eds), World Scientific, Singapore (1983).
[4] Dolan, L. *Physical Review Letters* **47**, 1371–4 (1981).
[5] Hirota, R. *Lecture Notes in Mathematics* **515**, 40–68 (1976).
[6] Kaufman, B. *Physical Review* **76**, 1232–43 (1949).
[7] Lax, P. D. *Communications in Pure and Applied Mathematics* **21**, 467–90 (1968).
[8] Lepowsky, J. and Wilson R. L. *Communications in Mathematical Physics* **62**, 43–53 (1978).
[9] Onsager, L. *Physical Review* **65**, 117–49 (1944).
[10] Sato, M. and Sato, Y. *RIMS Kokyuroku* **388**, 183–204 (1980).
[11] Sato, M., Miwa, T. and Jimbo, M. *Publications RIMS Kyoto University* **14**, 223–67 (1978).
[12] Sato, M., Miwa, T. and Jimbo, M. *Publications RIMS Kyoto University* **15**, 201–78, 577–629, 871–972 (1979).
[13] Sato, M., Miwa, T. and Jimbo, M. *Publications RIMS Kyoto University* **16**, 531–84 (1980).
[14] Ueno, K. *Publications RIMS Kyoto University* **19**, 59–82 (1981).
[15] Wu, T. T., McCoy, B. M., Tracy C. and Barouch E. *Physical Review* **B13**, 316–74 (1976).

14 *P. G. Grinevich and I. M. Krichever*

Algebraic-geometry methods in soliton theory

14.1 Introduction

The theory of 'finite-gap' or algebraic-geometry integration of nonlinear equations was founded as a natural analogue to the periodic case of the inverse scattering method in the rapidly decreasing case.

The general principle of the inverse problem method is to present the nonlinear equation as compatibility conditions of an overdetermined system of auxiliary linear equations. The transform to the 'spectral data' of the auxiliary linear equation 'linearises' the initial nonlinear equation.

This concept is common to both the rapidly decreasing and periodic cases, but the spectral theories of the related linear operators differ significantly.

In the rapidly decreasing case the apparatus corresponding to the needs of the initial stage of this method was well known and developed – it includes the direct and inverse scattering problems. The construction of the periodic and quasiperiodic solutions of the KdV (Korteweg–de Vries) equation required profound implementation of the Flochet spectral theory of the Sturm–Liouville operator with a periodic potentical. This programme was implemented in the series of papers by Novikov, Dubrovin, Matveev and Its (see their reviews in [11, 12]). These results were partly obtained later by Lax [35] and McKean and van Moerbeke [39].

The approach which was posed in these papers has brought about a new understanding of the spectral theory of an arbitrary, ordinary linear differential operator with periodic coefficients.

For any operator L with periodic coefficient the linear space $L(E)$ of the solutions of equation $L\psi = E\psi$ is invariant to the monodromy operator $\hat{T}:\psi(x) \to \psi(x + T)$, determining on this space the finite-dimensional linear operator $\hat{T}(E)$. The solutions of the equation $L\psi = E\psi$ which are eigenfunctions of this operator, $\psi_i(x + T) = w_i(E)\ \psi_i(x)$ are called the Bloch functions. Here $w_i(E)$ are the different eigenvalues of the operator $\hat{T}(E)$.

The statement that the Bloch solutions, which are multi-valued functions of E, become a single-valued function on the Riemann surface of the characteristic equation

$$\det(\hat{T}(E) - w \cdot 1) = F(w, E) = 0, \tag{14.1.1}$$

which seems quite evident now, had previously remained outside the frame of classical spectral Flochet theory.

It turned out that the analytical properties of Bloch's functions on this Riemann surface dominate in the solution of the inverse problem of reconstructing the coefficients of the operator L from the spectral data. When the genus of the surface is finite the solution of the inverse problem is based on the theory of the classic algebraic geometry and theta-functions. (The generalisation of the theories of Abelian differentials and theta-functions for the case of the hyperelliptic curves, which correspond to the general periodic potentials of the Sturm–Liouville operator, was obtained by McKean and Trubowitz [38].)

The overall significance of the algebraic geometry methods was revealed in the papers [24, 25], where the general construction of the periodic and quasiperiodic solutions was advanced for the $(2 + 1)$-systems, which allow the commutation representation. Such equations are called the Kadomtsev–Petviashvili type. The inverse problem for the operators of the form

$$M_i = \frac{\partial}{\partial t_i} - \sum_j u_{ij}(x, t_1, \ldots) \frac{\partial^j}{\partial x^j}. \tag{14.1.2}$$

was solved by this method. The coefficients u_{ij} of these operators, which are the scalar or matrix functions of their arguments, were uniquely determined by the data which characterised the analytical properties on the auxiliary algebraic curve (Riemann surface of the finite genus) of the common eigenfunctions $\psi(x, t_1, \ldots; Q)$, $Q \in \Gamma$ of these operators. The specific nature of these properties, which generalise the properties of Bloch functions, provides that for any function with such properties there exist the operators M_i of (14.1.2), such that

$$M_i\psi(x, t_1, \ldots; Q) = 0. \tag{14.1.3}$$

Nonlinear equations on the coefficients of these operators M_i, which are equivalent to the compatibility conditional of equations (14.1.3), are called the equations of KP type.

The different stages of the consequent development of the theory of finite-gap integration are presented in detail in the series of reviews by Dubrovin, Novikov and Krichever [12, 14, 26, 28]. That is why in the first two sections we confine ourselves to the presentation of the general ideas and methods of the theory and omit all applications of this construction to special equations other than the KP equation. The main objective of this chapter is to consider the problem, which has not been considered in the above mentioned reviews – the problem of the role and place of the 'finite-gap' solutions in the general periodic problem for integrable equations with two spatial dimensions.

The response to this question in the case of one-dimensional evolution equations is basically clear (though it has not always been developed up to the level of a mathematical theorem). The existence of the direct spectral transformation allows us to prove that the set of finite-gap solutions is dense in the space of all periodic solutions (for the KdV equation the theorem stating that an arbitrary periodic solution can be approximated by the finite-gap solutions with the same period was proved by Marchenko and Ostrowskii [37]).

The role of algebraic-geometry methods becomes still more evident in the case of finite-dimensional integrable systems, which were the main subject of the review [14]. In fact all such systems posses the commutation representation of the form

$$[\partial_t - U(t, \lambda), W(t, \lambda)] = 0, \tag{14.1.4}$$

where U, W are matrices which depend on λ as rational or elliptic functions. The *general solutions* of such equations can be represented in terms of theta-functions of the Riemann surface Γ which is determined by the equation

$$F(w, \lambda) = \det(W(t, \lambda) - w \cdot 1) = 0. \tag{14.1.5}$$

(For more details and examples see [14].) The coefficients of the equations and, consequently, the curve Γ are the integrals of the initial equation. It is worth mentioning that these exact expressions for solutions in terms of theta-functions demonstrate that algebraic-geometry methods are more efficient than the classical Liouville theory of completely integrable systems.

As was mentioned above, the problem of the role and place of the finite-gap solutions of the two-dimensional (2 + 1)-systems can be solved only if one can construct the complete (not only inverse) spectral theory for the two-dimensional linear differential operators with periodic coefficients. The corresponding theory for the operator

$$M = \sigma\partial_y - \partial_x^2 + u(x, y), \quad \sigma = 1, \tag{14.1.6}$$

was recently obtained (see [34]). It allows us to prove the integrability of the periodic problem for the KP-2 equation.

The Kadomtsev–Petviashvili equation

$$\tfrac{3}{4}\sigma^2 u_{yy} + (u_t - \tfrac{3}{2}uu_x + \tfrac{1}{4}u_{xxx})_x = 0, \quad \sigma^2 = \pm 1, \tag{14.1.7}$$

is one of the most fundamental equations of mathematical physics. In the papers by Zakharov and Shabat [21] and Druma [10] there was found for this equation the representation:

$$[\sigma\partial_y - L, \partial_t - A] = 0, \quad \sigma^2 = \pm 1, \tag{14.1.8}$$

$$L = \partial_x^2 - u(x, y, t); \quad A = \sigma_x^3 - \tfrac{3}{2}u\partial_x - w(x, y, t). \tag{14.1.9}$$

Two variants of this equation, corresponding to different signs of σ^2, describe the media with different dispersion laws, and are called the KP-1 ($\sigma^2 = -1$) and KP-2 ($\sigma^2 = 1$) equations, respectively. The methods proving the integrability of these equations in the rapidly decreasing case are quite different. For KP-1 it was proved by Manakov [36] with the help of the non-local Riemann problem. For KP-2 it was proved by Ablowitz and Fokas [1] with the help of the $\bar{\partial}$-problem.

In the periodic case the answers also differ. Zakharov and Shul'man [23] have proved that the periodic problem for the KP-1 equation is formally non-integrable and formulated the conjecture of integrability of this problem in the KP-2 case. As was mentioned above, this conjecture was proved by the present author. Moreover, in [16] it was proved that for any real, nonsingular periodic finite-gap solution of the KP-2 equation there exists a neighbourhood in which the finite-gap solutions form a dense set (see a more precise formulation in Theorem 14.5.3).

In addition to the investigations of the periodic problem, the construction of the perturbation theory and the averaging methods for (2 + 1)-systems, the spectral theory for the two-dimensional linear differential operators with periodic coefficients has some other physically important applications – they are the problems of the solid-state physics of the Peierls models type.

In the papers by Brasovskii, Gordunin, Kirova and Belokolos [2, 3, 5] the 'one-gap' extremums in different continual approximations of the Peierls model were found by means of the spectral theory of finite-gap Sturm–Liouville operators. These investigations were continued in the papers of Brasovskii, Dzaloshinskii and Krichever [6, 15, 16, 29]. The non-stationary Peierls models were investigated by means of a spectral theory of the operator (14.1.6) with $\sigma = i$ in [33].

Although the main part of this chapter is devoted to 'finite-gap' integration we present in the Section 14.7 a brief review of the 'fixed-energy spectral theory' of the two-dimensional Schrödinger operator with rapidly decreasing potential. This is done because it turned out that the algebraic-geometry ideas and the methods of the finite-gap theory have been very useful for the development of this theory also. Section 14.7 was written by P. G. Grinevich.

The other sections were written by I. M. Krichever. The presentation of the material in these sections is as follows. The general construction of the finite-gap solutions is formulated in Section 14.2. In the general case such solutions are complex meromorphic functions. The two main types of the conditions which separate the real and non-singular solutions are demonstrated in the Section 14.3 with special reference to the two variants of KP equation. At the beginning of this section we formulate the scheme of the finite-gap integration once again separately for the KP equation. This is done in order to omit, at the first reading some purely technical details which are connected with the matrix auxiliary systems of the linear equations (probably, to begin with the reader should become aquainted with the main construction with reference to the KP-equation).

For the KP case we expand our construction in such a way that it should include the multi-soliton fronts and rational solutions of this equation as particular cases. It is noteworthy that with the help of this example we shall try to demonstrate that algebraic-geometry methods are applicable to the construction of multi-soliton solution for any equation which allows the commutation representation. In this case the corresponding construction can be formulated in a closed way, using only the simplest concepts of linear algebra, though the ideas of this construction are common to the general periodic case.

In the Section 14.4 we present the spectral theory of the periodic operators (14.1.6). This theory is used in Section 14.5 for proving the integrability of the periodic problem for the KP-2 equation. In Section 14.6 the perturbation theory for the KP-2 equation is obtained.

14.2 The main construction of the theory of 'finite-gap' integration

Let us consider the system of nonlinear equations for the coefficients $u_i(x, y, t)$ and $v_j(x, y, t)$ of the operators

$$L = \sum_{i=0}^{n} u_i(x, y, t)\, \partial_x^i; \quad A = \sum_{j=0}^{m} v_j(x, y, t)\, \partial_x^j, \tag{14.2.1}$$

which is equivalent to the operator equation

$$[\partial_y - L, \partial_t - A] = 0 \tag{14.2.2}$$

Here u_i, v_j are the scalar or $(l \times l)$-matrix functions. Below we shall assume that the highest coefficients of L and A are diagonal constant matrices with different element on diagonals, $u_n^{\alpha\beta} = u_n^{\alpha}\, \delta_{\alpha\beta}$, $v_m^{\alpha\beta} = v_m^{\alpha}\, \delta_{\alpha\beta}$. In this case one can reduce the operators to the form, where $v_{m-1}^{\alpha\alpha} = 0$ with the help of transformation $L' = gLg^{-1}$, $A' = gAg^{-1}$ (where $g(x)$ is a diagonal matrix).

The operator equation (14.2.2) is equivalent to the system of the equations for u_i and v_j – the coefficients of both the operators L and A. It turns out, that if $n \leqslant m$, then this system can be reduced to the sheaf of systems for the coefficients of the operator A only. These systems are parametrised by the constants q_i^α, $i = 0, \ldots, M$, $\alpha = 1, \ldots, L$. (Similarly, if $n \geqslant m$ than the system (14.2.2) can be reduced to the sheaf of equations for the coefficients of L only.) For the sake of definiteness, we shall assume further on, that $n \leqslant m$. In order to express the coefficients of L through v_j it is enough to note that the operator $[L, A]$ (as it follows from (14.2.2)) has degree not higher than $m - 1$ and the diagonal elements of the coefficients at ∂_x^{m-1} must be equal to zero. The successive expressions for $\partial_x u_i^{\alpha\alpha}$ and $u_{i-1}^{\alpha\beta}$ $\alpha \neq \beta$ can be obtained by successively equalising the coefficients of $[L, A]$ at ∂_x^{m-1+i}, $i = n, n-1, \ldots, 1$ and the diagonal elements of the coefficient at ∂_x^{m-1} to zero. As was shown in [25], the matrix elements u_i are the differential polynomials of $v_j^{\alpha\beta}$, $j \geqslant i$, and $q_{i_1}^\alpha$, $i_1 \geqslant$ i – the constants of integration of the expressions for $\partial_x u_{i_1}^{\alpha\alpha}$. (In particular, $q_n^\alpha = u_n^\alpha$.)

The starting point of the present author's construction of algebraic geometry solutions of the equations (14.2.2) is the non-singular algebraic curve Γ of genus g with fixed points $P_1, \ldots, P_l$ and fixed local parameters $k_\alpha^{-1}(Q)$, $\alpha = 1, \ldots, l$, in their neighbourhoods, $k_\alpha^{-1}(P_\alpha) = 0$. Besides this, we fix the sets of polynomials $q_\alpha(k)$ of degree n, $R_\alpha(k)$ of degree m and polynomials $\sigma_{i\alpha}(k)$ of abitrary degree, $i = 1, \ldots, 2g$.

For an arbitrary set of points $\gamma_1, \ldots, \gamma_{g+l-1}$ in the general position on Γ there is a unique function $\psi_\alpha(x, y, t, \zeta, Q)$, $Q \in \Gamma$, $\zeta = (\zeta_1, \ldots, \zeta_{2g})$ satisfying the following conditions:

(i) ψ_α is meromorphic on Γ except for the points P_β with poles at the points γ_s (these poles are simple if all γ_s are different, and may be multifold if some of γ_s coincide; in the latter case the multiplicity of the pole must be no greater than the multiplicity with which the corresponding point enters the set $\gamma_1, \ldots, \gamma_{g+l-1}$).

(ii) In the neighbourhood of the point P_β the function ψ_α has the representation of the form

$$\psi_\alpha = \left(\sum_{S=0}^{\infty} \xi_S^{\alpha\beta} k_\beta^{-S}\right) \exp\Big(\kappa_\beta x + q_\beta(\kappa_\beta) y + R_\beta(\kappa_\beta) t + \langle \zeta, \sigma_\beta(\kappa_\beta)\rangle\Big) \tag{14.2.3}$$

where $\kappa_\beta = \kappa_\beta(Q)$, $\xi_S^{\alpha\beta} = \xi_S^{\alpha\beta}(x, y, t, \zeta)$, $\xi_0^{\alpha\beta} = e^{\Phi_\alpha} \delta_{\alpha\beta}$, Φ_α are fixed constants and

$$\langle \zeta, \sigma_\beta(\kappa_\beta)\rangle = \sum_i \zeta_i \, \sigma_{i\beta}(\kappa_\beta).$$

The functions of such a type are called the Clebsh–Gordan–Baker–Akhiezer functions (sometimes, briefly, Baker–Akhiezer functions).

The proof of this statement can be obtained from the exact formulae for ψ_α which we shall place below. Let us prove that from the analytical properties and the uniqueness of ψ_α the statement of Theorem 14.2.1. follows.

Let us denote by $\psi(x, y, t, \zeta, Q)$ the vector-function with coordinates ψ_α, $\alpha = 1, \ldots, l$.

Theorem 14.2 ([25])
There exist unique operators L and A of the form (14.2.1) with $(l \times l)$-matrix coefficients (which depend upon the parameters ζ), such that

$$(\partial_y - L)\psi(x, y, t, \zeta, Q) = 0; \quad (\partial_t - A)\psi(x, y, t, \zeta, Q) = 0. \quad (14.2.4)$$

Proof
Consider the formal series of the form

$$\Psi = \left(\sum_{S=0}^{\infty} \xi_S(x, y, t, \zeta)k^{-S}\right) \times \exp[kx + \hat{q}(k)y + \hat{R}(k)t + <\zeta, \hat{\sigma}(k)>], \quad (14.2.5)$$

where $\xi_S^{\alpha\beta}$ are the same as in (14.2.3); $\hat{q}$, $\hat{R}$, $\hat{\sigma}_i$ are the diagonal matrices with diagonal elements equal to $q_\alpha = \Sigma\, q_i^\alpha k^i$, $R_\alpha = \Sigma\, r_i^\alpha\, k^i$, $\sigma_{i\alpha}(k)$ respectively.

Lemma 14.2.1
For any formal series Ψ of the form (14.2.5) there exists the unique operator L of the form (14.2.1) such that

$$(\partial_y - L)\Psi = 0(k^{-1})\exp[kx + \hat{q}(k)y + \hat{R}(k)t + <\zeta, \hat{\sigma}(k)>]. \quad (14.2.6)$$

To prove this it is sufficient to put the series (14.2.6) in the left-hand side of equation (14.2.6) and equalise to zero the coefficients at k^{-S}, $S = -n, -n + 1, \ldots, 0$. We obtain

$$u_n^{\alpha\beta} = q_n^\alpha\, \delta_{\alpha\beta},\ u_{n-1} = q_{n-1} + [\xi_1, q_n]\xi_0^{-1} \quad (14.2.7)$$

and so on. Here $q_i^{\alpha\beta} = q_i^\alpha\, \delta_{\alpha\beta}$.

Consider now the vector-function $\tilde{\psi}(x, y, t, \zeta, Q) = (\partial_y - L)\psi$ where L is the operator which was constructed in Lemma 14.2.1. The components $\tilde{\psi}_\alpha$ of this vector have the same properties as ψ_α on Γ outside the points P_β. In the neighbourhood of P_β the expansion of $\tilde{\psi}_\alpha$ has the form (14.2.3), but the first term in the pre-exponential factor $\tilde{\xi}_0^{\alpha\beta}$ is equal to zero, $\tilde{\xi}_0^{\alpha\beta} = 0$, as follows from (14.2.6). From the uniqueness of ψ_α we conclude that $\tilde{\psi}_\alpha$ identically equals zero. Hence, the first equation (14.2.4) is proved. One can obtain the proof of the second equation, similarly.

Both equations (14.2.4) are fulfilled for all Q. Therefore we have the following.

Corollary
The operators L and A satisfy (for all ζ) the equation (14.2.2).

Let us determine the function $\psi_0(\zeta, Q)$ by the equality

$$\psi_0 = \sum_\alpha \exp(-\Phi_\alpha - \Phi'_\alpha)\, \psi_\alpha(0, 0, 0, \zeta, Q).$$

Then the functions $\psi'_\alpha = \psi_\alpha(x, y, t, \zeta, Q)\psi_0^{-1}(\zeta, Q)$ are the Baker–Akhiezer functions which correspond to the parameters φ'_α, $\zeta'_i = 0$ and have the poles at the points $\gamma'_1, \ldots, \gamma'_{g+l-1}$ which are zeros of ψ_0. As the vector-function ψ' with components ψ'_α satisfies the same equations (14.2.4) one can conclude that variation parameters are equivalent to the variation of the set γ_S. The set of γ_S is usually chosen as independent parameters determining finite-gap operators L and A on the assumption that $\zeta_i = 0$, $\varphi_\alpha = 0$. As will be seen below, in order to distinquish the real and non-singular solutions it is more convenient to fix in a special way the set $\gamma_1, \ldots, \gamma_{g+l-1}$ and choose $\Phi_1, \ldots, \Phi_{l-1}$ (below we shall always assume that $\varphi_l = 0$) and real constants ζ_i as independent parameters.

The group of substitution of local coordinates $k'_\alpha = h_\alpha^{-1}k_\alpha + \ldots$, $h_\alpha^{-1} \neq 0$, acts on the sets of polynomials q_α, R_α, $\sigma_{i\alpha}$ transforming them into the polynomials σ'_α, R'_α, $\sigma'_{i\alpha}$ so that $q'_\alpha(k'_\alpha) = q_\alpha(k_\alpha) + O(k_\alpha^{-1})$ and similarly for R_α, $\sigma_{i\alpha}$. The Baker–Akhiezer functions which correspond to the two sets k_α, q_2, R_α, $\sigma_{i\alpha}$ and k'_α, q'_α, R'_2, $\sigma'_{\dot\alpha}$ coincide with each other. Let us choose k_α in such a way that polynomial R_α is reduced in this parameter to the form $R_\alpha = v_m^\alpha\, k^m$. Such coordinates are unique up to the substitution of the form $k'_\alpha = k_\alpha + O(k_\alpha^{-m})$. Two coordinates k_α^{-1} and $(k'_\alpha)^{-1}$ are called equivalent if they are connected with each other by the latter relationship. The set of classes of the equivalences would be denoted by $[k_\alpha^{-1}]_m$. With such coordinates, for which $R_\alpha = v_m^\alpha\, k^m$, the coefficients of the polynomials $q_\alpha(k) = \Sigma\, q_i^\alpha\, k^i$ coincide with those introduced at the beginning of this section. Therefore, the polynomials q_α in our construction parametrise the equations, while the other parameters parametrise the solutions of the corresponding equation. The real dimension of the set of the parameters

$$(\Gamma, P_\alpha, [k_\alpha^{-1}]_m, \Phi_1, \ldots, \Phi_{l-1}, \zeta_i) \tag{14.2.8}$$

which determine the solution is equal to $N = 2 \times (4g + l(m + 3) - 4)$ (for each fixed genus g).

Now we shall obtain the exact formulae for the Baker–Akhiezer function in a form slightly different from the standard [12, 25], but more convenient for the 'real' theory and for some applications. (All the definitions from the theta-function theory and the theory of the Abelian differentials, which will be necessary below, can be found in [12, 25].)

Let us fix on Γ the canonical basis of the cycles a_i, b_j with the following intersection matrix $a_i \circ b_j = \delta_{ij}$, $a_i \circ a_j = b_i \circ b_j = 0$. As usual one can define: the basis of normalised holomorphic differentials w_k, vectors $B_k = (B_{ki})$ of their b-periods; the corresponding Riemann theta-function. This function is the entire function of g complex variables, which is transformed under the shifts on the basis vectors e_k in C^g and vectors B_k in the following way:

$$\theta(\tau + l_k) = \theta(\tau); \quad \theta(\tau + B_k) = \exp(-\pi i B_{kk} - 2\pi i \tau_k)\theta(\tau). \tag{14.2.9}$$

The lattice in C^g which is generated by the vectors e_k and B_k determines the torus $\mathcal{T}(\Gamma)$, the so-called Jacobian variety of the curve.

Let Q_0 be the distinguished point on Γ, then the correspondence $A_\kappa(Q) = \int_{Q_0}^{Q} w_k$ determines the well-defined mapping $A: \Gamma \to \mathcal{T}(\Gamma)$ which is called Abelian. The vector $A(Q)$ is defined to the accuracy of adding of the linear combinations of the vectors e_k, B_k with integer coefficients. Nevertheless the zeros $\tilde{\gamma}_S$ of the multi-valued functions $\theta(A(P) + Z)$ are correctly defined as follows from (14.2.9). The number of these zeros is equal to g. There is the following relation between vector Z and the images under the Abelian map of the corresponding points $\tilde{\gamma}_S$:

$$Z = -A(\tilde{\gamma}_1) - \ldots - A(\tilde{\gamma}_g) + \mathcal{K}, \tag{14.2.10}$$

where $\mathcal{K}$ is the vector of the Riemann constants.

Let $\gamma_1^0, \ldots, \gamma_{g+l-1}^0$ be an arbitrary fixed set of the points on Γ. According to the Riemann–Roch theorem there exists the unique function h_α, which has poles at the points γ_S^0 and satisfies the conditions $h_\alpha(P_\beta) = \delta_{\alpha\beta}$.

Let us determine the function $\varphi_\alpha(Z, Q)$, $Z = (Z_1, \ldots, Z_{2g})$ by the formula

$$\varphi_\alpha = h_\alpha(Q) \exp\left(2\pi i \sum_{k=1}^{g} (A_k(P) - A_k(P_\alpha))Z_{k+g}\right) \frac{\theta(A(Q) + Z_\alpha + \sum_{k=1}^{n} (Z_k e_k + Z_{k+g} B_k))\theta(A(P) + Z_\alpha)}{\theta(A(Q) + Z_\alpha)\theta(A(P_\alpha) + Z_\alpha + \sum_{k=1}^{g} (Z_k e_k + Z_{k+g} B_k))}, \tag{14.2.11}$$

where

$$Z_\alpha = \mathcal{K} + \sum_{\beta \neq \alpha} A(P_\beta) - Z_0, \quad Z_0 = \sum_S A(\gamma_S^0). \tag{14.2.12}$$

From (14.2.10) it follows that the functions φ_α are periodic functions with their periods equal to 1 for all variables Z_i.

Consider the differentials $i\,dp$, $i\,dE$, $i\,d\Omega$ which have the only singularities at the points P_α of the form dk_α, $dq_\alpha(k_\alpha)$, $dR_\alpha(k_\alpha)$ correspondingly. They may be uniquely determined by means of the conditions that their periods over all the cycles on Γ are purely imaginary. Let us define the real vector U with coordinates

$$U_k = \frac{1}{2\pi} \oint_{a_k} dp, \quad U_{k+g} = -\frac{1}{2\pi} \oint_{b_k} dp, \quad k = 1, \ldots, g. \tag{14.2.13}$$

Similarly, the differentials dE and $d\Omega$ determine the $2g$-dimensional real vectors V and W.

If we cut the curve Γ over the cycles a_i and b_j we obtain a simple-connected domain. That is why we can choose in this domain a single-valued branch of the integrals $p(Q)$, $E(Q)$, $\Omega(Q)$ of the corresponding differentials. In the neighbourhoods of the points P_α they have the form

$$ip = k_\alpha - a_\alpha + O(k_\alpha^{-1}), \quad iE = q_\alpha(k_\alpha) - b_x + O(k_\alpha^{-1}),$$
$$i\Omega = R_\alpha(k_\alpha) - c_\alpha + O(k_\alpha^{-1}). \tag{14.2.13}$$

The integrals p, E, Ω would be uniquely defined if we assume that $a_l = b_l = c_l = 0$.

Let us determine the differentials $d\sigma_1, \ldots, d\sigma_2 g$. They have the only singularities in P_α. The differentials $d\sigma_j$, $d\sigma_{j+g}$, $j = 1, \ldots, g$ have the only non-zero periods over cycles a_j, b_j respectively, which are equal to $\pm 2\pi$. The integrals of $d\sigma_1, \ldots, d\sigma_{2g}$ wouldbe denoted as $\sigma_1(Q), \ldots, \sigma_{2g}(Q)$.

Let the polynomials $\sigma_{j\alpha}$, which were introduced above into the definition of the Baker–Akhiezer functions, be equal to a singular part of the expansions of $i\sigma_j(Q)$ in the neighbourhoods of P_α.

Lemma 14.2.2

The Baker–Akhiezer vector-function, which has poles at the points $\gamma_1^0, \ldots, \gamma_{g+l-1}^0$, can be represented in the form

$$\psi = \exp[ax + by + ct + \Phi]\, \varphi(Ux + Vy + Wt + \zeta, Q)$$
$$\times \exp[i(px + Ey + \Omega t + \langle \Phi, \sigma \rangle)] \tag{14.2.14}$$

where φ is the vector-function with the above-determined coordinates $\varphi_\alpha(Z, Q)$; a, b, c, Φ are the diagonal matrices with the a_α, b_α, c_α, Φ_α on the diagonals (according to the above-made assumptions, $a_l = b_l = c_l = \varphi_l = 0$); and

$$p = p(Q), \ E = E(Q), \quad \Omega = \Omega(Q), \quad \sigma_j = \sigma_j(Q).$$

To prove the lemma it is sufficient to check that, although $\varphi_\alpha(Z, Q)$, $Q \in \Gamma$, is the multi-valued function on Γ, the function ψ which is determined by the formula (14.2.14), is single-valued. Its behaviour inside and outside the neighbourhood P_α follows from the definitions of φ_α, h_α and p, E, Ω, σ_j.

Let us denote the coefficient of the expansion of φ_α in the neighbourhood of P_β at κ_β^{-S} as $\varphi_S^{\alpha\beta}$. From (14.2.11) it follows that this coefficient is the quasiperiodic function of the variables x, y, t with the vector of the periods which are equal to U, V, W, respectively. The matrices ξ_S in (14.2.5) are equal to $\exp(ax + by + ct + \Phi)\hat{\varphi}_S$, $\hat{\varphi}_S = (\varphi_S^{\alpha\beta})$.

Corollary

The finite-gap operators L, A have the form

$$L = g\hat{L}g^{-1}, \quad A = g\hat{A}g^{-1}, \tag{14.2.15}$$

where $g = \exp(ax + by + ct + \Phi)$ and the coefficients of $\hat{L}$ and $\hat{A}$ are quasiperiodic functions of x, y, t, that is,

$$\hat{u}_i = \hat{u}_i(Ux + Vy + Wt + \zeta), \quad \hat{v}_j = \hat{v}_j(Ux + Vy + Wt + \zeta) \tag{14.2.16}$$

Here $\hat{u}_i(Z_1, \ldots, Z_{2g})$, $\hat{v}_j(Z_1, \ldots, Z_{2g})$ are the periodic functions with the periods equal to one. They are the differential polynomials of $\varphi_S(x, y, t, \zeta)$ and may be expressed in terms of the Riemann theta-functions. As was mentioned above, in this chapter we do not consider the special examples which can be found in the other reviews [11, 12].

For the construction of the perturbation theory of the equations (14.2.2), it is necessary to know the solutions not only of the system (14.2.4) but also of its conjugate. Let us define, following the paper of Cherednik [42], the dual Baker–Akhiezer functions. For any set of points $\gamma_1, \ldots, \gamma_{g+l-1}$ in general position on Γ, there exists a unique differential $\hat{\omega}$ of the second type with poles of second degree at the points P_α and zeros at the points γ_S. The remaining zeros of $\hat{\omega}$, $\gamma_1^+, \ldots, \gamma_{g+l-1}^+$ is said to be dual to the set $\gamma_1, \ldots, \gamma_{g+l-1}$. (Recall that any differential on Γ with poles of second degree at P_α has $2g + 2l - 2$ zeros.) From this definition it follows that the vectors Z_0 and Z_0^+ corresponding under the Abelian mapping to the divisors (γ_S^0) and (γ_S^{0+}) are connected to each other by the following relation:

$$Z_0 + Z_0^+ = \tilde{K} + 2\sum_\alpha A(P_\alpha),$$

where $\tilde{K}$ is the canonical class.

The row vector ψ^+ with coordinates $\psi_\alpha^+(x, y, t, \zeta, Q)$, which are meromorphic functions on Γ, except at points P_β, have poles at $\gamma_1^+, \ldots, \gamma_{g+l-1}^+$, and the form (14.2.18) defined in a neighbourhood of each of the points P_β, is called the dual Baker–Akhiezer function.

$$\psi_\alpha^+ = \exp[-k_\beta x - q_\beta(k_\beta)y - R_\beta(k_\beta)t - \langle\zeta, \sigma_\beta(k_\beta)\rangle] \times \left(\sum_{S=0}^{\infty} \xi_S^{+\alpha\beta}k^{-S}\right), \tag{14.2.18}$$

where $\xi_S^{+\alpha\beta} = \xi_S^{+\alpha\beta}(x, y, t, \zeta)$, $\xi_0^{+\alpha\beta} = e^{-\Phi_\alpha}\delta_{\alpha\beta}$.

This dual Baker–Akhiezer function can be represented in the form

$$\psi^+ = \exp[-i(px + Ey + \Omega t + <\zeta, \sigma>)]\ \varphi^+(-Ux - Vy - Wt - \zeta, Q) \times \exp[-ax - by - ct - \varphi] \qquad (14.2.19)$$

where the components φ_α^+ of the vector-row $\varphi^+(Z, Q)$ are given by the formula (14.2.11) in which we must replace the vector Z_α by the vector $Z_\alpha^+ = Z_\alpha + Z_0 - Z_0^+$ and replace the function h_α by the function h_α^+, which has poles in γ_S^+ and is such that: $h_\alpha^+(P_\beta) = \delta_{\alpha\beta}$.

In the paper [42] it was proved that ψ^+ satisfies the equations

$$\psi^+(x, y, t, \zeta, Q)(\partial_y - L) = 0; \quad \psi^+(x, y, t, \zeta, Q)(\partial_t - A) = 0 \qquad (14.2.20)$$

where L and A are the same as in (14.2.4).

In these equations and below the right action of any differential operator $\mathcal{D} = \Sigma\ \omega_i \partial_x^i$ on a row vector is defined as the action of the formal adjoint operator, given by

$$f^+\mathcal{D} = \sum_i (-\partial_x)^i (f^+\omega_i); \quad f^+\partial_y = -\partial_y f^+. \qquad (14.2.21)$$

The proof of (14.2.20) is based on the fact that according to the definition of ψ, ψ^+ the differentials

$$d\Lambda_{\alpha\beta} = \psi_\alpha(x, y, t, \zeta, Q)\psi_\beta^+(x', y', t', \zeta', Q)\ \hat{\omega} \qquad (14.2.22)$$

are holomorphic on Γ outside $P_1, \ldots, P_l$. Therefore,

$$\sum_\gamma \operatorname{res}_{P_\gamma} d\Lambda_{\alpha\beta} = 0. \qquad (14.2.23)$$

(The bilinear relations for the definition of the τ-function, which were used by Date *et al.* [7], are the further generalisation of (14.2.23).)

14.3 **Algebraic-geometry solutions of the KP equation**

The operators L and A (14.1.9), which enter the commutation representation for the KP equation, have scalar coefficients. That is why for the application of our general scheme for this equation we must consider the curves Γ with one fixed point $P_1 \subset \Gamma$. (i.e. $l = 1$). The polynomials q and R which determine the corresponding Baker–Akhiezer function must be equal to $q = \sigma k^2$, $R = k^3$.

As was mentioned above, we shall once again formulate the scheme of algebro-geometrical integration especially for the KP equation for the sake of simplicity (in this case we can omit some technical details which are necessary for the systems with matrix auxiliary operators). What is more, in this case we shall expand our construction in order to include into it practically all the other (multi-soliton, rational, etc.) sets of exact solutions.

Let Γ be the non-singular algebraic curve of the genus g with a fixed point P_1 and local coordinate $k^{-1}(Q)$ in its neighbourhood.

Lemma 14.3.1 [25]
For any set of points $\gamma_1, \ldots, \gamma_{g+N}$ the dimension of the linear space of functions $\psi(x, y, t, \zeta, Q)$, satisfying the following conditions is equal to $N + 1$:

(i) ψ is meromorphic on Γ outside P_1 and has poles at $\gamma_1, \ldots, \gamma_{g+N}$.
(ii) In the neighbourhood of P_1 it takes the form

$$\psi = \exp(kx + \sigma k^2 y + k^3 t + \Sigma\, \zeta_i k^i) \times \left(\sum_{S=0}^{\infty} \xi_S(x, y, t, \zeta) k^{-S} \right). \tag{14.3.1}$$

The function ψ can be determined up to a constant multiple by the system of N linear homogeneous equations.

(iii) $$\sum_{i,S} \alpha_{iS}^j \frac{\partial^S}{\partial Z_i^S} \psi(x, y, t\, \zeta, \varkappa_i) = 0, \; j = 1, \ldots, N, \tag{14.3.2}$$

(here the pairs i, S belong to a certain finite set). The points $\varkappa_i$ and constants α_{iS}^j are the added parameters of the solutions of the KP equation, $Z_i = Z_i(Q)$ are some local coordinates in the neighbourhood of P_1.

It should be emphasised that for the following statements it is essential only that equations (14.3.2) are linear and their coefficients are independent of x, y, t. Because of this, for any differential operator $\mathcal{D}$ in variables x, y, t the function $\mathcal{D}\psi$ satisfies the equations (14.3.2) if they are valid for ψ. Presumably, the most general form of such equation is

$$\text{(iii}') \quad \int_{\sigma j} \psi \, d\mu_j = 0, \quad j = 1, \ldots, N. \tag{14.3.3}$$

Here σ_j are arbitrary fixed cycles on Γ and $d\mu_j$ are the fixed differentials on σ_j. The constraints (iii) become the particular case (iii′) if one admits that $d\mu_j$ are the generalised differentials (in case (iii) $d\mu_j$ ought to be the linear combination of the δ-function and its derivatives).

Theorem 14.3.1
Let function $\psi(x, y, t, \zeta, Q)$ satisfying the constraints (i), (ii) and (iii) – or (iii′) – and which is normalised by the condition that ξ_0 in the expansion (14.3.1) equals one, $\xi_0 \equiv 1$.

Then the equations

$$(\sigma\partial_y - L)\psi(x, y, t, \zeta, Q) = 0; \quad (\partial_t - A)\psi(x, y, t, \zeta, Q) = 0 \tag{14.3.4}$$

are fulfilled. Here, L and A have the form (14.1.9) and their coefficients are equal to

$$u(x, y, t, \zeta) = 2\partial_x \xi_1, \quad w = \tfrac{3}{2}u_x + 3\,\xi_{2x}. \tag{14.3.5}$$

From (14.3.4) it follows that $u(x, y, t\,\zeta)$ determined by (14.3.5) is the solution of the KP equation.

Let us give a rough sketch of the proof, which repeats the proof of the general Theorem 14.2.1.

Consider the function $\tilde{\psi} = (\sigma\partial_y - L)\psi$. It satisfies condition (i) because the poles γ_S of ψ are independent of x, y. As has been explained above, $\tilde{\psi}$ satisfies the conditions (iii) – or (iii′) – as well as $\xi_0 \equiv 1$. From (14.3.1) and (14.3.5) it follows that the first coefficient $\tilde{\xi}_0$ in the expansion of $\tilde{\psi}$ in the neighbourhood of P_1 equals zero.

Therefore, $\tilde{\psi} = 0$ and the first equation of (14.3.4) is proved. The proof of the second equation of (14.3.4) is absolutely analogous.

At the level of the complex and meromorphic solutions, which we obtain in our construction in general, the difference between KP-1 and KP-2 equations is not essential.

The distinguishing of the real and non-singular solutions would be made at the beginning in the case of the KP-2 equation.

The KP-2 equation

Let us assume that there exists an anti-holomorphic involution τ (which is oftne briefly called an anti-involution) on Γ for which P_1 is stationary, $\tau(P_1) = P_1$.

Lemma 14.3.2
Let the set $\gamma_1, \ldots, \gamma_{g+N}$ and the conditions (iii) – or (iii′) – be invariant with respect to τ. Then, if $k(\tau(Q)) = \bar{k}(Q)$, $\zeta_i = \bar{\zeta}_i$ as well, the solution $u(x, y, t, \zeta)$ of the KP-2 equation, corresponding to these parameters according to Theorem 14.3.1, is real.

The invariance of the conditions (iii) means that there are two permutations $i'(i)$ and $j'(j)$ of the indices of the points $\varkappa_i$ and of equations (14.3.2), so that $\alpha_{i'S}^{j'} = \bar{\alpha}_{iS}^{j}$ and $\varkappa_{i'} = \tau(\varkappa_i)$.

Consider the function $\bar{\psi}(x, y, t, \zeta, \tau(Q))$. From the conditions of the lemma it follows that this function satisfies the same analytical properties which uniquely determine the function ψ. Therefore,

$$\psi(x, y, t, \zeta, Q) = \bar{\psi}(x, y, t, \zeta, \tau(Q)). \tag{14.3.6}$$

Consequently, $\xi_1 = \bar{\xi}_1$ and from (14.3.5) we obtain $u = \bar{u}$.

The further constraints on the parameters of the construction are necessary in order to separate the non-singular solutions. We shall formulate them for the case of conditions (iii) (for conditions of (iii′) type these

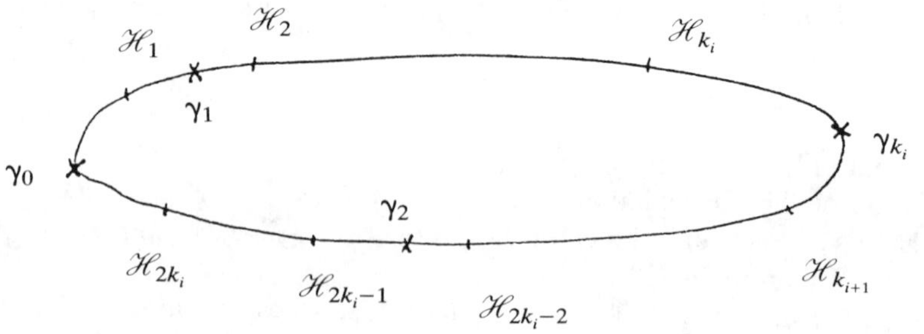

Figure 14.1

constraints can be found in [32]). It is known that the number of stationary cycles of the arbitrary anti-involution on the curve of genus g is not greater than $g + 1$. We shall consider the so-called M-curves – the curves with anti-involution τ, which have $g + 1$ stationary cycles $a_0, \ldots, a_g$ of τ. We shall assume that $P_1 \in a_0$. Let us fix on each cycle a_i, $i \neq 0$, the set (probably empty) of the points $\varkappa_{i,1}, \ldots, \varkappa_{i,2k_i}$ which are arranged in the order of some cyclic directions. The points $\gamma_{i,0}, \ldots, \gamma_{i,k_i}$ must be chosen in the following way. The point $\gamma_{i,S}$, $S = 1, \ldots, k_i$ must belong to the interval between the points $\varkappa_{i,S}, \varkappa_{i,S+1}$ or $\varkappa_{i,2k_i-S}, \varkappa_{i,2k_i-S+1}$. The point $\gamma_{i,0}$ belongs to the interval between $\varkappa_{i,1}, \varkappa_{i,2k_i}$. The possible arrangement of the points $\gamma_{i,0}, \ldots, \gamma_{i,k_i}$ is demonstrated in Figure 14.1. (If $k_i = 0$, then the point $\gamma_{i,0}$ is an arbitrary point of the cycle a_i.)

Consider the Baker–Akhiezer function $\psi(x, y, t, \zeta, Q)$ having the poles at $\gamma_{i,j}$, $i = 1, \ldots, g, j = 0, \ldots, k_i$. The total number of the poles equals $g + N$, $N = \Sigma\, k_i$. As in the previous lemma we assume that the parameters $\zeta_i = \bar{\zeta}_i$ are real. We take conditions (iii) in the following special form:

$$\psi(x, y, t, \zeta, \varkappa_{i,S}) = \psi(x, y, t, \zeta, \varkappa_{i,2k_i-S}) \tag{14.3.7}$$

(the total number of these equations equals N).

Theorem 14.3.2
The solution of the KP-2 equation corresponding to these parameters is real and regular.

Proof
The solution is real because all the points $\gamma_{i,j}$, $\varkappa_{i,S}$ are stationary points of τ. Therefore, the statement of Lemma 14.3.2 can be applied.

As follows from the main construction, the solution u has singularities only for such values x_0, y_0, t_0 of variables x, y, t for which we cannot normalise the corresponding Baker–Akhiezer function by the condition $\xi_0 = 1$. This means that for these values there exists the function $\tilde{\psi}(x_0, y_0, t_0, \zeta, Q)$ which satisfies the conditions (i), (ii) (14.3.7) and such that $\tilde{\xi}_0 = 0$ in the expansion (14.3.1).

The logarithmic differential $d\psi/\psi$ of any Baker–Akhiezer-type function is meromorphic on Γ, has residues ± 1 at zeros and at the poles of ψ. This differential has the pole of the second order at P_1 also. If $\xi_0 \neq 0$ in (14.3.1) then the residue of $d\psi/\psi$ at P_1 equals zero. If $\xi_0 = 0$ then $d\psi/\psi$ has non-zero positive residue at P_1. Therefore, in the latter case the function $\tilde{\psi}$ has less than $g + N$ zeros outside P_1. Now we shall prove that ψ cannot have less than $g + N$ zeros for any x, y, t outside P_1. This fact would complete the proof of the theorem. If we glue together the points $\varkappa_{i,s}$ and $\varkappa_{i,2k_i-s}$, the cycle a_i would be broken into $k_i + 1$ cycles. The conditions (14.3.7) mean that ψ defines a continuous (except $\gamma_{i,j}$) function on each of these cycle. Therefore, ψ has no less than one zero on each of these cycles. Consequently, there are no less than $g + N$ zeros of ψ outside P_1.

Remark

In the framework of the present chapter we leave out the proof of the statement that the solutions which have been obtained above are 'multi-solitons' at the background of finite-gap solutions. the latter means that for $x \to \pm\infty$ this solution tends to $u_{\pm}(x, y, t)$, where $u_{\pm}$ are the quasiperiodic finite-gap solutions which have the form of (14.4.1) with different values of the phases $\zeta_{\pm}$. For $g = 0$ these solutions coincide with the ordinary multi-soliton-front solutions of the KP-2 equation. The proof of this can be obtained by the same method as in [33].

The KP-1 equation ($\sigma = i$)

The reasons which guarantee the reality and non-singularity of the solutions which were obtained above for the KP-1 equation, are quite different from those which were considered in the previous section.

Let Γ, as earlier, be the curve with such an anti-involution τ that $\tau(P_1) = P_1$. In the case of the KP-1 equation it is sufficient that this anti-involution has a separating-type. This means that the stationary cycles $a_0, \ldots, a_l$ of this anti-involution separate Γ into the two non-connected domains $\Gamma^{\pm}$. (All the M-curves have this type. Therefore, the set of curves possible for the KP-1 case is much greater than in the KP-2 case.)

The local parameter in the neighbourhood of P_1 must satisfy the condition $k(\tau(Q)) = -\bar{k}(Q)$. Let $\mathrm{i}dp$, as it was in the previous paragraph, be the differential with the only pole which has the form of $dk(1 + 0(k^{-2}))$ in the neighbourhood of this point. This differential is uniquely determined by the condition that its periods over any cycle on Γ are purely imaginary. From this it follows, that Im $p(Q)$ (where p is the integral of dp) is a single-valued function. The cycles $a_0, \ldots, a_l$ coincide with the set of points Q such that Im $p(Q) = 0$. Consequently, the differential dp is positive at all a_i (in respect to the orientation of a_i which is induced on it as at the boundary of Γ^+). Therefore, its zeros can be separated into the two sets $\gamma_1, \ldots, \gamma_g$ and $\gamma_1^+ = \tau(\gamma_1), \ldots, \gamma_g^+ = \tau(\gamma_g)$.

Consider the Baker–Akhiezer functions ψ which have the poles at the points $\gamma_1, \ldots, \gamma_g, \mu_1, \tau(\mu_1), \ldots, \mu_L, \tau(\mu_L), \varkappa_1, \ldots, \varkappa_M$ where $\varkappa_i, \mu_j$ are the arbitrary points on Γ, which do not belong to $a_0, \ldots, a_g$.

The parameters ζ_i which were introduced for describing the form of ψ in the neighbourhood of P_1 must satisfy the following condition:

$$\zeta_i = (-1)^{i+1}\,\bar{\zeta}_i,\ i = 1, \ldots, g + N. \tag{14.3.8}$$

We shall take the linear conditions which determine ψ in the form:

$$\operatorname{res}_{\varkappa_i} \psi dp = \sum_{j=1}^{M} c_{ij}\psi(\bar{\varkappa}_j),\ i = 1, \ldots, M. \tag{14.3.9}$$

$$\psi_{j,1} = C_j\psi_{j,2},\quad \tilde{\psi}_{j,1} = \tilde{C}_j\,\tilde{\psi}_{j,2} \tag{14.3.10}$$

where $\psi_{j,1}$ and $\psi_{j,2}$ are the coefficients of the expansion ψ in the neighbourhood of μ_j:

$$\psi = \frac{\psi_{j,1}}{p - p(\mu_j)} + \psi_{j,2} + O(p - p(\mu_j)) \tag{14.3.11}$$

and, similarly, $\tilde{\psi}_{j,1}, \tilde{\psi}_{j,2}$ are the coefficients of the expansion ψ at the point $\bar{\mu}_j = \tau(\mu_j)$.

Below it will be assumed that $\varkappa_i$ is arranged in such a way that $\varkappa_1, \ldots, \varkappa_p$ belong to Γ^+ and $\varkappa_{p+1}, \ldots, \varkappa_M$ belong to Γ^-.

Theorem 14.3.3
Let the matrix $\hat{c} = (c_{ij})$ be skew-Hermitian, $c_{ij} = -\bar{c}_{ji}$, with the minors of $i\hat{c}$ with indices $1 \leqslant i, j \leqslant p$ and $p + 1 \leqslant i, j \leqslant M$ being respectively non-negative and non-positive. Let also $\bar{C}_j = -\tilde{C}_j$. Then the corresponding solution of the KP-1 equation is real and non-singular.

Proof
Let us define the function ψ^+ by the equality

$$\psi^+(x, y, t, \zeta, Q) = \bar{\psi}(x, y, t, \zeta, \tau(Q)). \tag{14.3.12}$$

From (14.3.8) it follows that the differential

$$d\hat{\Omega}(x, y, t, \zeta) = i\psi(x, y, t, \zeta, Q)\psi^+(x, y, t, \zeta, Q)\, dp \tag{14.3.3}$$

is meromorphic, because the exponential factors of ψ and ψ^+ in their expansions in the neighbourhood of P_1 cancel each other. The points γ_S, γ_S^+ are the zeros of dp. Therefore, $d\hat{\Omega}$ has only the simple poles at the points $\varkappa_i$, $\bar{\varkappa}_i$ and the poles of the second degree at μ_j and $\bar{\mu}_j$. From

(14.3.10) and the condition $\tilde{C}_j = -\bar{C}_j$ it follows that the residues of $d\hat{\Omega}$ at the points μ_j and $\bar{\mu}_j$ are equal to zero. The sum of residues of $d\hat{\Omega}$ at $\bar{\varkappa}_i$ and κ_i is equal to zero due to the skew-Hermiticity of $\hat{c}$. Consequently, the residue of $d\hat{\Omega}$ at P_1 equals zero (because the sum of all residues of any meromorphic differential on an algebraic curve must be equal to zero). The substitution of (14.3.1) in (14.3.13) leads to:

$$0 = \operatorname{res}_{P_1} d\hat{\Omega} = \xi_1 - \bar{\xi}_1. \tag{14.3.14}$$

From (14.3.5) we conclude that u is real, $\bar{u} = u$.

Assume that u has its singularity at x_0, y_0, t_0. As was explained above, this is possible only for such values of $x = x_0$, $y = y_0$, $t = t_0$ for which the corresponding pre-exponential factor in (14.3.1) has the form $O(k^{-1})$. Therefore, the differential $d\hat{\Omega}(x_0, y_0, t_0, \zeta)$ is regular at P_1 and it is possible to consider its integral over all the cycles

$$I = \frac{1}{2\pi i} \sum_j \oint_{a_j} d\hat{\Omega}(x_0, y_0, t_0, \zeta). \tag{14.3.15}$$

As follows from (14.3.12), there is the relation $\psi^+ = \bar{\psi}$ for any point which belongs to a_j. Hence, the integral (14.3.15) must be positive, $I > 0$ (recall that the differential dp is positive on a_j).

On the other hand, this integral is equal to the sum of the residues of $d\hat{\Omega}$ at the points $\varkappa_1, \ldots, \varkappa_p, \bar{\varkappa}_{p+1}, \ldots, \bar{\varkappa}_M$ which belong to Γ^+. The matrices (ic_{ij}), $1 \leqslant i, j \leqslant p$, and (ic_{ij}), $p+1 \leqslant i, j \leqslant M$, are non-negative and non-positive, respectively. From this we obtain that the residues of $d\hat{\Omega}$ in Γ^+ are all non-positive. Hence, $I \leqslant 0$. This contradiction proves the statement of the theorem.

The conditions on the parameters $(\Gamma, P_1, \gamma_1, \ldots, \gamma_g)$ from which the reality and regularity of the corresponding solution of the KP-1 equation follow was found by Dubrovin [13] for the case $N = 0$. The particular case of the conditions of Theorem 14.3 when the matrix $\hat{c}$ is diagonal was found by Krichever [33]. The general conditions on c_{ij} and C_j were recently formulated by these two workers for the case $g = 0$.

The conditions (14.3.9) and (14.3.10) are sufficient but not necessary for the reality and regularity of the corresponding solutions. The example of the conditions which involves the higher derivatives of ψ was considered in the paper [17].

Example

Case $g = 0$. The functions ψ which satisfy the conditions (i), (ii) in the case when Γ is an ordinary complex plane (i.e. $g = 0$) have the form

$$\psi = \frac{k^N + a_1 k^{N-1} + \ldots + a_N}{\prod_{i=1}^{M} (k - \varkappa_i) \prod_{j=1}^{L} (k - \mu_j)(k + \bar{\mu}_j)} \exp\left(kx + ik^2 y + k^3 t + \sum_{i=1}^{N} \zeta_i k^i\right) \tag{14.3.16}$$

(The anti-involution τ in this case has the form $\tau(k) = -\bar{k}$.)

The conditions (14.3.9), (14.3.10) are the system of linear equations on the $a_i(x, y, t, \zeta)$ in (14.3.16). If we denote the matrix of the coefficients at a_i in these equations by Δ then

$$a_1 = \frac{\partial_x(\det \Delta)}{\det \Delta} \Rightarrow u = 2\partial_x^2 \ln \det \Delta. \qquad (14.3.17)$$

For $L = 0$ we obtain the multi-soliton fronts of the KP-1 equation. For $M = 0$ the corresponding solutions are rational solutions of this equation which were first found by another method in the paper [4]. The general rational solutions for all equations of (14.2.2)-type were constructed in [27].

In conclusion of this section we would give the exact form of the matrix Δ in the case $L = 0$. For this it is convenient to present ψ in the form

$$\psi = \left(1 + \sum_{i=1}^{M} \frac{r_i}{k - \varkappa_i}\right)\exp(kx + \mathrm{i}k^2y + k^3t + \Sigma\zeta_j k^j). \qquad (14.3.18)$$

The variables a_i are the linear combination of r_i.

Let us denote

$$w_i = \varkappa_i x + \mathrm{i}\varkappa_i^2 y + \varkappa_i^3 t + \Sigma\zeta_j \varkappa_i^j.$$

Then the matrix of the coefficients at r_i in the equations which are equivalent to (14.3.9) equals

$$\tilde{\Delta}_{ij} = c_{ij}\frac{\mathrm{e}^{-w_i - \bar{w}_j}}{\varkappa_i + \bar{\varkappa}_j} + \delta_{ij}. \qquad (14.3.19)$$

This matrix is similar to Δ. Hence the function

$$u = 2\delta_x^2 \ln \det \Delta(x, y, t, \zeta)$$

is the real and regular solution of the KP-1 equation.

14.4 The spectral theory of the 'finite-gap' non-stationary Schrödinger operators

The construction [25], which was described above, in the case $N = 2L + M = 0$ provides the quasiperiodic solutions of the KP equations which have the form

$$u = 2\partial_x^2 \ln \theta(Ux + Vy + Wt + \zeta) \qquad (14.4.1)$$

(This formula [25] can be easily obtained from the exact expressions for the Baker–Akhiezer functions.)

The vectors U, V, W in (14.4.1) are the vectors of the periods of the differentials $\mathrm{d}p$, $\mathrm{d}E$ and $\mathrm{d}\Omega$ respectively. In the case when

$$U_k = \frac{n_k}{L_1}, \quad V_k = \frac{m_k}{L_2}, \quad n_k, m_k \text{ integers}, \tag{14.4.2}$$

the solution (14.4.1) is periodic in x, y. In this section we will consider the spectral theory of the corresponding operators (14.1.6) (see the details in [33]).

Let $\psi(x, y, t, \zeta, Q)$ and $\psi^+(x, y, t, \zeta, Q)$ be the Baker–Akhiezer function and the dual Baker–Akhiezer function, respectively. As was proved above,

$$(\sigma\partial_y - L)\psi = 0, \quad (\sigma\partial_y + L)\psi_1^+ = 0, \tag{14.4.3}$$

where $\psi = \psi(x, y, t, \zeta, Q)$, $\psi_1^+ = \psi^+(x, y, t, \zeta, Q_1)$.

From (14.4.3) it follows that

$$\sigma\partial_y(\psi\psi_1^+) = \partial_x(\psi'\psi_1^+ - \psi\psi_1^{+\prime}). \tag{14.4.4}$$

If Q_1 tends to Q, then with the help of averaging the equality (14.4.4), we can obtain that

$$\sigma dE<\psi\psi^+>_x = dp<\psi'\psi^+ - \psi\psi^{+\prime}>_y \tag{14.4.5}$$

where $<\cdot>_x$ and $<\cdot>_y$ are the averages of the functions in x and y, respectively.

Lemma 14.4.1
The differential

$$\hat{\omega} = \frac{dp}{<\psi\psi^+>_x} = \frac{\sigma dE}{<\psi'\psi^+ - \psi\psi^{+\prime}>_y} \tag{14.4.6}$$

is holomorphic on Γ outside P_1, where it has the poles of the second degree. The zeros of $\hat{\omega}$ coincide with the poles $\gamma_{S0}\gamma_S^+$ of the functions ψ and ψ^+.

The statement immediately follows from the equality (14.4.5).

Corollary
If the poles of ψ are the zeros of the differential dp, then

$$<\psi\psi^+>_x \equiv 1. \tag{14.4.7}$$

The following statement means that the set of the functions $\psi(x, y, t, \zeta, Q)$ for all $Q \in a_i$, is the basis in the space of square-integrable functions on the real axis of the variable x (this basis depends on y, t, ζ as on parameters).

Theorem 14.4.1
Let Γ, P_1, k, $\gamma_1, \ldots, \gamma_g$ be the parameters for which the corresponding solutions are real and regular. Then there exists the following equality:

$$\delta(x - x') = \sum_{i=0}^{l} \oint_{a_i} \psi(x, y, t, \zeta, Q)\psi^+(x', y, t, \zeta, Q)\hat{\omega} \quad (14.4.8)$$

where $a_0, \ldots, a_l$ are the stationary cycles of τ.

The proof of the theorem was obtained in a more general case in [33].

Let U_k satisfy the conditions (14.4.2), then the function

$$w_1(Q) = \exp\left(\mathrm{i}L_1 \int_{Q_0}^{Q} \mathrm{d}p\right) = \exp(\mathrm{i}L_1 p(Q)) \quad (14.4.9)$$

is well-defined on Γ (i.e. it does not depend on the path of the integration). This function is holomorphic on Γ outside the point P_1, where it has the essential singularity. This function gives the eigenvalues for the monodromy operator:

$$\psi(x + L_1, y, t, \zeta, Q) = w_1(Q)\, \psi(x, y, t, \zeta, Q),$$
$$\psi^+(x + L_1, y, t, \zeta, Q) = w_1^{-1}(Q)\, \psi^+(x, y, t, \zeta, Q). \quad (14.4.10)$$

The relations (14.4.10) follow from the obvious statement that both sides of (14.4.10) have the same analytical properties. For any complex number w we denote by $Q_n(w)$ the points of Γ for which we have

$$w_1(Q_n(w)) = w. \quad (14.4.11)$$

For $n \to \pm\infty$ the corresponding points Q_n tend to P_1. In the following formulae we shall briefly write $\psi_n = \psi_n(x, y, t, \zeta)$ instead of $\psi(x, y, t, \zeta, Q_n)$.

For any function $f(x)$ such that

$$f(x + L_1) = w\, f(x) \quad (14.4.12)$$

we can consider the series

$$S = \sum_n c_n(y, t, \zeta)\psi_n(x, y, t, \zeta) \quad (14.4.13)$$

where

$$c_n = \frac{1}{L_1} \int_0^{L_1} f(x') \frac{\psi_n^+(x, y, t, \zeta)}{<\psi_n\, \psi_n^+>_x} \mathrm{d}x'. \quad (14.4.14)$$

Theorem 14.4.2 [33]
If the function f(x) is continuously differentiable then the series (14.4.13) converges to $f(x)$.

In the case $g = 0$ Theorems 14.4.1 and 14.4.2 are usual statements about Fourier transformations of the square-integrable functions on the real axis and on the circle, respectively.

14.5 The periodic problem for the KP equation

Consider a curve Γ with a fixed point P_1 and local parameter $k^{-1}(Q)$ such that the periods of the corresponding differentials $\mathrm{d}p$ and $\mathrm{d}E$ satisfy the conditions (14.4.2). Then in addition to the function $w_1(Q)$ which was introduced earlier there exists a well-defined function

$$w_2(Q) = \exp\left(\mathrm{i}L_2 \int_{Q0}^{Q} \mathrm{d}E\right) = exp(iL_2E(Q)). \tag{14.5.1}$$

We have the relations

$$\psi(x, y + L_2, t, \zeta, Q) = w_2(Q)\psi(x, y, t, \zeta, Q), \tag{14.5.2}$$

$$\psi^+(x, y + L_2, t, \zeta, Q) = w_2^{-1}(Q)\psi^+(x, y, t, \zeta, Q). \tag{14.5.3}$$

In the following formulae we shall omit the dependence of ψ, u, ... on the parameters ζ as insignificant for the following results. (One can assume that we take $\zeta = 0$, for example).

Consider the equation

$$(\sigma\partial_y - \partial_x^2 + u_0(x, y, t) + \delta u(x, y, t))\tilde{\psi} = 0, \tag{14.5.4}$$

where $\delta u(x, y, t) = \delta u(x + L_1, y, t) = \delta u(x, y + L_2, t)$ is an arbitrary smooth periodic perturbation of the periodic finite-gap potential $u_0(x, y, t)$.

We shall find the Bloch solutions of the equation (14.5.4). Let Q_0 be an arbitrary point on Γ. The Baker–Akhiezer function $\psi(x, y, t, Q_0)$ is the Bloch function for an unperturbed equation (14.5.4) ($\delta u \equiv 0$) with the eigenvalues of monodromy operators equal to $w_1(Q)$ and $w_2(Q)$. The Bloch solution of equation (14.5.4) with the same eigenvalue $w_1(Q_0)$ for the shift of the variable x we shall find the form of:

$$\tilde{\psi} = \sum_{S=0}^{\infty} \varphi_S(x, y, t, Q_0), \quad \varphi_0 = \psi(x, y, t, Q_0), \tag{14.5.5}$$

where

$$\varphi_S(x + L_1, y, t, Q_0) = w_1(Q_0)\varphi_S(x, y, t, Q_0). \tag{14.5.6}$$

According to Theorem 14.4.2, the function φ_S can be represented in the form:

$$\varphi_S(x, y, t, Q_0) = \sum_n c_n^{(S)}(y, t)\psi_n(x, y, t). \tag{14.5.7}$$

Here $\psi_n = \psi(x, y, t, Q_n)$ and Q_n are determined by the relation (14.4.11), where $w = w_1(Q_0)$.

If φ_S satisfies the system of the recurrent equations

$$(\sigma\partial_y - \partial_x^2 + u_0)\varphi_S = -\delta u\ \varphi_{S-1} \tag{14.5.8}$$

then $\tilde{\psi}$ is the formal solution of equation (14.5.4).

The functions ψ_n are solutions of the homogeneous equation (14.5.8) and satisfy the relations $<\psi_n\ \psi_m^+>_x = <\psi_n\ \psi_n^+>\delta_{nm}$, which can be easily obtained from (14.4.4). If we substitute the series (14.5.7) in (14.5.8), we shall obtain that the coefficients c_n will be determined from the equation

$$\partial_y c_n = -\frac{<\psi_n^+\delta u\varphi_{S-1}>_x}{<\psi_n\ \psi_n^+>_x}. \tag{14.5.9}$$

The initial data for equation (14.5.9) can be uniquely determined from the condition that the corresponding $\tilde{\psi}$ would be the Bloch solution with respect to the shift of the variable y. This condition can be fulfilled if for any n

$$w_{2n} = w_2(Q_n) \neq w_{20} = w_2(Q_0). \tag{14.5.10}$$

Corresponding initial data for (14.5.9) are equal to

$$c_n^{(S)}(t) = \frac{\ll\psi_n^+\delta u\varphi_{S-1}\gg}{<\psi_n\ \psi_n^+>_x}\ \frac{w_{2n}}{w_{20} - w_{2n}} + \frac{w_{20}}{w_{2n} - w_{20}}\sum_{i=1}^{S-1} c_n^{(S-i)}e_i,$$

$$c_0^{(S)}(t) = -\sum_{n\neq 0} c_n^{(S)}(t), \tag{14.5.11}$$

where

$$e_S = e_S(t, Q_0) = \frac{\ll\psi_0^+\delta u\varphi_{S-1}\gg}{<\psi_0\psi_0^+>_x} + \sum_{i=1}^{S-1} c_0^{(S-i)}e_i. \tag{14.5.12}$$

(here and below, $\ll\cdot\gg$ denotes the average over x, y).

Lemma 14.5.2

The series (14.5.5) determined by equalities (14.5.7), (14.5.9)–(14.5.12) is the formal Bloch solution of equation (14.5.4). It is multiplied by $w_1(Q_0)$ and

$$w_2(t, Q_0) = w_{20}\left(1 + \sum_{S=1}^{\infty} e_S(t, Q_0)\right) \tag{14.5.13}$$

during the shift of variables x, y on L_1 and L_2 respectively.

Up to now there is no difference between the cases $\sigma^2 = \pm 1$. This difference becomes evident when we consider the structure of the set of the points for which the condition (14.5.10) is not fulfilled.

The points Q and Q' would be called the resonance pair, if $w_i(Q) = w_i(Q')$, $i = 1, 2$. In the KP-1 case the set of the resonance points on the corresponding curves is dense on the stationary cycles a_i. We omit the proof of this statement and consider in detail the KP-2 case ($\sigma = 1$).

In the latter case the corresponding curves are the M-curves. According to definition, these are the curves with $g + 1$ stationary cycles $a_0, \ldots, a_g$ of the anti-involution τ. These cycles cut Γ into the two unconnected domains $\Gamma^{\pm}$. We shall denote by Γ^+ the domain in which in the neighbourhood of P_1 one has $\operatorname{Re} p > 0$.

The single-valued branches of the multi-valued functions $\operatorname{Re} p$, $\operatorname{Re} E$ can be determined in each domain $\Gamma^{\pm}$.

Lemma 14.5.3
The mapping

$$Q \to (\operatorname{Re} p(Q), \operatorname{Re} E(Q)) \tag{14.5.14}$$

for any M-curve Γ is a real diffeomorphism of Γ^+ on the right half-plane in $R^2(\operatorname{Re} p > 0)$ with g deleted points. The solutions of the KP-2 equations corresponding to Γ and P_1 are periodic if the coordinates of these points are $(\pi n_S L_1^{-1}, \pi m_S L_2^{-1})$ where n_s, m_S are integers. All the resonance pairs on Γ are the pairs $P^{\pm}_{nm}$, such that $\tau(P^+_{nm}) = P^-_{nm}$ and $P^+_{nm} \subset \Gamma^+$ are the points which are uniquely determined by the relations $\operatorname{Re} p(P^+_{nm}) = \pi n L_1^{-1}$, $\operatorname{Re} E(P^+_{nm}) = \pi m L_2^{-1}$, $(n, m) \neq (n_S, m_S)$ are integers.

Proof
The differentials dp, dE are purely imaginary on the stationary cycles of τ. Therefore, from the reality of their periods over all the cycles on Γ it follows that

$$\oint_{a_S} dp = 0, \quad \oint_{a_S} dE = 0, \quad S = 1, \ldots, g. \tag{14.5.15}$$

The mapping (14.5.14) continuously expands on the cycles a_S. All the points of the cycle a_S are mapped on the same point in R^2 with the coordinate

$$\frac{1}{2}\oint_{b_S} dp = \pi U_S, \quad \frac{1}{2}\oint_{b_S} dE = \pi V_S. \tag{14.5.16}$$

Consider the level-line $\operatorname{Re} p = r$ of the function $\operatorname{Re} p$ on Γ^+. The function $\operatorname{Re} E$ has no extremums on this line. We shall prove it first for $r \neq \pi U_S$, $S = 1, \ldots, g$. Let us assume that $\operatorname{Re} E$ has the extremum at some point Q. Then we have that

$$\frac{dE}{dp} = \lambda, \quad \text{where } \lambda \text{ is real.} \tag{14.5.17}$$

Consider the differential $dE - \lambda dp$. It has $2g + 1$ zeros on Γ. From (14.5.15) and (14.5.17) it follows that the periods of this differential over $a_1, \ldots, a_g$ are equal to zero and it is purely imaginary on these cycles. Therefore, it has at least two zeros on $a_1, \ldots, a_g$. From the asymptotics of this differential near P_1 it follows that it has at least one zero on a_0. Hence, it cannot be equal to zero at Q, which contradicts (14.5.17). Similarly, it can be proved that Re E is monotone on all the connected components of the line Re $p = \pi U_S$. Thus the first statement of the lemma is proved. The second statement follows from (14.5.16) and (14.4.2). Now we shall consider the last statement of the lemma. Similarly to the proof of the first statement one can obtain that the function Im E cannot have the extremums on the level-line of the function Im $p = r$ in Γ^+. Therefore, the relations Im $p(Q) =$ Im $p(Q')$, Im $E(Q) =$ Im $E(Q')$ can be fulfilled only if $Q' = \tau(Q)$ because Im p and Im E are even functions with respect to τ. The resonance condition for the points Q and Q' means that as well as the above relations, the following must be fulfilled:

$$\text{Re } p(Q) - \text{Re } p(Q') = 2\pi n L_1^{-1}, \text{ Re } E(Q) - \text{Re } E(Q') = 2\pi m L_2^{-1} \tag{14.5.18}$$

for some integers n, m. The functions Re p and Re E are odd with respect to τ. Hence, from (14.5.18) it follows that Q, Q' coincide with the points $P_{nm}^{\pm}$.

Let us denote the zeros of dp by p_j, $j = 1, \ldots, 2g$. As follows from Lemma 14.5.3, the points $P_{nm}^{\pm}$ tend to P_1 as $|n|, |m| \to \infty$. Hence, we can choose the neighbourhoods $\mathscr{D}_j$ and $\mathscr{D}_{nm}^{\pm}$ of the points p_j and $P_{nm}^{\pm}$, respectively, so that for any point Q_0 which does not belong to these domains, there exists the uniform estimate $|w_{20}w_{2n}^{-1} - 1| \geqslant h_0 > 0$, where h_0 is a small constant. From this follow the estimates of all the terms of the series (14.5.5) and hence the proof of the statement.

Lemma 14.5.4

There exists the constant ε such that for $|\delta u| < \varepsilon$ the series (14.5.5) uniformly converges in $\Gamma \backslash (P_1 \cup \mathscr{D}_j \cup \mathscr{D}_{nm}^{\pm} \cup \tilde{\mathscr{D}}_S)$ and determines there the holomorphic Bloch solution $\tilde{\psi}(x, y, t, Q_0)$ of the equation (14.5.4). (Here $\tilde{\mathscr{D}}_S$ is the neighbourhood of the pole γ_S of the function ψ.) The function $\tilde{\psi}$ can be analytically expanded into the $\tilde{\mathscr{D}}_S$ where it has a simple pole. The images of $\mathscr{D}_j$ and $\mathscr{D}_{nm}^{\pm}$ in the w_1-plane for $w_1(Q)$ would be denoted by $\mathscr{D}_\alpha$ where α denotes $j = 1, \ldots, 2g$ or $(n, m) \neq (n_S, m_S)$ respectively. The series (14.5.5) does not converge into the domains $\mathscr{D}_j$ and $\mathscr{D}_{nm}^{\pm}$.

Similarly to the perturbation theory, for the multiple eigenvalue of operators we can construct for any $w_1 \in \tilde{\mathscr{D}}_\alpha$ two formal series $\tilde{\psi}_l^{(\alpha)}(x, y, t, w_1)$, $l = 1, 2$, which are the formal solutions of (14.5.4). They have the representation (14.5.5), (14.5.6). The initial data for determining the coefficients $c_n^{(S),\alpha}$ can be obtained from the condition that these series must satisfy the following relation:

$$\tilde{\psi}_l^{(\alpha)}(x, y + L_2, t, w_1) = \sum_{k=1}^{2} T_{lk}^{(\alpha)} \tilde{\psi}_k^{(\alpha)}(x, y, t, w_1) \quad (14.5.19)$$

where $T_{lk}^{(\alpha)}(t, w_1)$ are determined by the series which are similar to (14.5.12). For the sake of brevity we omit the exact formulae.

Lemma 14.5.5
For $|\delta u| < \varepsilon$ there exist the holomorphic (at w_1) solutions of (14.5.4) which satisfy the relation (14.5.19). The $T_{lk}^{(\alpha)}$ are also holomorphic functions of w_1.

This statement can be obtained from the estimates of the corresponding series for $\tilde{\psi}_l^{(\alpha)}$.

The Bloch solutions of the equation (14.5.4) for $w_1 \in \hat{\mathscr{D}}_\alpha$ are well-defined on the D_α which is a twofold cover of $\hat{\mathscr{D}}_\alpha$ and is determined by the characteristic equation of the matrix $T^{(\alpha)}$:

$$0 = \det(Z_\alpha - T_{lk}^{(\alpha)}) = Z_\alpha^2 - (\mathrm{Sp}T^{(\alpha)})Z_\alpha + \det T^{(\alpha)} \quad (14.5.20)$$

Let us denote the values $w_1(p_j)$ or $w_1(P_{nm}^{\pm})$ by $w_{1\alpha}$ respectively. The numbers $d_\alpha(t)$ would be determined by the equalities

$$d_\alpha(t) = (\mathrm{Sp}\ T^{(\alpha)}(t, w_{1\alpha}))^2 - 4\det T^{(\alpha)}(t, w_{1\alpha}). \quad (14.5.21)$$

The numbers d_{nm} are the values of 'splitting' the multiple eigenvalue of the shift operator of y. They are real and non-negative, $d_{nm} \geqslant 0$. The surface D_{nm} is irreducible iff $d_{nm} > 0$ (The pairs (n, m) such that $\mathrm{d_{nm}} > 0$ would be called the marked pairs).

Let us denote by $\hat{\Gamma}$ the Riemann surface which is obtained from $\Gamma \backslash P_1$ iff one cuts out the domains $\mathscr{D}_j$ and $\mathscr{D}_{nm}^{\pm}$ with $d_{nm} > 0$ and glues instead of them the corresponding domains D_{nm}. The anti-involution τ can be expanded on $\hat{\Gamma}$ where it has the stationary cycles $\tilde{a}_S$ which coincide with the old cycles a_S outside $\hat{\mathscr{D}}_S$ and stationary cycles $\tilde{a}_{nm} \subset D_{nm}$ for each marked pair (n, m).

From the above-formulated lemmas the statement of the theorem follows.

Theorem 14.5.1
The Bloch solutions of the equation (14.5.4) for $|\delta u| < \varepsilon$ determine the function $\tilde{\psi}(x, y, t, Q)$ on the surface Γ. This function has one simple pole on each stationary cycle $\tilde{\gamma}_S \in \tilde{a}_S$, $S = 1, \ldots, g$; $\tilde{\gamma}_{nm} \in \tilde{a}_{nm}$ and is holomorphic at the other points.

Remark

In the recent paper [41] by Taimanov it was proved that the Riemann surface of the Bloch solutions exists for the equation $(\partial_y - \partial_x^2 + u)\,\psi = 0$ with any smooth periodic potential u. There was no efficient description of the corresponding surface in [41], and so Taimanov has not obtained the approximation theorem. Up till now the variable t has played the role of a parameter (as well as the omitted variables ζ_l). If we take $t = 0$, the previous results provide for any periodic smooth function $\delta u(x, y)$ the following set of 'spectral data':

$$\delta u \rightarrow (d_{2j-1}, \tilde{\gamma}_j, j = 1, \ldots, g; d_{nm}, \tilde{\gamma}_{nm}) \qquad (14.5.22)$$

where (n, m) are marked.

Theorem 14.5.2

There exists the constant ε_1 depending on $u_0(x, y)$ only, such that for any set of data (14.5.22), satisfying the relation:

$$0 \leqslant d_{nm} \leqslant \varepsilon_1(|n| + |m|)^{-2l-1} \qquad (14.5.23)$$

there exists the unique l-differentiable function δu for which this set of data is the image for mapping (14.5.22).

We shall briefly formulate the main idea of the proof of this theorem in the following section. From this proof it follows that the potential $\tilde{u} = u_0 + \delta u$ is finite-gap iff it corresponds to only a finite number of non-zero d_{nm}. For any $\tilde{u}$ which belongs to ε_1-neighbourhood of u_0 we shall denote the finite-gap potential such that $\delta u_G = u_G - u_0$ corresponds (according to the theorem) to the data (14.5.23) with $|n| + |m| < G$ by u_G. If $G \rightarrow \infty$ then u_G tends to $\tilde{u}$.

For an arbitrary $\delta u(x, y, t)$ the surface $\hat{\Gamma}$ and $\tilde{\gamma}_\alpha$ depend on t as a parameter.

Lemma 14.5.6

The function $\tilde{u} = u_0 + \delta u$ is the solution of the KP-2 equation iff $\hat{\Gamma}$ and $\{\gamma\}$ do not depend on t.

The proof of the lemma can be obtained by using the fact that if $\tilde{u}$ is the solution of the KP equation, $\tilde{\psi}$ must satisfy the second of the equations (14.3.4).

Due to the previous results we obtain the following theorem (see remark at end of chapter).

Theorem 14.5.3

For any periodic regular finite-gap solution $u_0(x, y, t) = u_0(x + L_1, y, t) = u_0(x, y + L_2, t)$ of the KP-2 equation there exists the constant ε such that for an arbitrary real regular periodic function $v(x, y)$ satisfying $|u_0(x, y, 0) - v(x, y)| < \varepsilon$ there exists the unique solution $u(x, y, t)$

of this equation satisfying the initial condition $u(x, y, 0) = v(x, y)$. This solution is a quasiperiodic function of t. There exists a sequence of finite-gap solutions $u_G(x, y, t)$ (corresponding to curves of increasing genera) which tends to $u(x, y, t)$ with any number of derivatives uniformly on any compact region.

14.6 The perturbation theory for the finite-gap solutions of the KP-2 equation

Consider the problem of the construction of asymptotic solutions of the equation

$$\tfrac{3}{4} u_{yy} + (u_t - \tfrac{3}{2} uu_x + \tfrac{1}{4} u_{xxx})_x + \varepsilon K[u] = 0, \qquad (14.6.1)$$

where ε is the a small parameter, $K[u]$ is an arbitrary differential polynomial. There are two formulations of the problem. One of them is connected to the investigations of the influence of perturbation term $\varepsilon K[u]$ on the periodic solutions of the KP-2 equation. In this case we must consider the asymptotic series for the solution of the Cauchy problem for (14.6.1) with the initial data belonging to the neighbourhood of the finite-gap solution u_0 of the KP-2 equation. The second formulation of the problem has sense even when $\varepsilon = 0$. In this formulation one must find the asymptotic solution of the KP-2 equation with the main term $u_0(x, y, t|X, Y, T)$ which is determined by the formula

$$u_0(x, y, t) = \tilde{u}_0(\varepsilon^{-1}S(X, Y, T)|X, Y, T), \qquad (14.6.2)$$

where

$$\tilde{u}_0(Z|X, Y, T) = 2 \sum_{i=1}^{g} (U_i\partial_{Z_i})^2 \ln \theta(Z|X, Y, T)$$

is the periodic function of the variables $Z = (Z_1, \ldots, Z_g)$. The parameters of this function, which are the matrix of the b-periods of the holomorphic differential on Γ, depend on 'slow' variables $X = \varepsilon x$, $Y = \varepsilon y$, $T = \varepsilon t$. The vector-function $S(X, Y, T)$ is determined from the equations

$$\frac{\partial S}{\partial X} = U(X, Y, T); \ \frac{\partial S}{\partial Y} = V(X, Y, T); \ \frac{\partial S}{\partial T} = W(X, Y, T), \qquad (14.6.3)$$

where $2\pi U$, $2\pi V$, $2\pi W$ are the period vectors of the differentials dp, dE, $d\Omega$ and depend on X, Y, T through the dependence on these variables of the main parameters which are $(\Gamma, P_1, [k^{-1}]_3)$.

It must be noted, that for one-dimensional systems (in particular, for the KdV equation) mainly the second formulation of the problem has been considered [8, 9, 19, 20].

Unifying both of these problems we shall find the solutions of the equation (14.6.1) in the form:

$$u(x, y, t) = u_0(x, y, t|X, Y, T) + \sum_{i=1}^{\infty} \varepsilon^i w_i(x, y, t|X, Y, T). \tag{14.6.4}$$

In the case when $u_0(x, y, t)$ is the periodic function of x, y (14.6.4) for the construction of the formal series (14.6.4) it is sufficient to find the set of solutions of the 'linearised' equation

$$\tfrac{3}{4} v_{yy} + (v_t - \tfrac{3}{2} u_0 v_x - \tfrac{3}{2} u_{0x} v + \tfrac{1}{4} v_{xxx})_x = 0, \tag{14.6.5}$$

which form for each t the basis in the space of the periodic functions of x, y. We must also find the dual basis of the solutions of the adjoint linear equation

$$\tfrac{3}{4} \varphi_{yy} + (-\varphi_t - \tfrac{3}{2} u_0 \varphi_x + \tfrac{1}{4} \varphi_{xxx})_x = 0. \tag{14.6.6}$$

The solutions of equation (14.6.5) can be obtained by using the following obvious statement. If there is a set of solutions of the KP-2 equations depending on some parameters, the derivatives of these solutions with respect to these parameters satisfy the equation (14.6.5). First we shall consider the derivatives with respect to the parameters which determine the finite-gap solutions of the KP-2 equation which correspond to the curves Γ of the same genus g.

Consider the manifold of the parameters

$$M_g = (\Gamma, P_1, [k^{-1}]_1), \tag{14.6.7}$$

which together with the vector $\zeta = (\zeta_1, \ldots, \zeta_g)$ determine the finite-gap solutions of the KP-2 equation.

Remark

In M_g we consider the equivalence classes of the local coordinates which are defined up to substitutions of $k' = k + O(k^{-1})$. It slightly differs from the description of the parameters in Section 14.1. There for the KP case we considered the substitutions of the form $k' = k + O(k^{-3})$. The nature of this difference is a result of the following. The operator equation (14.1.8) is equivalent to the system of two equations for the two functions $u(x, y, t)$ and $w(x, y, t)$. This system is also often called the KP equation. Equation (14.1.7) is obtained by means of the elimination of w from the system. Thus, equation (14.1.7) is not completely equivalent to the system. That is why there is a slight difference between the manifold of the parameters in these two cases.

Parameters (14.6.7) are the parameters of the finite-gap solutions of equation (14.1.7), which satisfy the additional condition

$$\langle u \rangle_x = 0. \tag{14.6.8}$$

Now we shall give one of the possible systems of coordinates on M_g. Let $p(Q)$ be (as before) a single-valued branch on Γ^+ of the function

$$p(Q) = \int_{Q0}^{Q} \mathrm{d}p, \tag{14.6.9}$$

where the initial point $Q_0 \in a_0$ (the fixed point P_1 belongs to a_0 as well).

Lemma 14.6.1
The function $p(Q)$ maps the open domain Γ^+ onto the right half-plane (Re $p > 0$) with g vertical cuts between the points

$$(p_{2S-1}, p_{2S}), \quad \text{Re } p_{2S-1} = \text{Re } p_{2S}. \tag{14.6.10}$$

The edges of these 'cuts' are the images of a_S during the expansion of p from Γ^+ onto a_S.

The proof is obvious. The particular case of the statement of the lemma corresponding to hyperelliptic curves is well known (see [37]). The hyperelliptic case corresponds to the relation Im $p_{2S-1} = -$Im p_{2S}.

Lemma 14.6.2
The mapping

$$(\Gamma, P_1, [k^{-1}]_1) \to (p_1, \ldots, p_{2g}), \tag{14.6.11}$$

which follows from the previous lemma is a one-to-one correspondence. Here p_j are arbitrary points satisfying the condition (14.6.10) and such that the intervals $\{p\text{: Re } p_{2S-1} = \text{Re } p_{2S}, \text{Im } p_{2S-1} \leqslant \text{Im } p \geqslant \text{Im } p_{2S}\}$ do not intersect.

Consider the submanifold $M_g^0 \subset M_g$ of the data such that the conditions (14.4.2) are fulfilled for the fixed integers (n_S, m_S). The dimensions of this submanifold is equal to g. The parameters $\tau_S = \text{Tm } p_{2S}$, $S = 1, \ldots, g$ can be chosen as independent coordinates on M_g^0. As periods U, V do not change at the variations of τ_S and ζ_S (which are the components of the phase-vector ζ in (14.4.1), the functions

$$v_S^+(x, y, t) = \frac{\partial u_0}{\partial \tau_S}; \; v_S^-(x, y, t) = \frac{\partial u_0}{\partial \zeta_S} \tag{14.6.12}$$

are the periodic solutions of (14.6.5).

The other set of solutions of (14.6.5) can be obtained if we use the variation with which the curve of the genus g is transformed into the curve of the genus $g + 1$. The general type of such real transformation is as follows. Let us cut the neighbourhood of the two points Q_0 and $\tau(Q_0)$ out of Γ and 'glue' instead of them the corresponding piece of the surface determined by the equation

$$Z^2 = \nu^2 - \delta^2. \tag{14.6.13}$$

(The set of points of the surface (14.6.13) which correspond to the circle $|v| = v_0$ forms two circles if $v_0 > \delta$. These circles can be glued to the boundaries of the neighbourhood of Q_0 and $\tau(Q_0)$.) It should be noted that the transformations of the curves, which were introduced in Theorem (14.5.1), have the same type.

Let B' be the matrix of b-periods of the holomorphic differential on the curve Γ' which we have obtained. It has the form

$$B'_{ij} = B_{ij} + O(\delta^2), \quad i, j \leqslant g, \quad B'_{g+1,g+1} = \frac{1}{\pi i} \ln \delta + O(\delta^2),$$

$$B'_{i,g+1} = B'_{g+1,i} = A(Q_0) - A(\tau(Q_0)) + O(\delta^2), \quad (14.6.14)$$

where B is the matrix of b-periods of the initial curve Γ. The vector U' of b-periods of the differential $dp/2\pi$ on Γ' has the form

$$U'_i = U_i + O(\delta^2), \quad i = 1, \ldots, g, \quad U_{g+1} = \frac{1}{2\pi}(p(Q_0) - p(\tau(Q_0))).$$
$$(14.6.15)$$

Here U_i are the b-periods of $dp/2\pi$ on Γ. Similar relations are fulfilled for the vectors V' and W'. (The relations of (14.6.14)-type can be found in the review by Fay [18].)

The theta-function corresponding to the matrix B' and the vectors U', V', W' determine by means of formula (14.4.1) the exact solution of the KP-2 equation. If we use in this formula the relations (14.6.14) and (14.6.15), we shall obtain that the derivative of this solution with respect to δ is the linear combination of the functions $v(x, y, t, Q_0)$ and $v(x, y, t. \tau(Q_0))$, where

$$v(x, y, t, Q) = 2\partial_x^2 \frac{\theta(Ux + Vy + Wt + \zeta + A(Q) - A(\tau(Q)))}{\theta(Ux + Vy + Wt + \zeta)} \times$$
$$\exp[\pi i(x(p(Q) - p(\tau(Q))) + y(E(Q) - E(\tau(Q))) + t(\Omega(Q) - \Omega(\tau(Q))))].$$
$$(14.6.16)$$

The coefficients in this linear combination are expressed in terms of the coordinate ζ'_{g+1} of the phase-vector ζ' in (14.4.1). Since this coordinate is arbitrary we obtain the following lemma.

Lemma 14.6.3
The functions $v(x, y, t, Q)$ are the solutions of the equation (14.6.5).

Proof
As follows immediately from (14.6.16) the solutions $v_{nm}^{\pm} = v(x, y, t, P_{nm}^{\pm})$ are the periodic solutions of (14.6.5).

Let us define the functions

$$\hat{\varphi}_S^{+} = \hat{\varphi}(x, y, t, p_{2S}) - 1, \; \hat{\varphi}_{nm}^{\pm}(x, y, t) = \hat{\varphi}(x, y, t, P_{nm}^{\pm}), \tag{14.6.17}$$

where

$$\hat{\varphi}(x, y, t, Q) = \frac{\psi(x, y, t, Q)\psi^{+}(x, y, t, \tau(Q))}{<\psi(x, y, t, Q)\psi^{+}(x, y, t, Q)>_x}. \tag{14.6.18}$$

Theorem 14.6.1
The set of functions $v_S^{\pm}$, $v^{\pm}{}_{nm}$ is (for each t) the basis in the space $\mathscr{L}_0^2$ of the periodic functions of the variables x, y with zero average in relation to $x(f \in \mathscr{L}_0^2, <f>_x = 0)$. There are the following 'orthogonality' relations:

$$\ll v_S^{\pm}\, \hat{\varphi}_{nm}^{\pm} \gg = 0, \quad \ll v_S^{+}\, \hat{\varphi}_{S'}^{+} \gg = \delta_{SS'},$$
$$\ll v_S^{-}\, \hat{\varphi}_S^{+} \gg = 0, \quad \ll v_{nm}^{\pm}\, \hat{\varphi}_{n'm'}^{\pm} \gg = 0, \quad (n, m) \neq (n', m'),$$
$$\ll v_{nm}^{+}\, \hat{\varphi}_{nm}^{+} \gg = \ll v_{nm}^{-}\, \hat{\varphi}_{nm}^{-} \gg = 0, \quad \ll v_{nm}^{\pm}\, \hat{\varphi}_S^{+} \gg = 0,$$
$$\ll v_{nm}^{+}\, \hat{\varphi}_{nm}^{-} \gg = \ll v_{nm}^{-}\, \hat{\varphi}_{nm}^{+} \gg = e^{r_{nm}} \neq 0. \tag{14.6.19}$$

Here $r_{nm} = r(P_{nm}^{+})$ and $r(Q)$ is the real function which is determined in the following way. Let $\tilde{\omega}$ be the differential on Γ with the poles at Q, $\tau(Q)$ only. These poles are simple and the residues of $\tilde{\omega}$ at these points are equal to ± 1. This differential is uniquely defined if we assume that its a-periods are equal to zero. For $Q' \to Q$ we have

$$\int_{\tau(Q')}^{Q'} \tilde{\omega} = \ln|p(Q') - p(Q)|^2 + 2r(Q) + O(p(Q') - p(Q)). \tag{14.6.20}$$

From the theorem the following corollary can be obtained.

Corollary
The functions $\hat{\varphi}_S^{+}$, $\hat{\varphi}_{nm}^{\pm}$ are the periodic solutions of (14.6.6).

It follows from the theorem that there exist solutioons $\hat{\varphi}_S^{-}(x, y, t)$ of (14.6.6), $S = 1, \ldots, g$, such that

$$\ll \hat{\varphi}_S^{-} v_S^{+} \gg = \ll \hat{\varphi}_S^{-} v_{nm}^{\pm} \gg = 0, \quad \ll \hat{\varphi}_S^{-} v_{S'}^{-} \gg = \delta_{SS'}.$$

We shall omit their exact expression. In the available space we can give only the main idea of the proof of Theorem 14.6.1.

First we consider the periodic fucntion δu which is orthogonal to $\hat{\varphi}_S^{+}$ and $\hat{\varphi}_{nm}^{\pm}$. For such a function δu the series (14.5.7) for φ_1 can be determined for all points Q_0 including the resonance points $P_{nm}^{\pm}$. For a sufficiently small δu this series uniformly converges everywhere except small neighbourhoods $\tilde{\mathscr{D}}_S$ of the poles γ_S of ψ. The function $\psi(x, y, t, Q_0) + \varphi_1(x, y, t, Q_0)$ is holomorphic on Γ outside $\tilde{\mathscr{D}}_S$ and P_1. It can by analytically expanded on $\tilde{\mathscr{D}}_S$ where it would have the simple pole $\tilde{\gamma}_S$. Its behaviour near P_1 is the same as in (14.3.1). Therefore, $u_0 + \delta u$ is a finite-gap potential corresponding to the same curve Γ, though to another set of poles $\tilde{\gamma}_S$ of the Baker–Akhiezer function. Hence, the set of the solutions of (14.6.5) which are dual to $\hat{\varphi}_S^{\pm}$, $\hat{\varphi}^{\pm}{}_{nm}$ and the solutions $v_S^{\mp}(x, y, t)$ are the basis of the space of all the periodic solutions. The relations (14.6.19) which mean that $v_S^{\pm}$ and $v_{nm}^{\pm}$ are dual to $\hat{\varphi}_S$, $\hat{\varphi}_{nm}^{\pm}$ follow from the construction of $v(x, y, t, Q)$ and from the formulae for perturbation of $w_2(Q)$ and for d_{nm}. This completes the proof of the theorem.

Remark

The proof of Theorem (14.5.2) is based on the relations (14.6.19) as well, because in the first approximation, the values of d_j and d_{nm} are equal to

$$d_S \approx \langle\langle \delta u\ \hat{\varphi}_S^{\pm} \rangle\rangle, \quad d_{nm} \approx \langle\langle \hat{\varphi}_{nm}^{+} \delta u \rangle\rangle \ \langle\langle \delta u\ \hat{\varphi}_{nm}^{-} \rangle\rangle. \qquad (14.6.21)$$

The formulae for $v(x, y, t, Q)$ and $\hat{\varphi}(x, y, t, Q)$ allow us to obtain all the terms of the series (14.6.4) for the case of the periodic u_0. It turns out that the corresponding expressions can be expanded for arbitrary finite-gap solutions by means of their approximation by the periodic solutions with the periods L_1, L_2, tending to infinity. For such approximation the limit of the resonance points is the set of points Q, not belonging to $a_1, \ldots, a_g$ and such that there exist the integers $r_1, \ldots, r_g$ such that

$$\operatorname{Re} p(Q) = \pi(r_1 U_1 + \ldots + r_g U_g); \quad \operatorname{Re} E(Q) = \pi(r_1 V_1 + \ldots + r_g V_g). \qquad (14.5.22)$$

For U, V let us denote the group of the integer vectors $(r_1, \ldots, r_g)$ such that the right-hand side of (14.6.22) is equal to zero, by $R(U, V)$. For any integer vector $\rho = (r_1, \ldots, r_g)$ we shall denote the corresponding element in the factor-group $Z^g/R(U, V)$ by $\bar{\rho}$. Then the points Q in (14.6.22) are uniquely determined by $\bar{\rho}$ and would be further denoted by Q_ρ. *Here* $\rho \neq \rho_S^{\pm}$ where $\rho_S^{\pm}$ is a vecctor with the only non-zero coordinate $r_S = \pm 1$.

Let us denote by $F_i(x, y, t | X, Y, T) = F_i[u_0, \ldots, w_{i-1}]$ the coefficient at ε^i which is obtained by the substitution into (14.6.1) of the corresponding partial sum of (14.6.4).

Theorem 14.6.2

The term $w_i(x, y, t | X, Y, T)$ of the series (14.6.4) is given by the expression

$$w_i = \sum_{S=1}^{g} c_{iS}^{+}(t)\ v_S^{+}(x, y, t) + c_{iS}^{-}(t) v_S^{-}(x, y, t)$$
$$+ \sum_{\bar{\rho}}{}' \ c_i(t, Q_{\bar{\rho}}) v(x, y, t, Q_\rho). \qquad (14.6.23)$$

Here

$$c^{\pm}_{iS}(t) = \tilde{c}^{\pm}_{iS} - \int_0^t \ll\partial_x F_i(x, y, t')\, \hat{\varphi}^{\pm}_S(x, y, t')\gg \mathrm{d}t', \quad (14.6.24)$$

$$c_i(t, Q_{\bar{\rho}}) = \tilde{c}_i(Q_{\bar{\rho}}) - \int_0^t \mathrm{e}^{-r(Q_{\bar{\rho}})} \ll\partial_x\, F_i\, \hat{\varphi}\gg \,\mathrm{d}t. \quad (14.6.25)$$

It must be noted that these formulae give us the dependence of all the terms on the 'fast' variables x, y, t only, though these terms depend on X, Y, T as well. They depend on X, Y, T according to the dependence of $(\Gamma, P_1, [k^{-1}]_1)$ (which determine the $\hat{\varphi}^{\pm}_S$, $\hat{\varphi}$, $v^{\pm}_S$, v) on these variables. Also the constants $\tilde{c}^{\pm}{}_{iS}$ and $\tilde{c}_i(Q_{\bar{\rho}})$ in (14.6.24) and (14.6.25) depend on X, Y, T as well. Their dependence can be found from the condition that the term w_{i+1} in (14.6.4) must be uniformly bounded for all t.

In this chapter we shall restrict ourselves and formulate only the analogue of the Withem equation for the parameters Γ, P_1, $[k^{-1}]_1$ of the main term $u_0(x, y, t|X, Y, T)$ in (14.6.4).

Theorem 14.6.3

The dependence of the finite-gap solution $u_0(x, y, t|X, Y, T)$ on the parameters is uniquely determined from the equation for the corresponding differentials

$$\mathrm{d}p\left(\frac{\partial E}{\partial T} - \frac{\partial \Omega}{\partial Y}\right) - \mathrm{d}E\left(\frac{\partial p}{\partial T} - \frac{\partial \Omega}{\partial X}\right) + d\Omega\left(\frac{\partial p}{\partial Y} - \frac{\partial E}{\partial X}\right) = \ll K[u_0]\psi\psi^+\gg \frac{\mathrm{d}p}{<\psi\psi^+>_x}. \quad (14.6.26)$$

These equations are the necessary conditions for the existence of the asymptotic solution (14.6.4) with the main term $u_0(x, y, t|X, Y, T)$ such that the corresponding second term is uniformly bounded for all t.

Remark

Equation (14.6.26) is valid for any equation which has the representation (14.2.2). In the particular case of one-dimensional systems they coincide with the Withem equations in the form which was first found by Flaschka, Forest and McLaughlin [19] ($K \equiv 0$).

14.7 The two-dimensional Schrödinger operator with potential decreasing at infinity. Fixed-energy scattering problems

Let

$$L = -\Delta + u(\vec{x}),\ x = (x_1, \ldots, x_n), \quad n \geqslant 2 \quad (14.7.1)$$

be the n-dimensional time-independent Schrödinger operator with a real potential. No non-trivial evolution equations, connected with the scattering theory for L at all energies are at present known. For example, only linear equations can be presented in the Lax form [43]

$$\partial L/\partial t + [L, A] = 0. \tag{14.7.2}$$

(The multi-dimensional inverse scattering problem has been studied by Faddeev [45]. The modern approach, based on the $\bar{\partial}$-problem method is developed in [46–48]. Equations connected with the multi-dimensional first-order systems, are considered in [44].)

In contrast, Veselov and S. P. Novikov [49] have constructed a hierarchy of nonlinear equations, connected with the fixed-energy scattering problem for the two-dimensional Schrödinger operator. These equations have the form of the Manakov L-A-B triples [50]:

$$\partial L/\partial t + [L, A_k] = B_k L, \tag{14.7.3}$$

where $A_k = \partial^{2k+1} + \bar{\partial}^{2k+1} + \ldots$, and the degree of B_k is less than $2k + 1$. In the two-dimensional case we shall use the complex notation

$$L = -\partial\bar{\partial} + u(z, \bar{z}) \tag{14.7.4}$$

where $z = x + iy$, $\bar{z} = x - iy$, $\partial = \partial_x - i\partial_y$, $\bar{\partial} = \partial_x + i\partial_y$, $\partial_z = \frac{1}{2}\partial$, $\partial_{\bar{z}} = \frac{1}{2}\bar{\partial}$.

For $k = 1$ we have

$$\partial_t v = \partial^3 v + \bar{\partial}^3 v + \partial(vw) + \bar{\partial}(v\bar{w}), \quad \bar{\partial}w = -3\partial v, \quad \partial\bar{w} = -3\bar{\partial}v, \quad v = \bar{v}, \tag{14.7.5}$$

where $v = v(z, \bar{z}, t)$, $w = w(z, \bar{z}, t)$.

The Manakov L-A-B triples are the correct two-dimensional analogues of the Lax pairs (14.7.2) and coincide with them at the zero energy level of the operator L, that is, on the set of functions ψ such that $L\psi = 0$.

Konopelchenko [51] has proved that the Veselov–Novikov equations (14.7.3), (14.7.5) can be represented in the Lax form (14.7.2) with a first-order 2×2 matrix operator L. However, in this chapter we shall use the representation (14.7.3).

Equation (14.7.5) can be treated as a generalisation of the Kadomtsev–Petviashvili equation (14.1.7). Let v, w depend slowly on y.

$$v = u - E, \quad X = x + Vt, \quad Y = \epsilon y, \quad V = 6E, \quad E = \pm 1/(2\epsilon), \quad T = 2t,$$
$$\operatorname{Re} w = -3v + 0(\epsilon^2), \quad (\operatorname{Im} w)_X = -6\epsilon v_Y, \quad \tilde{w} = -(\operatorname{Im} w)/(6\epsilon).$$

Then for $\epsilon \to 0+$ we have KP-1 for $E > 0$ and KP-2 for $E < 0$:

$$u_T = u_{XXX} - uu_X + \tilde{w}_Y, \ \tilde{w}_X = u_Y. \tag{14.7.6}$$

The periodic inverse scattering problem for L on the basis of spectral data pertaining to a single energy level has been studied by S. P. Novikov, Dubrovin, Veselov and Krichever ([49, 52, 53] and references therein). For the so-called 'finite-zone at a fixed energy' potentials and the corresponding solutions of (14.7.3), explicit θ-functional formulae have been obtained [49]. S. P. Novikov pointed out that the methods of soliton theory can be useful in the decreasing case.

The 'classical' fixed-energy scattering problem is strongly underdetermined. The spherically symmetrical case has been studied by Newton, Redge and Sabatier (see [54]). The set of spherically symmetrical potentials with a fixed scattering amplitude at a given energy is parametrised by an arbitrary function of one variable.

So we shall consider 'generalised' scattering and inverse scattering problems for the two-dimensional time-independent Schrödinger operator without assuming the potential to be spherically symmetrical. The scattering data in our approach are connected not only with the bounded eigenfunctions of L, but also with those that are exponentially increasing at infinity. These data are formed from a fixed energy level ϵ_0, where ϵ_0 is a real number. They are complete and independent. As a consequence we obtain a simple direct parametrisation of the set of potentials with a fixed scattering amplitude at a given energy. The constructed scattering transform is applied to solve the Veselov–Novikov equations in an appropriate class of potentials. The main results of this section were obtained by R. G. Novikov, Manakov, S. P. Novikov, Grinevich [55–60].

Considering the direct problem we shall assume that $\epsilon_0 \neq 0$. The zero-energy scattering problem is more complicated, because in the two-dimensional case $\epsilon_0 = 0$ is a singular point for the scattering amplitude and the physical eigenfunctions. The inverse problem at zero energy level can be considered, but in this case the non-singular $\bar{\partial}$-problem gives rise to potentials which are not in general position.

Direct problem

We consider the scattering equation:

$$\begin{gathered} L\psi(\lambda, \bar{\lambda}, z, \bar{z}) = \epsilon_0 \psi(\lambda, \bar{\lambda}, z, \bar{z}), \\ \epsilon_0 = -4\varkappa^2, \quad \varkappa^2 \in R, \quad \varkappa \neq 0, \quad \sigma = \operatorname{sgn} \varkappa^2. \end{gathered} \tag{14.7.7}$$

If the potential $u(z, \bar{z})$ is sufficiently small (or $|\varkappa|$ is sufficiently large), then equation (14.7.7) has the unique solution $\psi(\lambda, \bar{\lambda}, z, \bar{z})$ such that

$$\begin{gathered} \psi(\lambda, \bar{\lambda}, z, \bar{z}) = \exp[-\varkappa(\lambda z + \bar{z}/\lambda)]\chi(\lambda, \bar{\lambda}, z, \bar{z}), \\ \chi(\lambda, \bar{\lambda}, z, \bar{z}) = 1 + 0(1) \quad \text{as} \quad |z| \to \infty. \end{gathered} \tag{14.7.8}$$

The function $\chi(\lambda, \bar{\lambda}, z, \bar{z})$ satisfies

$$L_0(\lambda)\chi(\lambda, \bar{\lambda}, z, \bar{z}) = -u(z, \bar{z})\chi(\lambda, \bar{\lambda}, z, \bar{z}),$$
$$L_0(\lambda) = -\partial\bar{\partial} + 2\varkappa\lambda\bar{\partial} + 2\varkappa/\lambda\partial$$

and can be obtained by solving the integral equation

$$\chi(\lambda, \bar{\lambda}, z, \bar{z}) = 1 - \iint G(\lambda, \bar{\lambda}, z - z', \bar{z} - \bar{z}')u(z', \bar{z}')\chi(\lambda, \bar{\lambda}, z', \bar{z}')dz'_R dz'_I \tag{14.7.9}$$

where the Green function for $L_0(\lambda)$ is given by

$$G(\lambda, \bar{\lambda}, Z, \bar{Z}) = \frac{1}{4\pi^2}\iint \frac{\exp[i(kZ + \bar{k}\bar{Z})]dk_R dk_I}{k\bar{k} + i\varkappa\bar{k}\lambda + i\varkappa k/\lambda}. \tag{14.7.10}$$

The function $G(\lambda, \bar{\lambda}, z, \bar{z})$ is nowhere holomorphic by λ:

$$\frac{\partial G}{\partial\bar{\lambda}} = -\frac{sgn(|\lambda|^2 - 1)}{4\pi\bar{\lambda}}\exp\{\varkappa[(\lambda + \sigma/\bar{\lambda})z + (1/\lambda + \sigma\bar{\lambda})\bar{z}]\} \tag{14.7.11}$$

The departure of $G(\lambda, \bar{\lambda}, z, \bar{z})$ from holomorphicity is connected with the presence of zeros in the denominator of (14.7.10). The relationship (14.7.11) can be derived from the identity

$$\partial_{\bar{\lambda}}(1/\lambda) = \pi\delta(\lambda_R)\delta(\lambda_I)$$

which follows from the generalised Cauchy formula

$$f(\lambda, \bar{\lambda}) = \frac{1}{2\pi i}\oint_{\partial S}\frac{f(\mu, \bar{\mu})d\mu}{\mu - \lambda} + \frac{1}{\pi}\iint_S \frac{\partial f(\mu, \bar{\mu})/\partial\bar{\mu}}{\lambda - \mu}d\mu_R d\mu_I.$$

If $\varkappa^2 > 0$, $G(\lambda, \bar{\lambda}, z, \bar{z})$ is a continuous function of λ; if $\varkappa^2 < 0$, $G(\lambda, \bar{\lambda}, z, \bar{z})$ has a jump across the unit circle $|\lambda| = 1$.

For us the following properties of $\psi(\lambda, \bar{\lambda}, z, \bar{z})$ are important.

Theorem 14.7.1

If equation (14.7.9) has a unique solution for all λ, then:

(i) $\psi(\lambda, \bar{\lambda}, z, \bar{z})$ has essential singularities (14.7.8) as $\lambda \to 0, \infty$.

(ii) $$\frac{\partial\psi(\lambda, \bar{\lambda}, Z, \bar{Z})}{\partial\bar{\lambda}} = T(\lambda, \bar{\lambda})\psi\left(-\frac{\sigma}{\bar{\lambda}}, -\frac{\sigma}{\lambda}, Z, \bar{Z}\right) \tag{14.7.13}$$

where

$$T(\lambda, \bar{\lambda}) = \frac{sgn(|\lambda|^2 - 1)}{4\pi\bar{\lambda}}\iint e^{-\varkappa}\,\sigma(Z/\bar{\lambda} + \bar{\lambda}\bar{Z}) \cdot u(Z, \bar{Z})\psi(\lambda, \bar{\lambda}, Z, \bar{Z})dZ_R dZ_I \tag{14.7.14}$$

(iii) (a) If $\varkappa^2 < 0$ then $\psi(\lambda, \bar{\lambda}, z, \bar{z})$ is a continuous function of λ.

(b) If $\varkappa^2 > 0$ then there exists a function $R(\lambda, \lambda')$, $|\lambda| = |\lambda'| = 1$ such, that

$$\psi_-(\lambda, z, \bar{z}) = \psi_+(\lambda, z, \bar{z}) + \oint_{|\lambda|=1} R(\lambda, \lambda')\, \psi_+(\lambda', z, \bar{z})|d\lambda'|, \ |\lambda| = 1, \tag{14.7.15}$$

where $\psi_\pm(\lambda, z, \bar{z})$ denotes the boundary values of $\psi(\lambda, \bar{\lambda}, z, \bar{z})$ at the unit circle $\psi_\pm(\lambda, z, \bar{z}) = \psi(\lambda(1 \pm 0), \lambda(1 \pm 0), z, \bar{z})$.

An explicit expression for $R(\lambda, \lambda')$ is obtained in [57].

The functions $R(\lambda, \lambda')$ and $T(\lambda, \bar{\lambda})$ for $\varkappa^2 < 0$ or $T(\lambda, \bar{\lambda})$ for $\varkappa^2 > 0$ are the scattering data in our approach.

For the scattering matrices, corresponding to purely real potential Schrödinger operators, the following relations hold:

$$S(-\vec{l}, -\vec{k}) = S(\vec{k}, \vec{l}), \quad S(\vec{k}, \vec{l}) + S(\vec{l}, \vec{k}) + \int_{|m|=\varepsilon_0} S(\vec{k}, \vec{m})\bar{S}(\vec{l}, \vec{m})d\vec{m} = 0,$$

the time-reversal invariance and the unitarity.

Analogues of these reductions are given by the following theorem.

Theorem 14.7.2

(i) The scattering data, constructed in Theorem 14.7.1, satisfy the constraints

$$T(\sigma/\bar{\lambda}, \sigma/\bar{\lambda}) = -\sigma\lambda\bar{\lambda}T(\lambda, \bar{\lambda}), \tag{14.7.16}$$

$$R(\lambda, \lambda') + R(-\lambda', -\lambda) + \oint_{|\lambda'|=1} R(\lambda, \lambda'')R(-\lambda', -\lambda'')|d\lambda''| = 0. \tag{14.6.17}$$

(ii) Let the potential $u(z, \bar{z})$ in Theorem 14.7.1 be a real-valued function $u(z, \bar{z}) = \bar{u}(z, \bar{z})$. Then,

$$R(\lambda', \lambda) = \bar{R}(\lambda, \lambda'), \quad T(-\sigma/\bar{\lambda}, -\sigma/\bar{\lambda}) = \sigma\lambda^2\, \bar{T}(\lambda, \bar{\lambda}). \tag{14.7.18}$$

(In the case $\epsilon_0 < 0$ we can assume $R(\lambda, \lambda') \equiv 0$.)

The connection between the scattering amplitude and the generalised 'scattering data' is given by the following theorem.

Theorem 14.7.3

Let $\varkappa = i\theta$, $\theta \in R$, $P^2 = 4\theta^2$, $\psi_\pm(p, z, \bar{z}) = \psi_\pm((ip_y - p_x)/2\theta, z, \bar{z})$, $\psi_-(p, z, \bar{z}) = (1 + \hat{R}')\psi_+(p, z, \bar{z})$. Let

$$(1 + \hat{W}^i_-) = (1 + \hat{R}')(1 + \hat{W}^i_+), \ (1 + \hat{W}^0_i) = (1 + \hat{R}')(1 + \hat{W}^0_+) \tag{14.7.19}$$

be the quasi-triangular factorisations for $1 + \hat{R}'$, that is,

$$W^i_-(p, q) = W^0_+(p, q) = 0 \quad \text{for} \quad p_x q_y - q_y p_x > 0,$$
$$W^i_+(p, q) = W^0_-(p, q) = 0 \quad \text{for} \quad p_x q_y - q_y p_x < 0.$$

Then for the scattering operator $1 + \hat{S}$ the following relations hold:

$$(1 + \hat{W}^0_-) = (1 + \hat{S})(1 + \hat{W}^i_-), \quad (1 + \hat{W}^0_+) = (1 + \hat{S})(1 + \hat{W}^i_+). \tag{14.7.20}$$

For the scattering amplitude in general position the formulae (14.7.19), (14.7.20) establish one-to-one correspondence between $R(\lambda, \lambda')$ and $S(k, l)$.

Considering the inverse scattering problem we shall prove that our scattering data are complete and independent. Hence, the set of potentials with a given scattering amplitude at a fixed energy is parametrised by the function $T(\lambda, \bar{\lambda})$.

Inverse problem

The inverse problem is reconstructing the potential from the scattering data. A formal solution of the inverse problem can be formulated as the following theorem.

Theorem 14.7.4

Let the 'scattering data' be given: arbitrary functions $R(\lambda, \lambda')$ and $T(\lambda, \lambda)$ for $\epsilon_0 > 0$ or $T(\lambda, \bar{\lambda})$ for $\epsilon_0 < 0$ such that $T(\lambda, \bar{\lambda}) \to 0$ sufficiently fast as $\lambda \to 0, \infty$. Then

(i) For data in general position there exists a unique function $\psi(\lambda, \bar{\lambda}, z, \bar{z})$ satisfying equations (14.7.13) and (14.7.15), or (14.7.13) for $\sigma = \pm 1$ respectively, such that

$$\psi(\lambda, \bar{\lambda}, z, \bar{z}) = (1 + 0(1)) \exp(-\varkappa\lambda z) \quad \text{as} \quad \lambda \to \infty,$$
$$\psi(\lambda, \bar{\lambda}, z, \bar{z}) = (c(z, \bar{z}) + 0(1)) \exp(-\varkappa\bar{z}/\lambda) \quad \text{as} \quad \lambda \to 0,$$

where $c(z, \bar{z})$ is some unknown function of $z, \bar{z}$.

(ii) Consider the asymptotic expansion for $\chi = \psi\exp[\varkappa(\lambda z + \bar{z}/\lambda)]$ as $\lambda \to \infty$:

$$\chi(\lambda, \bar{\lambda}, z, \bar{z}) = 1 + \chi_1(z, \bar{z})/\lambda + 0(1/\lambda) \tag{14.7.21}$$

(we use $T(\lambda, \bar{\lambda}) \to 0$ as $\lambda \to \infty$).

Let $$A(z, \bar{z}) = \partial \ln(c(z, \bar{z})), \quad u(z, \bar{z}) = -2\varkappa\bar{\partial}\chi_1(z, \bar{z}). \tag{14.7.22}$$

Then $\psi(\lambda, \bar{\lambda}, z, \bar{z})$ is an eigenfunction of the Schrödinger operator L:

$$L = -\partial\bar{\partial} + A(z, \bar{z})\bar{\partial} + u(z, \bar{z}), \quad L\psi(\lambda, \bar{\lambda}, z, \bar{z}) = \epsilon_0\psi(\lambda, \bar{\lambda}, z, \bar{z}),$$
$$\epsilon_0 = -4\varkappa^2,$$

in general complex and with magnetic field.

The function $\psi(\lambda, \bar{\lambda}, z, \bar{z})$ can be obtained by solving the linear integral equation:

$$\chi(\lambda, \bar{\lambda}, z, \bar{z}) = 1 - \frac{1}{\pi}\iint \frac{d\mu_R d\mu_I}{\mu - \lambda} T(\mu, \bar{\mu}) \exp\left\{\varkappa\left[\left([(\mu + \frac{\sigma}{\bar{\mu}}\right)z + \left(\sigma\bar{\mu} + \frac{1}{\mu}\right)\bar{z}\right]\right\} + \frac{1}{2\pi i}\oint\oint_{|\lambda'|=|\mu|=1} \frac{|d\lambda'| d\mu}{\mu - \lambda} R(\mu, \lambda')$$
$$\times \exp\{\varkappa[(\mu - \lambda')z + (1/\mu - 1/\lambda')z']\}. \qquad (14.7.23)$$

(In the case $\epsilon_0 < 0$, the second term vanishes, i.e. $R(\lambda, \lambda') = 0$.) If the scattering data R, T are sufficiently small, equation (14.7.23) has unique solution for all z and the potentials $A(z, \bar{z})$, $u(z, \bar{z})$ are non-singular smooth functions.

An important question is that of distinguishing 'scattering data' corresponding to real purely potential operators ($A(z, \bar{z}) = 0$, $u(z, \bar{z}) = \bar{u}(z, \bar{z})$. We have shown that the scattering data shall satisfy the constraints (14.7.16)–(14.7.18). These conditions are not only necessary, but also sufficient.

Theorem 14.7.5

(i) Let the scattering data satisfy the constraints (14.7.16), (14.7.17) for $\epsilon_0 > 0$ or (14.7.16) for $\epsilon_0 < 0$. Then they give rise to purely potential operators $L = -\partial\bar{\partial} + u(z, \bar{z})$.

(ii) Let the scattering data satisfy also (14.7.18). Then the potential $u(z, \bar{z})$ is a real function.

Condition (14.7.17) means that the operator $1 + \hat{R}$ preserves the scalar product $\langle f, g\rangle$ in the space $L^2(S^2)$:

$$\langle f, g\rangle = \oint_{|\lambda|=1} f(\lambda)g(-\lambda)|d\lambda|, \quad \langle(1 + \hat{R})f, (1 + \hat{R})g\rangle = \langle f, g\rangle.$$

In the simplest case, $T(\lambda, \bar{\lambda}) = 0$, we have

$$(c(z, \bar{z}))^2 = \oint_{|\lambda|=1} \psi_-(\lambda, z, \bar{z})\psi_-(-\lambda, z, \bar{z})|d\lambda|$$
$$= \oint_{|\lambda|=1} \psi_+(\lambda, z, \bar{z})\psi_+(-\lambda, z, \bar{z})|d\lambda| = 1.$$

Applying $A(z, \bar{z}) = \partial\ln(c(z, \bar{z}))$ we obtain $A(z, \bar{z}) = 0$.

The non-local Riemann problem can be represented in a more symmetric form:

$$\psi_-(\lambda, z, \bar{z}) - \psi_+(\lambda, z, \bar{z})$$
$$= \oint_{|\lambda'|=1} W(\lambda, \lambda')(\psi_-(\lambda', z, \bar{z}) + \psi+(\lambda', z, \bar{z}))|d\lambda'|,$$

where $\hat{W}$ is the Cayley transform for $(1 + R)$:

$$\hat{W} = \hat{R}(2 + \hat{R})^{-1}.$$

Reductions (14.7.17), (14.7.18) at R are equivalent to linear constraints at W:

$$W(\lambda', \lambda) = \bar{W}(\lambda, \lambda'), \quad W(-\lambda', -\lambda) = -W(\lambda, \lambda').$$

Remark

The non-local Riemann problem and the $\bar{\partial}$-problem can be considered as particular cases of the non-local $\bar{\partial}$-problem [61]:

$$\partial_{\bar{\lambda}}\psi(\lambda, \bar{\lambda}, z, \bar{z}) = \iint d\lambda'_R \, d\lambda'_I \, T(\lambda, \bar{\lambda}, \lambda', \bar{\lambda}') \, \psi(\lambda', \bar{\lambda}', z, \bar{z})$$

or equivalently,

$$\partial_{\bar{\lambda}}\chi(\lambda, \bar{\lambda}, z, \bar{z}) = \iint d\lambda'_R \, d\lambda'_I \, \tilde{T}(\lambda, \bar{\lambda}, \lambda', \bar{\lambda}', z, \bar{z})\chi(\lambda', \bar{\lambda}', z, \bar{z})$$

where

$$\tilde{T}(\lambda, \bar{\lambda}, \lambda', \bar{\lambda}', z, \bar{z}) = T(\lambda, \bar{\lambda}, \lambda', \bar{\lambda}') \exp[-\varkappa((\lambda - \lambda')z + (1/\lambda - 1/\lambda')\bar{z})].$$

If the kernel $T(\lambda, \bar{\lambda}, \lambda', \bar{\lambda}')$ satisfies the conditions

$$T(\lambda, \bar{\lambda}, \lambda', \bar{\lambda}')/\lambda = T(-\lambda', -\bar{\lambda}', -\lambda, -\bar{\lambda})/\lambda',$$
$$T(\sigma/\bar{\lambda}, \sigma/\lambda, \sigma/\bar{\lambda}', \sigma/\lambda') = -\sigma\bar{T}(\lambda, \bar{\lambda}, \lambda', \bar{\lambda}')\lambda^2\bar{\lambda}^2\lambda'^2$$

then the scattering data gives rise to real purely potential operators (at least at the formal level).

Assuming T to be a bounded function of z we obtain $T(\lambda, \bar{\lambda}, \lambda', \bar{\lambda}') \neq 0$ only for $\lambda = -\sigma/\bar{\lambda}$ or $|\lambda| = |\lambda'| = 1$ for $\epsilon_0 > 0$. Hence we obtain the 'pure' $\bar{\partial}$-problem (14.7.13) for $\epsilon_0 < 0$ or the $\bar{\partial}$-problem with the non-local Riemann problem for $\epsilon_0 < 0$. This fact has been pointed out by Zakharov.

The problem of characterising the 'scattering data' corresponding to non-singular potentials is more complicated. The solution in some particular cases is given by the following theorem.

Theorem 14.7.6

Assume that the hypotheses of Theorem 14.7.4 hold.

(i) Let at least one of the following assumptions be valid:
   (a) $\varkappa^2 > 0$, $T(\lambda, \bar{\lambda}) = 0$, the integral operator $\hat{R}$ has L^2-norm less than 1.
   (b) $R(\lambda, \lambda') = 0$, $T(\lambda, \bar{\lambda})$ is a non-singular function, decreasing rapidly as $\lambda \to 0, \infty$.

   Then $u(z, \bar{z})$ is a smooth non-singular function, decreasing at infinity like $O(1/|z|)$.

(ii) Let $\varkappa^2 < 0$, $T(\lambda, \bar{\lambda})$ has the same properties as in the point (b).

Then the operator $L - \epsilon_0$ is strictly positive, that is, $((L - \epsilon_0)f, f) \geqslant C(f, f)$, $C > 0$ for any $f \in L^2(R^2)$.

The proof of this theorem in the case $R(\lambda, \lambda') = 0$ is based upon the theory of generalised analytic functions [62]. The positivity of $L - \epsilon_0$ *is derived from the existence of a strictly positive real eigenfunction* $F(z, \bar{z})$ at the level ϵ_0 [59]:

$$LF(z, \bar{z}) = \epsilon_0 F(z, \bar{z}), \quad F(z, \bar{z}) = \oint_{|\lambda|=1} \psi(\lambda, \bar{\lambda}, z, \bar{z})|d\lambda|.$$

If the potential $u(z, \bar{z})$ decreases sufficiently fast as $|z| \to \infty$, then the boundary values of $T(\lambda, \bar{\lambda})$ at the unit circle satisfy some 'continuity' equations. These have been discussed in [60] for $\epsilon_0 > 0$ and in [59] for $\epsilon_0 < 0$. For example, assuming $u(z, \bar{z}) = 0(1/|z|^{2+\delta})$ and $\epsilon_0 < 0$, we obtain

$$\pi U(\lambda, \bar{\lambda}) - \iint U(1/\bar{\mu}, 1/\mu) T(\mu, \bar{\mu})(\lambda - \mu)^{-1}\, d\mu_R\, d\mu_I|_{|\lambda|=1} = \text{const}.$$

where $U(\lambda, \bar{\lambda}) = \operatorname{sgn}(|\lambda|^2 - 1)\pi\bar{\lambda} T(\lambda, \bar{\lambda})$.

Theorem 14.7.7

Let the scattering data depend on an infinite set of extra variables, $t_1, \ldots, t_k, \ldots$:

$$R(\lambda, \lambda', \ldots, t_k, \ldots) = R(\lambda, \lambda')\exp\left[\sum_k (2\varkappa)^{2k+1}(\lambda^{2k+1} + 1/\lambda^{2k+1} - (\lambda')^{2k+1} - 1/(\lambda')^{2k+1})t_k\right],$$

$$T(\lambda, \bar{\lambda}, \ldots, t_k, \ldots) = T(\lambda, \bar{\lambda})\exp\left[\sum_k (2\varkappa)^{2k+1}(\lambda^{2k+1} + 1/\lambda^{2k+1} + \sigma\bar{\lambda}^{2k+1} + \sigma/\bar{\lambda}^{2k1})t_k\right].$$

Then the function $v(z, \bar{z}, \ldots, t_k, \ldots) = u(z, \bar{z}, \ldots, t_k, \ldots) - \epsilon_0$ satisfies the hierarchy of the Veselov–Novikov equations (14.7.3).

Remark

In our treatment of the scattering problems we assume that equation (14.7.9) has no homogeneous solutions. In fact, if L has negative eigenvalues, such solution exist for some real ϵ and correspond to the signularities of $\psi(\lambda, \bar{\lambda}, z, \bar{z})$. In the case of general position $\psi(\lambda, \bar{\lambda}, z, \bar{z})$ has singular contours at the λ-plane. Some necessary conditions at the scattering data in this case are discussed in [59].

Explicit solutions

(i) Let $T(\lambda, \bar{\lambda}) = 0$, $R(\lambda, \lambda')$ be degenerate, that is, $R(\lambda, \lambda') = \sum_k g_k(\lambda) f_k(\lambda')$. Then an explicit expression for the potential can be obtained. The simplest case is $n = 1$, $\epsilon_0 < 0$. (The non-local Riemann problem at a negative energy can be considered, but it gives rise to slowly decreasing potentials.) The kernel

$$R(\lambda, \lambda') = f(-\lambda)f(\lambda')/ \oint_{|\lambda''|=1} f(\lambda'')f(-\lambda'')|d\lambda''|$$

where $f(\lambda)$ is any real positive function $f(\lambda) = \bar{f}(\lambda)$, $f(\lambda) > 0$, gives rise to purely real potential operator with smooth potential, decreasing at infinity like $0(1/|z|)$. The corresponding solutions of (14.7.3) are given by:

$$v(z, \bar{z}, t) = 4\varkappa^2 - 2\partial\bar{\partial}F(z, \bar{z}, t), \quad w(z, \bar{z}, t) = 6\partial^2 F(z, \bar{z}, t),$$

$$F(z, \bar{z}, t) = \ln \oint_{|\lambda|=1} f(\lambda) \exp[-\varkappa(\lambda z + \bar{z}/\lambda) - 8\varkappa^3 (\lambda^3 + 1/\lambda^3)t]|d\lambda|.$$

(ii) The simplest example of the function $\psi(\lambda, \bar{\lambda}, z, \bar{z})$ with singular points corresponds to $\epsilon_0 > 0$, $R(\lambda, \lambda') = 0$, $T(\lambda, \bar{\lambda}) = 0$, $\chi(\lambda, \bar{\lambda}, z, \bar{z})$ is a rational function of λ with the simple poles λ_j. The standard procedure with respect to the reductions (see [58]) gives rise to real non-singular rational potentials. They are reflectionless at the level $\epsilon_0(S(k, l) = 0$ for $|k| = |l| = \epsilon_0)$ and decrease at infinity like $0(1/|z|^2)$. The corresponding multi-soliton solutions ('lumps') of (14.7.3) are given by

$$v(z, \bar{z}, t) = 4\varkappa^2 - \partial\bar{\partial}\ln \det A, \; w(z, \bar{z}, t) = 3\partial^2 \ln \det A,$$
$$A_{jk} = 1/(\lambda_k - \lambda_j) \text{ for } j \neq k,$$

$$A_{kk} = q_k(z, \bar{z}, t) + c_k, \; q_k(z, \bar{z}, t) = \varkappa(z - z/\lambda^2) + 24\varkappa^3(\lambda^2 - 1/\lambda^4)t), \; j, k = 1, \ldots, 4n,$$

$$\lambda_{2k+1} = -\lambda_{2k+2}, \; \lambda_{4k+3} = 1/\bar{\lambda}_{4k+1},$$
$$c_{2k+1} - c_{2k+1} = 1/\lambda_{2k+1}, \; c_{4k+3} = \bar{\lambda}^2_{4k+1}\bar{c}_{4k+1}.$$

Remark

The non-local Riemann problem method was proposed by Manakov [63] in the theory of the time-dependent Schrödinger operator. The $\bar{\partial}$-method was introduced in the study of inverse scattering problems on the line by Beals and Coifman (for references see [47]), and successfully extended to two-dimensional problems by Ablowitz, Bar Yaacov and Fokas ([23] and references therein). Multi-dimensional problems on the basis of the $\bar{\partial}$-method have been studied by Ablowitz, Nachman and Fokas [44, 46], Beals and Coifman [47], and R. G. Novikov and Henkin [48]. The partial solution of the two-dimensional inverse scattering problem has been obtained in [55, 56], the general solution for sufficiently small potentials in [57] for $\epsilon_0 > 0$. The case $\epsilon_0 < 0$ has been studied in [59]. Explicit formulae (14.7.19), (14.7.20) for the scattering amplitude have been obtained in [60].

References

[1] Fokas, A. and Ablowitz, M. and Lectures on the inverse scattering transform for multi-dimensional (2 + 1) problems. *Nonlinear Phenomena*, Lecture Notes in Physics **189**, 137–83, Springer, Berlin (1983).

[2] Belokolos, E. D. Peierls–Frölich problems and finite-gap potentials I, *Teoret. Mat. Fiz.* **45**, 2, 268–80 (1980); English transl. in *Theoret. Math. Phys.* **45** (1980).

[3] Belokolos, E. D. Peierls–Frölich's problems and finite-gap potentials II, *Teoret. Mat. Fiz.* **48**, 1; 60–9 (1981); English transl. in *Theoret. Math. Phys.* **48** (1981).

[4] Bordag, L. A., Its, A. R., Matveev, V. B., Manakov, S. V. and Zakharov, V. E. Two-dimensional solitons of Kadomtzev–Petviashvili equation *Phys. Lett.* 205–7 (1977).

[5] Brasovskii S. A., Gordjunin, S. A. and Kirova, N. N. Exact solution of Peierls model with arbitrary number of electrons on elementary cell *Pis'ma v ZETF* **31**, 8, 486–90 (1980); English transl. in *JETP Letters* **31** (1980).

[6] Brasovskii, S. A., Dzjaloshinskii, I. E. and Krichever, I. M. Exactly solvable discrete Peierls models *Z. Eksper. Teoret. Fiz* **83**, 2, 389–415 (1982); English transl. in *JETP* **56** (1982).

[7] Date, E., Jimbo, M., Kashiwara, M. and Miwa, T. Transformation groups for soliton equations, *Phys. Soc. Japan* **50**, 3806–12 (1981).

[8] Dobrokhotov, S. Y. and Maslov, V. P. Finite-gap quasi-periodic solutions in WKB-approximations, *Itogi Nauki: Sovremennye Problemy Mat.* **15**, 3–94 (1980); English transl. in *J. Soviet Math* **16**, no. 6 (1981).

[9] Dobrokhotov, S. Y. and Maslov, V. P. Multiphase asymptotics of nonlinear partial differential equations with a small parameter, Soviet Scientific Reviews. Math. Phys. Reviews v.3: OPA, Amsterdam (1982).

[10] Druma, V. S. Analytical solution of two-dimensional Korteweg–de Vries equation, *Pis'ma v ZETF* **19**, 12, 219–25 (1974).

[11] Dubrovin, B. A., Matveev, V. B. and Novikov, S. P. Non-linear equation of Korteweg–de Vries type, finite-gap linear operators, and Abelian varieties, *Uspekhi Mat. Nauk* **31**, 1, 55–136 (1976); English transl. in *Russian Math. Surveys* **31** (1976).

[12] Dubrovin, B. A. Theta-functions and non-linear equations, *Uspekhi Mat. Nauk* **36**, 2, 11–80 (1981); English transl. in *Russian Math. Surveys* **36** (1981).

[13] Dubrovin, B. A. Matrix finite-gap operators, *Itogi Nauki: Sovremennye Problemy Mat.* **23**, 33–78 (1983).

[14] Dubrovin, B. A., Krichever, I. M. and Novikov, S. P. Integrable systems 1, *Itogi Nauki: Sovremennye Problemy Mat., Fundamentalniye napravleniya* **4**, 220–326 (1985).

[15] Dzyaloshinskii, I. E. and Krichever, I. M. Sound and charge density wave in Peierls discrete model, *Z. Esper. Teoret Fiz.* **85**, 11, 1771–89 (1983).

[16] Dzyaloshinskii, I. E. and Krichever, I. M. Commensurability effect in discrete Peierls model, *Z. Eksper. Teoret. Fiz.* **83**, 5, 1576–81 (1982).

[17] Eleonskii, V. M., Krichever, I. M. and Kulagin, N. E. Rational multi-soliton solutions of the non-linear Schrödinger equation, *Dokl. Akad. Nauk SSSR* **287**, 3, 606–10 (1986).

[18] Fay, J. *Theta-functions on Riemann Surfaces*, Lecture Notes in Mathematics 352 Springer, Berlin (1973).

[19] Flaschka, H., Forest, M. and McLaughlin, D. The multiphase averaging and the inverse spectral solution of the Korteweg–de Vries equation, *Comm. Pure and Appl. Math.* **33**, 6, 739–84 (1980).

[20] Gurevitch, A. V. and Pitaievskii, L. P. Unstationary structure of the non-collision shock wave, **Z. Eksper. Teoret. Fiz. 65**, 3, 590–604 (1973).

[21] Zakharov, V. E. and Shabat, A. B. A scheme for the integration of non-linear equations of mathematical physics by the method of the inverse scattering problem 1, *Funktional. Anal. i Prilozhen* **8**, 3, 43–53 (1974); English transl. in *Functional Anal. Appl.* **8**, 226–35 (1974).

[22] Zakharov, V. E., Manakov, S. V., Novikov, S. P. and Pitaievskii, L. P. *The Theory of Solitons. The Method of Inverse Scattering*, Plenum, New York (1984).

[23] Zakharov, V. E. and Shul'man, E. I. The problems of the integrability of two-dimensional systems, *Dokl. Akad. Nauk SSSR* **283**, 6, 1325–9 (1985).

[24] Krichever, I. M. An algebraic-geometrical construction of the Zakharov–Shabat equations and their periodic solutions, *Dokl. Akad. Nauk SSSR* **227**, 2, 291–4 (1976); English transl. in *Soviet Math. Dokl.* **17**, 394–7 (1976).

[25] Krichever, I. M. The integration of non-linear equations by methods of algebraic geometry, *Funktsional. Anal. i Prilozhen.* **11**, 1, 15–31 (1977); English transl. in *Functional Anal. Appl.* **11**, 11–26 (1977).

[26] Krichever, I. M. Methods of algebraic geometry in the theory of non-linear equations, *Uspehi Mat. Nauk* **32**, 6, 183–208 (1977); English transl. in *Russian Math. Surveys* **33** (1978).

[27] Krichever, I. M. About rational solutions of the Kadomtsev–Petviashvili equation and integrable systems of particles on line, *Funktional. Anal. i Prilozhen.* **12**, 1, 76–8 (1978); English transl. in *Function. Anal. Appl.* **12** (1978).

[28] Krichever, I. M. and Novikov, S. P. Holomorphic bundles on Riemann surfaces and non-linear equations, *Uspehi Mat. Nauk* **35**, 6, 47–68 (1980).

[29] Krichever, I. M. The Peierls model, *Funktional. Anal. i Prilozhen*, **16**, 4, 10–26 (1982); English transl. in *Function. Anal. Appl.* **16** (1982).

[30] Krichever, I. M. 'Hessians' of integrals of the Korteweg–de Vries equation and perturbations of finite-zone solutions, *Dokl. Akad. Nauk SSSR* **270**, 6, 1312–17 (1983); English transl. in *Soviet Math. Dokl.* **27**, no. 3 (1983).

[31] Krichever, I. M. Non-linear equation and elliptic curves, *Itogi Nauki: Sovremennye Problemy Mat.* **23**, 79–136 (1983).

[32] Krichever, I. M. Laplace method, algebraic curve and non-linear equations *Funktsional. Anal. i Prilozhen.* **18**, 3, 43–56 (1984); English transl. in *Functional Anal. Appl.* **18** (1984).

[33] Krichever, I. M. Spectral theory of finite-gap non-stationary Schrödinger operators, non-stationary Peierls model, *Funktsional. Anal. i Prilozhen.* **20**, 3, 42–54 (1986); English transl. in *Function. Anal. Appl.* **20** (1986).

[34] Krichever, I. M. Periodic problem for the equation of Kadomtsev–Petviashvili, *Dokl. Akad. Nauk SSSR* **298**, 4 (1988).

[35] Lax, P. D. Periodic solutions of Korteweg–de Vries equation, *Comm. Pure and Appl. Math.* **28**, 141–88 (1975).

[36] Manakov, S. V. Nonlocal Riemann problem and KP-equation, *Physica*, **3D**, 1–2, 420–67 (1981).

[37] Marchenko, V. A. *The Sturm–Liouville Operator and its Applications*, Naukova Dumka, Kiev (1977).
[38] McKean, H. and Trubowitz, E. Hill's operator and hyperelliptic function in the presence of infinitely many branch points, *Comm. Pure and Appl. Math* **29**, 2, 142–226 (1976).
[39] McKean, H. and Moerbeke, P. van, The spectrum of Hill's equation, *Invent. Math.* **30**, 3, 217–74 (1975).
[40] Novikov, S. P. Two-dimensional Schrödinger operators in periodic fields, *Itogi Nauki: Sovremennye Problemy Mat.* **23**,2–32 (1983).
[41] Taimanov, I. K. Bloch's functions for two-dimensional periodic linear operators, *Dokl. Akad. Nauk.* **289**, 5 (1986).
[42] Cherednik, I. V. Elliptic curve and matrix soliton differential equations, *Itogi Nauki: Algebra, Geom. Topologiya* **22**, (1984).
[43] Perelomov, A. M. Nonlinear evolution equations that leave the spectrum of multidimensional Schrödinger operator invariant do not exist, *Lett. Math. Phys.* **1**, 175 (1976).
[44] Fokas, A. S. Inverse scattering and integrability in multi-dimensions, *Phys. Rev. Lett* **57**, 159–62 (1986) and references therein.
[45] Faddeev, L. D. Inverse problems of quantum scattering theory, *Itogi Nauki: Sovremennye Problemy Math.* **3**, 93–180 (1974); English trans. in *J. Soviet Math.* **5**, no. 3 (1976).
[46] Nachman, A. J. and Ablowiotz, M. J. A multidimensional inverse-scattering method, *Stud. Appl. Math.* **71**, 243–50 (1984).
[47] Beals, R. and Coifman, R. G. Multidimensional inverse scattering and nonlinear P. D. E., *Proc. Symp. Pure Math.*, **43** (1985).
[48] Novikov, R. G. and Henkin, G. M. $\bar{\partial}$-equation in the multidimensional inverse scattering theory, *Uspekhi Mat. Nauk* **42**, 27–55 (1987) (Russian).
[49] Veselov, A. P. and Novikov, S. P. Finite-zone two-dimensional Schrödinger operators. Explicit formulas and evolution equations, *Dokl. Akad. Nauk SSSR* **279**, 20–4 (1984); English trans. in *Soviet Math. Dokl.* **30** (1984).
[50] Manakov, S. V. Inverse scattering problem method and two-dimensional evolution equations, *Uspekhi Mat. Nauk* **31**, 245–6 (1976) (Russian).
[51] Konopelchenko, B. G. Two-dimensional second order differential spectral problem: compatibility conditions, general BTs and integrable equations, preprint 86–163, Institute of Nuclear Physics, Novosibirsk (1986).
[52] Dubrovin, B. A., Krichever, I. M. and Novikov, S. P. Schrödinger equation in a periodic field and Riemann surfaces, *Dokl. Akad. Nauk SSSR*, **229**, 15–18 (1976); English trans. in *Soviet Math. Dokl.* **17** (1976).
[53] Veselov, A. P. and Novikov, S. P. Finite-zone two-dimensional Schrödinger operators. Potential operators, *Dokl. Akad. Nauk SSSR* **279**, 784–8 (1984); English trans. in *Soviet Math. Dokl.* **30** (1984).
[54] Chadan, K. and Sabatier, P. G. *Inverse Scattering Problems in Quantum Scattering Theory*, Berlin, Springer (1977).
[55] Grinevich, P. G. and Novikov, R. G. Analogues of multisoliton potentials for the two-dimensional Schrödinger operators, *Funct. analis i ego prilog.* **19**, 4, 32–42 (1985) (Russian).
[56] Grinevich, P. G. and Novikov, R. G. Analogues of multisoliton potentials for the two-dimensional Schrödinger operators, and a nonlocal Riemann problem, *Dokl. Akad. Nauk SSSR* **226**, 19–22 (1986); English trans. in *Soviet Math. Dokl.* **33** (1986).
[57] Grinevich, P. G. and Manakov, S. V. Inverse scattering problem for the two-dimensional Schrödinger operator $\bar{\partial}$-method and nonlinear equations, *Funct. analis i ego prilog.*, **20**, 3, 14–24 (1986) (Russian).

[58] Grinevich, P. G. Rational solitons of the Veselov–Novikov equations are potentials reflectionless at fixed energy, *Theor. Math. Phys.* **69**, 307–10 (1986).
[59] Grinevich, P. G. and Novikov, S. P. Two-dimensional inverse scattering problem at negative energies and generalized analytic functions. 1 Energies below the ground state, *Funct. analis i ego prilog.* (to appear) (Russian).
[60] Novikov, R. G. Reconstructing the two-dimensional Schrödinger operator from the fixed-energy scattering amplitude, *Funct. analis i ego prilog.*, **20**, 90–1, 3, (1986) (Russian).
[61] Zaharov, V. E. and Manakov, S. V. Constructing the multidimensional integrable systems and their solution, *Funct. analis i ego prilog.*, **19**, 11–25, 2, (1985) (Russian).
[62] Vekua, N. I. *Generalized Analytic Functions*, Phismatgis, Moscow (1959) (Russian).
[63] Manakov, S. V. The inverse scattering transform for the time-dependent Schrödinger equation and Kadomtsev–Petviashvily equation, *Physica* **3D**, 420–7 (1981).
[64] Krichever I. M. The spectral theory of two-dimensional periodic differential operators and its applications to nonlinear equations, *Uspeki Mat. Nauk* **44**, 2 (1989).

Note added in proof

Further developments of these results are to be found in [64] which, in particular, contains a proof of the following statement: the general periodic problem for KP-2 is integrable and the set of the finite-gap solutions is dense in the space of all periodic solutions.

Part VI

Testing for complete integrability

15 *A. P. Fordy*

Prolongation structures of nonlinear evolution equations

15.1 Introduction

Most of this book has been concerned with the mathematical properties of exactly soluble nonlinear evolution equations (NLEE). A recurrent theme has been that with a knowledge of the linear spectral problem associated with our NLEE we can deduce most of the interesting properties of our equation. We can construct exact solutions and solve the initial value problem; we can generate infinite families of commuting flows, which are themselves isospectral to the same linear problem, and construct the recursion operator which acts as a ladder between these flows; we can use asymptotic expansions to generate infinite hierarchies of conserved densities which can be used as Hamiltonians to generate these same commuting flows; we can construct Miura maps, modified equations and deformations and use these to generate second Hamiltonian structures.

In the final part of the book we are going to pose a different type of question. Suppose we were given an equation (by a physicist or engineer, for instance). How do we test this equation for complete integrability? When it passes this test how do we discover or construct the associated linear spectral problem? For most of the better-known (and physically interesting) equations the spectral problem was either stumbled upon by accident or, at best, constructed by *ad hoc* means.

An interesting related question is how to isolate integrable families in larger classes of differential equation. This is of great interest to the soliton theorist, who often tries to classify and characterise integrable systems. Since physicists often have a general class rather than a single model for a physical process, they are interested in isolating specific integrable models. These are not only useful in themselves but also in the realm of perturbations. Testing classes of equation is much more difficult than testing individual equations. If our class merely has a few free parameters, then the process is straightforward. However, when arbitrary functions are involved, calculations can easily grind to a halt. It is therefore a remarkable fact that such tests have been carried out.

In this chapter we will be concerned with one particular method, Wahlquist–Estabrook prolongation, which simultaneously tests an equation for complete integrability and constructs the associated linear spectral problem or Bäcklund transformation. However, I shall, first, briefly describe some of the other approaches to the question. These are basically indicators, rather than decisive tests.

Numerical test

Here, some initial data is allowed to evolve (numerically) according to the given equation. In the case of the KdV (Korteweg–de Vries) equation, solitons emerge 'asymptotically' in one direction while radiation goes in the other. Alternatively, one can obtain a travelling wave solution, which entails solving a nonlinear ODE. For nonlinear equations the wave speed depends upon the amplitude, so that larger waves travel faster. Starting with a pair of separated solitary waves, with the faster 'behind' the slower, as initial conditions it is possible to numerically observe their interaction. If these pass through each other/exchange identities this indicates the existence of solitons. Such numerical tests are only indicators. Indeed, if the equation is close to being integrable it is difficult to distinguish between a true soliton interaction with fictitious 'numerical radiation' and a non-soliton interaction with true, but small, emitted radiation as in the case of the BBM (Benjamin, Bona and Mahony) equation [9]. Such problems can only be resolved analytically [24]. However, this was the method by which solitons were discovered [31].

Constants of motion

As previously discussed (Chapter 1) integrable equations possess infinitely many conservation laws and (usually) constants of the motion. In the case of the KdV equation, $u_t = u_{xxx} + 6uu_x$, the first four conserved densities are:

$$\mathscr{H}_0 = u, \quad \mathscr{H}_1 = \tfrac{1}{2}u^2, \quad \mathscr{H}_2 = u^3 - \tfrac{1}{2}u_x^2, \quad \mathscr{H}_3 = \tfrac{1}{2}u_{xx}^2 - 5uu_x^2 + \tfrac{5}{2}u^4. \tag{15.1.1}$$

The first three of these just correspond to mass, momentum and energy, so are no surprise. The existence of the fourth is a little suspicious. 'Too many' conserved densities is a good indication that an equation is worthy of further investigation. Hirota and Satsuma [14] used this approach to isolate a coupled KdV system, which was subsequently shown to be integrable [5] by the prolongation technique (see Example 15.2.2).

Generalised symmetries

We know that exactly soluble nonlinear evolution equations belong to infinite hierarchies of isospectral flows, associated with a single spectral problem. Furthermore, these flows commute with respect to a natural Lie bracket of vector fields (Chapter 1). In the Hamiltonian case these are generated by the sequence of conserved densities discussed in the previous section. It is thus possible to search for evolution equations which commute with our given equation. Integrable second- and third-order equations have been isolated by this method [11, 16].

Hirota's direct method

Hirota first presented this method in [13] and he, his collaborators and many others have exploited this to great effect. The method is described in Chapter 4. If it is possible to construct an N-soliton solution (arbitrary N) then complete integrability is effectively 'proved'. However, the less positive result of the existence of 3 or more (but not general N) is still a good indicator. Hirota and Satsuma discovered a 3-soliton solution for the coupled KdV system mentioned above. The direct method is one of the most powerful tests when in the right hands. Thus, we have the so-called 'postcards test': to test an equation, just send it on a postcard to Hirota. By return of post you will receive an answer regarding its integrability!

Painlevé test

In [1] it was noticed that the similarity solutions of various well-known soliton equations satisfy very special, second-order, ordinary differential equations, discovered by Painlevé and Gambier between 1893 and 1906 (see [17] for a description of the classical work). These equations possess a very special singularity structure. There exist ODEs of all orders which share this property – equations of Painlevé type. Ablowitz Ramani and Segur (ARS) [2] devised a method to test equations for this property, based on Kowaleskaya's method of isolating integrable tops. ARS [2] conjectured that all reductions of integrable NLEEs to ODEs should have this property. Weiss *et al.* [30] have extended the method to act directly on partial differential equations, thus avoiding the need to check *all* reductions to ODEs. This method is described in Chapter 16.

The last three methods have been successfully used to test nonlinear evolution equations for exact solubility. The Painlevé approach seems to be the current favourite, although, through a lack of supporting theorems, it is difficult to make strict statements regarding classification within this framework. On the other hand Shabat recently proved a theorem enabling him and his co-workers [22, 23, 27] to give an exhaustive list of integrable equations of nonlinear diffusion and of NLS (nonlinear Schrödinger) type. However, none of these methods directly leads to the construction of the associated linear spectral problem, although the Painlevé approach helps here [29, 30].

In the present chapter we concentrate on a sixth method: the Wahlquist–Estabrook prolongation technique [10, 28]. This method immediately proceeds with an attempt to construct the linear spectral problem (or another device, such as the Bäcklund transformation). If we are successful we have not only tested our equation for complete integrability but also constructed a device with which to integrate the equation, generate Hamiltonians and commuting flows, construct the recursion operator, and so on. If our attempt fails it suggests that the equation is not integrable. However, there are several starting assumptions so that failure does not guarantee non-integrability. Since prolongation is asking for a stronger result than the previous five listed methods, it is not surprising that the technique can often be more difficult to apply. Nevertheless, it is an interesting and powerful method which has been successfully applied to several new equations [5–8].

Wahlquist and Estabrook originally formulated their technique in the framework of differential forms and Cartan exterior differential systems. This approach was more fully developed in the language of jet-bundles in [25]. This geometric approach gives a nice conceptual picture (for those who like geometry and differential forms) but appears not to be useful from the practical point of view. The basic process of constructing a linear spectral problem is algebraic, so the geometry is inessential baggage.

In this chapter we shall consider prolongation theory from the point of view expounded in [6], to which the reader is referred for a more detailed discussion. For a simple algebraic approach, albeit limited to equations ultimately associated with $sl(2, \mathrm{C})$, see also [4, 19].

15.2 **Prolongation theory**

To avoid the use of indices we shall develop the general theory for a single component equation and, furthermore, consider only equations of evolution type. Systems of equations are considered in Example (15.2.2) and Section 15.5.

We start with an equation of type:

$$u_t = K(u, u_x, u_{xx}, \ldots, u_{nx}) \qquad (15.2.1)$$

where $u(x, t)$ is a real function of two variables and $u_{nx} = \partial^n u/\partial x^n$. We assume that there exist matrices $\mathrm{F}(u, u_x, \ldots)$, $\mathrm{G}(u, u_x, \ldots)$, depending upon u and its x-derivatives, such that:

$$\psi_x = \mathbb{F}\psi, \tag{15.2.2a}$$

$$\psi_t = \mathbb{G}\psi, \tag{15.2.2b}$$

and such that equation (15.2.1) is represented as the integrability conditions of (15.2.2a, b):

$$\mathbb{F}_t - \mathbb{G}_x + [\mathbb{F}, \mathbb{G}] = 0 \tag{15.2.3}$$

where $[\mathbb{F}, \mathbb{G}] = \mathbb{F}\mathbb{G} - \mathbb{G}\mathbb{F}$ is the matrix commutator (see Chapter 1 for examples of $\mathbb{F}$ and $\mathbb{G}$). From (15.2.1) and (15.2.3) we need to derive the structure of the matrices $\mathbb{F}$ and $\mathbb{G}$. This is not easy is general. There are two very important facts which we would like to know:

(i) The dependence of $\mathbb{F}$ and $\mathbb{G}$ upon the variables u, u_x, ... In particular, precisely which of these variables enter each of these matrices.
(ii) The size of the vector ψ and, therefore, the matrices $\mathbb{F}$ and $\mathbb{G}$.

The first of these is largely a guess, but not totally arbitrary. The second is denied us until towards the end of the prolongation calculation. During the calculation we introduce a partially defined Lie algebra. The size of the matrices comes from the representation of this algebra. At some stage during the calculation we need to introduce a spectral parameter into our matrices. Before discussing the method in detail, we look at the simple, but non-trivial, example of the KdV equation.

Example 15.2.1 KdV equation

$$u_t = u_{xxx} + 6uu_x. \tag{15.2.4}$$

Let us assume that $\mathbb{F}$ is a function of u only. The integrability conditions (15.2.3) then lead to $\mathbb{G}$ being a function of only u, u_x, u_{xx}, as a consequence of (15.2.4). The first step is to find the functional dependence of $\mathbb{F}$ and $\mathbb{G}$ upon these variables. Thus:

$$\mathbb{F}(u),\ \mathbb{G}(u, u_x, u_{xx}) \Rightarrow u_{xxx}(\mathbb{F}_u - \mathbb{G}_{u_{xx}}) + 6uu_x\mathbb{F}_u - u_x\mathbb{G}_u - u_{xx}\mathbb{G}_{u_x} + [\mathbb{F}, \mathbb{G}] = 0. \tag{15.2.5a}$$

Since u_{xxx} does not appear in either $\mathbb{F}$ or $\mathbb{G}$ we can equate coefficients of this variable to zero:

$$\mathbb{G}_{u_{xx}} = \mathbb{F}_u, \tag{15.2.5b}$$

$$6uu_x\mathbb{F}_u - u_x\mathbb{G}_u - u_{xx}\mathbb{G}_{u_x} + [\mathbb{F}, \mathbb{G}] = 0. \tag{15.2.5c}$$

We can solve (15.2.5b) to obtain:

$$\mathbb{G} = u_{xx}\mathbb{F}_u + \mathbb{L}^1(u, u_x) \tag{15.2.6a}$$

where $\mathbb{L}^1(u, u_x)$ is a new matrix function, independent of u_{xx}. On substitution of this, it is possible to read off the coefficients of u_{xx} in (15.2.5c):

$$\mathbb{L}^1_{u_x} = [\mathbb{F}, \mathbb{F}_u] - u_x\mathbb{F}_{uu}, \tag{15.2.6b}$$

$$u_x(6u\mathbb{F}_u - \mathbb{L}^1_u) + [\mathbb{F}, \mathbb{L}^1] = 0. \tag{15.2.6c}$$

We can now integrate (15.2.6b) to get:

$$\mathbb{L}^1(u, u_x) = \mathbb{L}^0(u) + u_x[\mathbb{F}, \mathbb{F}_u] - \tfrac{1}{2}u_x^2\mathbb{F}_{uu} \tag{15.2.7a}$$

where $\mathbb{L}^0(u)$ is a matrix function of u only. On substitution of this, (15.2.6c) decouples into four equations:

$$\mathbb{F}_{uuu} = [\mathbb{F}, \mathbb{F}_{uu}] = 0, \tag{15.2.7b}$$

$$\mathbb{L}^0_u = 6u\mathbb{F}_u + [\mathbb{F}, [\mathbb{F}, \mathbb{F}_u]], \tag{15.2.7c}$$

$$[\mathbb{F}, \mathbb{L}^0] = 0. \tag{15.2.7d}$$

At this stage a new feature arises: we obtain some *constraints* (15.2.7b) which involve *only* $\mathbb{F}$. These define $\mathbb{F}$ up to the *matrix constants of integration* $\mathbb{X}_i$:

$$\mathbb{F} = \mathbb{X}_1 + \mathbb{X}_2u + \mathbb{X}_3u^2 \tag{15.2.8a}$$

subject to

$$[\mathbb{X}_1, \mathbb{X}_3] = [\mathbb{X}_2, \mathbb{X}_3] = 0. \tag{15.2.8b}$$

This enables us to integrate (15.2.7c):

$$\mathbb{L}^0(u) = \mathbb{X}_0 + [\mathbb{X}_1, [\mathbb{X}_1, \mathbb{X}_2]]u + \tfrac{1}{2}(6\mathbb{X}_2 + [\mathbb{X}_2, [\mathbb{X}_1, \mathbb{X}_2]])u^2 + 4\mathbb{X}_3u^3 \tag{15.2.8c}$$

where X_0 is a further constant of integration. Piecing together (15.2.6a), (15.2.7a) and (15.2.8c), and using (15.2.8a), we have:

$$\begin{aligned}\mathbb{G} = \mathbb{X}_0 &+ (u_{xx} + 3u^2)\mathbb{X}_2 + (2uu_{xx} - u_x^2 + 4u^3)\mathbb{X}_3 \\ &+ u_x[\mathbb{X}_1, \mathbb{X}_2] + u[\mathbb{X}_1, [\mathbb{X}_1, \mathbb{X}_2]] + \tfrac{1}{2}u^2[\mathbb{X}_2, [\mathbb{X}_1, \mathbb{X}_2]].\end{aligned} \tag{15.2.9a}$$

To make the formulae less messy we define some new elements $\mathbb{X}_i$:

$$[\mathbb{X}_1, \mathbb{X}_2] = \mathbb{X}_4, \quad [\mathbb{X}_1, \mathbb{X}_4] = \mathbb{X}_5, \quad [\mathbb{X}_2, \mathbb{X}_4] = \mathbb{X}_6. \tag{15.2.9b}$$

The constraint (15.2.7d) can now be written explicitly:

$$\begin{aligned}&[\mathbb{X}_2, \mathbb{X}_6] = [\mathbb{X}_0, \mathbb{X}_1] = 0, \quad [\mathbb{X}_0, \mathbb{X}_2] = [\mathbb{X}_1, \mathbb{X}_5].\\ &[\mathbb{X}_0, \mathbb{X}_3] = 3\mathbb{X}_4 + \tfrac{1}{2}[\mathbb{X}_1, \mathbb{X}_6] + [\mathbb{X}_2, \mathbb{X}_5].\end{aligned} \tag{15.2.9c}$$

Thus, the problem of finding the linear spectral problem for the KdV equation (15.2.4) is reduced to the problem of finding a set of matrices $\mathbb{X}_0, \mathbb{X}_1, \ldots, \mathbb{X}_6$ which satisfy the commutation relations (15.2.8b),

(15.2.9b, c). We can consider these to be a set of generators and relations used to abstractly define a Lie algebra. In such circumstances we have algebraic machinery for finding a matrix representation of our elements $\mathbb{X}_i$ and thus our spectral problem (15.2.2). This machinery will be described in Section 15.3. For the moment we just present a solution:

$$\mathbb{X}_1 = \begin{pmatrix} 0 & 1 \\ \lambda^2 & 0 \end{pmatrix}, \quad \mathbb{X}_2 = \begin{pmatrix} 0 & 0 \\ 1 & 0 \end{pmatrix},$$

$$\mathbb{X}_3 = \begin{pmatrix} 0 & 0 \\ 0 & 0 \end{pmatrix}, \quad \mathbb{X}_0 = 2\mathbb{X}_1^3. \tag{15.2.10a}$$

Equation (15.2.2a) is a first-order system of equations equivalent to the scalar equation:

$$(\partial^2 + u)\psi = \lambda^2\psi \tag{15.2.10b}$$

which is the usual Schrödinger equation associated with the KdV equation.

Remark
The solution (15.2.10a) is not unique, since any similarity transformation will give another. However, all 2 × 2 matrix solutions will lead to the same scalar equation (15.2.10b). A solution such as (15.2.10a), in canonical form, is a feature of the algebraic methods described in Section 15.3. The matrices (15.2.10a) are elements of $sl(2, \mathrm{C})$, written in the fundamental representation. Any other representation, such as the 3 × 3 adjoint, would also provide us with a solution to our problem. However, such a solution would be equivalent to a scalar differential equation different from (15.2.10b). For instance, the adjoint representation leads to the 'squared eigenfunction' equation.

Remark
When solving (15.2.5a) we first found the dependence of $\mathbb{G}$ upon the highest derivative u_{xx} and then worked down to its dependence upon u. Along the way we found some constraints (15.2.7b) which enabled us to find expression (15.2.8a) for $\mathbb{F}$. These are typical features of a prolongation calculation.

The reader who works through the above example will find the calculation straightforward. A natural question arises:

Is it always as easy as for the KdV?

At present there exist no algorithms for testing nonlinear partial differential equations for complete integrability and constructing the associated linear spectral problem. The prolongation method is the nearest thing we have to such an algorithm, but even here there are four possible sticking points:

(i) Choice of variables entering $\mathbb{F}$ (and therefore $\mathbb{G}$): that is, $\mathbb{F}(u)$ or $\mathbb{F}(u, u_x)$ or $\mathbb{F}(u, u_x, u_{xx})$ and so on.

(ii) We cannot always solve for $\mathbb{F}$ and $\mathbb{G}$ in terms of the variables u, $u_x, u_{xx}, \ldots$, to obtain:

$$\mathbb{F} = \sum_1^n \mathbb{X}_i \mathbb{F}_i(u, u_x, \ldots)$$

and a similar expression for $\mathbb{G}$.

(iii) We need to 'close off' the Lie algebra generated by the elements $\mathbb{X}_i$. This means finding a finite-dimensional Lie algebra $\mathscr{g}$ (with parameter) consistent with the relations satisfied by the generators $\mathbb{X}_i$ defining $\mathbb{F}$ and $\mathbb{G}$.

(iv) We need a linear representation of the Lie algebra g, preferably the 'fundamental' one.

For equations of evolution-type, problems (i) and (ii) cause little trouble (see [6] for further discussion). However, problems (iii) and (iv) are universally considered to be the most difficult ones to overcome. Wahlquist and Estabrook devised a clever but *ad hoc* procedure which worked well for the KdV and other simple equations. To represent the algebra associated with the Hirota–Satsuma example below, however, requires much more sophisticated machinery.

It is important to note:

These sticking points are intrinsic to the problem, not the method.

They will appear in other methods albeit, perhaps, in disguise.

Example 15.2.2 Hirota–Satsuma equation

Hirota and Satsuma [14] considered the following coupled KdV system:

$$u_t = a(u_{xxx} + 6uu_x) - 6vv_x,$$
$$v_t = -v_{xxx} - 3uv_x. \tag{15.2.11}$$

They found that for the particular value of $a = 1/2$ there exists a fourth conserved density and a 3-soliton solution. In [5] and [6] Dodd and Fordy also isolated the case $a = 1/2$ and presented the prolongation structure and linear spectral problem, associated with this equation. We consider these results here to act as motivation and example for the 'heavy' algebraic machinery of the next section. 'Sticking points' (i) and (ii) present no difficulty here, but the four generators $\mathbb{X}_0, \ldots, \mathbb{X}_3$ satisfy a complicated set of commutation relations which ultimately lead to the matrix algebra $sl(4, \mathbb{C})$. We find:

$$\mathbb{F} = \mathbb{X}_1 + \mathbb{X}_2 u + \mathbb{X}_3 v,$$

$$\begin{aligned}\mathbb{G} = \mathbb{X}_0 &+ \mathbb{X}_2(au_{xx} + 3au^2 - 3v^2) - \mathbb{X}_3(v_{xx} + 3uv) + a\mathbb{X}_4 u_x + a\mathbb{X}_5 u \\ &+ \tfrac{1}{2}a\mathbb{X}_6 u^2 - \mathbb{X}_7 v_x + \mathbb{X}_8 v - \tfrac{1}{2}\mathbb{X}_9 v^2 - \mathbb{X}_{10} uv,\end{aligned} \tag{15.2.12a}$$

where $\mathbb{X}_0$, $\mathbb{X}_1$, $\mathbb{X}_2$ and $\mathbb{X}_3$ are constants of integration and

$$\begin{aligned}&[\mathbb{X}_1, \mathbb{X}_2] = \mathbb{X}_4, \quad [\mathbb{X}_1, \mathbb{X}_4] = \mathbb{X}_5, \quad [\mathbb{X}_2, \mathbb{X}_4] = \mathbb{X}_6, \\ &[\mathbb{X}_1, \mathbb{X}_3] = \mathbb{X}_7, \quad [\mathbb{X}_1, \mathbb{X}_7] = \mathbb{X}_8, \quad [\mathbb{X}_2, \mathbb{X}_7] = \mathbb{X}_{10}, \quad [\mathbb{X}_3, \mathbb{X}_7] = \mathbb{X}_9.\end{aligned} \tag{15.2.12b}$$

As a consequence of the integrability conditions, these elements are subject to the further restrictions:

$$\begin{gathered}[\mathbb{X}_2, \mathbb{X}_3] = 0, \quad (a + 1)[\mathbb{X}_3, \mathbb{X}_4] = -3\mathbb{X}_3, \quad [\mathbb{X}_3, \mathbb{X}_9] = 0, \\ [\mathbb{X}_2, \mathbb{X}_9] + 2[\mathbb{X}_3, \mathbb{X}_{10}] = 0, \quad \tfrac{1}{2}a[\mathbb{X}_3, \mathbb{X}_6] = [\mathbb{X}_2, \mathbb{X}_{10}], \quad [\mathbb{X}_2, \mathbb{X}_6] = 0, \\ \tfrac{1}{2}[\mathbb{X}_1, \mathbb{X}_9] + [\mathbb{X}_3, \mathbb{X}_8] = -3\mathbb{X}_4, \quad \tfrac{1}{2}[\mathbb{X}_1, \mathbb{X}_6] + [\mathbb{X}_2, \mathbb{X}_5] = -3\mathbb{X}_4, \\ [\mathbb{X}_1, \mathbb{X}_{10}] + [\mathbb{X}_2, \mathbb{X}_8] + 3\mathbb{X}_7 = a[\mathbb{X}_3, \mathbb{X}_5], \quad [\mathbb{X}_0, \mathbb{X}_1] = 0, \\ [\mathbb{X}_0, \mathbb{X}_2] = a[\mathbb{X}_1, \mathbb{X}_5], \quad [\mathbb{X}_0, \mathbb{X}_3] = -[\mathbb{X}_1, \mathbb{X}_8].\end{gathered} \tag{15.2.12c}$$

The Jacobi identities imply further restrictions and we are able to deduce

$$a = \tfrac{1}{2} \qquad \mathbb{X}_6 = \mathbb{X}_9 = -2\mathbb{X}_2, \quad \mathbb{X}_{10} = -2\mathbb{X}_3, \tag{15.2.13a}$$

$$[\mathbb{X}_4, \mathbb{X}_2] = 2\mathbb{X}_2, \quad [\mathbb{X}_1, \mathbb{X}_2] = \mathbb{X}_4, \quad [\mathbb{X}_4, \mathbb{X}_3] = 2\mathbb{X}_3. \tag{15.2.13b}$$

We can use (15.2.13) to find a very simple (canonical) representation for our generators:

$$\mathbb{X}_1 = \left[\begin{array}{cc|cc} 0 & -1 & 0 & 0 \\ \lambda^2 & 0 & -1 & 0 \\ \hline 0 & 0 & 0 & -1 \\ 2\lambda^4 & 0 & -\lambda^2 & 0 \end{array}\right] \quad \mathbb{X}_2 = \left[\begin{array}{cc|cc} 0 & 0 & 0 & 0 \\ 1 & 0 & 0 & 0 \\ \hline 0 & 0 & 0 & 0 \\ 0 & 0 & 1 & 0 \end{array}\right],$$

$$\mathbb{X}_3 = \left[\begin{array}{cc|cc} 0 & 0 & 0 & 0 \\ 1 & 0 & 0 & 0 \\ \hline 0 & 0 & 0 & 0 \\ 2\lambda^2 & 0 & -1 & 0 \end{array}\right], \quad \mathbb{X}_0 = 2\mathbb{X}_1^3. \tag{15.2.14a}$$

The remaining elements can be found from the definitions (15.2.12b). The resulting linear spectral problem is:

$$\begin{bmatrix} \psi_1 \\ \psi_2 \\ \psi_3 \\ \psi_4 \end{bmatrix}_x = \left[\begin{array}{cc|cc} 0 & -1 & 0 & 0 \\ \lambda^2 + u + v & 0 & -1 & 0 \\ \hline 0 & 0 & 0 & -1 \\ 2\lambda^4 + 2\lambda^2 v & 0 & -\lambda^2 + u - v & 0 \end{array}\right] \begin{bmatrix} \psi_1 \\ \psi_2 \\ \psi_3 \\ \psi_4 \end{bmatrix} \tag{15.2.14b}$$

which is equivalent to the fourth-order scalar Lax equation:

$$\mathbb{L}\psi = -\lambda^4\psi \tag{15.2.14c}$$

with $\psi = \psi_1$ and

$$\begin{aligned} \mathbb{L} &= \partial^4 + 2u\partial^2 + 2(u_x - v_x)\partial + u_{xx} - v_{xx} + u^2 - v^2 \\ &= (\partial^2 + u + v)(\partial^2 + u - v). \end{aligned}$$

The corresponding time evolution of ψ can be calculated from (15.2.12a) and is equivalent to:

$$\psi_t = 2(\partial^3 + \tfrac{3}{2}u\partial + \tfrac{3}{4}u_x - \tfrac{3}{2}v_x)\psi. \tag{15.2.14d}$$

15.3 **Lie algebras**

In this section we suppose we are given a free Lie algebra with constraints; that is, we are given a set of vectors $\mathbb{X}_i$ that satisfy certain commutation relations. Such a free algebra would, in general, be infinite-dimensional (even in this context). However, we wish to find a non-trivial homomorphism of this set into a finite-dimensional Lie algebra; in the prolongation literature this is usually called 'closing off' the algebra [28].

It is worth noting that without further information there can exist no systematic method of finding such a homomorphism. Algebraists refer to this as a word problem, which is known to be insoluble in such a general context. This does not mean that for a specific set of generators subject to a specific set of constraints one cannot find such a homomorphism; indeed, in the context of prolongation such homomorphisms are explicitly constructed. However, it does mean that there can exist no algorithm that is guaranteed to work for any example that may arise.

However, for prolongation algebras we do have further information, which allows us to contemplate an algorithm. We know that in almost all examples the spectral problem is governed by a simple Lie algebra. There are degenerate examples known to be related to solvable algebras, such as the Burgers equation.

This argument is given further credence by the following discussion.

General Lie algebras

Suppose we have a linear spectral problem

$$\psi_x = \mathbb{F}\psi, \quad \psi_t = \mathbb{G}\psi, \tag{15.3.1a}$$

where F and G are elements of a Lie algebra g with coefficients in some differential algebra. The integrability conditions are:

$$\mathbb{F}_t - \mathbb{G}_x + [\mathbb{F}, \mathbb{G}] = 0. \tag{15.3.1b}$$

We can use Levy's theorem [18] to write g as the semi-direct sum of a semi-simple algebra and a solvable ideal:

$$g = s \oplus_s r \tag{15.3.2a}$$

where s is a semi-simple and r is the maximal solvable ideal of g, known as the radical. The algebra g satisfies the following commutation rules:

$$[s, s] = s, \quad [s, r] \subset r, \quad [r, r] \subset r. \tag{15.3.2b}$$

The last of these relations is strict inclusion, since r is solvable. Because r is an ideal, s has the quotient structure $s \cong g/r$. As a vector space g is the direct sum of s and r. Consequently, there exists a basis of g such that $\mathbb{F}$ and $\mathbb{G}$ can be decomposed as:

$$\mathbb{F} = \mathbb{F}_s + \mathbb{F}_r, \quad \mathbb{G} = \mathbb{G}_s + \mathbb{G}_r, \tag{15.3.3a}$$

and the integrability condition (15.3.1b) takes the form:

$$\mathbb{F}_{st} - \mathbb{G}_{sx} + [\mathbb{F}_s, \mathbb{G}_s] = 0, \tag{15.3.3b}$$

$$\mathbb{F}_{rt} - \mathbb{G}_{rt} + [\mathbb{F}_s, \mathbb{G}_r] + [\mathbb{F}_r, \mathbb{G}_s] + [\mathbb{F}_r, \mathbb{G}_r] = 0. \tag{15.3.3c}$$

This follows immediately from the commutation relation (15.3.2b) and the existence of the above vector space direct sum.

Notice that the semi-simple part of the algebra gives rise to a self-contained system of differential equations, with no dependence on the variables occurring in the radical. The only coupling between these equations is in the second of them, and then the semi-simple components can be thought of as 'known' from (15.3.3b). It can be seen that, unless we have a system of equations with this type of coupling, we can further restrict our attention to semi-simple algebras.

The further restriction to simple algebras follows from the standard theorem that any semi-simple Lie algebra is the direct sum (both as a vector space and as a Lie algebra) of a number of simple ideals:

$$s = s_1 \oplus s_2 \oplus \ldots \oplus s_n, \quad s_i \text{ simple}, \quad [s_i, s_j] = \delta_{ij} s_j. \tag{15.3.3c}$$

Thus there exists a basis of s such that:

$$\mathbb{F}_s = \sum_{i=1}^{n} \mathbb{F}_{s_i}, \quad \mathbb{G}_s = \sum_{i=1}^{n} \mathbb{G}_{s_i}, \tag{15.3.3d}$$

and the integrability condition (15.3.3b) completely decouples into n separate systems:

$$\mathbb{F}_{s_i t} - \mathbb{G}_{s_i x} + [\mathbb{F}_{s_i}, \mathbb{G}_{s_i}] = 0, \qquad (15.3.3e)$$

each being governed by a simple Lie algebra.

In what follows we shall be mainly interested in simple Lie algebras. However, since most of the general theory is true for semi-simple algebras, we often work with these.

Semi-simple Lie algebras

Having reduced our problem to the study of (semi-)simple Lie algebras we use the structure and representation theory of these algebras to simplify the calculations.

The usual approach to 'closing off' a prolongation algebra is to find a complete multiplication table and then to find a representation in terms of finite matrices. The method we shall use is to embed the elements $\mathbb{X}_i$ in a simple Lie algebra, with known (fundamental) matrix representation.

There is not, of course, a unique way of doing this, since any similarity transformation will give rise to an equivalent embedding. We are thus interested in conjugacy classes of elements, and so are at liberty to choose a representative matrix for at least one element, and sometimes two or more simultaneously. Most of what follows is concerned with identifying elements with particularly simple representatives. This is particularly useful when the element is one of the generators used to define $\mathbb{F}$.

In particular, we concentrate on nilpotent elements and the embedding of them in copies of $sl(2, \mathbb{C})$. It is well known [15] that every complex, semi-simple Lie algebra contains at least one (and usually many) subalgebras isomorphic to $sl(2, \mathbb{C})$. By using Weyl's theorem and the standard irreducible representations of $sl(2, \mathbb{C})$ it is then possible to give each such subalgebra a canonical form within a given (semi)-simple algebra. For $sl(n, \mathbb{C})$, this is essentially the Jordan normal form of the three elements.

For completeness we briefly give the basic representations of $sl(2, \mathbb{C})$. The standard basis for $sl(2, \mathbb{C})$ in the fundamental representation is:

$$h = \begin{bmatrix} 1 & 0 \\ 0 & -1 \end{bmatrix}, \quad e_+ = \begin{bmatrix} 0 & 1 \\ 0 & 0 \end{bmatrix}, \quad e_- = \begin{bmatrix} 0 & 0 \\ 1 & 0 \end{bmatrix}. \qquad (15.3.4a)$$

They satisfy the commutation relations:

$$[h, e_\pm] = \pm 2 e_\pm, \quad [e_+, e_-] = h. \qquad (15.3.4b)$$

The irreducible $(m + 1)$-dimensional representations of $sl(2, \mathbb{C})$ are well known [15, 20]:

$$e_+^{(m)} = \begin{bmatrix} 0 & m & & & \\ & \ddots & m-1 & & \\ & & \ddots & \ddots & \\ & 0 & & \ddots & 1 \\ & & & & 0 \end{bmatrix}, \quad e_-^{(m)} = \begin{bmatrix} 0 & & & & \\ & \ddots & & 0 & \\ 1 & & \ddots & & \\ & 2 & & \ddots & \\ 0 & & \ddots & & \\ & & & m & 0 \end{bmatrix}, \quad h^{(m)} = \begin{bmatrix} m & & & & & \\ & m-2 & & & & \\ & 0 & \ddots & & & \\ & & & \ddots & 0 & \\ & & & & -(m-2) & \\ & & & & & -m \end{bmatrix}. \tag{15.3.4c}$$

Notice that all of the representations of e_+ and e_- are triangular, and therefore nilpotent matrices, while those of h are diagonal.

All matrix representations of $sl(2, \mathbb{C})$ can be constructed out of the irreducibles (15.3.4c). This is a special case of the following theorem.

Theorem 15.3.1 (Weyl)
Let ϕ: $g \to gl(\mathcal{V})$ be a (finite-dimensional) representation of a semi-simple Lie algebra g. Then ϕ is completely reducible.

This means that there exists a basis of $\mathcal{V}$ such that the representation of g takes block-diagonal form, each block being an irreducible representation of g.

For $sl(2, \mathbb{C})$ this means that all the elements can simultaneously be brought to (almost) Jordan normal form.

Given a prolongation algebra we wish to identify certain of its elements as basis elements of $sl(2, \mathbb{C})$. This copy of $sl(2, \mathbb{C})$, if not the whole prolongation algebra, has then to be embedded in a larger simple Lie algebra. We equate our copy of $sl(2, \mathbb{C})$ with the simplest possible representation of one of the conjugacy classes. Section (3.3) of [6] gives a rather abstract discussion of this topic, to which the interested reader is referred.

Prolongation algebras

To be able to use the above canonical forms we need to be able to locate nilpotent and semi-simple elements using only commutation relations. The following theorem is very useful [18, 20].

Theorem 15.3.2 (Jacobson)
Let $\mathbb{X}$ and $\mathbb{Y}$ be any two (non-commuting) elements of g such that $[\mathbb{Y}, [\mathbb{Y}, \mathbb{X}]] = 0$. Then $[\mathbb{Y}, \mathbb{X}]$ is nilpotent. In particular, if $[\mathbb{X}, Y] = \alpha\mathbb{Y}$, $\alpha \neq 0$, then $\mathbb{Y}$ is nilpotent.

Corollary 15.3.3
Let $\mathbb{X}$ and $\mathbb{Y}$ be any two elements of g such that $[\mathbb{X}, \mathbb{Y}] = \alpha\mathbb{Y}$, $\alpha \neq 0$, and $\mathbb{X} \in$ range ad$\mathbb{Y}$ where $(\text{ad}\mathbb{Y})\mathbb{Z} = [\mathbb{Y}, \mathbb{Z}]$. Then we may identify $\mathbb{Y}$ with $e_\pm$ and $\mathbb{X}$ with $\pm\frac{1}{2}\alpha h$.

Given a prolongation algebra our strategy will be as follows:

(1) Locate elements of the centre of the algebra. Assuming the algebra to be (semi-)simple we may equate these elements to zero.
(2) Use Corollary (15.3.3) to locate a nilpotent and a semi-simple element.
(3) Embed these elements in a simple Lie algebra g: identify the elements with the canonical form of one of the appropriate conjugacy classes. For $sl(n+1, \mathbb{C})$ these are block-diagonal, where each block is of the form (15.3.4c).
(4) Express the remaining generators of our prolongation algebra as linear combinations of a suitable basis of g.
(5) Use the fundamental representation of g to generate a linear spectral problem.

Some comments are in order:

(1) This enables us to reduce the number of independent elements $\mathbb{X}_i$ in the given prolongation algebra. Elements of the centre are associated with conservation laws. In Example (15.2.1) the element $\mathbb{X}_3$ corresponds to the conservation law of the KdV equation (15.2.4).
(2) There is no guarantee that we can locate nilpotent elements, but in all known examples this can be done. However, locating semi-simple elements is much more difficult. In many examples we do not have the crucial property $\mathbb{X} \in$ range ad$\mathbb{Y}$.
(3) The method is easiest to apply when $g = sl(n+1, \mathbb{C})$. There do exist, however, equations for which the appropriate algebras are symplectic or orthogonal (or even exceptional) (see Chapter 12). Of course, these algebras can always be embedded in the special linear algebras of higher rank and dimensions, but this would make the calculations much longer. However, we shall always assume $g = sl(n+1, \mathbb{C})$, starting with $n = 1$. If not successful we move on to $n = 2$, and so on.

    There is no *a priori* way of knowing the conjugacy class of our nilpotent element. The number of classes in $sl(n+1, \mathbb{C})$ is the number of ways (minus one), of partitioning $n + 1$ with a monotonic sequence of positive integers. Without divine inspiration we start with the simplest form.
(4) The advantages of expressing generators in terms of a known algebra are manifest. Furthermore, it is only the generators that need to be determined since all other coefficients are derived from them.
(5) The representation is a bonus. Closing off in the usual fashion would still leave the problem of finding a representation.

The above method is very general and most of the strategy works most of the time. However, it is not a strict algorithm and problems do arise. Most notably, we do not always know that $\mathbb{X} \in$ range ad$\mathbb{Y}$ when wishing to apply Corollary (15.3.3). If we know $\mathbb{Y} = e_+$ and $[\mathbb{X}, \mathbb{Y}] = 2\mathbb{Y}$, then we can deduce $\mathbb{X} - h \in \text{Ker ad}e_+$. Only if we also know that $\mathbb{X} - h \in$ range ad$\mathbb{Y}$ can we actually bring $\mathbb{X}$ onto h. However, even this result greatly simplifies the calculation.

At present we do not have a single technique that is applicable to all examples. We mainly use the above strategy, but some other methods are described in the examples of [6]. However, even these few technique are very powerful in practice.

Equations of evolution type

For equations of evolution type, with dispersion, the form of $\mathbb{F}$ and $\mathbb{G}$ are found to be:

$$\mathbb{F} = \sum_{k=1}^{n} \mathbb{F}_i \mathbb{X}_i, \quad \mathbb{G} = \mathbb{X}_0 + \mathbb{G}_F, \tag{15.3.5a}$$

where $\mathbb{F}_i$ are functionally independent and $\{\mathbb{X}_i\}_0^n$ are constants of integration that generate a free Lie algebra with constraints. We assume that there is a homomorphism of these generators into a simple Lie algebra g. The elements $\{\mathbb{X}_i\}_0^n$ generate a subalgebra of g, which we call g_F. Then $\mathbb{G}_F$ is an element of this subalgebra. The element $\mathbb{X}_0$ is the only one that is seemingly not an element of g_F. However, we have:

$$D_t\mathbb{F} - D_x\mathbb{G}_F + [\mathbb{F}, \mathbb{G}_F] = [\mathbb{X}_0, \mathbb{F}]. \tag{15.3.5b}$$

The left-hand side is an element of g_F, so that $[\mathbb{X}_0, \mathbb{F}] \in g_F$. We deduce that

$$[\mathbb{X}_0, \mathbb{X}_i] \in g_F \ \forall \text{ generators } \mathbb{X}_i \in g_F, \tag{15.3.5c}$$

so that $g_F \subset g$ is an ideal. Since g is simple we have $g_F = g$.

Scaled flows

A symmetry of a system of differential equations is any transformation of independent and dependent variables that permutes solutions of the system. Assuming the linear equations (15.3.1) to be invariant under such a transformation, we induce a symmetry of the elements $\mathbb{F}$ and $\mathbb{G}$, and consequently of the generators $\mathbb{X}_i$.

In particular, a scale symmetry of the form:

$$x \to \lambda^{-1}x, \quad t \to \lambda^{-n}t, \quad q_i \to \lambda^{m_i}q_i \tag{15.3.6a}$$

induces

$$\mathbb{F} \to \lambda \mathbb{F}, \quad \mathbb{G} \to \lambda^n \mathbb{G}, \tag{15.3.6b}$$

which induces a scale symmetry of the elements $\{\mathbb{X}_i\}$.

Each copy of $sl(2, \mathbb{C})$ has an automorphism that corresponds to such a scale symmetry:

$$e_- \to \lambda^{-1} e_-, \quad h \to h, \quad e_+ \to \lambda e_+. \tag{15.3.7a}$$

In all the examples of [6] it is this symmetry that gives rise to the eigenvalue.

Other useful automorphisms of $sl(2, \mathbb{C})$ are:

$$e_- \to -e_-, \quad h \to h, \quad e_+ \to -e_+, \tag{15.3.7b}$$

$$e_- \to \mathrm{e}^{-\mathrm{i}\phi} e_-, \quad h \to h, \quad e_+ \to \mathrm{e}^{\mathrm{i}\phi} e_+, \tag{15.3.7c}$$

$$e_- \to e_+, \quad h \to -h, \quad e_+ \to e_-. \tag{15.3.7d}$$

The first two of these are special cases of (15.3.7a) but are especially useful; the first corresponds to discrete symmetries such as $q \to -q$ of the MKdV equation while the second occurs when dealing with complex equations.

Example 3.4 Hirota–Satsuma equation

We finish this section by indicating how the above methods were used in [6] to obtain the representation (15.2.14) for the algebra (15.2.12). The equalities (15.2.13a) are a consequence of such equations as $[\mathbb{X}_i, \mathbb{X}_6 + 2\mathbb{X}_2] = 0$, $i = 1, 2, 3$, thus showing that $\mathbb{X}_6 + 2\mathbb{X}_2$ lies in the centre of the algebra. The first two equations of (15.2.13b) show that we can choose:

$$\mathbb{X}_2 = e_-, \quad \mathbb{X}_4 = -h, \tag{15.3.8}$$

for some copy of $sl(2, \mathbb{C})$. It is easy to see that neither $sl(2, \mathbb{C})$ nor $sl(3, \mathbb{C})$ can be the *whole* algebra. It is fortunately not necessary to go beyond $sl(4, \mathbb{C})$. Of the four conjugacy classes the correct choice is $\mathbb{X}_4 = \mathrm{diag}(-1, 1, -1, 1)$ and $\mathbb{X}_2$ given by (15.2.14a). The choice of $\mathbb{X}_4$ dictates the eigenvalues of $\mathrm{ad}\mathbb{X}_4$ and thus the value of a: $3/(a + 1) = 2 \Rightarrow a = \frac{1}{2}$. This largely fixes $\mathbb{X}_3$. The form of $\mathbb{X}_1$ is restricted, but not determined, by the second of (15.2.13b). The precise form of the solution (15.2.14) was obtained by using some scaling arguments [6], which will not be repeated here.

Remark

It cannot be over-emphasised that the greatest labour-saving device was the choice of canonical forms for $\mathbb{X}_2$ and $\mathbb{X}_4$ (saving 15 parameters !). The remaining calculation is straightforward but tedious, but could be easily carried out with the aid of an algebraic manipulator such as REDUCE [12].

15.4 Testing classes of equation: generalised KdV equations

We now turn to the identification of integrable families belonging to some given class of equation. In this section we discuss a class of generalised KdV equation. Here we restrict ourselves to a single-component generalisation turning to 2-component systems in the next section.

Consider the equation:

$$u_t = u_{xxx} + K(u, u_x, u_{xx}). \tag{15.4.1}$$

Let us first assume that $\mathbb{F}$ depends upon u and u_x and decouple (15.2.3) according to the coefficients of u_{xxxx}:

$$\mathbb{F}(u, u_x), \quad \mathbb{G}(u, u_x, u_{xx}, u_{xxx}), \tag{15.4.2a}$$

$$\mathbb{G}_{u_{xxx}} = \mathbb{F}_{u_x}, \tag{15.4.2b}$$

$$\mathbb{F}_u(u_{xxx} + K) + \mathbb{F}_{u_x}(\mathrm{K_u u_x} + \mathrm{K_{u_x} u_{xx}} + \mathrm{K_{u_{xx}} u_{xxx}}) \\ - \mathbb{G}_u u_x - \mathbb{G}_{u_x} u_{xx} - \mathbb{G}_{u_{xx}} u_{xxx} + [\mathbb{F}, \mathbb{G}] = 0. \tag{15.4.2c}$$

Equation (15.4.2b) can be integrated and used to decouple (15.4.2c):

$$\mathbb{G} = \mathbb{F}_{u_x} u_{xxx} + \mathbb{L}^2(u, u_x, u_{xx}), \tag{15.4.3a}$$

$$\mathbb{L}^2_{u_{xx}} = \mathbb{F}_u + \mathbb{F}_{u_x} K_{u_{xx}} - \mathbb{F}_{uu_x} u_x - \mathbb{F}_{u_x u_x} u_{xx} + [\mathbb{F}, \mathbb{F}_{u_x}], \tag{15.4.3b}$$

$$\mathbb{F}_u K + u_x \mathbb{F}_{u_x} K_u + u_{xx} \mathbb{F}_{u_x} K_{u_x} - u_x \mathbb{L}^2_u - u_{xx} \mathbb{L}^2_{u_x} + [\mathbb{F}, \mathbb{L}^2] = 0. \tag{15.4.3c}$$

We can next integrate (15.4.3b), but the decoupling procedure is more complicated at this stage:

$$\mathbb{L}^2 = \mathbb{L}^1(u, u_x) + (\mathbb{F}_u - u_x \mathbb{F}_{uu_x} + [\mathbb{F}, \mathbb{F}_{u_x}]) u_{xx} - \tfrac{1}{2} u_{xx}^2 \mathbb{F}_{u_x u_x} + \mathbb{F}_{u_x} K, \tag{15.4.4a}$$

$$[\mathbb{F}, \mathbb{L}^1] - u_x \mathbb{L}^1_u - u_{xx}(\mathbb{L}^1_{u_x} + u_x \mathbb{F}_{uu} + u_x^2 \mathbb{F}_{uuu_x} - [\mathbb{F}, \mathbb{F}_u] + u_x[\mathbb{F}_u, \mathbb{F}_{u_x}] \\ + 2u_x[\mathbb{F}, \mathbb{F}_{uu_x}] - [\mathbb{F}, [\mathbb{F}, \mathbb{F}_{u_x}]]) + \tfrac{3}{2} u_{xx}^2 (u_x \mathbb{F}_{uu_x u_x} - [\mathbb{F}, \mathbb{F}_{u_x u_x}]) \\ + \tfrac{1}{2} u_{xx}^3 \mathbb{F}_{u_x u_x u_x} + (\mathbb{F}_u - u_x \mathbb{F}_{uu_x} - u_{xx} \mathbb{F}_{u_x u_x} + [\mathbb{F}, \mathbb{F}_{u_x}]) K = 0. \tag{15.4.4b}$$

As this point the precise dependence of K upon u_{xx} is important. Let K be polynomial in u_{xx}:

$$K = \sum_{m=1}^{N} \mathrm{K_m}(u, u_x) u_{xx}^m, \quad K_N \neq 0. \tag{15.4.5}$$

If $N > 2$ then $K_N \mathbb{F}_{u_x u_x} = 0$. Thus, $K_N \neq 0$ implies $\mathbb{F}_{u_x u_x} = 0$ so that

$$\mathbb{F} = \mathbb{F}_0(u) + \mathbb{F}_1(u) u_x. \tag{15.4.6a}$$

We can then use a gauge transformation $\mathbf{T}(u)$ to remove the u_x-dependence of $\mathbb{F}$:

$$\mathbf{T}_u + \mathbf{T}\mathbb{F}_1 = 0, \quad \hat{\mathbb{F}} = \mathbf{T}\mathbb{F}_0\mathbf{T}^{-1}. \tag{15.4.6b}$$

Next we note that $\mathbb{F}_{u_x} = 0 \Rightarrow K_m = 0 \ \forall \ m \geq 2$, contradicting our assumption $N > 2$ and $K_N \neq 0$. We conclude that $N = 2$, so that (after a change of notation):

$$K(u, u_x, u_{xx}) = p(u, u_x) + q(u, u_x)u_{xx} + \tfrac{1}{2}r(u, u_x)u_{xx}^2. \tag{15.4.7}$$

It is now possible to decouple (15.4.4b). It is convenient to define $\mathbb{H}$ as below:

$$\mathbb{H} = \mathbb{F}_u - u_x \mathbb{F}_{uu_x} + [\mathbb{F}, \mathbb{F}_{u_x}], \tag{15.4.8a}$$

$$\mathbb{F}_{u_x u_x u_x} = r\mathbb{F}_{u_x u_x}, \tag{15.4.8b}$$

$$\mathbb{H}_{u_x} - \tfrac{1}{3}r\mathbb{H} = \tfrac{2}{3}q\mathbb{F}_{u_x u_x}, \tag{15.4.8c}$$

$$\mathbb{L}^1_{u_x} = q\mathbb{H} - u_x\mathbb{H}_u + [\mathbb{F}, \mathbb{H}] - p\mathbb{F}_{u_x u_x}, \tag{15.4.8d}$$

$$u_x \mathbb{L}^1_u = p\mathbb{H} + [\mathbb{F}, \mathbb{L}^1]. \tag{15.4.8e}$$

Given an equation of the form (15.4.1) it is a relatively simple matter to substitute p, q, r into (15.4.8) and solve for $\mathbb{F}$ and $\mathbb{G}$. It is considerably more difficult, however, to use these equations to isolate *all* integrable cases of (15.4.1). Nevertheless, this *is* possible [26], but the details are too cumbersome to include here. To illustrate the general approach the restricted class (15.4.11) will be analysed below.

Remark

We have shown above that $r = 0$ when $\mathbb{F}_{u_x} = 0$ (as can be seen by (15.4.8c)). It is also possible to prove that if $r = 0$, there exists a gauge such that $\mathbb{F}_{u_x} = 0$. We can thus restrict ourselves to a fairly well defined subclass of (15.4.1):

Restricted class

$$r = 0, \quad \mathbb{F}_{u_x} = 0$$

Equations (15.4.8) now reduce to:

$$\mathbb{H} = \mathbb{F}_u, \quad \mathbb{L}^1_{u_x} = q\mathbb{F}_u - u_x\mathbb{F}_{uu} + [\mathbb{F}, \mathbb{F}_u], \quad u_x\mathbb{L}^1_u = p\mathbb{F}_u + [\mathbb{F}, \mathbb{L}^1], \tag{15.4.9}$$

which can be partially solved to give:

$$q = q_0 + q_1 u_x, \quad p = p_0 + p_1 u_x + p_2 u_x^2 + p_3 u_x^3, \tag{15.4.10a}$$

$$(\mathbb{F}_{uu} - q_1 \mathbb{F}_u)_u + 2p_3 \mathbb{F}_u = 0, \tag{15.4.10b}$$

$$(\tfrac{3}{2}[\mathbb{F}, \mathbb{F}_u] + q_0 \mathbb{F}_u)_u - p_2 \mathbb{F}_u - \tfrac{1}{2} q_1 [\mathbb{F}, \mathbb{F}_u] = 0, \tag{15.4.10c}$$

$$\mathbb{L}_u^0 = p_1 \mathbb{F}_u + q_0 [\mathbb{F}, \mathbb{F}_u] + [\mathbb{F}, [\mathbb{F}, \mathbb{F}_u]], \tag{15.4.10d}$$

$$[\mathbb{F}, \mathbb{L}^0] + p_0 \mathbb{F}_u = 0, \tag{15.4.10e}$$

where q_i and p_i are functions of u only.

Although this is a subcase of the general system (15.4.8), the detailed analysis is still rather cumbersome. We thus restrict once more.

Equations in conservation form

A particularly interesing case of the above class is that of equations in conservation form:

$$u_t = (u_{xx} + a_0(u) + a_1(u) u_x + \tfrac{1}{2} a_2(u) u_x^2]_x. \tag{15.4.11}$$

In these circumstances (15.4.10) reduce to:

$$\mathbb{F}_{uuu} = a_2 \mathbb{F}_{uu}, \tag{15.4.12a}$$

$$\tfrac{3}{2}[\mathbb{F}, \mathrm{F}_{uu}] + a_1 \mathbb{F}_{uu} - \tfrac{1}{2} a_2 [\mathbb{F}, \mathbb{F}_u] = 0, \tag{15.4.12b}$$

$$\mathbb{L}_u^0 = a_{0u} \mathbb{F}_u + a_1 [\mathbb{F}, \mathbb{F}_u] + [\mathbb{F}, [\mathbb{F}, \mathbb{F}_u]], \tag{15.4.12c}$$

$$[\mathbb{F}, \mathbb{L}^0] = 0. \tag{15.4.12d}$$

The first two of these equations enable us to write $\mathbb{F}$ in the form (15.3.5a) and to derive some constraints on the constants $\mathbb{X}_i$:

$$\mathbb{F} = \mathbb{X}_1 + \mathbb{X}_2 u + \mathbb{X}_3 g(u), \quad g''' = a_2 g'', \tag{15.4.13a}$$

$$(2a_1 \mathbb{X}_3 + 3[\mathbb{X}_1, \mathbb{X}_3] + 3u[\mathbb{X}_2, \mathbb{X}_3]) g'' - a_2([\mathbb{X}_1, \mathbb{X}_3] + u[\mathbb{X}_2, \mathbb{X}_3]) g' + a_2 [\mathbb{X}_2, \mathbb{X}_3] g = a_2 [\mathbb{X}_1, \mathbb{X}_2]. \tag{15.4.13b}$$

We can first dismiss the case $g'' = 0$, since this is equivalent to $\mathbb{X}_3 = 0$ and the only integrable case is the mixed KdV/MKdV equation. We may assume that the linear part of g has been absorbed in the definitions of $\mathbb{X}_1$ and $\mathbb{X}_2$. There are some other special cases, in which the equation can be transformed into the KdV/MKdV or into the third-order member of the Burgers hierarchy.

The function g must satisfy both the third-order differential equation (15.4.13a) *and* the second-order differential equation (with *matrix coef-*

ficients) (15.4.13b). It is this which restricts g (with $g'' \neq 0$) to one of only four functions.

Write (15.4.13b) as:

$$Ag'' = a_2([\mathbb{X}_1, \mathbb{X}_2] - g[\mathbb{X}_2, \mathbb{X}_3] + g'([\mathbb{X}_1, \mathbb{X}_3] + u[\mathbb{X}_2, \mathbb{X}_3])) \quad (15.4.13c)$$

where $A = 2a_1\mathbb{X}_3 + 3[\mathbb{X}_1, \mathbb{X}_3] + 3u[\mathbb{X}_2, \mathbb{X}_3]$. Differentiating (15.4.13b) with respect to u gives (using $g''' = a_2 g''$):

$$(a_2(A - [\mathbb{X}_1, \mathbb{X}_3] - u[\mathbb{X}_2, \mathbb{X}_3]) + A')g'' = a_2'([\mathbb{X}_1, \mathbb{X}_2] - g[\mathbb{X}_2, \mathbb{X}_3] + g'([\mathbb{X}_1, \mathbb{X}_3] + u[\mathbb{X}_2, \mathbb{X}_3])). \quad (15.4.13d)$$

Comparing (15.4.13c, d), with $a_2' \neq 0$, we get

$$2(a_1a_2' - a_2a_1' - a_1a_2^2)\mathbb{X}_3 + (3a_2' - 2a_2^2)[\mathbb{X}_1, \mathbb{X}_3] + (3ua_2' - 2ua_2^2 - 3a_2)[\mathbb{X}_2, \mathbb{X}_3] = 0. \quad (15.4.14)$$

Remark

The case $a_2' = 0$ is one of the above mentioned special cases which leads to a soluble prolongation algebra and equation of Burgers type.

Equation (15.4.14) (having *matrix coefficients*) is very restrictive and possesses four non-trivial solutions of a_1 and a_2, giving in turn the four solutions for g. Only one of these, the deformed MKdV, is an interesting equation and corresponds to setting the coefficient of $[\mathbb{X}_2, \mathbb{X}_3]$ to zero:

$$3ua_2' - 2ua_2^2 - 3a_2 = 0, \quad (15.4.14b)$$

leading to:

$$a_2 = \frac{3\varepsilon^2 u}{1 - \varepsilon^2 u^2}, \quad g = \sqrt{(1 - \varepsilon^2 u^2)}. \quad (15.4.14c)$$

The complete equation is:

$$u_t = \left(u_{xxx} + \frac{3}{2}\frac{\varepsilon^2 u u_x^2}{1 - \varepsilon^2 u^2} + 2u^3\right)_x. \quad (15.4.15)$$

and is studied in detail in [6]. The prolongation algebra can be represented in terms of $sl(2, \mathbb{C})$, with

$$\mathbb{F} = \begin{pmatrix} \lambda\sqrt{(1 - \varepsilon^2 u^2)} & u\sqrt{(1 - \lambda^2\varepsilon^2)} \\ -u\sqrt{(1 - \lambda^2\varepsilon^2)} & -\lambda\sqrt{(1 - \varepsilon^2 u^2)} \end{pmatrix}. \quad (15.4.16)$$

It is thus seen that for certain simple classes of equations it is possible to use the prolongation method to isolate *all* integrable cases.

15.5 Coupled KdV systems

We saw in Section 15.4 that the single-component, generalised KdV equation (15.4.1), with an $\mathbb{F}$ which depends upon u and u_x, gives rise to a prolongation structure of the form (15.4.8). By assuming $\mathbb{F}$ to be of simpler form, depending upon u only, the system (15.4.8) reduced to (15.4.10), which was further reduced to (15.4.12) by assuming the equation to be in conservation form.

In this section we generalise equation (15.4.11) for a system of two coupled KdV equations in conservation form:

$$u_t = (u_{xx} + \rho_{10}u_x + \rho_{01}v_x + a)_x,$$
$$v_t = (v_{xx} + \sigma_{10}u_x + \sigma_{01}v_x + b)_x. \qquad (15.5.1)$$

We start with the assumption that $\mathbb{F}$ depends upon just u and v:

$$\mathbb{F}(u, v), \quad \mathbb{G}(u, v, u_x, v_x, u_{xx}, v_{xx}). \qquad (15.5.2a)$$

After some manipulation we obtain:

$$\mathbb{F} = \mathbb{X}_1 + \mathbb{X}_2 u + \overline{\mathbb{X}}_2 v, \qquad (15.5.2b)$$

$$\mathbb{G} = (u_{xx} + \rho_{10}u_x + \rho_{01}v_x)\mathbb{X}_2 + (v_{xx} + \sigma_{10}u_x + \sigma_{01}v_x)\overline{\mathbb{X}}_2 + u_x\mathbb{X}_3 + v_x\overline{\mathbb{X}}_3$$
$$+ (uv_x - vu_x)[\mathbb{X}_2, \overline{\mathbb{X}}_2] + \mathbb{L}^0(u, v), \qquad (15.5.2c)$$

where $\mathbb{X}_1$, $\mathbb{X}_2$ and $\overline{\mathbb{X}}_2$ are (matrix) constants of integration and $\mathbb{X}_3 = [\mathbb{X}_1, \mathbb{X}_2]$, $\overline{\mathbb{X}}_3 = [\mathbb{X}_1, \overline{\mathbb{X}}_2]$. Terms quadratic in u and v do occur in $\mathbb{F}$, but their coefficients turn out to be in the centre of the prolongation algebra, so can be set equal to zero. The function $\mathbb{L}^0(u, v)$ must satisfy the equations

$$\mathbb{L}^0_u = \mathbb{X}_2 a_u + \overline{\mathbb{X}}_2 b_u + \mathbb{X}_3\rho_{10} + \overline{\mathbb{X}}_3\sigma_{10} + [\mathbb{X}_2, \overline{\mathbb{X}}_2](u\sigma_{10} - v\rho_{10}) + [\mathbb{X}_1, \mathbb{X}_3]$$
$$+ u[\mathbb{X}_2, \mathbb{X}_3] + v([\overline{\mathbb{X}}_2, \mathbb{X}_3] - [\mathbb{X}_1, [\mathbb{X}_2, \overline{\mathbb{X}}_2]]) - uv[\mathbb{X}_2, [\mathbb{X}_2, \overline{\mathbb{X}}_2]]$$
$$- v^2[\overline{\mathbb{X}}_2, [\mathbb{X}_2, \overline{\mathbb{X}}_2]], \qquad (15.5.3a)$$

$$\mathbb{L}^0_v = \mathbb{X}_2 a_v + \overline{\mathbb{X}}_2 b_v + \mathbb{X}_3\rho_{01} + \overline{\mathbb{X}}_3\sigma_{01} + [\mathbb{X}_2, \overline{\mathbb{X}}_2](u\sigma_{01} - v\rho_{01}) + [\mathbb{X}_1, \overline{\mathbb{X}}_3]$$
$$+ u([\mathbb{X}_1, [\mathbb{X}_2, \overline{\mathbb{X}}_2]] + [\mathbb{X}_2, \overline{\mathbb{X}}_3]) + v[\overline{\mathbb{X}}_2, \overline{\mathbb{X}}_3]$$
$$+ uv[\overline{\mathbb{X}}_2, [\mathbb{X}_2, \overline{\mathbb{X}}_2]] + u^2[\mathbb{X}_2, [\mathbb{X}_2, \overline{\mathbb{X}}_2]], \qquad (15.5.3b)$$

together with the constraint:

$$[\mathbb{F}, \mathbb{L}^0] = 0. \qquad (15.5.3c)$$

The integrability conditions of (15.5.3a, b) give us constraints upon ρ_{ij}, σ_{ij}:

$$R\mathbb{X}_3 + S\overline{\mathbb{X}}_3 - 3[\mathbb{X}_1, [\mathbb{X}_2, \overline{\mathbb{X}}_2]] - 3u[\mathbb{X}_2, [\mathbb{X}_2, \overline{\mathbb{X}}_2]] - 3v[\overline{\mathbb{X}}_2, [\mathbb{X}_2, \overline{\mathbb{X}}_2]]$$
$$+ (uS - vR - \Omega)[\mathbb{X}_2, \overline{\mathbb{X}}_2] = 0 \qquad (15.5.4a)$$

where

$$R(u, v) = \rho_{10v} - \rho_{01u}, \quad S(u, v) = \sigma_{10v} - \sigma_{01u}, \Omega(u, v) = \rho_{10} + \sigma_{01}. \tag{15.5.4b}$$

This prolongation structure is studied in detail in [3]. The constraint (15.5.4) enables us to broadly classify such structures, but still leaves us with a complicated algebra. The number of free functions in (15.5.1) can be reduced by assuming some sort of Hamiltonian structure. We just consider one example here:

Example

$$\begin{aligned} u_t &= (u_{xx} - 3vu_x + 3uv^2 - 3u^2)_x, \\ v_t &= (v_{xx} + 3vv_x + v^3 - 6uv)_x. \end{aligned} \tag{15.5.5}$$

This equation was first discussed by Levi *et al.* [21] and is the third-order flow of the dispersive water waves hierarchy (see Chapter 11). In this case, (15.5.4a) decouples to give:

$$[\overline{\mathbb{X}}_2, [\mathbb{X}_2, \overline{\mathbb{X}}_2]] = [\mathbb{X}_2, \overline{\mathbb{X}}_2], \quad [\mathbb{X}_2, [\mathbb{X}_2, \overline{\mathbb{X}}_2]] = 0, \quad [\mathbb{X}_1, [\mathbb{X}_2, \overline{\mathbb{X}}_2]] = -\mathbb{X}_3. \tag{15.5.6a}$$

Thus, $[\mathbb{X}_2, \overline{\mathbb{X}}_2] + \mathbb{X}_2$ lies in the centre of the prolongation algebra, so that

$$[\overline{\mathbb{X}}_2, \mathbb{X}_2] = \mathbb{X}_2. \tag{15.5.6b}$$

It is then possible to integrate (15.5.3a, b) to give:

$$\begin{aligned} \mathbb{L}^0 = \mathbb{X}_0 &+ (uv^2 - 3u^2)\mathbb{X}_2 + (v^3 - 6uv)\overline{\mathbb{X}}_2 + [\mathbb{X}_1, \mathbb{X}_3]u + [\mathbb{X}_1\overline{\mathbb{X}}_3]v \\ &- \tfrac{3}{2}uv\,\mathbb{X}_3 + \tfrac{3}{2}v^2\overline{\mathbb{X}}_3 + \tfrac{1}{2}u^2[\mathbb{X}_2, \mathbb{X}_3] + \tfrac{1}{2}v^2[\overline{\mathbb{X}}_2, \overline{\mathbb{X}}_3] \\ &+ \tfrac{1}{2}uv[\overline{\mathbb{X}}_2, \mathbb{X}_3] + [\mathbb{X}_2, \overline{\mathbb{X}}_3]). \end{aligned} \tag{15.5.7}$$

The remaining equation (15.5.3c) gives rise to further commutation relations. The details can be found in [3]. The prolongation algebra can be represented in terms of $sl(2, \mathbb{C})$ so that the linear spectral problem is:

$$\mathbb{X}_1 = \begin{pmatrix} \lambda & 0 \\ 1 & -\lambda \end{pmatrix}, \quad \mathbb{X}_2 = \begin{pmatrix} 0 & 1 \\ 0 & 0 \end{pmatrix}, \quad \overline{\mathbb{X}}_2 = \begin{pmatrix} \frac{1}{2} & 0 \\ 0 & -\frac{1}{2} \end{pmatrix}, \tag{15.5.8a}$$

$$\begin{pmatrix} \psi_1 \\ \psi_2 \end{pmatrix}_x = \begin{pmatrix} \frac{1}{2}(\lambda + v) & u \\ 1 & -\frac{1}{2}(\lambda + v) \end{pmatrix} \begin{pmatrix} \psi_1 \\ \psi_2 \end{pmatrix}. \tag{15.5.8b}$$

Conclusions

This chapter has been concerned with the Wahlquist–Estabrook prolongation method of constructing the linear spectral problem associated with an integable, nonlinear evolution equation. For a given equation, this construction is a straightforward, but possibly arduous task [5]. On the other hand, it is possible to use the prolongation method to test classes of NLEE for exact solubility, to isolate the integrable cases and calculate the corresponding spectral problems. This approach was illustrated in Sections 15.4 and 15.5 and is further employed in [3, 7, 26].

Acknowledgements

I thank Mario Pytka and Marek Antonowicz for their contributions to the calculation of results in Sections 15.4 and 15.5. We thank the SERC for financial support.

References

[1] Ablowitz, M. J. and Segur, H. Exact linearisation of a Painlevé transcendent, *Phys. Rev. Lett.* **38**, 1103–6 (1977).
[2] Ablowitz, M. J. Ramani, A. and Segur, H. A connection between nonlinear evolution equations and ordinary differential equations of P-type, I. *J. Math. Phys.* **21**, 715–21 (1980).
[3] Antonowicz, M. and Fordy, A. P. Prolongation structure of coupled KdV equations (in preparation).
[4] Corones, J. Solitons and simple pseudo-potentials, *J. Math. Phys.* **17**, 756–9 (1976).
[5] Dodd, R. K. and Fordy, A. P. On the integrability of a system of coupled KdV equations, *Phys. Letts. A* **89**, 168–70 (1982).
[6] Dodd, R. K. and Fordy, A. P. The prolongation structures of quasi-polynomial flows, *Proc. R. Soc. Lond.* **A385**, 389–429 (1983).
[7] Dodd, R. K. and Fordy, A. P. Prolongation structures of complex quasi-polynomial evolution equations, *J. Phys. A* **17**, 3249–66 (1984).
[8] Dodd, R. K. and Gibbon, J. D. The prolongation structure of some higher order Korteweg–de Vries equations, *Proc. R. Soc. Lond.* **A358**, 287–96 (1977).
[9] Eilbeck, J. C. and McGuire, G. R. Numerical study of the regularised long wave equation. I. Interaction of solitary waves, *J. Computational Phys.* **23**, 63–73 (1977).
[10] Estabrook, F. B. and Wahlquist, H. D. Prolongation structures of nonlinear evolution equations II, *J. Math. Phys.* **17**, 1293–7 (1976).
[11] Fokas, A. S. A symmetry approach to exactly solvable evolution equations, *J. Math. Phys.* **21**, 1318–25 (1980).
[12] Gragert, P. K. H. Symbolic computations in prolongation theory, Ph D Thesis, Twente University of Technology (1981).
[13] Hirota, R. Exact solution of the Korteweg–de Vries equation for multiple collision of solitons, *Phys. Rev. Lett.* **27**, 1192–4 (1972).
[14] Hirota, R. and Satsuma, J. Soliton solutions of a coupled Korteweg–de Vries equation, *Phys. Letts. A* **85**, 407–8 (1981).
[15] Humphreys, J. E. *Introduction to Lie Algebras and Representation Theory*, Springer, Berlin (1972).
[16] Ibragimov, N. Kh. and Shabat, A. B. Evolutionary equations with nontrivial Lie–Bäcklund group, *Funct. Anal. Appls.* **14**, 19–28 (1980).

[17] Ince, E. L. *Ordinary Differential Equations*, Dover New York (1956).
[18] Jacobson, N. *Lie Algebras*, Dover New York (1979).
[19] Kaup, D. J. The Estabrook–Wahlquist method with examples of applications, *Physica* **1D**, 391–411 (1980).
[20] Kostant, B. The principal three dimensional subgroup and the Betti numbers of a complex simple Lie group, *Am. J. Math.* **81**, 973–1032 (1959).
[21] Levi, D. Sym, A. and Wojciechowski, S. A hierarchy of coupled Korteweg-de Vries equations and the normalisation conditions of the Hilbert–Riemann problem, *J. Phys. A* **16**, 2423–32 (1983).
[22] Mikhailov, A. V. and Shabat, A. B. Integrability conditions for the system of two equations of the form $u_t = A(u)u_{xx} + F(u, u_x)$ II, *Theor. Math. Phys.* **66**, 47–65 (1986).
[23] Mikhailov, A. V. Shabat, A. B. and Yamilov, R. I. Extension of the module of invertible transformations. Classification of integrable systems, *Commun. Math. Phys.* **115**, 1–19 (1988).
[24] Olver, P. J. Euler operators and conservation laws of the BBM equation, *Math. Proc. Camb. Phil. Soc.* **85**, 143–60 (1979).
[25] Pirani, F. A. E. Robinson, D. C. and Shadwick, W. F. Local jet bundle formulation of Bäcklund transformations, *Math. Phys. Stud.* **1**, 1–132 (1979).
[26] Pytka, M. Testing nonlinear evolution equations for exact solubility, PhD Thesis, Leeds University, to be submitted.
[27] Sokolov, V. V. and Shabat, A. B. Classification of integrable evolution equations, *Math. Phys. Rev.* **4**, 221–80 (1986).
[28] Wahlquist, H. D. and Estabrook, F. B. Prolongation structures of nonlinear evolution equations, *J. Math. Phys.* **16**, 1–7 (1975).
[29] Weiss, J. The Painlevé property for partial differential equations, II: Bäcklund transformations, Lax pairs and the Schwarzian derivative, *J. Math. Phys.* **24**, 1405–13 (1983).
[30] Weiss, J. Tabor, M. and Carnevale, G. The Painlevé property for partial differential equations, *J. Math. Phys.* **24**, 522–6 (1983).
[31] Zabusky, N. J. and Kruskal, M. D. Interaction of 'solitons' in a collisionless plasma and the recurrence of initial states, *Phys. Rev. Letts.* **15**, 240–3 (1965).

16 *M. Tabor*

Painlevé property for partial differential equations

16.1 Introduction and historical background

Determining whether or not a given system of nonlinear ordinary or partial differential equations is integrable raises many fundamental issues. In particular there is the issue of what is meant by 'integrability' and the issue of how that integrability can best be determined without having to resort to a complete solution of the problem. For Hamiltonian systems the notion of integrability, as discussed in Chapter 10, is well-defined, that is, the existence of as many involutive first integrals as there are degrees of freedom. For non-Hamiltonian systems things are less clear-cut. Clearly the existence of integrals (which can be time-dependent) can lead to a reduction of the order of the system and hence to a solution in terms of an 'integration by quadratures'. However, this is clearly not the whole story since there are simple, apparently 'integrable', equations such as the Painlevé transcendents which do not have algebraic integrals and for which an integration by quadratures is not possible. None the less, despite these difficulties, it now seems that it is possible to identify many different classes of integrable system on the basis of their analytic structure, that is, the types of singularities exhibited by their solutions in the complex domain. The techniques for doing this can be applied to both ordinary and partial differential equations and furthermore can be extended to provide, in some cases, explicit solutions to the systems in question. It also turns out that the complex domain of non-integrable systems also contains much valuable information and overall it would seem that the study of analytic structure can provide a wide-ranging and unified treatment for a large class of nonlinear problems.

Historically, the first application of these ideas is due to Kovalevskaya [1] in her famous work on the rigid body problem. This concerned the then popular problem of trying to find a general solution to the Euler–Poisson equations which describe the motion of a top spinning about a fixed point. Kovalevskaya's approach was to determine the conditions under which the only *movable singularities*, that is, those singularities whose positions are initial-condition-dependent, exhibited by the solutions to the equations of motion, in the complex time plane, are ordinary poles. She found that this only occurred for four special combinations of the adjustable system parameters (the moments of inertia and the position of the centre of gravity). Three of these cases were already known as special solutions and the fourth one was a new case for which she was able to effect an integration of the equations of motion by means of some virtuoso work with hyperelliptic functions.

Following Kovalevskaya came the extensive work of Painlevé [2] and co-workers to determine the categories of second-order equations whose only movable singularities are ordinary poles. (This is the analytic property of ordinary differential equations that is now referred to as the 'Painlevé property'. In retrospect, especially for dynamical applications, it seems fairer to call it the Kovalevskaya or Kovalevskaya–Painlevé property.) Working with the general class of equations

$$\frac{\mathrm{d}^2y}{dx^2} = F\left[\frac{\mathrm{d}y}{\mathrm{d}x}, y, x\right], \tag{16.1.1}$$

where F is analytic in x and rational in y and $\mathrm{d}y/\mathrm{d}x$, Painlevé found that there were 50 types which had the desired analytical property. (A complete list is given in [3].) Forty-four of these equations are soluble in terms of known functions, for example, elliptic functions. The remaining six equations, referred to as the Painlevé transcendents, have transcendental meromorphic solutions for which convergent expansions are explicity known. The first two transcendents are

$$\text{PI: } y'' = 6y^2 + x, \tag{16.1.2}$$

$$\text{PII: } y'' = 2y^3 + xy + \alpha, \tag{16.1.3}$$

where primes denote derivative with respect to x. (Some additional historical details have been given elsewhere [4].)

Despite their mathematical interest it appeared, for many years, that the Painlevé transcendents were devoid of physical content. However, in the last ten years or so they have started to reappear in various important physical contexts such as the work of Wu *et al.* [5] which showed that two-point correlation function of the Ising model satisfied the third Painlevé transcendent. Particularly significant for our discussion here is their appearance as various reductions of some of the integrable partial differential equations soluble by IST (inverse spectral transform). This latter work has given a fresh impetus to Kovalevskaya's ideas and there is now a growing literature on the Painlevé property as a means of identifying integrable cases of both ordinary and partial differential equations. In the field of ordinary differential equations there have been, among others, detailed studies of the analytic structure of the Lorenz equations [6], the Henon–Heiles system [7, 8] and the three-particle Toda lattice [9]. The reader is also referred to the attempts by Ramani *et al.* [10] to formulate a 'weak' Painlevé property (which allows for rational branch points) and the investigations by Yoshida [11]. More geometrical perspectives have been given by Adler and van Moebeke [12] and Ercolani and Siggia [13]. In the last section of this chapter we will show how the techniques for determining the Painlevé property of partial differential equations can also be used to find the Lax pairs for ordinary differential equations.

16.2 Painlevé property for partial differential equations

As mentioned in Section 16.1 it had been noted some time ago that various of the IST soluble partial differential equations could be reduced to nonlinear ODEs possessing the Painlevé property. For example the KdV (Korteweg–de Vries) equation

$$q_t + 6qq_x + q_{xxx} = 0 \tag{16.2.1}$$

under the travelling wave reduction $q(x, t) = f(z = x - ct)$ becomes

$$-cf_z + 6ff_z + f_{zzz} = 0 \tag{16.2.2}$$

which, after one integration, is exactly soluble in terms of elliptic functions. The modified KdV equation,

$$q_t - 6q^2q_x + q_{xxx} = 0, \tag{16.2.3}$$

under the similarity reduction $q(x, t) = (3t)^{-1/3}f(z = x(3t)^{-1/3})$ becomes, after one integration, the second Painlevé transcendent (16.1.3). These observations led various workers (Ablowitz *et al.* [14], Nakach [15], McLeod and Olver [16]) to conjecture that all ODEs obtained from completely integrable PDEs should have the Painlevé property – thereby providing a simple analytical test for integrability. This conjecture has proved to be quite valuable – although there are clearly certain drawbacks such as the difficulty of identifying all possible reductions of the PDE to ODEs and the inability of the conjecture to provide further information about actual solutions to the equation in hand.

In order to overcome these difficulties Weiss, Tabor and Carnevale [17] (henceforth referred to as WTC) suggested a generalisation of the Painlevé property that could be applied directly to PDEs, without reference to associated ODEs, and which furthermore contains valuable information about the associated IST solutions. The underlying idea is the recognition that the singularities of functions of many complex variables, $z_1, \ldots, z_n$, are not isolated but lie on some analytic manifold, the *singular manifold*, of dimension 2n-2 determined by conditions of the form [18]:

$$\varphi(z_1, \ldots, z_n) = 0 \tag{16.2.4}$$

where φ is analytic in the neighbourhood of that manifold. To show that a given evolution equation, say

$$q_t = K(q), \quad q = q(x, t), \tag{16.2.5}$$

possesses the generalised Painlevé property requires a demonstration that the solutions may be expanded locally in a Laurent-like series about the singular manifold in the complex hyperspace, that is, an expansion of the form

$$q(x, t) = \frac{1}{\varphi(x, t)^\alpha} \sum_{j=0}^{\infty} q_j(x, t)\varphi^j(x, t), \tag{16.2.6}$$

where α is some (integer) leading order and the $q_j(x, t)$ a set of expansion coefficients analytic in the neighbourhood of the singular manifold $\varphi(x, t) = 0$. Such an expansion will have certain values of j, termed 'resonances', at which the corresponding q_j should be arbitrary. One resonance always occurs at $j = -1$ and corresponds to the arbitrariness of φ itself and the positions of the other resonances are determined by the nonlinearities in $K(q)$. If the expansion (16.2.6) is a local representation of the general solution it will, in keeping with the Cauchy–Kovalevskaya theorem, have as many arbitrary functions as the order of the equation. Furthermore, depending again on the form of $K(q)$, the equation may also have different leading orders – each one of which will have its own resonance structure. We usually call the leading order corresponding to the general solution the 'principle branch' and the others (with less arbitrary functions, that is, representing singular solutions) as 'lower branches'. Integrability requires that all branches possess the Painlevé property. If, at a resonance (of any branch) the associated q_j fails to be arbitrary, terms of the form $\varphi^j \ln \varphi$ must be included in the expansion. This makes the solution multi-valued about the singular manifold and hence the Painlevé property is lost.

In order that the solution (16.2.6) can be constructed in the neighbourhood of $\varphi = 0$ it is clearly necessary that the flow directions do not lie along that surface, that is, *neither* φ_x *nor* φ_t *is zero along* $\varphi(x, t) = 0$. As we shall describe in a later section on the KdV hierarchy there seems to be an important connection between the vanishing of the components of $\nabla\varphi$ on $\varphi = 0$ and the lower branches of the Painlevé expansion.

Expansions of the form (16.2.6) appear to contain much valuable information. If the expansion for the general solution (principal branch) is truncated at $0(\varphi^0)$, that is,

$$q = \frac{q_0}{\varphi^\alpha} + \frac{q_1}{\varphi^{\alpha-1}} + \ldots q_\alpha, \tag{16.2.7}$$

one obtains a system of equations involving φ and the q_j $(0 \le j \le \alpha)$ with both q and q_α satisfying the same evolution equation, that is, $q_t = K(q_t)$ and $q_{\alpha,t} = K(q_\alpha)$. For integrable systems this approach can (sometimes) yield a non-trivial auto-Bäcklund transformation with the equations involving φ being transformable into the associated Lax pair.

Many of the above ideas are well illustrated by the example of the KdV equation (16.2.1). Demonstration that this equation possesses the Painlevé property is by now a standard exercise [17] and the solutions have the local expansion

$$q = \sum_{j=0}^{\infty} q_j(x, t)\varphi^{j-2}(x, t) \tag{16.2.8}$$

where $q_0 = -2\varphi_x{}^2$, $q_1 = 2\varphi_{xx}$ and q_4 and q_6 (along with φ itself) are the arbitrary functions. If one sets $q_4 = q_6 = 0$ and requires that $q_3 = 0$ the series can be truncated in a self-consistent manner at $0(\varphi^0)$, that is,

$$q = 2\frac{\partial^2}{\partial x^2}\ln\varphi + u, \tag{16.2.9}$$

where for notational convenience, we set $u \equiv q_2$. Direct substitution of (16.2.9) into (16.2.1) gives

$$\begin{aligned} q_t &+ 6qq_x + q_{xxx} \\ &= \left[\left(\frac{2\varphi_x}{\varphi}\right)_t + \frac{(6\varphi_{xx}^2 - 8\varphi_x\varphi_{xxx} - 12u\varphi_x^2)}{\varphi^2} + \frac{(2\varphi_{xxxx} + 12u\varphi_{xx})}{\varphi}\right]_x \\ &\quad + u_t + 6uu_x + u_{xxx} = 0. \end{aligned} \tag{16.2.10}$$

The usual WTC approach is to set each order of φ equal to zero separately, leading to the pair of equations

$$\varphi_x\varphi_t + 3\varphi_{xx}^2 - 4\varphi_x\varphi_{xxx} - 6u\varphi_x^2 = 0, \tag{16.2.11a}$$

$$\varphi_{xt} + \varphi_{xxxx} + 6u\varphi_{xx} = 0, \tag{16.2.11b}$$

which, under the remarkable substitution

$$\varphi_x = \psi^2 \tag{16.2.12}$$

transform to the standard Lax pair for the KdV equation. At this stage however we prefer, for reasons that will soon become apparent, to make this substitution directly into the composite form (16.2.10) and obtain

$$\begin{aligned} q_t &+ 6qq_x + q_{xxx} \\ &= \left[\frac{-4\psi^2}{\varphi^2}\int^x \psi(\psi_t + 4\psi_{xxx} + 6u\psi_x + 3u_x\psi)\mathrm{d}x\right. \\ &\quad \left. + \frac{4\psi}{\varphi}\left\{\psi(\psi_t + 4\psi_{xxx} + 6u\psi_x + 3u_x\psi) - 3\psi\left(\frac{\psi_{xx}}{\psi} + u\right)_x\right\}\right]_x \\ &\quad + u_t + 6uu_x + u_{xxx}. \end{aligned} \tag{16.2.13}$$

From this we immediately deduce the Lax pair

$$\psi_{xx} + (\lambda + u)\psi = 0 \tag{16.2.14a}$$

and

$$\psi_t + 4\psi_{xxx} + 6u\psi_x + 3u_x\psi = c\psi \tag{16.2.14b}$$

with

$$u_t + 6uu_x + u_{xxx} = 0 \tag{16.2.14c}$$

where λ is the spectral parameter (and c an arbitrary constant).

Originally it seemed that the squared eigenfunction substitution (16.2.12) only applied to the KdV equation and this led Weiss [19] to develop an alternative and elegant procedure, involving the Schwarzian derivative of φ, to find the Lax pairs. For example, for the pair of equations (16.2.11) one may eliminate u and obtain the single equation for φ:

$$\frac{\varphi_t}{\varphi_x} + \{\varphi; x\} = \lambda \tag{16.2.15}$$

where

$$\{\varphi; x\} = \frac{\varphi_{xxx}}{\varphi_x} - \frac{3}{2}\left(\frac{\varphi_{xx}}{\varphi_x}\right)^3 \tag{16.2.16}$$

is the Schwarzian derivative of φ. A linearisation of (16.2.15) then enables one to find the Lax pair. Weiss has developed this approach and systematically applied it to the KdV hierarchy [20], the Boussinesq hierarchy [21] and the KP hierarchy [22].

That the WTC method provides a useful test of integrability (apart from some obvious counterexamples involving ration branch point, see e.g. [22]) and a route to the Lax pairs, has now been demonstrated for quite a large number of integrable systems. It is now highly desirable to have a better understanding of why and how the method actually works and what its connections are with other techniques for completely integrable systems. In recent work Tabor, Newell and others [23, 24] have started to address these issues. Some of these can be summarised in the following list of questions:

(i) What is the function φ and how might it be characterised?
(ii) Why can one identify $\varphi_x = \psi^2$ and how general is this relationship?
(iii) Can one always truncate, in a self-consistent manner, the Painlevé expansion of the principal branch?
(iv) What is the pattern and significance of the resonances for all branches?

(v) What is the role of the lower branches?
(vi) How does one find the Lax pair and algebraic curve for integrable systems of ODEs?
(vii) What is the connection with other techniques such as Hirota's method?

In the sections that follow we will attempt to provide some answers to these questions through a variety of examples and discussions.

16.3 Painlevé property for the KdV hierarchy

When using the WTC method to find the Bäcklund transformation (16.2.9) for the KdV equation (16.2.1) we should recall that this equation is just the first member of the infinite hierarchy

$$q_{t_{2n+1}} = \frac{\partial}{\partial x}(L^n q), \tag{16.3.1}$$

where t_{2n+1} is the 'time' associated with the $(2n+1)$th flow, and

$$L = \frac{1}{4}\frac{\partial^2}{\partial x^2} - q + \frac{1}{2}\int^x \mathrm{d}x' q_{x'}. \tag{16.3.2}$$

The first three members of this hierarchy are:

$$4q_{t_3} = -(q_{xx} + 3q^2)_x, \tag{16.3.3}$$

$$16q_{t_5} = +(q_{xxxx} + 5q_x^2 + 10qq_{xx} + 10q^3)_x, \tag{16.3.4}$$

$$\begin{aligned} 64q_{t_7} = &-(q_{6x} + 14qq_{4x} + 28q_x q_{xxx} + 21q_{xx}^2 \\ &+ 70q^2 q_{xx} + 70q_x^2 + 35q^4)_x. \end{aligned} \tag{16.3.5}$$

Thus we should think of φ as a function of all the time coordinates $t_1 = x$, $t_3, t_5, \ldots, t_{2n+1} \ldots$ and therefore write (16.2.9) as

$$q(x, t_3, t_5, \ldots) = 2\frac{\partial^2}{\partial x^2}\ln\varphi(x, t_3, t_5, \ldots) + u(x, t_3, t_5, \ldots) \tag{16.3.6}$$

which, as we shall see, is indeed the Bäcklund transformation for all members of the hierarchy (16.3.1).

For all members of the family it is easy enough to show that the leading order is -2 and hence one makes the *Ansatz*

$$q = \sum_{j=0}^{\infty} q_j(x, t_3, \ldots)\varphi^{j-2}(x, t_3, t_5 \ldots). \tag{16.3.7}$$

At $0(\varphi^{-2n-2})$ one finds that there are n possible solutions

$$q_0 = -m(m+1)\varphi_x^2 \quad m = 1, 2, \ldots, n \tag{16.3.8}$$

and at $0(\varphi^{-2n-1})$ that

$$q_1 = m(m+1)\varphi_{xx} \quad m = 1, 2, \ldots, n. \tag{16.3.9}$$

Thus for the $(2n+1)$th flow there are n branches with the expansions

$$q = m(m+1)\frac{\partial^2}{\partial x^2}\ln\varphi(x, t_3, t_5, \ldots) + \sum_{j=2}^{\infty} q_j\varphi^{j-2}. \tag{16.3.10}$$

For each branch there are $2n$ resonances and for the principal branch, $m = 1$, $2n - 1$ of them occur for $j \geq 0$ and therefore this branch gives rise to the full complement (which includes φ itself) of $2n$ arbitrary functions. The lower branches ($m > 1$) have only $2n - m$ resonances for $j \geq 0$ and the corresponding expansion (16.3.10) has only $2n + 1 - m$ arbitrary functions.

For now we shall concentrate on the principal branch. Rather than attempt to demonstrate that this branch (or the others, for that matter) explicitly possess the Painlevé property we have proved [24] that the truncated expansion (16.3.6) leads to a self-consistent system of equations (the natural generalisation of the set (16.2.11)) which yield the BT (Bäcklund transformation) and Lax pair for each member of the KdV hierarchy. An important feature of our analysis is that the squared eigenfunction substitution (16.2.12) may still be used for all members of the hierarchy. That this is so should not be too surprising since the equation for q_1, occurring at $0(\psi^{-1})$ in the truncated expansion, always satisfies the PDE linearised about the solution u. For example, for the t_3 flow this linearisation is just

$$4q_{1t} + q_{1xxx} + 6uq_{1x} + 6q_1u_x = 0. \tag{16.3.11}$$

Thus q_1 is a symmetry of the KdV equation and from previous studies it is known that the associated infinitesimal BTs are generated by the derivative of the squared eigenfunction. Hence q_1, which is $2\varphi_{xx}$, should be identified with $(\psi^2)_x$ which immediately leads to the identification $\varphi_x = \psi^2$. Similar arguments apply to other types of integrable PDE such as the AKNS hierarchy to be discussed in the next section.

Working with the integrated KdV hierarchy

$$-\int q_{t_{2n+1}} + L^n q \tag{16.3.12}$$

and defining the set of coefficients $A_k^{(n)}$ as

$$L^n q = \sum_{k=0}^{2n+2} \frac{A_k^{(n)}}{\varphi^k} \tag{16.3.13}$$

we have proved [23, 24] that the coefficients in the truncated expansion for (16.3.12) are, at each order of φ,

$$\varphi^{-2n-2}\colon \frac{(-1)^n(2n+1)!!}{2^n(2n+1)!}\prod_{m=0}^{n}(u_0 + m(m+1)\varphi_x^2), \tag{16.3.14a}$$

$$\varphi^{-2n-1}\colon \frac{(-1)^{n+1}(2n+3)n!(2n-3)!!}{2^n}\varphi_x^{2n}(u_1 - m(m+1)\varphi_{xx}), \tag{16.3.14b}$$

$$\varphi^{-2n}\colon A_{2n}^{(n)} = 0, \tag{16.3.14c}$$

$$\varphi^{-2n+1}\colon A_{2n-1}^{(n)} \propto \psi^{4n-2}\left(\frac{\psi_{xx}}{\psi} + u\right)_x, \tag{16.3.14d}$$

$$\varphi^{-m}\colon A_m^{(n)} = 0, \quad m = 2n, \ldots, 3, \tag{16.3.14e}$$

$$\varphi^{-2}\colon -\psi^2 \int^x (-4\psi\psi_{t_{2n+1}} + A_1^{(n)})\mathrm{d}x, \tag{16.3.14f}$$

$$\varphi^{-1}\colon (-4\psi\psi_{t_{2n+1}} + A_1^{(n)}), \tag{16.3.14g}$$

$$\varphi^0\colon -\int^x u_{t_{2n+1}} + A_0^{(n)} = -\int^x u_{t_{2n+1}} + L^n u, \tag{16.3.14h}$$

where

$$A_1^{(n)} = 4\psi[\tfrac{1}{2} B_x^{(n)}\psi - B^{(n)}\psi_x]$$

with

$$B^{(n)} = \sum_{r=0}^{n} B_{n-r}\lambda^r, \quad 2B_{k+1} = L^k q, \quad B_0 = -1.$$

All coefficients can be self-consistently set to zero by choosing $u(x, t)$ to satisfy

$$-\int u_{t_{2n+1}} + L^n u = 0, \tag{16.3.15a}$$

where

$$L^n u = \left[-\tfrac{1}{4}\frac{\partial^2}{\partial x^2} - u + \tfrac{1}{2}\int^x \mathrm{d}x' u_x\right]^n u,$$

and ψ to satisfy

$$\psi_{xx} + (\lambda + u)\psi = 0, \tag{16.3.15b}$$

$$\psi_{t_{2n+1}} = -B^{(n)}\psi_x + \tfrac{1}{2} B_x^{(n)}\psi, \tag{16.3.15c}$$

the latter two equations being precisely the Lax pair for the associated flow. A different treatment of the KdV hierarchy has been given by Weiss [20] in terms of his Schwarzian derivative formalism.

A little calculation shows that the resonances (for $j > -1$) for the stationary equations of the KdV hierarchy,

$$L^n q = 0 \tag{16.3.16}$$

where

$$q = m(m+1)\frac{\partial^2}{\partial x^2}\ln\varphi + \sum_0^\infty q_{j+2}\varphi^j, \quad m = 1, 2, \ldots, n \tag{16.3.17}$$

occur at the zeros of a polynomial $P(n, m; k)$ in k. It satisfies the recursion relation [23, 24]:

$$P(n, m; k) = -\frac{(k - 2n - 1)(k - 2n + 2m)(k - 2n - 2m - 2)}{4(k - 2n - 2)} \times P(n - 1, m; k) + \frac{(2n - 1)!!(k - 4n - 2)}{2^n n!(k - 2n - 2)} \times \prod_{p=0}^{n-1} (m - p)(m + p + 1), \quad (16.3.18)$$

with $P(0, m; k) = 1$. For $n > m$,

$$P(n, m; k) = -\frac{(k - 2n - 1)(k - 2n + 2m)(k - 2n - 2m - 2)}{4(k - 2n - 2)} \times P(n - 2, m; k) \quad (16.3.19)$$

and for $m = 1$, the principal Painlevé expansion,

$$P(n, 1; k) = [-\tfrac{1}{4}]^n (k + 1)\{(k - 5)(k - 7) \ldots (k - 2n - 1)\} \times \{(k - 2)(k - 4) \ldots (k - 2n + 2)\}(k - 2n - 4). \quad (16.3.20)$$

For the time-dependent flows,

$$q_{t_{2n+1}} = (L^n q)_x, \quad (16.3.21)$$

the resonances occur at the zeros of the polynomial $(k - 2n - 2)P(n, m; k)$. Table 16.1 lists the resonances of each of the Painlevé expansions for the first three stationary and time-dependent flows in the KdV hierarchy.

Now, observe the pattern for the principal balance. The resonance at $j = -1$ is simply a reflection of the arbitrary nature of φ and the leading nonlinear balance. The resonance at $j = 2$ does not occur for the t_3 flow because at that stage the time-dependent term has already come into play. On the other hand, the resonance at $j = 2$ is present for the t_{2n+1} flow, $n \geq 2$. Hence u_2 is not determined at this stage but appears at the next level φ^{-2n+1} as an x-derivative. When we make $u_3 = 0$, the remaining part of the coefficient of φ^{-2n+1} is a perfect x-derivative which, when set to zero and integrated, gives rise to a free parameter λ which is the *spectral parameter* in the Lax pair.

For the principal balance in the t_{2n+1} flow, the resonances which produce undetermined functions occur at

$$2, 4, 5, 6, \ldots, 2n - 2, 2n - 1, 2n + 1, 2n + 4$$

for the *stationary* equation $L^n q = 0$. For the time-dependent flow $-q_{t_{2n+1}} + \partial/\partial x\, L^n q = 0$ there is one more at the position $2n + 2$. In the former case, therefore, there are, including φ, $2n$ arbitrary functions, exactly the

Table 16.1 Resonances of Painlevé expansions

Branch	t_3	t_5	t_7
Stationary flows			
$m = 1$	$-1, 6$	$-1, 2, 5, 8$	$-1, 2, 4, 5, 7, 10$
$m = 2$		$-3, -1, 8, 10$	$-3, -1, 2, 7, 10, 12$
$m = 3$			$-5, -3, -1, 10, 12, 14$
Time-dependent flows			
$m = 1$	$-1, 4, 6$	$-1, 2, 5, 6, 8$	$-1, 2, 4, 5, 7, 8, 10$
$m = 2$		$-3, -1, 6, 8, 10$	$-3, -1, 2, 7, 8, 10, 12$
$m = 3$			$-5, -3, -1, 8, 10, 12, 14$

number one expects for the stationary equation $L^n q = 0$, an ordinary differential equation of order $2n$ in x. The only gap in the series occurs at $j = 3$. With u_3 set equal to zero, the remaining terms in the coefficient form a perfect x-derivative. As we have already pointed out, this gives rise to the introduction of the arbitrary spectral parameter.

In reference [24] we consider in some detail the role of the lower branches and the significance of the pattern of resonances. What we are able to show, on the basis of the properties of the rational solutions to the KdV hierarchy, is that the lower branches are a re-expansion of the principal Painlevé expansion about those points on the singular manifold at which the zeros of $\varphi(x, t_{2n+1})$ become multiple. Furthermore there is reason to believe that the resonances are associated with the fluxes of the system – the lowest branch containing information about the fluxes of $(2n + 1)$th flow and the other branches containing information about the fluxes of lower flows in the hierarchy.

16.4 **Painlevé property for the AKNS hierarchy**

Most of the results to date concerning the connection between the WTC method and inverse scattering transforms have been for those systems with scalar Lax pairs. Although some connections with the AKNS scheme, for example, for the NLS and MKdV equations, have been noted [25], we are now able to give a more complete account for this family. The first three members of this hierarchy ('NLS hierarchy') are the NLS equation,

$$q_{t_2} = \frac{i}{2}(q_{xx} - 2q^2 r), \quad r_{t_2} = -\frac{i}{2}(r_{xx} - 2qr^2), \qquad (16.4.1)$$

the MKdV equation,

$$q_{t_3} = -\frac{1}{4}(q_{xxx} - 6qrq_x), \quad r_{t_3} = -\frac{1}{4}(r_{xxx} - 6qrr_x), \qquad (16.4.2)$$

and the NLS4 equation,

$$q_{t_4} = -\frac{\mathrm{i}}{8}\,(q_{xxxx} - 8qrq_{xx} - 6q_x^2 r - 4q_x r_x q - 2q^2 r_{xx} + 6q^3 r^2),$$

$$r_{t_4} = \frac{\mathrm{i}}{8}\,(r_{xxxx} - 8qrr_{xxx} - 6r_x^2 q - 4q_x r_x r - 2r^2 q_{xx} + 6r^3 q^2). \tag{16.4.3}$$

The NLS equation is easily shown [26] to possess the Painlevé property with both q and r having the expansions

$$q = \sum_{j=0}^{\infty} u_j \varphi^{j-1}, \quad r = \sum_{j=0}^{\infty} v_j \varphi^{j-1},$$

about the same singular manifold with resonances at $j = -1, 0, 3, 4$. Working with the truncated expansions

$$q = \frac{u_0}{\varphi} + u, \quad r = \frac{v_0}{\varphi} + v, \tag{16.4.4}$$

the following overdetermined systems of equations is obtained:

$$0(\varphi^{-3}):\ u_0 v_0 = \varphi_x^2, \tag{16.4.5}$$

$$0(\varphi^{-2}):\ 2\mathrm{i}\varphi_t = -\left[\varphi_{xx} + \frac{2u_{0x}}{u_0}\varphi_x\right] - 2u_0 v - 4v_0 u, \tag{16.4.6a}$$

$$-2\mathrm{i}\varphi_t = -\left[\varphi_{xx} + \frac{2v_{0x}}{v_0}\varphi_x\right] - 2v_0 u - 4u_0 v, \tag{16.4.6b}$$

which on addition yield

$$(u_0 v + v_0 u) = -\varphi_{xx} \tag{16.4.6c}$$

and on subtraction yield

$$4\mathrm{i}\varphi_t = \frac{2(u_0 v_{0x} - v_0 u_{0x})}{\varphi_x} + 2(u_0 v - v_0 u). \tag{16.4.6d}$$

$$0(\varphi^{-1}):\ -2\mathrm{i}u_{0t} = u_{0xx} - 2u^2 v_0 - 4u_0 uv, \tag{16.4.7a}$$

$$2\mathrm{i}v_{0t} = v_{0xx} - 2v^2 u_0 - 4v_0 uv, \tag{16.4.7b}$$

$$0(\varphi^0):\ u_t = \frac{\mathrm{i}}{2}\,(u_{xx} - 2u^2 v), \tag{16.4.8a}$$

$$v_t = -\frac{\mathrm{i}}{2}\,(v_{xx} - 2v^2 u). \tag{16.4.8b}$$

By noting that $(v_0 u_{0t} + u_0 v_{0t}) = (u_0 v_0)_t = 2\varphi_x \varphi_{xt}$ for equation (16.4.7) and using the expression for φ_t in (16.4.6d) one may, after a little manipulation, obtain

$$\left[\frac{u_{0x}}{u_0} + \frac{2u\varphi_x}{u_0}\right]_x = \left[\frac{v_{0x}}{v_0} + \frac{2v\varphi_x}{v_0}\right]_x = 0. \tag{16.4.9}$$

Treating the constant of integration as the spectral parameter $\pm 2\mathrm{i}\xi$ one obtains

$$u_{0x} + 2\mathrm{i}\xi u_0 = -2u\varphi_x, \tag{16.4.10a}$$

$$v_{0x} - 2\mathrm{i}\xi v_0 = -2v\varphi_x, \tag{16.4.10b}$$

for which the squared eigenfunction substitution

$$u_0 = w_1^2, \quad v_0 = w_2^2, \quad \varphi_x = -w_1 w_2, \tag{16.4.11}$$

gives the scattering problem

$$w_{1x} + i\xi w_1 = u w_2, \tag{16.4.12a}$$

$$w_{2x} - \mathrm{i}\xi w_2 = v w_1. \tag{16.4.12b}$$

The time-dependent part of the problem may be obtained directly from equations (16.4.7) by repeated use of (16.4.10) and (16.4.12), giving

$$w_{1t} = -w_1\left[\mathrm{i}\xi^2 + \frac{\mathrm{i}}{2}uv\right] + w_2\left[\xi u + \frac{\mathrm{i}}{2}u_x\right], \tag{16.4.13a}$$

$$w_{2t} = -w_1\left[\xi v + \frac{\mathrm{i}}{2}v_x\right] + w_2\left[\mathrm{i}\xi^2 + \frac{\mathrm{i}}{2}uv\right]. \tag{16.4.13b}$$

An amusing side-product of this derivation of the AKNS scheme for the NLS equation from the WTC method are some apparently novel rational solutions whose derivation is given in [24].

Exactly the same type of analysis can be carried through for the MKdV equation (16.4.2). It is easily shown to pass the Painlevé test [26] with the expansions

$$q = \sum_{j=0}^{\infty} u_j \varphi^{j-1}, \quad r = \sum_{j=0}^{\infty} v_j \varphi^{j-1}, \tag{16.4.14}$$

with resonances at $j = -1, 0, 1, 3, 4, 5$. Again working with the truncated expansions (16.4.4) one can derive the standard AKNS scheme for the MKdV equation by noting the squared eigenfunction relations (16.4.11). The NLS4 equation (16.4.3) has two possible leading orders, and both branches are again easily shown to pass the Painlevé test and the scattering problem can be deduced from the principal branch.

Finally we mention the results for the system of equations

$$p_x = \tfrac{1}{2}(qr)_t, \tag{16.4.15a}$$

$$q_{xt} = -4\mathrm{i}pq, \tag{16.4.15b}$$

$$r_{xt} = -4\mathrm{i}pr, \tag{16.4.15c}$$

from which integrable systems such as the sine-Gordon equation can be obtained. This system is easily shown [26] to pass the Painlevé test with the expansions

$$p = \sum_{j=0}^{\infty} p_j \varphi^{j-2}, \quad q = \sum_{j=0}^{\infty} q_j \varphi^{j-1}, \quad r = \sum_{j=0}^{\infty} r_j \varphi^{j-1}, \tag{16.4.16}$$

and resonances at −1, 0, 2, 3, 4. Using the truncated expansions

$$p = \frac{p_0}{\varphi^2} + \frac{p_1}{\varphi} + p_2, \quad q = \frac{q_0}{\varphi} + q_1, \quad r = \frac{r_0}{\varphi} + r_1, \tag{16.4.17}$$

and making the squared eigenfunction substitutions $q_0 = w_1^2$, $r_0 = w_2^2$, $\varphi_x = -w_1 w_2$ again gives the standard scattering problem

$$w_{1x} + \mathrm{i}\xi w_1 = q_1 w_2, \tag{16.4.18a}$$

$$w_{2x} - \mathrm{i}\xi w_2 = r_1 w_1, \tag{16.4.18b}$$

and

$$w_{1t} = w_1(p_2) + w_2[-\tfrac{1}{2}\,\xi q_{1t}], \tag{16.4.19a}$$

$$w_{2t} = w_1[\tfrac{1}{2}\,\xi r_{1t}] + w_2(p_2). \tag{16.4.19b}$$

The choice of $q_1 = -r_1 = u_x/2$ and $p_2 = \frac{1}{4}\cos u$ then gives the scattering scheme for the sine-Gordon equation $u_{xt} = \sin u$. This route, from the truncated Painlevé expansions (16.4.17) to the AKNS scheme (equations (16.4.18) and (16.4.19)), is far less cumbersome than that proposed in [27] which attempts to find the Lax pair directly from the sine-Gordon equation itself.

16.5 The nature of φ and the connection with Hirota's method

Hirota's method, which has been reviewed in Chapter 4, for constructing n-soliton solutions has proved to be one of the most enduring and useful techniques for studying integrable systems – even if the workings of the method are still somewhat mysterious. Recent studies by Gibbon *et al.* [25] on the connection between the Painlevé property and Hirota's method have cast some light on the latter and also provided some insight into the nature of the singular manifold function φ.

A simple example is provided by the KdV equation (16.2.1). The substitution

$$q = \frac{\partial^2}{\partial x^2} \ln f \tag{16.5.1}$$

leads, after one integration, to the equation

$$f f_{xxxx} - 4 f_x f_{xxx} + 3 f_{xx}^2 + f f_{xt} - f_x f_t = 0, \tag{16.5.2}$$

which can be written in terms of Hirota's bilinear operators as

$$(\mathrm{D}_x^4 + \mathrm{D}_x \mathrm{D}_t)\, f \cdot f = 0, \tag{16.5.3}$$

where $D_x^m(f \cdot f) = (\partial_x - \partial_{x'})^m f(x)f(x')|_{x=x'}$. As is well known, f can be developed in a series expansion which 'self-truncates' at each order yielding 1, 2, 3, ..., n-soliton solutions. The connection between this approach and the Painlevé analysis is demonstrated as follows. With each new, that is, n-soliton, solution of (16.2.1) we can associate a corresponding Hirota function $f^{(n)}$, that is,

$$q^{(n)} = \frac{\partial^2}{\partial x^2} \ln f^{(n)}. \qquad (16.5.4)$$

From the auto-Bäcklund transformation (16.2.9) determined by the Painlevé analysis we can regard the q and u in that equation as a pair of adjacent solutions $q^{(n)}$ and $q^{(n-1)}$. Thus, with each iteration of the BT, a new singular manifold function φ_{n-1} is produced, that is,

$$q^{(n)} = \frac{\partial^2}{\partial x^2} \ln \varphi_{n-1} + q^{(n-1)}. \qquad (16.5.5)$$

Comparison of (16.5.5) and (16.5.4) immediately yields

$$\varphi_{n-1} = \frac{f^{(n)}}{f^{(n-1)}}. \qquad (16.5.6)$$

In fact, one can prove the following theorem [25]: if the functions $f^{(n)}$, $n = 1, 2, \ldots$ satisfy the Hirota equation

$$(D_x^4 + D_x D_t) f^{(n)} \cdot f^{(n)} = 0 \qquad (16.5.7)$$

for every n and if $f^{(n)} = \varphi_{n-1} f^{(n-1)}$, then the resulting equation in φ_{n-1} and $u^{(n-1)}$ is satisfied by the Painlevé relations (16.2.11a, b) and, furthermore,

$$f^{(n)} = \prod_{i=0}^{n-1} \varphi_i. \qquad (16.5.8)$$

Theorems such as the above can be proved for other integrable systems for which the initial transformation to bilinear form proceeds by means of a second logarithmic derivative. Analogous results can also be obtained for members of the AKNS hierarchy which proceed via the initial substitution $q = g/f$ [25].

The crucial relationship is (16.5.6) which, since the Hirota $f^{(n)}$ functions are nothing more than τ-functions in disguise [23], tells us that φ *is a ratio of* τ*-functions* with the additional property that the singular manifold itself, that is, $\varphi_{n-1} = 0$, is a manifold on which $f^{(n)}$ has zeros but on which $f^{(n-1)}$ does not vanish. Further, the condition that $\nabla\varphi$, the gradient of φ_{n-1} with respect to x and the flow times, has no zero components on the zero manifold, says that, on $\varphi_{n-1} = 0$, $f^{(n)}$ has a *simple* zero when we consider x or any particular time t_{2n+1} as a function of all the other times. (As we have indicated, the lower branches for the equations in the KdV hierarchy reflect the fact that at certain special locations on the $\varphi = 0$ manifold in $(x, t_3, \ldots)$ space, these zeros become multiple.) However in $x, t_3, t_5, \ldots$ space, the zeros of φ are not isolated but lie on smooth manifolds. They are the infinite-dimensional analogues of the *divisors* that arise for finite-dimensional, algebraically completely integrable systems. In that context, the divisors are codimension-1 subvarieties on an Abelian variety on which certain of the original dependent variables can have pole singularities. They are located at the zeros of the Riemann τ-function which, up to a non-vanishing factor, is the τ-function for that case. In the infinite-dimensional case, the set of allowable τ-functions is considerably larger and so therefore is the set of allowable zero manifolds; the simplest subset of zero manifolds being given by the zeros of the rational solutions.

16.6 Lax pairs for ordinary differential equations

By means of a simple extension, the WTC method can be used to obtain BTs and Lax pairs for systems of integrable ordinary differential equations. Consider, as a simple example, the stationary part of the lowest member of the KdV hierarchy, that is,

$$q_{xx} + 3q^2 = 0 \tag{16.6.1}$$

which can be integrated in terms of elliptic functions. This ODE is easily shown to possess the Painlevé property with the local Laurent expansion (with constant coefficients a_j)

$$q(x) = \sum_{j=0}^{\infty} a_j(x - x_0)^{j-2} \tag{16.6.2}$$

with resonances at $j = -1$ and 6. As it stands, the series (16.6.2) contains no further information about the actual solutions to (16.6.1). However, if by analogy with the PDE method, we write the series as

$$q(x) = \sum_{j=0}^{\infty} q_j(x)(\varphi(x))^{j-2}, \tag{16.6.3}$$

interesting results can be obtained by working with the truncated expansion:

$$q(x) = \frac{q_0}{\varphi^2} + \frac{q_1}{\varphi} + u(x). \tag{16.6.4}$$

On direct substitution into (16.6.1), it is easy enough to show, on equating the powers of φ^{-4} and φ^{-3} to zero, that $q_0 = -2\varphi_x^2$ and $q_1 = 2\varphi_{xx}$, that is,

$$q(x) = 2\,\frac{\partial^2}{\partial x^2}\ln\varphi(x) + u(x). \tag{16.6.5}$$

Thus one obtains

$$q_{xx} + 3q^2 = \frac{1}{\varphi^2}[-8\varphi_x\varphi_{xxx} - 12\varphi_x^2 u + 6\varphi_{xx}^2] + \frac{1}{\varphi}[2\varphi_{xxxx} + 12\varphi_{xx}u] + u_{xx} + 3u^2. \quad (16.6.6)$$

Introducing the squared eigenfunction substitution $\varphi_x = \psi^2$ one obtains, bar the t-dependent terms, a result virtually identical to (16.2.13), that is,

$$\begin{aligned} q_{xx} + 3q^2 = {} & \frac{-4\psi^2}{\varphi^2}\left[\int^x \psi\,(4\psi_{xxx} + 6u\psi_x + 3u_x\psi)\,dx'\right] \\ & + \frac{4\psi}{\varphi}\left[(4\psi_{xxx} + 6u\psi_x + 3u_x\psi) - 3\psi\left(\frac{\psi_{xx}}{\psi} + u\right)_x\right] \\ & + u_{xx} + 3u^2. \end{aligned} \quad (16.6.7)$$

Now by setting the different orders of φ equal to zero, according to the usual WTC prescription, one obtains

$$\psi_{xx} + (\lambda + u)\psi = 0, \quad (16.6.8a)$$

$$4\psi_{xxx} + 6u\psi_x + 3u_x\psi = 0, \quad (16.6.8b)$$

$$u_{xx} + 3u^2 = 0. \quad (16.6.8c)$$

However, as it stands, this system of equations is not a Lax pair for equation (16.6.8c) since, on eliminating u between equations (16.6.8a) and (16.6.8b), an explicit, indeed, trivial solution for ψ is obtained. Clearly an additional parameter has to be incorporated into the system. The crucial point is that the requirement that each order of φ be set to zero is too restrictive. Rather, *each order of φ can be set to zero modulo some function of φ*. Thus all one has to do is add an amount $4y\varphi\varphi_x$ to the terms at order φ^{-2} and subtract the identical amount at order φ^{-1}, namely $4y\varphi_x$, where y represents the separation parameter, that is, $\Psi(x, t) = \psi(x)e^{yt}$. Equation (16.6.7) now reads

$$\begin{aligned} q_{xx} + 3q^2 = {} & \frac{-4\psi^2}{\varphi^2}\left[\int^x \psi\,(4\psi_{xxx} + 6u\psi_x + 3u_x\psi - y\psi)dx'\right] \\ & + \frac{4\psi}{\varphi}\left[(4\psi_{xxx} + 6u\psi_x + 3u_x\psi - y\psi) - 3\psi\left(\frac{\psi_{xx}}{\psi} + u\right)_x\right] \\ & + u_{xx} + 3u^2. \end{aligned} \quad (16.6.9)$$

from which one obtains the desired Lax pair

$$\psi_{xx} + (\lambda + u)\psi = 0, \quad (16.6.10a)$$

$$4\psi_{xxx} + 6u\psi_x + 3u_x\psi = y\psi. \quad (16.6.10b)$$

Exactly the same procedure holds for all other members of the stationary KdV hierarchy

$$L^n q = 0. \tag{16.6.11}$$

By adding $4y\varphi\varphi_x$ at order φ^{-2} and subtracting $4y\varphi_x$ at order φ^{-1}, one obtains

$$-B^{(n)}\,\psi_x + \tfrac{1}{2}\,B_x^{(n)}\psi = y\psi, \tag{16.6.12}$$

which combined with the eigenvalue problem obtained at order φ^{-2n+1} gives the Lax pair for the nth member of the hierarchy.

This approach is not restricted to the KdV equations. Another illustration is provided by one of the integrable cases of the Henon–Heiles system, which has the equations of motion

$$q_{1xx} = -\tfrac{1}{2}\,q_2^2 - 3q_1^2, \tag{16.6.13a}$$

$$q_{2xx} = -q_1 q_2. \tag{16.6.13b}$$

The expansions for the general, that is, four-parameter, solution truncated at order φ^0 are found [24] to be:

$$q_1 = 2\,\frac{\partial^2}{\partial x^2}\log\varphi + u_1, \tag{16.6.14a}$$

$$q_2 = -\frac{\psi}{\varphi}\int^x \psi u_2 \mathrm{d}x' + u_2, \tag{16.6.14b}$$

where again the squared eigenfunction relation $\varphi_x = \psi^2$ is used. Direct substitution of (16.6.14) into (16.6.13) yields

$$\begin{aligned} q_{1xx} &+ \frac{1}{2}\,q_2^2 + 3q_1^2 \\ &= \frac{4\psi^2}{\varphi^2}\left\{-\int\psi\left[4\psi_{xxx} + 6\psi_x u + 3\psi u_x - \frac{1}{4}\,u_2\int^x\psi u_2\mathrm{d}x'\right]\right\} \\ &+ \frac{4\psi}{\varphi}\left\{\left(4\psi_{xxx} + 6\psi_x u_1 + 3\psi u_{1x} - \frac{1}{4}\,u_2\int^x\psi u_2\mathrm{d}x'\right) - 3\left[\frac{\psi_{xx}}{\psi} + u_1\right]\right\} \\ &+ u_{1xx} + \frac{1}{2}\,u_2^2 + 3u_1^2. \end{aligned} \tag{16.6.15}$$

Again, by adding an amount $4y\varphi\varphi_x$ at order φ^{-2} and subtracting $4y\varphi_x$ at order φ^{-1} gives the Lax pair

$$\psi_{xx} + (u_1 + \lambda)\psi = 0, \tag{16.6.16a}$$

$$4\psi_{xxx} + 6\psi_x u_1 + 3\psi u_{1x} - \tfrac{1}{4}\,u_2\int^x\psi u_2\mathrm{d}x' = y\psi. \tag{16.6.16b}$$

Further details of this analysis and the calculation of the associated algebraic curve are given elsewhere [24] along with some results for the second and fourth Painlevé transcendents.

Another class of ordinary differential equations, for which a variation of the WTC method can be applied, is the Toda lattice, that is, the differential-difference system of equations

$$\ddot{Q}_n = \exp(Q_{n-1} - Q_n) - \exp(Q_n - Q_{n+1}) \quad n = 1, 2, \ldots, N. \tag{16.6.17}$$

In terms of the variables $q_n = Q_n$, $p_n = \exp(Q_n - Q_{n+1})$ these can be written as

$$\dot{p} = p_n(q_n - q_{n+1}), \tag{16.6.18a}$$

$$\dot{q} = p_{n-1} - p_n. \tag{16.6.18b}$$

The usual Painlevé analysis involves expanding each of the dependent variables about the same common pole position t_0. As N becomes large this analysis becomes very cumbersome although for the three-particle case the analysis was carried out by Bountis *et al.* [9] who were thus able to identify certain integrable cases depending on the particle masses and the boundary conditions.

A different approach has been proposed by Gibbon and Tabor [28] whereby each of the dependent variables is expanded about its own singular manifold $\varphi_n(t)$. The truncated expansions are thus written as

$$p_n = \frac{p_n^{(0)}}{\varphi_n^2} + \frac{p_n^{(1)}}{\varphi_n} + p_n^{(2)}, \tag{16.6.19a}$$

$$q_n = \frac{q_{n-1}^{(0)}}{\varphi_{n-1}} - \frac{q_n^{(0)}}{\varphi_n} + q_n^{(1)}. \tag{16.6.19b}$$

A little analysis then shows that

$$p_n = \frac{\partial^2}{\partial t^2} \ln \varphi_n + p_n^{(2)}, \tag{16.6.20a}$$

$$q_n = \frac{\partial}{\partial t} \ln (\varphi_{n-1}/\varphi_n) + q_n^{(1)} \tag{16.6.20b}$$

and

$$\frac{\partial^2}{\partial t^2} \ln \varphi_n = p_n^{(2)} \left[\frac{\varphi_{n+1}\varphi_{n-1}}{\varphi_n^2} - 1\right], \tag{16.6.20c}$$

where $p_n^{(2)}$ and $q_n^{(1)}$ satisfy the Toda system (16.6.18). It is shown in [28] that the Lax pair

$$\varphi_{n,t} = -\lambda^{-1}\, \varphi_{n-1}\, p_n^{(2)}, \tag{16.6.21a}$$

$$q_{n+1}^{(1)}\varphi_n = \lambda\varphi_{n+1} + \lambda^{-1}\varphi_{n-1}p_n^{(2)} \tag{16.6.21b}$$

can be deduced from (16.6.20c). Thus in this case the φ_n are themselves playing the role of the scattering eigenfunctions.

References

[1] Kovalevskaya, S. *Acta Math.* **12**, 177–232 (1889); **14**, 81–3 (1889).
[2] Painlevé, P. *Oeuvres*, Vol. I, CNRS, Paris (1973).
[3] Ince, E. L. *Ordinary Differential Equations*, Dover, New York (1956).
[4] Tabor, M. *Nature* **310**, 277–82 (1984).
[5] Wu, T. T., McCoy, B. M., Tracy, C. A. and Barouch, E. *Phys. Rev. B* **13**, 316–74 (1976).
[6] Tabor, M. and Weiss, J. *Phys. Rev. A* **24**, 2157–67 (1981).
[7] Chang, Y. F., Tabor, M. and Weiss, J. *J. Math. Phys.* **24**, 531–8 (1982).
[8] Chang, Y. F., Greene, J. M., Tabor, M. and Weiss, J. *Physica* **8D**, 183–207 (1983).
[9] Bountis, T., Sequr, H. and Vivaldi, F. *Phys. Rev. A* **25**, 1257–64 (1982).
[10] Ramani, A., Dorizzi, B. and Grammaticos, B. *Phys. Rev. Letts.* **49**, 1539–41 (1982).
[11] Yoshida, H. *Celest. Mech.* **31**, 363–79 (1983).
[12] Adler, M. and van Moebeke, P. *Adv. in Math.* **38**, 267–317 (1980).
[13] Ercolani, N. and Siggia, E. The Painlevé property and integrability, *Phys. Lett.* **119**, 112–16 (1986).
[14] Ablowitz, M. J., Ramani, A. and Segur, H. J. *J. Math. Phys.* **21**, 715–21 (1980).
[15] Nakach, R. Plasma physics, *Proc. 36th Nobel Symp.* (ed. Wilhelmsson) Plenum, New York (1977).
[16] McLeod, J. B. and Olver, P. J. *SIAM J. Math. Anal.* **14**, 488–506 (1983).
[17] Weiss, J., Tabor, M. and Carnevale, G. *J. Math. Phys.* **24**, 522–6 (1983).
[18] Osgood, W. F. *Topics in the Theory of Functions of Several Complex Variables*, Dover, New York (1966).
[19] Weiss, J. *J. Math. Phys.* **24**, 1405–13 (1983).
[20] Weiss, J. *J. Math. Phys.* **25**, 13–24 (1984).
[21] Weiss, J. *J. Math. Phys.* **26**, 258–69 (1985).
[22] Weiss, J. *J. Math. Phys.* **26**, 2174–80 (1985).
[23] Newell, A. C., Ratiu, T., Tabor, M. and Zeng, Y. *Soliton Mathematics*, to be published by Center for Research in Applied Mathematics, University of Montreal.
[24] Newell, A. C., Tabor, M. and Zeng, Y. B. A unified approach to Painlevé expressions, *Physica* **29D**, 1–68 (1987).
[25] Gibbon, J. D., Radmore, P., Tabor, M. and Wood, D. *Stud. Appl. Math.* **72**, 39–63 (1985).
[26] Chudnovsky, D. V., Chudnovsky, G. V. and Tabor, M. *Phys. Letts* **97A**, 268–74 (1983).
[27] Weiss, J. *J. Math. Phys.* **25**, 2226–35 (1984).
[28] Gibbon, J. D. and Tabor, M. *J. Math. Phys.* **26**, 1956–60 (1985).

Index